Numbers

$\gcd(a, b)$	the greatest common divisor of integers a and b
$\lfloor x \rfloor$	the greatest integer less than or equal to the real number x
$\lceil x \rceil$	the smallest integer greater than or equal to the real number x
$a \equiv b \pmod{m}$	a is equivalent to b modulo m: m divides $a - b$
$n!$	n factorial: $0! = 1$, $n! = n(n - 1) \ldots (2)(1)$
$P(n, r)$	the number of permutations of n objects taken r at a time: $n!/(n - r)!$
$C(n, r) = \binom{n}{r}$	the number of r element subsets of an n element set: $n!/[r!(n - r)!]$

Graph Theory

$V(G)$	the set of vertices of graph G
$E(G)$	the set of edges of graph G
K_n	the complete graph on n vertices
$\bar{G}$	the complement of graph G
$\delta(v)$	the degree of vertex v
$\delta_i(v)$	the in-degree of vertex v of a directed graph
$\delta_o(v)$	the out-degree of vertex v of a directed graph
$w(e)$	the weight of edge e in a weighted graph
$W(G)$	the weight of a weighted graph G

Automata and Languages

Σ	a finite set of symbols, an alphabet
ε	the empty word
$L_1 L_2$	the concatenation of languages L_1 and L_2: $\{w_1 w_2 \mid w_1 \in L_1, w_2 \in L_2\}$
L^n	the language L concatenated with itself n times
L^0	the set consisting of the empty word: $\{\varepsilon\}$
$L*$	the Kleene closure of language L: $\bigcup_{n=0}^{\infty} L^n$
$ac(M)$	the set of words accepted by machine M
$R(M)$	the dual of machine M
$D(M)$	a DFA with $ac(D(M)) = ac(M)$ where M is an NFA

Discrete Mathematics

Paul F. Dierker
University of Idaho

William L. Voxman
University of Idaho

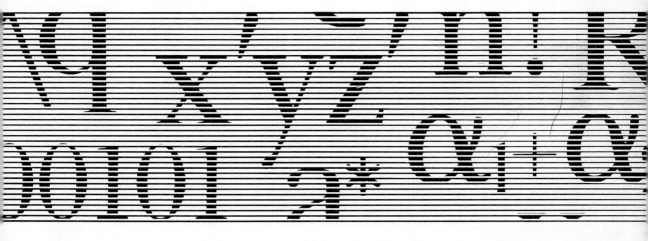

HBJ

Harcourt Brace Jovanovich, Publishers
San Diego   New York   Chicago   Atlanta   Washington, D.C.
London   Sydney   Toronto

Requests for permission to make copies of any part of the work should
be mailed to: Permissions, Harcourt Brace Jovanovich, Publishers,
Orlando, Florida 32887.

ISBN: 0-15-517691-9
Library of Congress Catalog Card Number: 85-60870
Printed in the United States of America

Preface

There is a growing consensus that a course in discrete mathematics should be offered at the freshman-sophomore level. Such a course provides a solid foundation for computer science students and is increasingly important in mathematics and in many mathematically oriented disciplines. But discrete mathematics has a number of other virtues as well. Topics covered in discrete mathematics are easily motivated and students usually find the material interesting and accessible. Moreover, introductory discrete mathematics helps develop computational skills and provides a first look at the theory underlying more advanced topics. The only prerequisite for this book is college algebra.

Generally, our goal has been to present a discrete mathematics text that offers several interesting and useful topics having solid mathematical content. We have worked toward a blend that enables the reader to see the close relation between theory and practice. The computational aspects of discrete mathematics are stressed, but not at the expense of the underlying theory. Students have ample opportunity to develop their own creative abilities.

We begin with an introduction to algorithms, material that is generally new to the reader yet is easily understood. Students soon appreciate the usefulness of algorithms, and this appreciation is reinforced by the algorithmic approach to problems taken throughout the book. Rather than employ a particular programming language for presenting algorithms, we have used commands that are common to many languages. Each algorithm in the book includes glossed explanatory statements that clarify the individual steps.

We have emphasized graph theory early in the text. Graphs have many interesting applications well within the grasp of students. In addition, graph theory provides a convenient vehicle for introducing elementary proofs that complement the more computational aspects of this subject. In studying graphs, students quickly realize why algorithms are necessary for solving problems of any significant magnitude.

Boolean algebra, the subject of Chapter 5, introduces an abstract mathematical structure which, although similar in some respects to previously learned structures, is sufficiently different to enable the student to see that operations such as addition and multiplication can take on many forms. Applications of Boolean algebra to computer science are obvious, and the material found in Chapter 5 illustrates the connection between theory and practice.

We give a fairly thorough presentation of difference equations in Chapter 7, emphasizing the application of these equations to various modeling situations. Difference equations have many other uses as well and, like calculus, provide students an excellent opportunity to develop and sharpen their algebraic skills.

Enumeration and probability form an integral component of a course in discrete mathematics, and in our presentation we stress basic counting concepts. The material in Chapters 9 and 10 gives students sufficient background to analyze the complexity of certain algorithms as well as to pursue more advanced courses in combinatorics and probability.

We take up generating functions in Chapter 10. Generating functions demonstrate the importance of infinite series and introduce students to an entirely different approach to solving difference equations than that found in Chapter 7. At a slightly more sophisticated level, we see how generating functions can be used to solve some difficult combinatorial problems.

This book concludes with a brief introduction to automata and formal languages. These topics help students appreciate the importance of abstraction in the design of computers and software. Type 3 languages offer a natural approach to the study of formal languages, and we use them as the focus for our discussion of these languages.

There is sufficient material in this book for a semester or a year course in discrete mathematics. We have arranged it to provide maximum flexibility in choice of topics. The following chart indicates the dependencies of the various chapters:

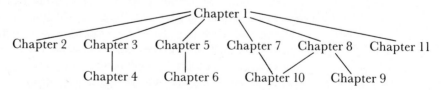

Many people have contributed to the development of this text. We wish to thank our colleagues Arie Bialostocki, Ya-Yen Wang, Willy Brandal, and

Tom Miller for class-testing the manuscript and offering many helpful comments. Erol Barbut and Jim Calvert suggested a number of improvements, and Leo Boron did his usual excellent job of proofreading.

Input from students, essential for the success of any textbook, has been a great benefit to us, and we would particularly like to acknowledge Greg Stenback and Mark Wilkins for their constructive ideas and for verifying answers to selected problems.

Finally, many thanks are due to the Harcourt Brace Jovanovich staff: to Richard Wallis for his sound advice and continued encouragement; to Barbara Girard for her handling of proof on a tight schedule; to Dean Reed for a book design that is both functional and appealing; and to Audrey Thompson for her exceptionally careful editing of the final manuscript.

PAUL F. DIERKER
WILLIAM L. VOXMAN

Contents

Preface   v

1 A First Look at Algorithms   1

1.1   Algorithms   1
1.2   Two Sorting Algorithms   8
1.3   Mathematical Induction   12
1.4   The Alternate Form of the Principle of Induction   21
1.5   Greedy Algorithms   28
*1.6 Backtracking Algorithms   35
1.7   Complexity of Algorithms and the Big O Notation   44
      Review   51

2 Number Systems and Modular Arithmetic   55

2.1   Numbers in Other Bases   55
2.2   Horner's Algorithm: Converting to the Decimal System   59
2.3   Converting from the Decimal System to Other
         Number Systems   62
2.4   Operations with Binary Numbers   69
2.5   Modular Arithmetic   73
*2.6 A Modular Code   76
      Review   83

*An asterisk indicates an optional section.

3 An Introduction to Graph Theory   85

3.1 Graphs   85
3.2 Connectedness   94
3.3 The Components of a Graph   99
3.4 Trees   103
3.5 Spanning Trees   107
3.6 Weighted Graphs and Minimal Spanning Trees   113
*3.7 A Proof of Kruskal's Algorithm   120
3.8 Minimum Weight Spanning Cycles   122
3.9 Directed Graphs and Relations   131
       Review   143

4 Applications of Graph Theory   147

4.1 Minimal Path Trees   147
4.2 The Critical Path Method   157
4.3 Euler Tours and Paths   167
4.4 Maximal Shift Register Sequences   174
4.5 Gray Codes   178
       Review   184

5 Boolean Algebra and Switching Systems   187

5.1 From Switching Systems to Boolean Algebra   187
5.2 Switching Systems, Boolean Functions, and
       Boolean Expressions   193
5.3 An Introduction to Simplifying Boolean Expressions   201
5.4 Karnaugh Maps   206
5.5 Prime Implicants   219
5.6 The Quine–McCluskey Algorithm: How It Works   224
*5.7 The Quine–McCluskey Algorithm: Why It Works   228
5.8 The Covering Problem   230
       Review   236

6 Symbolic Logic and Logic Circuits   238

6.1 Compound Statements and Logical Equivalence   238
6.2 Implication   242
6.3 Logic Circuits   246
6.4 Gate Implementation of Binary Addition   252
       Review   258

7 Difference Equations   260

7.1 Some Examples of Difference Equations   260
7.2 Linear Difference Equations with Constant Coefficients 266

7.3 The General Solution of a Homogeneous Linear
Difference Equation   271
7.4 Characteristic Equations with Repeated Roots   276
7.5 Characteristic Equations with Complex Roots   280
7.6 Nonhomogeneous Difference Equations   285
7.7 The Method of Undetermined Coefficients   290
7.8 Some Additional Applications of Difference Equations   296
Review   300

8   An Introduction to Enumeration   **302**

8.1 Two Basic Principles   302
8.2 Permutations   307
8.3 Combinations   313
8.4 The Binomial Theorem   319
8.5 Pascal's Triangle and Some Basic Identities   322
8.6 Counting with Repetitions   328
8.7 The Principle of Inclusion and Exclusion   334
Review   340

9   Elementary Probability Theory   **342**

9.1 Basic Concepts   342
9.2 Some Basic Properties of the Probability Measure   348
9.3 Conditional Probability   355
9.4 Independent Events   360
9.5 Bernoulli Processes   366
*9.6 Bayes' Theorem   373
9.7 Random Variables and Probability Density Functions   376
9.8 Expected Values of Random Variables   381
*9.9 Poisson Distributions   387
*9.10 Introduction to Queueing Theory   393
Review   401

10   Generating Functions   **404**

10.1 The Limit of a Sequence   404
10.2 Computation of Limits   411
10.3 Infinite Sums and Series   415
10.4 Generating Functions   422
10.5 Generating Functions and Difference Equations   429
10.6 Partial Fractions and the Problem of Inversion   432
10.7 Convolution   438
10.8 A Combinatorial View of Generating Functions   444
10.9 Some Applications of Generating Functions to
Combinatorial Problems   448
Review   453

11   An Introduction to Automata and Formal Languages   **456**

11.1   An Introduction to Machines   456
11.2   Nondeterministic Automata   464
11.3   Comparing Pattern Recognition Capabilities of
           NFA's and DFA's   471
11.4   Regular Sets   476
11.5   Kleene's Theorem   479
11.6   Minimization of Automata   486
11.7   Formal Languages   489
11.8   Type 3 Languages and Automata   494
           Review   499

Appendix   **503**

A.1   Set Theory   503
A.2   Functions   512
A.3   Matrices and Matrix Operations   518
A.4   Sigma Notation   528
A.5   Relations   531

Answers to Selected Odd-Numbered Problems   536
Index   583

Discrete Mathematics

1 A First Look at Algorithms

In the past twenty years there has been a remarkable upsurge in interest in the study and analysis of algorithms. Although the concept of an algorithm (a step-by-step procedure for solving a problem) goes back some 4000 years to the Babylonians, the advent of computers has stimulated much of the current emphasis on algorithms. The word *algorithm* is derived from the last part of the name of the ninth century Persian scholar Abu Ja'far Mohammed ibn Musa al Khuwarizmi, who was one of the most influential mathematicians of his time.

In this chapter we shall look at a number of basic, though relatively simple, algorithms. We begin with a comparison of two standard search algorithms. Next we consider two sorting algorithms, and we introduce mathematical induction as a method of analyzing certain algorithms. (We use mathematical induction throughout the text in a variety of other contexts as well.) We also briefly examine two general types of algorithms—greedy algorithms and backtracking algorithms. The chapter concludes with an introduction to the problem of describing an algorithm's efficiency.

1.1 ALGORITHMS

In this text we view an algorithm as a well-defined, computer executable set of instructions that allows us to obtain a unique output from a given input. We assume that an algorithm will terminate in a finite number of steps. It is essential to have instructions precise enough to ensure that the output depends only on the input and the algorithm used.

Two fundamental questions arise about any algorithm:

1. Does the algorithm accomplish what it purports to accomplish?

2. Is there another algorithm that accomplishes the same task more efficiently?

From time to time we will consider these questions to assess the effectiveness of a particular algorithm.

The concept of an algorithm is perhaps best illustrated through specific examples. The first algorithm we consider is used to deal with the following problem. We have a computer which stores an alphabetical listing of the registered voters from a particular Congressional district. We must determine whether or not a given name is in the list, and if so, indicate its location within the ordered list.

This problem represents a large number of problems that involve locating a particular item in some ordered list of items. The simplest method, or algorithm, for solving this kind of a problem is to have the computer start at the top of the list and compare each item in the list to the given input. If at some stage in the process the computer recognizes the input, the procedure stops, and the output indicates that the given input has been found and gives the item's position in the list. If the computer does not recognize the input, the algorithm indicates that the stored list does not contain the entry that is sought.

To apply this algorithm to the problem of the registered voters, we use the following notation. Let $A(K)$ denote the name occupying the Kth position in the stored list of names. Because the list is stored in alphabetical order, we may have $A(1) = $ Adams, $A(2) = $ Arthur, $A(3) = $ Buchanan, and so on. The formal statement of this algorithm is as follows:

Algorithm 1.1   SEQUENTIAL SEARCH   To determine if a given name is in a stored list of names $A(1), \ldots, A(N)$, and if so, to print out its location within the list.

Input: *NAME*, the given name

Output: The position K of *NAME* if *NAME* is in the list;
"NOT FOUND" if *NAME* is not in the list

1. FOR $K = 1$ TO N	Each name $A(K)$ is examined in succession to determine if $NAME = A(K)$.
a. IF $NAME = A(K)$ THEN OUTPUT K AND STOP	If *NAME* is the Kth name in the list, the location of *NAME* is printed out, and the algorithm terminates.
2. OUTPUT "NOT FOUND"	*NAME* was not found in the list, and "NOT FOUND" is printed out.

Algorithm 1.1 illustrates the style we will use to express algorithms. A block of statements (such as 1a. in Algorithm 1.1) which is repeated several times during the execution of an algorithm will be indented. Such a block is called a *loop*. A control statement, given immediately before a loop, states the conditions (loop control) under which the loop is to be executed (in Algorithm 1.1, statement 1 is a control statement).

We use multiple indentations to indicate nested loops. When a loop is ended, control passes to the next statement of the algorithm (for example, in Algorithm 1.1, when the loop 1a. is ended control passes to statement 2). When appropriate, we provide comments on specific steps of the algorithm to the right of the given statement.

There is a better algorithm than SEQUENTIAL SEARCH for finding a particular name in a given list. By "better" we mean an algorithm that yields the same results but is more efficient because it involves fewer operations. The next algorithm we consider, frequently called BINARY SEARCH, is more efficient than SEQUENTIAL SEARCH because, unlike SEQUENTIAL SEARCH, this algorithm utilizes the alphabetical order of the stored list.

In SEQUENTIAL SEARCH we compared $NAME$ to successive entries in the stored list of N names. In BINARY SEARCH, however, we first compare $NAME$ to the "middlemost" name in the list. More explicitly, we first compare $NAME$ to

$$A\left(\left\lfloor \frac{N+1}{2} \right\rfloor\right)$$

where $\lfloor (N+1)/2 \rfloor$ is defined as the greatest integer that is less than or equal to $(N+1)/2$. For instance, if $N = 6$, then $A(\lfloor (6+1)/2 \rfloor) = A(3)$ is the middlemost entry in a list of six names $A(1)$, $A(2)$, $A(3)$, $A(4)$, $A(5)$, and $A(6)$; if $N = 11$, then $A(\lfloor (11+1)/2 \rfloor) = A(6)$ is the middlemost name in the list of the names $A(1)$, $A(2)$, ..., $A(11)$.

In this procedure, if

$$NAME = A\left(\left\lfloor \frac{N+1}{2} \right\rfloor\right)$$

then the output will be the integer

$$K = \left\lfloor \frac{N+1}{2} \right\rfloor$$

If $NAME$ precedes the name $A(\lfloor (N+1)/2 \rfloor)$ alphabetically, we write

$$NAME < A\left(\left\lfloor \frac{N+1}{2} \right\rfloor\right)$$

and we next compare $NAME$ with the middlemost name lying between $A(1)$

and $A([N + 1)/2])$. Note that in this case it will not be necessary to compare *NAME* with any name that follows $A([[(N + 1)/2])$.

If *NAME* follows the name $A(\lfloor (N + 1)/2 \rfloor$ alphabetically, we write

$$NAME > A\left(\left\lfloor \frac{N + 1}{2} \right\rfloor\right)$$

and we next compare *NAME* to the middlemost name that lies between $A(\lfloor (N + 1)/2 \rfloor)$ and $A(N)$. Note that in this case it will no longer be necessary to compare *NAME* with any name preceding $A(\lfloor (N + 1)/2 \rfloor)$.

We continue to "halve" the given list of names until we have either found *NAME* or have determined that *NAME* is not in the list. At each step of the procedure, at least one half of the remaining names are eliminated from further consideration. To see how this algorithm works when $N = 16$, consider the alphabetical list:

$$A(1) = \text{Adams}$$
$$A(2) = \text{Arthur}$$
$$A(3) = \text{Buchanan}$$
$$A(4) = \text{Cleveland}$$
$$A(5) = \text{Coolidge}$$
$$A(6) = \text{Eisenhower}$$
$$A(7) = \text{Fillmore}$$
$$A(8) = \text{Grant}$$
$$A(9) = \text{Harding}$$
$$A(10) = \text{Harrison}$$
$$A(11) = \text{Hayes}$$
$$A(12) = \text{Hoover}$$
$$A(13) = \text{Jackson}$$
$$A(14) = \text{Johnson}$$
$$A(15) = \text{Lincoln}$$
$$A(16) = \text{Madison}$$

and suppose that we are searching for *NAME* = Hayes. We first compare Hayes to the middlemost name of this list, which is

$$A\left(\left\lfloor \frac{1 + 16}{2} \right\rfloor\right) = A(8) = \text{Grant}$$

Since Hayes > Grant (follows Grant alphabetically), we next compare Hayes to the middlemost name between $A(9)$ and $A(16)$, that is, to $A(12) = $ Hoover. Since Hayes < Hoover (precedes Hoover alphabetically), we compare Hayes to the middlemost name lying between $A(9)$ and $A(11)$, that is, to $A(10) = $ Harrison. Finally, since Hayes > Harrison, we compare Hayes to the only remaining name to be considered in the original list (Hayes) and find that the given input is the eleventh name in the list.

~~$A(1), A(2), A(3), A(4), A(5), A(6), A(7), \boxed{A(8),}$~~ $A(9), A(10), A(11), A(12), A(13), A(14), A(15), A(16)$
~~$A(1), A(2), A(3), A(4), A(5), A(6), A(7), A(8),$~~ $A(9), A(10), A(11), \boxed{A(12),}$ ~~$A(13), A(14), A(15), A(16)$~~
~~$A(1), A(2), A(3), A(4), A(5), A(6), A(7), A(8), A(9), \boxed{A(10),}$~~ $A(11),$ ~~$A(12), A(13), A(14), A(15), A(16)$~~

Figure 1.1

Figure 1.1 illustrates this elimination process. We have crossed out entries that were no longer considered at each step and have circled the middlemost entries at each stage.

It is important to note that in this case BINARY SEARCH makes only three comparisons to locate *NAME*, whereas SEQUENTIAL SEARCH would require eleven comparisons to locate the same given input.

To write BINARY SEARCH and subsequent algorithms, we adopt the "assignment" notation

$$k \leftarrow j$$

to indicate that the current value of k is to be replaced with j.

Algorithm 1.2   BINARY SEARCH   To determine if a given name is in an alphabetically stored list of names $A(1), A(2), \ldots, A(N)$, and if so, to print out its location in the list.

Input: *NAME*, the given name
Output: The position K of *NAME* if *NAME* is in the list;
"NOT FOUND" if *NAME* is not in the list

1. $F \leftarrow 1; L \leftarrow N$

   F and L will designate the first and last names on the sublist being considered. We use the notation $F \leftarrow 1$ and $L \leftarrow N$ to indicate that F is assigned the value 1, and L is assigned the value N. Thus we initially have $F = 1$ and $L = N$.

2. WHILE $F \leq L$ DO

   This loop control indicates that the next steps are to be executed as long as $F \leq L$. Otherwise, control passes to 3.

   a. $K \leftarrow \left\lfloor \dfrac{F + L}{2} \right\rfloor$

   The position of the middlemost entry between $A(F)$ and $A(L)$ is assigned to K.

   b. IF *NAME* = $A(K)$ THEN
      1. OUTPUT K
      2. STOP

   If *NAME* is the middlemost entry, then we output its location and terminate the algorithm.

c. IF *NAME* > $A(K)$ THEN
 1. $F \leftarrow K + 1$

> If *NAME* follows the middlemost entry, then the list will be searched between $A(K + 1)$ and $A(L)$.

d. IF *NAME* < $A(K)$ THEN
 1. $L \leftarrow K - 1$

> If *NAME* precedes the middlemost entry, then the list will be searched between $A(F)$ and $A(K - 1)$.

3. OUTPUT "NOT FOUND"

> The entire list has been searched without finding *NAME*; the output is "NOT FOUND."

You should trace through the steps of this algorithm for the data discussed above for *NAME* = Hayes.

Both Algorithms 1.1 and 1.2 accomplish the same task. To determine which algorithm is "better," however, we consider what is called the *complexity* of an algorithm. Complexity is a measure of an algorithm's efficiency and may be expressed in terms of the time or the storage space required to execute the algorithm. All other factors being equal, the more efficient the algorithm, the more it is preferred. In Section 1.7 we shall investigate the notion of complexity further, and we will find that from the standpoint of efficiency, BINARY SEARCH is a better algorithm than SEQUENTIAL SEARCH.

1.1 Problems

1. Show that if *n* is even, then $[(n + 1)/2] = n/2$.
2. Show that $\lfloor x - 1 \rfloor = \lfloor x \rfloor - 1$.
3. True or false: $[2x] = 2[x]$. Prove your assertion.

A function similar to the greatest integer function is the *least integer function*: $[x]$ denotes the least integer greater than or equal to *x*.

4. Show that if *n* is odd, then $[(n + 1)/2] = (n + 1)/2$.
5. Show that $\lceil x - 1 \rceil = \lceil x \rceil - 1$.
6. Construct a table of the values of all variables that occur in Algorithm 1.2 when this algorithm is applied to the following list of "names":
 $A(1) = AB, A(2) = AC, A(3) = BE, A(4) = BG, A(5) = CF,$
 $A(6) = DE, A(7) = DF, A(8) = EK, A(9) = GR, A(10) = HI$
 Input: *NAME* = GR
7. Repeat Problem 6 for *NAME* = EK.
8. Repeat Problem 6 for *NAME* = AA.

In Problems 9–12, describe what the given algorithm accomplishes.

9. Input: *X*, *Y*
 Output: *Z*

1. $Z \leftarrow X$
2. IF $X < Y$ THEN
   a. $Z \leftarrow Y$
3. OUTPUT Z

10.   Input: X

   Output: Y

1. IF $X < 0$ THEN
   a. $Y \leftarrow -X$
2. ELSE $Y \leftarrow X$
3. OUTPUT Y

11.   Input: $A(1), A(2), \ldots, A(N)$
   Output: CT

1. $CT = 0$
2. FOR $I = 1, \ldots, N$
   a. IF $A(I) > 0$ THEN
      1. $CT \leftarrow CT + 1$
3. OUTPUT CT

12.   Output: S
1. $S \leftarrow 0$
2. $I \leftarrow 0$
3. WHILE $I \leq N$ DO
   a. $S \leftarrow S + I$
   b. $I \leftarrow I + 1$
4. OUTPUT S

13. Write an algorithm that will compute the sum of the first N positive integers.

14. Write an algorithm that will compute the sum of the first N positive even integers.

15. Write an algorithm that will compute the sum of the first N positive odd integers.

16. Write an algorithm that will compute the sum of the cubes of the first N positive integers.

17. Write an algorithm that finds the first entry that is out of order if a list of numbers $A(1), A(2), \ldots, A(N)$ is to be in decreasing order.

18. Write an algorithm that reverses the order of a sequence of entries $A(1)$, $A(2), \ldots, A(N)$.

19. Recall that the *greatest common divisor* of two integers a and b, $\gcd(a, b)$, is the largest integer that divides both a and b. For instance, it is easy to see that $\gcd(48, 84) = 12$. The following algorithm will find $\gcd(a, b)$ for any two integers a and b.

   Input: Positive integers A and B
   Output: $\gcd(A, B)$

1. WHILE $A \neq B$ DO
   a. IF $A < B$ THEN
      1. TEMP $\leftarrow A$
      2. $A \leftarrow B - A$
      3. $B \leftarrow$ TEMP
   b. ELSE
      1. TEMP $\leftarrow B$
      2. $B \leftarrow A - B$
      3. $A \leftarrow$ TEMP
2. OUTPUT A

Apply this algorithm to find
(a) gcd(64, 120)
(b) gcd(360, 1000)
(c) gcd(144, 896)

20. Why must the algorithm in Problem 19 terminate after a finite number of steps?

1.2   TWO SORTING ALGORITHMS

In this section we develop two sorting algorithms that reorder a given list of numbers from the largest number to the smallest. This problem is representative of a variety of problems involving lists of numbers, names, words, and so on.

One approach to the problem of sorting a list of numbers is to determine the largest number in the list and then determine the largest number among the remaining numbers. Continuing in this manner, we will eventually obtain the desired order. The following algorithm finds the largest number in any list of numbers.

Algorithm 1.3   To determine the largest number in a list of N numbers.

Input: $A(1), A(2), \ldots, A(N)$, the list of numbers
Output: $LARGE$, the largest of the numbers $A(1), A(2), \ldots, A(N)$;
M, the position of $LARGE$

1. $LARGE \leftarrow A(1)$;
   $M \leftarrow 1$

   Initialize: $LARGE$ designates the current largest value.

2. FOR $K = 2$ TO N

   All of the numbers $A(1), A(2), \ldots A(N)$ must be considered.

a. IF $A(K) > LARGE$
THEN
   1. $LARGE \leftarrow A(K)$
   2. $M \leftarrow K$
3. OUTPUT $LARGE$, M

If $A(K)$, the number being considered, is larger than $LARGE$, then $LARGE$ is replaced by $A(K)$, and M is assigned the value K.

By repeatedly applying Algorithm 1.3, we can sort a list of N numbers into decreasing order. The following algorithm implements this technique.

Algorithm 1.4 To sort a list of N numbers into decreasing order.

Input: $A(1)$, $A(2)$,..., $A(N)$, the list of numbers

Output: The list of numbers in decreasing order

1. FOR $K = 1$ TO $N - 1$
   a. Use ALGORITHM 1.3 to determine $LARGE = A(M)$, the largest of the numbers $A(K)$, $A(K + 1)$, ..., $A(N)$

       The computer makes successive passes through a list; for each pass it registers the largest number in the list and removes it from the list. Note that this step is unnecessary for $K = N$.

   b. $A(M) \leftarrow A(K)$;
      $A(K) \leftarrow LARGE$

       The largest number in the list is interchanged with the first number in the list.

2. FOR $I = 1$ TO N
   a. OUTPUT $A(I)$

Example 1 Applying Algorithm 1.4 to the list $A(1) = 4$, $A(2) = 1$, $A(3) = 6$, $A(4) = 10$, $A(5) = 2$ yields the lists in Table 1.1. □

	Initial Order	Pass 1	Pass 2	Pass 3	Pass 4
$A(1)$	4	10	10	10	10
$A(2)$	1	1	6	6	6
$A(3)$	6	6	1	4	4
$A(4)$	10	4	4	1	2
$A(5)$	2	2	2	2	1

Table 1.1

An alternate approach to this sorting problem is to start at the end of the given list and compare each two successive entries during each pass through the list. If two successive entries are not in the proper order, then the two numbers are interchanged. For instance, to apply this approach to the list from Example 1, we start with the numbers in the following order:

I	$A(I)$
1	4
2	1
3	6
4	10
5	2

During the first pass, the computer begins at the bottom of the list and makes the following changes:

4	4	4	4	10
1	1	1	10	4
6	6	10	1	1
10	10	6	6	6
2	2	2	2	2

Thus, the next pass has the task of reordering the list

I	$A(I)$
1	10
2	4
3	1
4	6
5	2

Note that the largest entry in the list, the number 10, has risen to the top. During the next pass, the next largest entry, 6, will rise to the position immediately below 10. This algorithm is generally known as BUBBLESORT because the large items continue to "bubble up" to the top of the list during the various passes.

Algorithm 1.5   BUBBLESORT   For sorting a list of N numbers into decreasing order.

Input: $A(1), A(2), \ldots, A(N)$, the list of numbers

Output: The list of numbers in decreasing order

1. FOR $J = 1$ TO $N - 1$
   a. FOR $I = 1$ TO $N - J$ — The first $J - 1$ entries are in order; only the entries $A(J), \ldots, A(N)$ need be sorted.

      1. IF $A(N + 1 - I) > A(N - I)$ THEN — The list is checked from the bottom up and the necessary interchanges are made.

   a. TEMP $\leftarrow A(N - I)$
   b. $A(N - I) \leftarrow A(N + 1 - I)$
   c. $A(N + 1 - I) \leftarrow$ TEMP
2. FOR $K = 1, 2, \ldots, N$
   a. OUTPUT $A(K)$

Example 2   Table 1.2 illustrates the application of Algorithm 1.5 to the list $A(1) = 4, A(2) = 1, A(3) = 6, A(4) = 10, A(5) = 2.$ $\square$

	Initial Order	Pass 1				Pass 2			Pass 3		Pass 4
J	—	1	1	1	1	2	2	2	3	3	4
I	—	1	2	3	4	1	2	3	1	2	1
$A(1)$	4	4	4	4	10	10	10	10	10	10	10
$A(2)$	1	1	1	10	4	4	4	6	6	6	6
$A(3)$	6	6	10	1	1	1	6	4	4	4	4
$A(4)$	10	10	6	6	6	6	1	1	2	2	2
$A(5)$	2	2	2	2	2	2	2	2	1	1	1

Table 1.2

BUBBLESORT can be improved by terminating the procedure whenever no changes are made during a pass. For example, the last pass in Example 2 is unnecessary.

We will consider the relative efficiencies of Algorithms 1.4 and 1.5 in Section 1.7.

1.2   Problems

1. Follow Algorithm 1.3 through step-by-step for the list of numbers 5, 12, -3, -6, 4, $\frac{1}{2}$.
2. Follow Algorithm 1.3 through step-by-step for the list of numbers 4, -6, 3, 10, 6, 8.
3. Follow Algorithm 1.4 through step-by-step for the list of numbers 5, -2, 7, 1, 3.
4. Follow Algorithm 1.4 through step-by-step for the list of numbers 3, 2, 4, 3, 1.

In Problems 5–8 give the successive lists produced during an application of BUBBLESORT (Algorithm 1.5).

5. I	$A(I)$	6. I	$A(I)$	7. I	$A(I)$	8. I	$A(I)$
1	10	1	9	1	9	1	4
2	9	2	10	2	10	2	5
3	6	3	6	3	8	3	6
4	8	4	8	4	3	4	8
5	5	5	5	5	4	5	9
6	4	6	4	6	5	6	10

9. Write an algorithm that locates the smallest entry in a list of numbers.

10. Write an algorithm that orders a list of numbers from the smallest number to the largest.

11. Write an algorithm that determines both the largest and the smallest number in a given list of numbers.

12. Explain why BUBBLESORT can be terminated if no changes are made during a pass.

13. Explain why any algorithm used to determine the largest number in a list of n numbers must make at least $n - 1$ comparisons.

14. Under what conditions will the maximum number of interchanges take place in BUBBLESORT?

1.3 MATHEMATICAL INDUCTION

Inductive reasoning is the process of drawing *general* conclusions after examining *particular* cases or observations. This process can be very useful in mathematics, but care must be taken when asserting the validity of a given proposition based on a few examples. For instance, in view of the observations in Table 1.3, what general conclusion might you be tempted to draw?

$$
\begin{array}{lll}
4 = 2 + 2 & 18 = 5 + 13 & 40 = 11 + 29 \\
6 = 3 + 3 & 20 = 7 + 13 & 50 = 13 + 37 \\
8 = 3 + 5 & 22 = 3 + 19 & 100 = 3 + 97 \\
10 = 3 + 7 & 24 = 7 + 17 & 200 = 19 + 181 \\
12 = 5 + 7 & 26 = 7 + 19 & 300 = 119 + 181 \\
14 = 3 + 11 & 28 = 5 + 23 & \\
16 = 5 + 11 & 30 = 13 + 17 & \\
\end{array}
$$

Table 1.3

The inferred pattern is that any even integer can be written as the sum of two prime numbers. Based on the particular examples in Table 1.3, however, can we be assured of the validity of this general conclusion? The Goldbach conjecture, that every even integer greater than 2 can be written as the sum of two primes, has tantalized mathematicians for more than 200 years. No one has yet found an even number that cannot be written as the sum of two prime numbers. Although inductive reasoning strongly suggests that the Goldbach conjecture is valid, we have no way at present of proving whether it is true or false.

Fortunately, many of the problems we deal with are more accessible and we can often apply *mathematical induction* to prove conjectures based on observations of particular cases. Mathematical induction is based intuitively on the well-known domino principle: If the first domino in a row of dominos falls, and if whenever one domino falls the next domino in the row also falls, then all of the dominos in the row will fall.

To use the domino principle, we must first translate it into mathematical terms. The domino principle, applied to the set $\mathbf{Z}^+$ of positive integers is expressed as follows:

Property P: Suppose that A is a subset of $\mathbf{Z}^+$ such that:

a.' $1 \in A$ (the first domino falls), and

b.' if $k \in A$, then $k + 1 \in A$ (whenever one domino falls so does the next one)

Then $A = \mathbf{Z}^+$.   (all the dominos fall)

Property P is an intrinsic property of the positive integers, and is the basis for the *Principle of Induction.*

Principle of Induction: Suppose that $S_1, S_2, \ldots$ is an infinite sequence of statements such that

a. The first statement S_1 is true, and

b. If the statement S_k is true, then the next statement S_{k+1} is also true.

Then all of the statements $S_1, S_2, \ldots$ are true; in other words, statement S_n is true for each positive integer n.

The Principle of Induction is a consequence of Property P. To see this, let $A = \{n | \text{statement } S_n \text{ is true}\}$, and suppose that **a.** and **b.** of the Principle of Induction hold. Then we have from **a.** that $1 \in A$, and from **b.** that if $k \in A$, then $k + 1 \in A$. But these are precisely the conditions necessary to satisfy the hypothesis of Property P, and therefore we can conclude that $A = \mathbf{Z}^+$. This is equivalent to saying that the statement S_n is true for each positive integer n.

For our first example of the Principle of Induction, we consider the following equations:

$$1 = 1$$
$$1 + 3 = 4$$
$$1 + 3 + 5 = 9$$
$$1 + 3 + 5 + 7 = 16$$
$$1 + 3 + 5 + 7 + 9 = 25$$

From these equations, we might conjecture that the sum of the first n odd integers is equal to n^2 for every positive integer n. In other words, for each $n \in \mathbf{Z}^+$,

$$1 + 3 + 5 + \cdots + (2n - 1) = n^2 \tag{1}$$

To apply the Principle of Induction to prove (1), for each positive integer n, let S_n be the statement

$$1 + 3 + 5 + \cdots + (2n - 1) = n^2 \tag{2}$$

We want to show that S_n is true for all n.

By the Principle of Induction it suffices to show

a. S_1 is true.

b. If S_k is true, then S_{k+1} is true.

To verify **a.** we substitute $n = 1$ into (2) and observe that S_1 is the statement $1 = 1^2$, which is clearly true.

Establishing **b.** is a bit trickier. We must show that if S_k is true, that is, if

$$1 + 3 + \cdots + (2k - 1) = k^2 \tag{3}$$

then S_{k+1} is true, that is,

$$1 + 3 + \cdots + (2k - 1) + (2(k + 1) - 1) = (k + 1)^2 \tag{4}$$

In other words, we must show that if equation (2) is true for $n = k$, then this equation is also true for $n = k + 1$. To establish (4) we use assumption (3) and some elementary algebra to obtain

$$
\begin{aligned}
1 + 3 + 5 + \cdots + (2k - 1) &+ (2(k + 1) - 1) \\
&= 1 + 3 + 5 \cdots + (2k - 1) + (2k + 1) \\
&= [1 + 3 + 5 + \cdots + (2k - 1)] + (2k + 1) \\
&= k^2 + (2k + 1) \\
&= k^2 + 2k + 1 \\
&= (k + 1)^2
\end{aligned}
\tag{5}
$$

which is what we wanted to show. Since we have established **a.** and **b.** of the Principle of Induction, we can conclude that the statement S_n defined by (2) is true for each positive integer n.

Example 1   We use the Principle of Induction to show that for each positive integer n,

$$1 + 2 + \cdots + n = \frac{n(n + 1)}{2} \tag{6}$$

For each $n \in \mathbf{Z}^+$, let S_n be the statement given in equation (6). We must show that each of the statements $S_1, S_2, \ldots$ is true.

By the Principle of Induction it suffices to show that

a. S_1 is true.

b. If S_k is true, then S_{k+1} is true.

To verify **a.**, substitute $n = 1$ into (6) and observe that

$$1 = \frac{1(1 + 1)}{2}$$

so S_1 is true.

As is frequently the case, establishing **b.** is somewhat more difficult. We assume that the statement S_k is true, that is, we suppose that equation (6) is valid if $n = k$. This produces

$$1 + 2 + \cdots + k = \frac{k(k + 1)}{2} \tag{7}$$

We must show that under this assumption, statement S_{k+1} is true. That is, we must show that if we substitute $n = k + 1$ into (6), then we still obtain a valid equation. Note that if we let $n = k + 1$, then the left-hand side of (6) becomes

$$1 + 2 + \cdots + k + (k + 1) \tag{8}$$

and the right-hand side becomes

$$\frac{(k + 1)(k + 2)}{2} \tag{9}$$

We must verify that (8) and (9) are equal.

Using assumption (7) and some elementary algebra we obtain

$$
\begin{aligned}
1 + 2 + \cdots + k + (k + 1) &= [1 + 2 + \cdots + k] + (k + 1) \\
&= \frac{k(k + 1)}{2} + (k + 1) \\
&= \frac{k(k + 1) + 2(k + 1)}{2} \\
&= \frac{(k + 1)(k + 2)}{2}
\end{aligned}
$$

which is what we wanted to show. Since we have established **a.** and **b.** of the Principle of Induction, we can conclude that the statement S_n defined by (6) is true for each positive integer n. □

Mathematical induction is often applied to the analysis of algorithms, as the next example illustrates.

Example 2   We use induction to show that a total of $n(n-1)/2$ comparisons are made when Algorithm 1.4 is used to sort a list of n numbers.

Let S_n denote the statement:

The number of comparisons that are made when Algorithm 1.4 is used to sort a list of n numbers is $n(n-1)/2$.

Observe that if $n = 1$, no comparisons are necessary; the list consists of only one member. Consequently, since

$$\frac{1(1-1)}{2} = 0$$

it follows that S_1 is true.

Suppose now that S_k is true. That is, suppose that $[k(k-1)]/2$ comparisons are made when we use Algorithm 1.4 to sort a list of k numbers. We must show that under this assumption S_{k+1} is true. In other words, we must show that $[k(k+1)]/2$ comparisons are made when we use Algorithm 1.4 to sort a list of $k+1$ numbers.

Observe that when we apply Algorithm 1.4 to a list of $k+1$ numbers, Algorithm 1.3 is called, and step 2 of Algorithm 1.3 is executed k times. That is, Algorithm 1.3 makes k comparisons. Algorithm 1.4 then places the largest element (found by Algorithm 1.3) first in the list and proceeds to sort the remaining k numbers on the list. By the induction hypothesis, Algorithm 1.4 will require $k(k-1)/2$ comparisons to sort the remaining k numbers. Thus Algorithm 1.4 requires a total of

$$k + \frac{k(k-1)}{2}$$

comparisons to sort a list of $k+1$ numbers. Since

$$k + \frac{k(k-1)}{2} = k(1 + \frac{k-1}{2})$$

$$= \frac{k(k+1)}{2}$$

statement S_{k+1} is true, and it follows from the Principle of Induction that for each positive integer n, Algorithm 1.4 uses $n(n-1)/2$ comparisons to sort a list of n numbers. □

The number of comparisons a sort algorithm makes is a key factor in determining its efficiency. We will return to a more detailed examination of this idea in Section 1.7.

Next we apply the Principle of Induction in a geometric context.

Example 3   We use the Principle of Induction to establish the following proposition:

Let $P_1, P_2, \ldots, P_n$ be n points in the plane such that no three of these points are collinear (lie on the same line). Then the total number of line segments connecting pairs of these points is $(n^2 - n)/2$. For instance, if $n = 5$, we have line segments as shown in Figure 1.2.   □

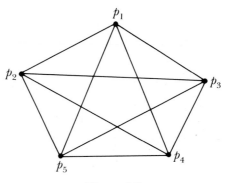

Figure 1.2

Note that there are $(5^2 - 5)/2 = 10$ line segments connecting the points.

To show that this is true in general, we verify conditions **a.** and **b.** of the Principle of Induction.

If $n = 1$, then there is only one point and there are no line segments. Note that in this case we have

$$\frac{1^2 - 1}{2} = 0$$

line segments; therefore, condition **a.** of the Principle of Induction is satisfied.

Suppose now that the formula $(n^2 - n)/2$ holds if $n = k$. We need to show that under this assumption this formula is also valid if $n = k + 1$. That is, we need to show that $k + 1$ points (no three of which lie in a straight line) are connected by

$$\frac{(k + 1)^2 - (k + 1)}{2} = \frac{k^2 + 2k + 1 - k - 1}{2} = \frac{k^2 + k}{2} \tag{10}$$

straight line segments. To see this, consider $k + 1$ points in the plane, and isolate one of these points, as shown in Figure 1.3.

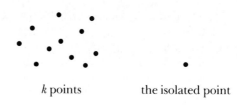

k points   the isolated point

Figure 1.3

Since we are assuming that the formula holds in the case of k points, the k points indicated in Figure 1.3 are connected with

$$\frac{k^2 - k}{2}$$

line segments. To connect the isolated point to each of the other points, we obviously need k line segments. Thus, to connect all pairs of the $k + 1$ points we need

$$\frac{k^2 - k}{2} + k$$

line segments. But

$$\frac{k^2 - k}{2} + k = \frac{k^2 - k + 2k}{2} = \frac{k^2 + k}{2}$$

which from (10) is what we wanted to show.

Consequently, conditions **a.** and **b.** of the Principle of Induction are satisfied and thus for each $n \in \mathbf{Z}^+$

$$\frac{n^2 - n}{2}$$

is the number of line segments joining noncollinear points.   □

We conclude this section with a word of caution. Many beginning students, once having "mastered" the induction technique, make the mistake of trying to solve *all* problems with this method. Though induction is very useful, it can be misused. You must be careful to apply it only when appropriate and to apply it correctly, as the following example illustrates.

Example 4   We will use the Principle of Induction to show that you and all of your classmates will receive an A in this course (under the likely assumption that at least one of you receives an A).

Let S_n be the statement that any set of exactly n students in your class will

Numbers

$\gcd(a, b)$ the greatest common divisor of integers a and b

$\lfloor x \rfloor$ the greatest integer less than or equal to the real number x

$\lceil x \rceil$ the smallest integer greater than or equal to the real number x

$a \equiv b \pmod{m}$ a is equivalent to b modulo m: m divides $a - b$

$n!$ n factorial: $0! = 1$, $n! = n(n-1)\ldots(2)(1)$

$P(n, r)$ the number of permutations of n objects taken r at a time: $n!/(n-r)!$

$C(n, r) = \binom{n}{r}$ the number of r element subsets of an n element set: $n!/[r!(n-r)!]$

Graph Theory

$V(G)$ the set of vertices of graph G

$E(G)$ the set of edges of graph G

K_n the complete graph on n vertices

$\bar{G}$ the complement of graph G

$\delta(v)$ the degree of vertex v

$\delta_i(v)$ the in-degree of vertex v of a directed graph

$\delta_o(v)$ the out-degree of vertex v of a directed graph

$w(e)$ the weight of edge e in a weighted graph

$W(G)$ the weight of a weighted graph G

Automata and Languages

Σ a finite set of symbols, an alphabet

ε the empty word

$L_1 L_2$ the concatenation of languages L_1 and L_2: $\{w_1 w_2 | w_1 \in L_1, w_2 \in L_2\}$

L^n the language L concatenated with itself n times

L^0 the set consisting of the empty word: $\{\varepsilon\}$

L^* the Kleene closure of language L: $\bigcup_{n=0}^{\infty} L^n$

$ac(M)$ the set of words accepted by machine M

$R(M)$ the dual of machine M

$D(M)$ a DFA with $ac(D(M)) = ac(M)$ where M is an NFA

receive the same grade. Then clearly S_1 is true, that is, condition **a.** of the Principle of Induction is satisfied.

To see that condition **b.** also holds, suppose that statement S_k is true. We show that under this assumption the statement S_{k+1} is also true. To see this, merely observe that if $\{z_1, z_2, \ldots, z_k, z_{k+1}\}$ is any set of $k+1$ students, and if statement S_k is true, then we have that all members of the set $\{z_1, \ldots, z_k\}$ will receive the same grade as will the members of the set $\{z_2, \ldots, z_{k+1}\}$ (there are k students in each of these sets). Since student z_2 is common to both of these sets, it follows that all of the students in the set $\{z_1, z_2, \ldots, z_k, z_{k+1}\}$ will receive the same grade, and, hence, statement S_{k+1} is true.

This completes the induction argument and we can conclude that for each positive integer n, any set of n students will receive the same grade. Since for some integer n, there are n students in your class, it is clear that if one member of your class receives an A, then every student in the class must receive this grade. From a practical standpoint then, only one of you needs to master this text.

$\square$

What is the problem in this application of induction? Simply that S_k does *not* imply S_{k+1} for all k. In fact, S_1 does not imply S_2 . To see this, check the above argument for $k = 1$.

1.3   Problems

In Problems 1–14 use induction to prove that the given statement is true for all positive integers.

1. $2 + 6 + 10 + \cdots + (4n - 2) = 2n^2$

2. $1 + 5 + 9 + 13 + \cdots + (4n - 3) = \dfrac{n(4n - 2)}{2}$

3. $1 + 3 + 9 + 27 + \cdots + 3^{n-1} = \dfrac{3^n - 1}{2}$

4. $1 + 2 + 4 + 8 + 16 + \cdots + 2^{n-1} = 2^n - 1$

5. $2 + 7 + 12 + 17 + 22 + \cdots + (5n - 3) = \dfrac{n(5n - 1)}{2}$

6. $1^3 + 2^3 + 3^3 + \cdots + n^3 = \dfrac{n^2(n + 1)^2}{4}$

7. $1^2 + 2^2 + 3^2 + \cdots + n^2 = \dfrac{n(n - 1)(2n + 1)}{6}$

8. $1 \cdot 3 + 3 \cdot 5 + 5 \cdot 7 + \cdots + (2n - 1)(2n + 1) = \dfrac{n(4n^2 + 6n - 1)}{3}$

9. $1 \cdot 2 + 2 \cdot 3 + 3 \cdot 4 + \cdots + n(n + 1) = \dfrac{n(n + 1)(n + 2)}{3}$

10. $\dfrac{1}{1 \cdot 2} + \dfrac{1}{2 \cdot 3} + \dfrac{1}{3 \cdot 4} + \cdots + \dfrac{1}{n(n + 1)} = \dfrac{n}{n + 1}$

11. $a + (a + d) + (a + 2d) + \cdots + [a + (n - 1)d] = \dfrac{n[2a + (n - 1)d]}{2}$

12. $a + ar + ar^2 + \cdots + ar^{n-1} = \dfrac{a(r^n - 1)}{r - 1}, r \neq 1$

13. $n < 2^n$

14. $n! \geq 2^{n-1}$

15. Show that $2n - 3 \leq 2^{n-2}$ for all $n \geq 5$.

16. Prove that 3 divides $(n + 1)^3 + 3n^2 + 2n + 2$ for all nonnegative integers n. (Recall that an integer n divides an integer m if $m = kn$ where k is also an integer.)

17. Prove that 6 divides $n(n^2 + 5)$ for all positive integers n.

18. In order to execute a particular algorithm on n data points, the same algorithm must be executed twice on $n - 1$ data points. The execution of this algorithm on one point requires one unit of time. Use induction to show that the execution of the algorithm on n points requires 2^{n-1} time units for all positive integers n.

19. Prove that the sum of the cubes of any three consecutive positive integers is divisible by 9.

20. Every pair of entries in a list of n entries must be compared. How many comparisons are necessary? (*Hint:* Use the result of Example 3.)

21. Use the fact that

$$1 + 2 + \cdots + (n - 1) = \frac{n(n - 1)}{2}$$

(established in Example 1) to give an alternate proof of the result obtained in Example 3.

22. What is the maximum number of interchanges that will be made when BUBBLESORT is applied to a list of n numbers?

23. Use induction to prove that the output of the following algorithm is 2^M for $M = 1, 2, \ldots$.
    1. $P \leftarrow 1$
    2. FOR $N = 1$ TO M
       a. $P \leftarrow 2P$
    3. OUTPUT P

1.4   THE ALTERNATE FORM OF THE PRINCIPLE OF INDUCTION

At times we will make use of a slightly different, though equivalent, formulation of the Principle of Induction.

Alternate Form of the Principle of Induction: Suppose that $S_1, S_2, \ldots$ is a sequence of statements such that

a. statement S_1 is true, and

b. if statements $S_1, S_2, \ldots, S_k$ are true, then so is statement S_{k+1}.

Then statement S_n is true for *each* positive integer n.

Note that the difference between this alternate form and the Principle of Induction lies in condition **b.** In the alternate form, we assume that *all* of the statements $S_1, S_2, \ldots, S_k$ are true. In the Principle of Induction, however, we assumed as our induction hypothesis that only statement S_k was true. In spite of this difference, it can be shown that these versions of induction are equivalent in the sense that each one implies the other.

We shall use the alternate form of the Principle of Induction to establish the validity of another sorting algorithm, generally known as QUICKSORT. As with the sorting algorithms discussed in Section 1.2, the purpose of QUICKSORT is to arrange a given list of numbers in decreasing order. QUICKSORT is an example of a *recursive algorithm*, an algorithm that calls on itself one or more times in executing a procedure. Although QUICKSORT can be implemented in a nonrecursive fashion for computer languages that do not permit recursive calls, its recursive form is particularly simple.

The idea behind QUICKSORT is easy: to QUICKSORT a list L of numbers, select an arbitrary number (call it X) from L. Then rearrange the list so that all of the entries in L that are greater than X precede X, and all of the entries in L that are less than or equal to X follow X. As a result of this rearrangement, the number X appears in its proper position.

Next let L_1 be the list of entries in L that are greater than X, and let L_2 be the list of entries in L that are less than or equal to X. Since X is in its proper position, we need only sort lists L_1 and L_2. Make a recursive call to QUICKSORT for each sublist and continue this process until the original list is arranged in decreasing order.

A formal description of this algorithm is fairly involved and is deferred to this section's Problem set.

Table 1.4 indicates the result of the application of QUICKSORT to a specific list of numbers.

i	$A(i)$	After the initial run of QUICKSORT
1	$5 = X$	8 ⎫
2	8	12 ⎪
3	2	6 ⎪
4	12	9 ⎬ L_1 Use QUICKSORT
5	6	14 ⎪
6	9	7 ⎭
7	14	5 ← Now in its proper position; this
8	1	2 ⎫ entry will never be moved
9	4	1 ⎬ L_2 Use QUICKSORT
10	7	4 ⎭

Table 1.4

The proof of QUICKSORT's validity is based on a very natural application of the alternate form of the Principle of Induction. Let S_n denote the statement:

QUICKSORT correctly sorts a list of n numbers.

First observe that statement S_1 is true, since in this case the list consists of a single entry and no sorting is necessary.

Now assume that all of the statements $S_1, S_2, \ldots, S_k$ are true. We must show that under this assumption, statement S_{k+1} is also true. In other words, we must show that under the assumption that QUICKSORT correctly sorts all lists with fewer than $k + 1$ numbers, it will also correctly sort any list containing $k + 1$ numbers.

Suppose then that we apply QUICKSORT to a list containing $k + 1$ numbers. In the initial step, QUICKSORT selects a number X and places it in its proper position. QUICKSORT subsequently processes the *shorter* lists on either side of X. By our induction assumption, QUICKSORT will correctly sort these sublists because their lengths are less than or equal to k. Thus, we find that QUICKSORT will correctly sort the entire list of $k + 1$ entries — in short, that it will accomplish what was intended. This concludes the inductive argument.

Recursive algorithms can sometimes be wildly inefficient. We illustrate this with a recursive algorithm that can be used to generate the Fibonacci numbers.

The Fibonacci numbers[1] have long held a particular fascination for the mathematician and nonmathematician alike. These numbers tend to occur naturally in a wide variety of contexts including plant growth, rabbit reproduction (see Problem 14), and optimization. The Fibonacci numbers,

---

[1] Named after Leonardo Fibonacci, an outstanding thirteenth-century Italian mathematician.

$F(i)$, are defined by

$$F(1) = 1$$
$$F(2) = 1$$

and for $n \geq 3$,

$$F(n) = F(n - 1) + F(n - 2) \tag{1}$$

Thus, the first two Fibonacci numbers are equal to one, and each successive Fibonacci number is the sum of the previous two numbers. For instance,

$$F(3) = F(2) + F(1) = 2$$
$$F(4) = F(3) + F(2) = 3$$
$$F(5) = F(4) + F(3) = 5$$
$$\vdots$$

Equation (1) provides a natural basis for a recursive algorithm.

Algorithm 1.6   To compute the Nth Fibonacci number, $F(N)$.

Input: N

Output: $F(N)$, the Nth Fibonacci number

1. $F(1) \leftarrow 1$	The first two Fibonacci numbers are given.
2. $F(2) \leftarrow 1$	
3. $F(N) \leftarrow F(N - 1) + F(N - 2)$	Each Fibonacci number is the sum of its two predecessors. Here the algorithm will call itself to compute both $F(N - 1)$ and $F(N - 2)$.

4. OUTPUT $F(N)$

As indicated above, this algorithm, calls itself in step 3, which makes it a recursive algorithm. Figure 1.4 illustrates the fifteen calls necessary to compute $F(6)$.

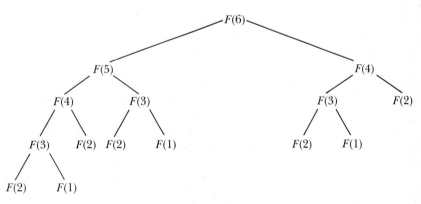

Figure 1.4

We can use the alternate form of the Principle of Induction to show that Algorithm 1.6 must make more than $\left(\frac{3}{2}\right)^{n-1} - 1$ calls to compute $F(n)$. Let $C(n)$ denote the number of calls of Algorithm 1.6 requires to compute $F(n)$. (Recall from Figure 1.4 that $C(6) = 15$.) Let S_n be the statement

$$C(n) > \left(\tfrac{3}{2}\right)^{n-1} - 1$$

Since only a single call of Algorithm 1.6 is necessary to compute $F(1)$ and $F(2)$, we have $C(1) = C(2) = 1$. Note that

$$C(1) = 1 > \left(\tfrac{3}{2}\right)^{0} - 1 = 0$$

and

$$C(2) = 1 > \left(\tfrac{3}{2}\right)^{1} - 1 = \tfrac{1}{2}$$

Therefore both S_1 and S_2 are true. (Since the computation of $F(1)$ and $F(2)$ are special cases, we check both to avoid the type of error illustrated in Example 4 of Section 1.3.)

Now we assume that statements $S_1, S_2, \ldots, S_k$ are all true, and under this assumption we show that statement S_{k+1} is also true. Observe that in order to compute $F(k + 1)$ we must first call Algorithm 1.6, which in turn calls itself to compute $F(k)$ and $F(k - 1)$. Hence we have

$$C(k + 1) = \underset{\substack{\uparrow \\ \text{the first} \\ \text{call of the} \\ \text{algorithm}}}{1} + \underset{\substack{\uparrow \\ \text{the number of} \\ \text{calls necessary} \\ \text{to compute } F(k)}}{C(k)} + \underset{\substack{\uparrow \\ \text{the number of} \\ \text{calls necessary} \\ \text{to compute } F(k - 1)}}{C(k - 1)} \qquad (2)$$

We are assuming that $S(k)$ and $S(k - 1)$ are true; thus we assume that

$$C(k) > \left(\tfrac{3}{2}\right)^{k-1} - 1 \qquad (3)$$

and

$$C(k - 1) > \left(\tfrac{3}{2}\right)^{k-2} - 1 \qquad (4)$$

Substituting (3) and (4) into (2), we obtain

$$C(k + 1) > 1 + \left(\tfrac{3}{2}\right)^{k-1} - 1 + \left(\tfrac{3}{2}\right)^{k-2} - 1$$

Consequently, we have

$$C(k + 1) > \left(\tfrac{3}{2}\right)^{k-1} + \left(\tfrac{3}{2}\right)^{k-2} - 1 = \left(\tfrac{3}{2}\right)^{k-2}\left(\tfrac{3}{2} + 1\right) - 1$$
$$= \left(\tfrac{3}{2}\right)^{k-2}\left(\tfrac{5}{2}\right) - 1$$

Since

$$\tfrac{5}{2} > \tfrac{9}{4} = \left(\tfrac{3}{2}\right)^{2}$$

it follows that

$$C(k + 1) > (\tfrac{3}{2})^{k-2}(\tfrac{3}{2})^2 - 1 = (\tfrac{3}{2})^k - 1$$

so statement S_{k+1} is also true. This concludes the induction argument.

In view of this result you can check that it would require in excess of 7,000,000 calls of Algorithm 1.6 to compute the fortieth Fibonacci number!

1.4   Problems

In Problems 1–4 give the intermediate lists that result from an application of QUICKSORT to the given list of numbers. In each case place the first number in the list in its proper position.

1. i	$A(i)$
1	5
2	8
3	3
4	1
5	7
6	2

2. i	$A(i)$
1	10
2	−3
3	2
4	4
5	11
6	12

3. i	$A(i)$
1	8
2	2
3	1
4	4
5	6
6	0
7	3

4. i	$A(i)$
1	10
2	2
3	3
4	2
5	4
6	5
7	4

In Problems 5–10 use the alternate form of the Principle of Induction to prove the given statement.

5. If $u_{n+2} = 5u_{n+1} - 6u_n$ for all n, and $u_1 = 2$ and $u_2 = 4$, then $u_n = 2^n$ for all positive integers n.

6. If $u_{n+2} = 5u_{n+1} - 6u_n$ for all n, and $u_1 = 5$ and $u_2 = 13$, then $u_n = 2^n + 3^n$ for all positive integers n.

7. If $u_{n+2} = 3u_n - 2u_{n+1}$ for all n, and $u_1 = u_2 = 2$, then $u_n = 2$ for all positive integers n.

8. If $u_{n+2} = 3u_n - 2u_{n+1}$ for all n, and $u_1 = -2$ and $u_2 = 10$, then $u_n = 1 + (-3)^n$ for all positive integers n.

9. If $u_{n+3} = 3u_{n+2} - 3u_{n+1} + u_n$ for all n, and $u_1 = 3, u_2 = 7$, and $u_3 = 13$, then $u_n = n^2 + n + 1$ for all positive integers n.

10. If $u_{n+3} = 2u_{n+2} + u_{n+1} - 2(u_n + 1)$ for all n and $u_1 = 5$, $u_2 = 8$, and $u_3 = 13$, then $u_n = 2 + n + 2^n$.

11. The average number of comparisons, $C(n)$, made by a particular sorting algorithm for a list of n entries is known to satisfy the inequality:

$$C(n) \le 1 + C(n/2) \qquad C(1) = 0$$

Show that $C(n) \le \log_2 n$ if n is a power of 2.

12. The computing time of an algorithm is known to satisfy the inequality:

$$T(n) < 8T(n/2) \qquad T(1) = 0$$

Show that $T(n) < n^3$ if n is a power of 2.

13. Write a nonrecursive algorithm for the computation of the nth Fibonacci number.

14. The Fibonacci numbers originated from a now-famous rabbit problem. Suppose that each pair of a certain breed of long-lived and amorous rabbits produces one new pair of rabbits every month. The rabbits must be at least two months old to breed, and they mate exclusively. At the beginning of month one, there is one pair of rabbits, and at the beginning of month two there is still one pair of rabbits. At the beginning of month three, however, there will be two pairs of rabbits, and at the beginning of month four there will be three pairs of rabbits. (The chart below visualizes this multiplying concept.)

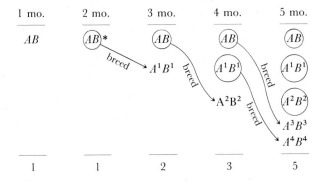

* A circle indicates the pair is old enough to breed.

Show that at the beginning of month n there will be $F(n)$ pairs of rabbits, where $F(n)$ is the nth Fibonacci number.

*15. Use the alternate form of the Principle of Induction to show that for each positive integer n, the nth Fibonacci number, $F(n)$, is defined by

$$F(n) = \frac{\alpha^n - \beta^n}{\sqrt{5}}$$

where $\alpha = (1 + \sqrt{5})/2$ and $\beta = (1 - \sqrt{5})/2$ are the solutions of the equation

$$x^2 = x + 1$$

*16. To write a more formal description of QUICKSORT to apply to a list of numbers $A(M)$, $A(M + 1), \ldots, A(K)$, we make use of two "pointers," I and J. The pointer I moves to the right through the given list in a search for some $A(I)$ such that $A(I) < A(M)$ (called X in the table below). After I is found such that $A(I) < A(M)$, the pointer J moves to the left through the list searching for an $A(J)$ such that $A(J) > A(M)$. If $I < J$, then the positions of $A(I)$ and $A(J)$ are interchanged. The procedure is illustrated in the following table, where $M = 1$ and $K = 6$.

	$A(1)$	$A(2)$	$A(3)$	$A(4)$	$A(5)$	$A(6)$
Original List	5	7	3	6	8	4
	↑		↑		↑	
	X		I		J	
New List	5	7	8	6	3	4

After several such interchanges, we would eventually obtain $J \leq I$. For example, if we apply the procedure to the new list in the above table, we have $J = 4$ and $I = 5$. Note that if P is a positive integer and if $P \leq J$, then $A(P) \geq X$; if $P > J$, then $A(P) \leq X$. Consequently, X should be at the position $A(J)$. To accomplish this, the computer interchanges $A(J)$ and $A(M) = X$.

Algorithm QUICKSORT   To sort a list $A(M)$, $A(M + 1), \ldots, A(K)$ into decreasing order.

Input: $A(M)$, $A(M + 1), \ldots, A(K)$

Output: The list in decreasing order

1. $X \leftarrow A(M)$ — $X = A(M)$ will be placed in its proper position.

2. $I \leftarrow M + 1$; $J \leftarrow K$ — The pointers I and J are initialized.
3. FOR $I < J$ — Interchanges will continue until $J \leq I$.

   a. WHILE $A(I) \geq X$ DO — I increases until we find $A(I)$ such that $A(I) < X$.

      1. $I \leftarrow I + 1$
   b. WHILE $A(J) \leq X$ DO — J decreases until we find $A(J)$ such that $A(J) > X$.

      2. $J \leftarrow J - 1$

    c. TEMP ← $A(J)$; $A(I)$ and $A(J)$ are interchanged.
      $A(J)$ ← $A(I)$;
      $A(I)$ ← TEMP
4. TEMP ← $A(J)$; $A(J)$ and $A(M)$ are interchanged.
  $A(J)$ ← X;   This places $A(M) = X$ in its proper
  $A(M)$ ← TEMP   position.
5. QUICKSORT
  $A(M), \ldots, A(J - 1)$
6. QUICKSORT
  $A(J + 1), \ldots, A(K)$
7. FOR $I = M$ TO K
  a. OUTPUT $A(I)$

Apply this algorithm step-by-step to Problems 1–4.

1.5 GREEDY ALGORITHMS

The following type of problem arises in a variety of situations. A hiker must decide which items, from a list of n items, she will carry in her backpack, and which items she will leave behind. The items are denoted by Item 1, Item 2, ..., Item n. For each i, Item i has weight, $w(i)$, and is assigned a value, $v(i)$. The backpacker wishes to carry no more than W pounds while maximizing the total value of items she carries.

In one approach to this problem, the backpacker first selects the item that has the greatest value and whose weight is less than or equal to the weight limit W. This item goes into the pack. Then she selects the item that has the next greatest value such that the total weight of this item and the first item selected is less than or equal to W. She continues this process until she cannot add any more items to the pack without surpassing the weight restriction W.

This procedure for making a selection is an example of a *greedy algorithm*. The algorithm is "greedy" in the sense that at each step, the most valuable item possible is selected with no regard to the items already chosen or to those that subsequently might be selected. This algorithm may be described more formally as follows.

Algorithm 1.7 A greedy selection of items to be included in a backpack.

      Input: $W(I)$, the weight of item I, $I = 1, 2, \ldots, N$;
           $V(I)$, the value of item I, $I = 1, 2, \ldots, N$;
           W, the weight restriction
     Output: B, a list of the items to be included;
            TW, the total weight to be carried;
            TV, the total value of the items to be carried

1. Sort the items so that
   $$V(1) \geq V(2) \geq \cdots \geq V(N)$$
2. $B \leftarrow \varnothing;\ TW \leftarrow 0;\ TV \leftarrow 0$
3. FOR $I = 1$ TO N
   a. IF $TW + W(I) \leq W$
      THEN
      1. $B \leftarrow B \cup \{I\}$
      2. $TW \leftarrow TW + W(I)$
      3. $TV \leftarrow TV + V(I)$
4. OUTPUT B, TW, TV

Use Algorithm 1.5, BUBBLESORT, for example.

Initialize; $\varnothing$ is the empty set.

Consider the items one by one, beginning with the first item. Include item I if its inclusion does not exceed the weight limit. If item I is included, then increment the total weight and the total value accordingly.

Example 1 In Table 1.5 we apply Algorithm 1.7 to the given data.

$$W = 35$$

$V(1) = 10$	$W(1) = 21$
$V(2) = 8$	$W(2) = 7$
$V(3) = 6$	$W(3) = 13$
$V(4) = 6$	$W(4) = 4$
$V(5) = 5$	$W(5) = 9$
$V(6) = 4$	$W(6) = 2$
$V(7) = 2$	$W(7) = 8$
$V(8) = 1$	$W(8) = 4$

I	—	1	2	3	4	5	6	7	8
B	$\varnothing$	$\{1\}$	$\{1, 2\}$	$\{1, 2\}$	$\{1, 2, 4\}$	$\{1, 2, 4\}$	$\{1, 2, 4, 6\}$	$\{1, 2, 4, 6\}$	$\{1, 2, 4, 6\}$
TW	0	21	28	28	32	32	34	34	34
TV	0	10	18	18	24	24	28	28	28

Table 1.5

We see that the computer selected items 1, 2, 4, and 6; they have a total weight of 34 and a total value of 28. □

Because Algorithm 1.7 started by choosing the highest-value item, it may overlook other high-value items that would cause excessive weight. As a result, this greedy algorithm does not always yield the best possible solution to this problem. In Example 1, for instance, if the backpacker were to choose items 2, 3, 4, 5, and 6, the items would have a total value of 29 and a total weight of 35. Thus in this instance, the greedy algorithm yields only an approximate solution to the problem.

This illustrates an important point about algorithms: In order to obtain the best possible solution to a given problem, it is sometimes necessary to use an extremely inefficient algorithm. At times, however, it may be better to settle for an approximate solution obtained from a more efficient—and less expensive—algorithm. (Of course, what is an acceptable approximation will depend on the particular problem under consideration. Such problems are treated in more advanced texts on algorithms.)

We can also use a greedy algorithm in the following context. Suppose a salesman is based in city x_1 and is to visit $n - 1$ other cities exactly once, and then return to city x_1. The salesman knows the costs of traveling between any two of these n cities, and he must find the route he should take in order to minimize the cost of the trip.

Suppose the salesman is to cover five cities, including his base city x_1. Figure 1.5 indicates the cost c_{ij} of travel between the cities x_i and x_j. In Figure 1.6 we indicate two possible routes and their respective costs.

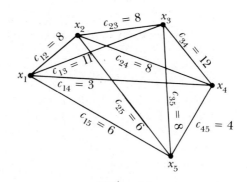

Figure 1.5

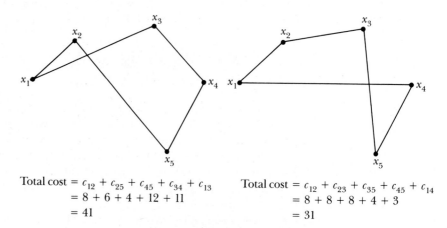

Total cost $= c_{12} + c_{25} + c_{45} + c_{34} + c_{13}$
$= 8 + 6 + 4 + 12 + 11$
$= 41$

Total cost $= c_{12} + c_{23} + c_{35} + c_{45} + c_{14}$
$= 8 + 8 + 8 + 4 + 3$
$= 31$

Figure 1.6

To determine a route using a greedy method, start with x_1 and select the city that is the least expensive to visit from x_1. For instance, from Figure 1.5 we have

$$c_{12} = 8$$
$$c_{13} = 11$$
$$c_{14} = 3$$
$$c_{15} = 6$$

Therefore, using the greedy procedure, we would choose city x_4 because the cost of going from x_1 to x_4 is less than the cost of going from x_1 to any other city.

Next we check the costs between x_4 and cities not yet visited and choose whichever city is the least expensive to reach. In the example in Figure 1.5, this would be city x_5. Continuing in this fashion, our greedy algorithm yields the route $x_1 x_4 x_5 x_2 x_3 x_1$. Note that the cost associated with the route is

$$3 + 4 + 6 + 8 + 11 = 32$$

As with the backpack problem, we have not obtained an optimal solution. (Recall that one of the routes indicated in Figure 1.6 had a total cost of 31.) Unfortunately, there is no efficient algorithm to solve the traveling salesman problem. To do so, a computer would have to examine all possible routes and compare the costs of these routes. As you are asked to show in Problem 10, there are $(n - 1)!$ possible routes if n cities are to be visited. Thus if $n = 30$ cities, the computer would have to check the costs of $29! \approx 8.8418 \times 10^{30}$ possibilities—an impossible task, even for a computer. In such a situation, you would probably settle for an algorithm that yields an approximate solution after a reasonable number of steps. Two such algorithms are discussed in Section 3.8.

Although in the preceding two examples the greedy algorithm did not lead to the best possible solution, this will not always be the case. As we see next, a greedy algorithm, when applied to a slightly different backpack problem, does result in an optimal solution.

The backpack problem discussed earlier is often called the *0–1 backpack problem* because an item either does not go into the pack or it does. Suppose, however, that we allow *portions* of the items to go into the pack. If we denote that portion or fraction of item i that goes into the pack by x_i ($0 \le x_i \le 1$), then our problem becomes that of maximizing

$$x_1 v(1) + x_2 v(2) + \cdots + x_n v(n)$$

under the condition that

$$x_1 w(1) + x_2 w(2) + \cdots + x_n w(n) \le W$$

where, as before, for each i, $v(i)$ is the value of the ith item, $w(i)$ is the weight of the ith item, and W is the total weight that the backpacker is willing to carry.

It can be shown that the following greedy algorithm yields an optimal solution to this problem.

Algorithm 1.8   A greedy algorithm for maximizing the value of items carried in a backpack when fractional parts of items are allowed.

Input: $W(I)$, the weight of item I, $1 \leq I \leq N$;
$V(I)$, the value of item I, $1 \leq I \leq N$;
W, the weight restriction.

Output: B, a list of the items to be included;
F, the weight of the portion of the last item included;
TV, the total value of the items carried

1. Sort the items so that $V(1)/W(1) \geq V(2)/W(2) \geq \ldots \geq V(N)/W(N)$

    For example use Algorithm 1.5, BUBBLESORT.

2. $B \leftarrow \varnothing$; $TW \leftarrow 0$; $TV \leftarrow 0$

    Initialize.

3. FOR $I = 1$ TO N

    a. IF $TW + W(I) \leq W$ THEN
        1. $B \leftarrow B \cup \{I\}$
        2. $TW \leftarrow TW + W(I)$
        3. $TV \leftarrow TV + V(I)$

    Examine the items in succession and include all of item I if possible; update total weight and total value.

    b. ELSE
        1. $B \leftarrow B \cup \{I\}$
        2. $F \leftarrow W - TW$
        3. $TV \leftarrow TV + (F)(V(I)/W(I))$

    Otherwise, include only a part $(W - TW)$ of item I. Update the total value.

4. OUTPUT B, F, TV

Example 2   In Table 1.6 the data of Example 1 is sorted so that $V(1)/W(1) \geq V(2)/W(2) \geq \ldots \geq V(N)/W(N)$.

Number of the item from Example 1	I	$V(I)$	$W(I)$	$V(I)/W(I)$
6	1	4	2	2
4	2	6	4	3/2
2	3	8	7	8/7
5	4	5	9	5/9
1	5	10	21	10/21
3	6	6	13	6/13
7	7	2	8	1/4
8	8	1	4	1/4

Table 1.6

In Table 1.7, we apply Algorithm 1.8 to the data from Table 1.6. ☐

I	—	1	2	3	4	5
B	$\varnothing$	$\{1\}$	$\{1, 2\}$	$\{1, 2, 3\}$	$\{1, 2, 3, 4\}$	$\{1, 2, 3, 4, 5\}$
TW	0	2	6	13	22	$F = 13$
TV	0	4	10	18	23	29 4/21

Table 1.7

1.5 Problems

In Problems 1–3, apply Algorithm 1.7 to the following sets of data.

1.

Item	Weight	Value
1	$w_1 = 8$	$v_1 = 15$
2	$w_2 = 16$	$v_2 = 13$
3	$w_3 = 4$	$v_3 = 7$
4	$w_4 = 2$	$v_4 = 7$
5	$w_5 = 9$	$v_5 = 4$
6	$w_6 = 1$	$v_6 = 2$
7	$w_7 = 3$	$v_7 = 2$

Maximum weight $W = 20$.

2.

Item	Weight	Value
1	$w_1 = 8$	$v_1 = 2$
2	$w_2 = 4$	$v_2 = 9$
3	$w_3 = 11$	$v_3 = 7$
4	$w_4 = 2$	$v_4 = 6$
5	$w_5 = 4$	$v_5 = 6$
6	$w_6 = 5$	$v_6 = 3$

Maximum weight $W = 21$

3.

Item	Weight	Value
1	$w_1 = 5$	$v_1 = 5$
2	$w_2 = 3$	$v_2 = 9$
3	$w_3 = 17$	$v_3 = 6$
4	$w_4 = 4$	$v_4 = 5$
5	$w_5 = 9$	$v_5 = 5$
6	$w_6 = 6$	$v_6 = 8$
7	$w_7 = 2$	$v_7 = 3$
8	$w_8 = 4$	$v_8 = 4$

Maximum weight $W = 32$

In Problems 4 and 5 apply a greedy algorithm to the following data for the traveling salesman problem. In each case start and end the route with city x_1.

4.

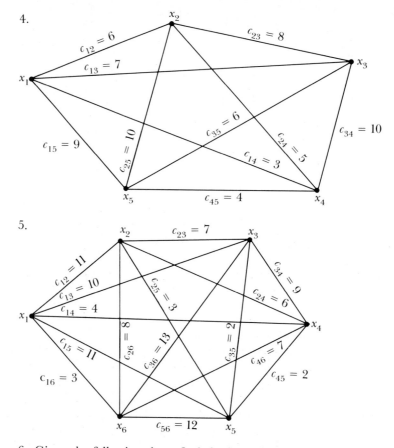

5.

6. Given the following data, find the least expensive route for the traveling salesman problem. Apply a greedy algorithm to the data and compare your results. Start and end with city x_1.

(a)

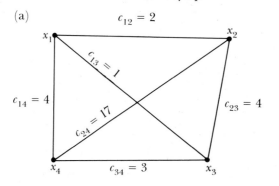

(b)

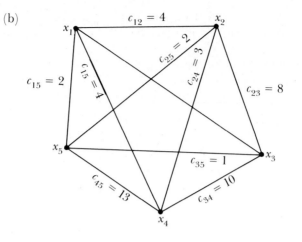

7. Apply Algorithm 1.8 to the data given in Problem 1.

8. Apply Algorithm 1.8 to the data given in Problem 2.

9. Apply Algorithm 1.8 to the data given in Problem 3.

10. Show that there are $(n - 1)!$ routes that a traveling salesman might take to visit each of n cities. Explain why only $(n - 1)!/2$ routes need be checked to find a route with minimal cost.

11. Write a greedy algorithm for the 0–1 backpack problem by placing items into the pack in order of increasing weights; that is, the lightest item is placed first, then the next lightest, and so on, until the weight capacity W is reached. Apply your algorithm to the data given in Problems 1, 2, and 3. Do you obtain an optimal solution?

12. Let $c_1, c_2, \ldots, c_n$ be a set of distinct coin values such that $c_1 > c_2 > c_3 > \cdots > c_n$. For instance, using U.S. coins, $c_1 = 25¢, c_2 = 10¢, c_3 = 5¢$, and $c_4 = 1¢$. You want to use the minimal number of coins to make up an exact amount A. For example, if $A = 63¢$, then 2 quarters, 1 dime, and 3 pennies would be the optimal solution. Describe a greedy algorithm for this problem. Does your algorithm generate an optimal solution? If not, can you find one that does?

*1.6   BACKTRACKING ALGORITHMS

One important type of algorithm involves *backtracking*. Basically, a backtracking algorithm makes some initial choices and then, as the algorithm proceeds, goes back and changes some of these choices in order to improve the final output.

To illustrate a backtracking algorithm, we consider a communications network that ties n cities together. We assume that there are enough direct links between pairs of cities to ensure that all of the cities are connected by some sequence of links. Our goal is to connect all of the cities using a *minimal* number of links in the network.

In Figure 1.7 we describe a network connecting six cities: $x_1, x_2, x_3, x_4, x_5,$ and x_6. A line segment connecting two cities indicates a direct link between the two cities. Note that only a series of links connects some cities, such as x_2 and x_5. Figure 1.8 presents two possible networks that connect all of the cities using a minimal number of links.

A network is a minimal network if it does not contain any cycles, or closed routes. (We define cycles formally in Chapter 3, but your intuitive idea of a cycle suffices here; in Figure 1.9, for instance, the links joining cities $x_3, x_4, x_5,$ and x_6 form a cycle.)

An algorithm that will produce a minimal network must generate a network that includes all of the given cities but contains no cycles. Suppose we wish to determine a minimal communications network M using the available links given in Figure 1.10.

We begin with city x_1 and find the smallest integer k for which there is a direct link $\langle x_1, x_k \rangle$ between x_1 and x_k. The first such link is with city x_3; thus $k = 3$, and we add the link $\langle x_1, x_3 \rangle$ to M. Next, we must find the smallest integer

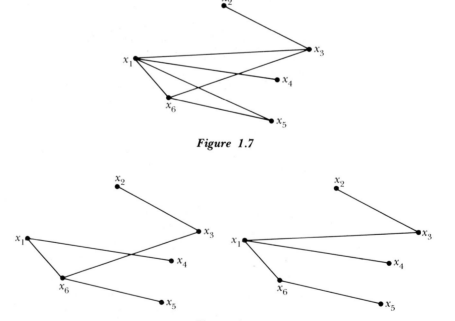

Figure 1.7

Figure 1.8

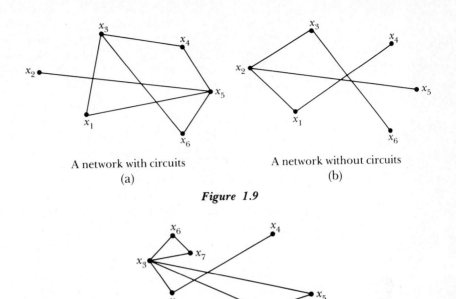

A network with circuits
(a)

A network without circuits
(b)

Figure 1.9

Figure 1.10

$k(k \neq 1)$ such that city x_k is directly connected to x_3. In our example, this is city x_2, and we assign the link $\langle x_3, x_2 \rangle$ to M. We then try to connect x_2 to a new city x_k without forming a cycle in M. From Figure 1.10 we see that we can add the link $\langle x_2, x_5 \rangle$ to M without creating such a cycle. Now we try to find a link $\langle x_5, x_k \rangle$ that we can add to M without forming a circuit in M. This, however, is impossible and we must backtrack. We return to city x_2 and look for a link $\langle x_2, x_k \rangle$—other than $\langle x_2, x_5 \rangle$—which still avoids creating a cycle in M. There is no such city, so we continue to backtrack to city x_3. Here we find that we could add either link $\langle x_3, x_6 \rangle$ or $\langle x_3, x_7 \rangle$ to M without creating a cycle; we choose the first of these and include the link $\langle x_3, x_6 \rangle$ in M. Next we add the link $\langle x_6, x_7 \rangle$ to M and then discover we cannot tie x_7 to any additional cities without forming a cycle. We backtrack to x_3, but find that we cannot use another link with x_3, either. Finally, we backtrack to city x_1 to obtain the link $\langle x_1, x_4 \rangle$, and all cities are serviced. Figure 1.11 illustrates the steps we used to achieve a minimal network M.

To describe this algorithm more formally, we let $\langle I, J \rangle$ denote a link between city I and city J, and use two functions, F and P, to keep track of the "follower" $F(K)$ and the "predecessor" $P(K)$ of a city K. More specifically $F(K)$ is an integer such that the link $\langle K, F(K) \rangle$ is under consideration for inclusion in the minimal network M; $P(K)$ is an integer such that $\langle P(K), K \rangle$ has been placed in M. We must keep track of the predecessor of a city so that we may backtrack when necessary.

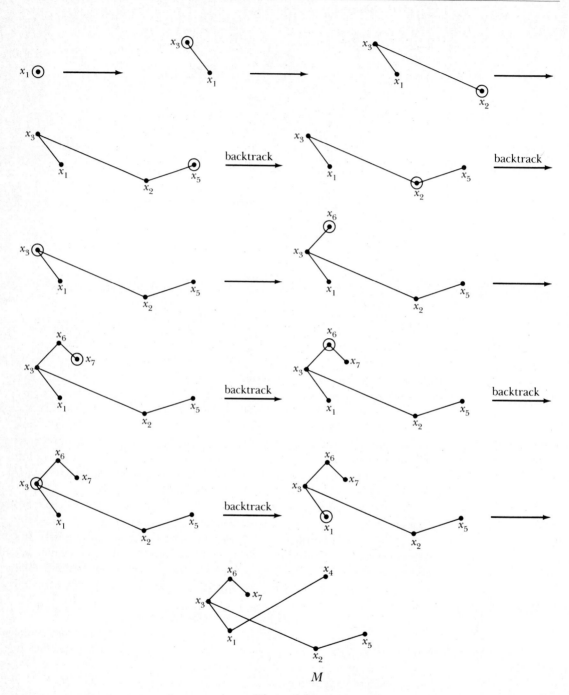

Figure 1.11

Algorithm 1.9 A backtracking algorithm for constructing a minimal network *M* contained in a network of *N* cities.

> Input: *N*, the number of cities in the original network;
>   *LINK*, a list of links connecting the cities
>
> Output: *MINNET*, a list of links comprising a minimal network containing the *N* cities

1. $K \leftarrow 1$; $MINNET \leftarrow \varnothing$	Initialize; *K* designates the city under consideration.
2. FOR $I = 1$ TO N a. $F(I) \leftarrow 1$	Initialize $F(I)$.

3. While at least one city is not included in some link in *M*
   a. IF $\langle K, F(K) \rangle \in$
   $LINK \backslash MINNET$ AND
   $MINNET \cup \langle K, F(K) \rangle$
   has no cycles
   THEN
     1. $MINNET \leftarrow MINNET$
     $\cup \langle K, F(K) \rangle$
     2. $P(F(K)) \leftarrow K$
     3. $K \leftarrow F(K)$
	If $\langle K, F(K) \rangle$ is in *LINK*, and has not been previously included in *MINNET*, and does not create any cycles in *MINNET*, then $\langle K, F(K) \rangle$ is added to *MINNET*. Replacing *K* by $F(K)$ initiates the process of searching for links that emanate from $F(K)$.

b. ELSE
     1. IF $F(K) = N$ THEN
       a. $K \leftarrow P(K)$
     2. ELSE
       a. $F(K) \leftarrow F(K) + 1$
	If $\langle K, F(K) \rangle$ cannot be included in *MINNET* and $F(K) = N$, then backtracking is necessary. Replace *K* by its predecessor, $P(K)$. If $F(K) \neq N$, then increment $F(K)$ and repeat the process.

4. OUTPUT *MINNET*

We apply Algorithm 1.9 to the network given in Figure 1.12. To trace this algorithm, we must keep a list of the current values assigned to all the variables in the algorithm: *K*, *MINNET*, $F(1)$, $F(2)$, $F(3)$, $F(4)$, $F(5)$,

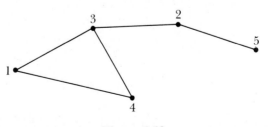

Figure 1.12

$P(2)$, $P(3)$, $P(4)$, and $P(5)$. Table 1.8 records the initial steps of Algorithm 1.9 for

$$N = 5 \quad \text{and} \quad L = \{\langle 1, 3 \rangle, \langle 1, 4 \rangle, \langle 3, 4 \rangle, \langle 3, 2 \rangle, \langle 2, 5 \rangle\}$$

In Problem 1 you are asked to complete this table.

K	MINNET	$F(1)$	$F(2)$	$F(3)$	$F(4)$	$F(5)$	$P(2)$	$P(3)$	$P(4)$	$P(5)$
1	$\varnothing$	1	1	1	1	1	—	—	—	—
1	$\varnothing$	2	1	1	1	1	—	—	—	—
1	$\varnothing$	3	1	1	1	1	—	—	—	—
3	$\langle 1, 3 \rangle$	3	1	1	1	1	—	1	—	—
3	$\langle 1, 3 \rangle$	3	1	2	1	1	—	1	—	—
2	$\langle 1, 3 \rangle, \langle 3, 2 \rangle$	3	1	2	1	1	3	1	—	—
2	$\langle 1, 3 \rangle, \langle 3, 2 \rangle$	3	2	2	1	1	3	1	—	—
2	$\langle 1, 3 \rangle, \langle 3, 2 \rangle$	3	3	2	1	1	3	1	—	—
2	$\langle 1, 3 \rangle, \langle 3, 2 \rangle$	3	4	2	1	1	3	1	—	—
2	$\langle 1, 3 \rangle, \langle 3, 2 \rangle$	3	5	2	1	1	3	1	—	—
5	$\langle 1, 3 \rangle, \langle 3, 2 \rangle, \langle 2, 5 \rangle$	3	5	2	1	1	3	1	—	2
5	$\langle 1, 3 \rangle, \langle 3, 2 \rangle, \langle 2, 5 \rangle$	3	5	2	1	2	3	1	—	2
5	$\langle 1, 3 \rangle, \langle 3, 2 \rangle, \langle 2, 5 \rangle$	3	5	2	1	3	3	1	—	2
5	$\langle 1, 3 \rangle, \langle 3, 2 \rangle, \langle 2, 5 \rangle$	3	5	2	1	4	3	1	—	2
5	$\langle 1, 3 \rangle, \langle 3, 2 \rangle, \langle 2, 5 \rangle$	3	5	2	1	5	3	1	—	2
2	$\langle 1, 3 \rangle, \langle 3, 2 \rangle, \langle 2, 5 \rangle$	3	5	2	1	5	3	1	—	2
3										

Table 1.8

For a second illustration of a backtracking procedure, we consider the *four queens* problem, stemming from the game of chess. The problem is to place four queens on a 4×4 portion of a chess board in such a way that no one queen can capture any of the other three queens. For the non-chess player, this means situating four objects, say coins, on a 4×4 board in such a way that no two of the coins lie in the same row, the same column, or the same diagonal. Two solutions to this problem are indicated in Figure 1.13.

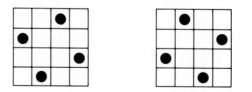

Figure 1.13

A backtracking algorithm leads to a solution of the four queens problem. First, place a coin in the square at the intersection of the first row and the first column of the board. Place another coin in the first available position of the next column so that the two coins do not lie in either the same row, the same column, or the same diagonal. Continue this process until you can no longer place a coin in the next column without violating at least one of the constraints of having no two coins in the same row, column, or diagonal. At this stage, it is necessary to backtrack. In the last column which contains a coin, try to move the coin to a different position in the same column so that

    a. the required conditions are still met, and

    b. you can place a coin in the next column without violating the given constraints.

If you cannot find such a new location, then you must backtrack further. In the next to the last column containing a coin, move the coin so that

    a. the coins in the columns up to and including this column satisfy the given requirements, and

    b. you can place coins in the next two columns without violating the restrictions.

If this proves to be impossible, then you must backtrack even further to try to improve on the earlier set-up. If the problem can be solved, repeated backtracking will eventually produce the desired solution.

The following notation will help describe the backtracking algorithm for solving the four queens problem:

R: The number of the row under consideration; $R = 1, 2, 3, 4$

C: The number of the column under consideration; $C = 1, 2, 3, 4$

$R(C)$:   The *row* in which a coin in column C is placed; for example, if $R(3) = 2$, then a coin is placed in the second row of the third column.

Algorithm 1.10   For finding a solution to the four queens problem.

       Output: $R(C)$, $C = 1, 2, 3, 4$

1. FOR $C = 1$ TO 4 a. $R(C) \leftarrow 0$ 2. $R \leftarrow 1; C \leftarrow 1$	In steps 1 and 2, the various variables are initialized.

3. WHILE $C < 4$ DO
   a. IF (R, C) is
     acceptable THEN
	At $C = 5$ this procedure will pass to step 4 where the completed results are printed out.

1. $R(C) \leftarrow R$
2. $C \leftarrow C + 1$
3. $R \leftarrow 1$

b. ELSE
    1. IF $R \geq 4$ THEN
        a. $C \leftarrow C - 1$
        b. $R \leftarrow R(C) + 1$

    2. ELSE
        a. $R \leftarrow R + 1$

4. FOR $C = 1$ TO 4
    a. OUTPUT $R(C)$

If (R, C) is an acceptable position, then place a coin there and proceed to the next column.

Otherwise determine if you are in the last row. If so, backtrack to the previous column and begin searching for the next available position. Note that if no position remains in this column, then $R = 5$, and backtracking to the previous column results.

Finally, if possible, proceed to check the next row.

You should verify that this algorithm generates the coin placements shown in Figure 1.14.

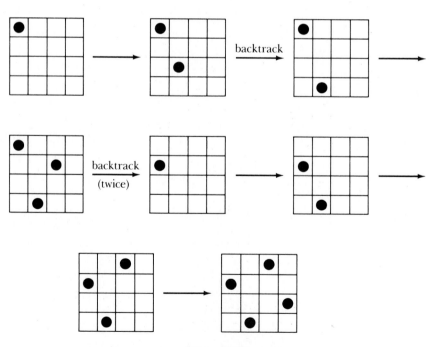

Figure 1.14

1.6   Problems

1. Complete Table 1.8 on page 40.

In Problems 2–4 apply Algorithm 1.9 to the given network. Arrange your results in a table similar to Table 1.8.

2.

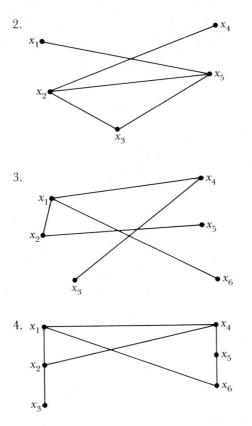

3.

4.

5. Explain how to use a backtracking algorithm for the four queens problem that will allow you to start with an entry in any square.

6. One step of Algorithm 1.10 requires the computer to determine whether (R, C) is an acceptable position. Write an algorithm to accomplish this task.

7. Using an 8×8 board, write and apply an algorithm for an eight queens problem. (The same rules apply to an 8×8 board as to the 4×4 board of the four queens problem.)

8. A magic square is an $n \times n$ array of the numbers $1, 2, \ldots, n^2$, such that the sum of the entries in any row or column is the same as the sum of the entries

in either of the two diagonals. For instance,

2	9	4
7	5	3
6	1	8

is a 3×3 magic square in which each column, row, and diagonal adds up to fifteen.

(a) Explain how to use a backtracking algorithm to obtain a 3×3 magic square.

* (b) Show that the common sum of an $n \times n$ magic square is $n(n^2 + 1)/2$.

9. Let N be a network of cities. A path in N is a sequence of cities in which successive cities are joined by a link. Explain how to use a backtracking algorithm to determine whether a network has a path that includes each city in the network exactly once. For example, the network illustrated in Figure 1.15(a) contains such a path, while the network in Figure 1.15(b) contains no such path.

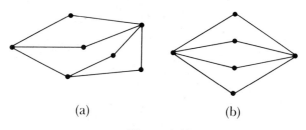

(a)   (b)

Figure 1.15

10. Explain how to use a backtracking algorithm to find a sequence of links between two specified cities of a network. Assume that the network contains sufficient links to ensure that such a path exists.

1.7 COMPLEXITY OF ALGORITHMS AND THE BIG *O* NOTATION

As we indicated earlier, one of the most important questions concerning algorithms is that of efficiency. Efficiency is frequently described in terms of what is called the *complexity* of an algorithm. This is essentially a measure of the computer time or computer storage required to carry out a given algorithm. In this text we treat complexity primarily in terms of time.

There are many ways to estimate the amount of time a computer needs to implement an algorithm. For instance, we may count the total number of operations (additions, multiplications, comparisons, and so on) the algorithm must perform. We might also express time in terms of the number of times the algorithm must execute a given instruction. In this text, we shall generally describe an algorithm's complexity by a function $S(n)$ where, for example, n is the number of items in a list or the number of cities to be visited; the value $S(n)$ may be the number of comparisons necessary to sort a list of n items or the number of possible routes to consider between n cities.

To illustrate the idea of complexity, we consider SEQUENTIAL SEARCH and BINARY SEARCH (Algorithms 1.1 and 1.2). Both algorithms find the location of a given name in an alphabetical list of n names. To measure their respective complexities, we will determine how many comparisons each algorithm requires in a *worst-case* situation (when the most comparisons are necessary).

Let $S(n)$ be the number of comparisons that SEQUENTIAL SEARCH makes in a worst-case situation when it is applied to an alphabetized list of n names. From the description of this algorithm, the worst case occurs if the input is not in the stored list, because then the computer compares the input to each item in the list (makes n comparisons) before it prints out "NOT FOUND." Therefore, we can describe the worst-case complexity of this algorithm by

$$S(n) = n \tag{1}$$

To calculate the worst-case complexity of BINARY SEARCH, we let $B(n)$ be the maximum number of comparisons the computer must make when BINARY SEARCH is applied to an alphabetized list of n names. (The worst case will occur when the input is not in the list.) Recall that BINARY SEARCH searches by eliminating half of the eligible items at each pass. If the computer has not located the input after the first comparison, then the computer will need to consider, at most, one-half of the names in future comparisons. This is true because the input $NAME\ N$ either precedes or follows the middlemost name of the list (see Figure 1.16.)

From Figure 1.16, observe that

$$B(n) \le 1 + B\left(\frac{n-1}{2}\right) \qquad \text{if } n \text{ is odd} \tag{2}$$

Total number   First   Maximum number of
of comparisons   comparison   comparisons needed after
the first comparison is done

$$B(n) \le 1 + B\left(\frac{n}{2}\right) \qquad \text{if } n \text{ is even} \tag{3}$$

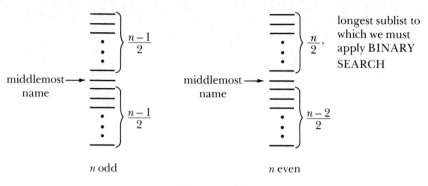

Figure 1.16

Observe, too, that if there is only one name in the stored list ($n = 1$), then only one comparison is necessary, and

$$B(1) = 1 \qquad (4)$$

Using the alternate form of the Principle of Induction we show that for each positive integer n

$$B(n) \leq 1 + \log_2 n \qquad (5)$$

If $n = 1$, then $\log_2 n = 0$, so (5) becomes $B(1) \leq 1$, which, in view of (4), is true.

Now assume that (5) is true for $n = 1, 2, \ldots, k$. Thus, we suppose that

$$B(1) \leq 1, \, B(2) \leq 1 + \log_2 2, \ldots, B(k) \leq 1 + \log_2 k$$

We must show that, under this assumption, (5) is also true for $n = k + 1$; that is,

$$B(k + 1) \leq 1 + \log_2 (k + 1) \qquad (6)$$

If $k + 1$ is odd, then it follows from (2) that

$$B(k + 1) \leq 1 + B(k/2)$$

By the induction hypothesis, we have $B(k/2) \leq 1 + \log_2 (k/2)$, and consequently,

$$B(k + 1) \leq 1 + 1 + \log_2 (k/2) = 1 + 1 + \log_2 k - \log_2 2$$

However, since $\log_2 2 = 1$, and since $\log_2 k < \log_2 (k + 1)$, we see that if $k + 1$ is an odd integer, then

$$B(k + 1) \leq 1 + \log_2 (k + 1)$$

Thus, (6) is true whenever $k + 1$ is odd.

If $k + 1$ is even, then (3) implies that

$$B(k + 1) \leq 1 + B\left(\frac{k + 1}{2}\right)$$

By the induction hypothesis, we have

$$B\left(\frac{k + 1}{2}\right) \leq 1 + \log_2\left(\frac{k + 1}{2}\right)$$

and consequently,

$$B(k + 1) \leq 1 + 1 + \log_2\left(\frac{k + 1}{2}\right) = 1 + 1 + \log_2(k + 1) - \log_2 2$$

Thus, if $k + 1$ is even, we have

$$B(k + 1) \leq 1 + \log_2(k + 1)$$

so (6) is true when $k + 1$ is even.

Since (6) is true for both the even and odd cases, we have completed the inductive proof of (5).

Moreover, since for $n > 2$,

$$1 + \log_2 n < n$$

we see in the worst case that BINARY SEARCH requires fewer comparisons than does SEQUENTIAL SEARCH, so it is the more efficient algorithm.

How do the two sorting algorithms given in Section 1.2 compare in terms of their complexity? Recall from Example 2 of Section 1.3 that if we define the complexity $S(n)$ of Algorithm 1.4 as the maximum number of comparisons the algorithm must make to order a list of n numbers, then

$$S(n) = \frac{(n - 1)n}{2}$$

In Chapter 7, you will be asked to use difference equations to show that the worst-case complexity of BUBBLESORT (Algorithm 1.5) is also given by

$$S(n) = \frac{(n - 1)n}{2}$$

Thus, rather surprisingly, Algorithms 1.4 and BUBBLESORT have the same worst-case complexity and, therefore, are equally efficient. There are, however, sorting algorithms that are more efficient than either Algorithm 1.4 or BUBBLESORT. For instance, one such algorithm (called HEAPSORT) has worst-case complexity defined by

$$S(n) = 2n \log_2 n$$

where $S(n)$ is the number of comparisons the algorithm must make in a worst-case situation.

For small values of n, $2n\log_2 n$ and $[(n-1)n]/2$ do not differ greatly; however, if, say, $n = 512$, then $2n\log_2 n = 9216$, while $[(n-1)n]/2 = 130{,}816$. In this case, Algorithm 1.3 and Algorithm 1.4 would each make fourteen times more comparisons than HEAPSORT.

From a practical standpoint, a worst-case analysis may not fairly describe an algorithm's efficiency. If the chances that a worst case will actually occur are very low, then it may be misleading to consider the efficiency of the algorithm in these terms. For instance, one of the most frequently used algorithms, the simplex method, is inefficient under a worst-case analysis. In practice, however, the algorithm is amazingly efficient, and usually only artificial examples yield the worst-case results.

Instead of a worst-case analysis, an *average-case* analysis—or even a *best-case* analysis—sometimes yields a more appropriate evaluation of an algorithm. As the name suggests, in an average-case analysis we try to measure the complexity or efficiency of an algorithm in terms of what happens "on the average." This kind of analysis requires some concepts from probability theory, however, and we defer further discussion of this topic to later chapters.

The "Big O" notation is often employed in describing the complexity of algorithms. This notation provides a convenient way to compare the "size" of various functions.

Definition 1.11   Suppose that f and g are two functions, each with domain in $\mathbf{Z}^+$. If there is a positive constant c and an integer N such that

$$|f(n)| \leq c\,|g(n)|$$

for each $n \geq N$, then f is said to be of *order g*. If f is of order g, then we say f is $O(g)$, or $f \in O(g)$, or (rather inappropriately) $f = O(g)$. Frequently, f and g are replaced by $f(n)$ and $g(n)$.

Example 1

(a) Let $f(n) = n^2 + 2n$, and $g(n) = n^2$. It is easy to verify (see Problem 3) that

$$n^2 + 2n \leq 2n^2 \tag{7}$$

for each $n \geq 2$. Therefore, $f \in O(g)$ since if we let $N = 2$ and $c = 2$, we have from (7)

$$|f(n)| \leq c\,|g(n)|$$

for each $n \geq N$.

(b) Any kth degree polynomial

$$f(n) = a_k n^k + a_{k-1} n^{k-1} + \cdots + a_1 n + a_0$$

is of order $g(n) = n^k$ since for sufficiently large values of n it can be shown that

$$|a_k n^k + a_{k-1} n^{k-1} + \cdots + a_1 n + a_0| \leq 2|a_k|n^k$$

(c)   The function $f(n) = 2^n$ is not of order n^2, since it can be shown that for any positive constant c,

$$2^n > cn^2$$

for sufficiently large values of n.

(d) 2^n is of order $n!$, since

$$2^n \leq n!$$

if $n \geq 4$. In Problem 19 you are asked to show that $n!$ is not of order a^n for any fixed positive integer a.   □

An algorithm is of *polynomial complexity* if the complexity function $S(n)$ is of polynomial order, that is, if $S(n) \in O(n^k)$ for some positive integer k. An algorithm is of *exponential complexity* if the complexity function $S(n) \in O(a^n)$ for some constant $a > 1$.

As a general rule, algorithms of polynomial complexity are "good" algorithms and are preferable to algorithms of exponential complexity. The reason is fairly obvious: An exponential function grows much more quickly than a polynomial function does. Hence, for large values of n, algorithms of polynomial complexity are considerably more efficient than those of exponential complexity. More generally, there is a hierarchy of increasing orders:

$$O(1), \; O(\log_2 n), \; O(n), \; O(n^2), \; O(n^3), \; \ldots, \; O(2^n), \; O(3^n), \; \ldots, \; O(n!), \; O(n^n)$$

Table 1.9 compares execution times for algorithms of various complexities. In this table we assume that each step of the algorithm takes one microsecond of computer time to execute.

Complexity Function	Execution time when				
	$n = 5$	$n = 10$	$n = 50$	$n = 100$	$n = 1000$
$S(n) = \log_2 n$	.000003 sec	.000004 sec	.000006 sec	.000007 sec	.00001 sec
$S(n) = 1$	.000005 sec	.00001 sec	.00005 sec	.0001 sec	.001 sec
$S(n) = n^2$	.000025 sec	.0001 sec	.0025 sec	.01 sec	1 sec
$S(n) = n^5$	.003125 sec	.1 sec	5.2 min	2.8 hrs	31.7 yrs
$S(n) = 2^n$	.000032 sec	.001 sec	35.7 yrs	4×10^{14} centuries	3×10^{285} centuries
$S(n) = 3^n$	.00024 sec	.059 sec	2×10^8 centuries	1.6×10^{32} centuries	Eons!
$S(n) = n!$	.00012 sec	3.63 sec	9.6×10^{48} centuries	Eons!	Eons!!

Table 1.9

For smaller values of n, an algorithm of exponential complexity may prove to be more efficient than an algorithm of polynomial complexity. For instance, if the complexities of two algorithms A and B are

$$f(n) = 96n^4 + 4n^3 - 2n$$

and

$$g(n) = 3^n$$

respectively, then for $n = 10$, we have

$$f(10) = 963,980$$

and

$$g(10) = 59,049$$

and, in this case, algorithm B is preferable. If $n = 30$, however, then

$$f(30) = 77,867,940$$

but

$$g(30) \approx 205,890,000,000,000$$

and clearly in this case, algorithm A is more efficient.

1.7   Problems

1. Show that in a worst-case analysis the complexity $S(n)$ of Algorithm 1.4 is

$$S(n) = \frac{(n-1)n}{2}$$

where $S(n)$ is the number of comparisons the computer must make.

2. Find the complexity in terms of the number of additions performed in order to add two $n \times n$ matrices of the form

$$\begin{pmatrix} a_{11} & a_{12} & a_{13} & \cdots & a_{1n} \\ 0 & a_{22} & a_{23} & \cdots & a_{2n} \\ 0 & 0 & a_{33} & \cdots & a_{3n} \\ \vdots & \vdots & \vdots & & \vdots \\ 0 & 0 & 0 & \cdots & a_{nn} \end{pmatrix}$$

3. Show that $n^2 + 2n \leq 2n^2$ for $n \geq 2$.

4. Show that for any c, $2^n > cn^2$ for sufficiently large n.

In Problems 5–14, state if the given statement is true or false.

5. $2n^2 - 7n \in O(n^2)$

6. $3^{n-1} \in O(2^n)$

7. $n^3 - 2n^5 \in O(n^5)$

8. $n^4 \in O(2^n)$

9. $n! \in O(n^n)$

10. $n! \in O(2^n)$

11. $(3n + 2)^2 \in O(n^2)$

12. $\log n \in O(n)$

13. $n \in O(\log n)$

14. $2^n \in O(n!)$

15. Show that if $f \in O(g)$ and k is a constant, then $kf \in O(g)$.

16. Show that if $f_1 \in O(g_1)$ and $f_2 \in O(g_2)$, then $f_1 f_2 \in O(g_1 g_2)$.

17. Show that if $f_1 \in O(g)$ and $f_2 \in O(g)$, then $f_1 + f_2 \in O(g)$.

18. Show that any kth degree polynomial $f(n) = a_k n^k + a_{k-1} n^{k-1} + \cdots + a_1 n + a_0$ is of order n^k. (*Hint:* Use induction and Problems 15–17.)

19. Show that $n!$ is not of order a^n for any positive integer a.

20. Use the results of Problems 15–17 to show that
$$(3n + n^2)(1 + \log n + n^2) \in O(n^4)$$

21. Show that $1 + 2 + \cdots + n \in O(n^2)$.

22. Show that if $f \in O(g)$ and $g \in O(h)$, then $f \in O(h)$.

Determine how many times the statement
$$1.\ \text{SUM}\ (I, J) = A(I) + A(J)$$
is executed in the portions of the algorithms given in Problems 23 and 24.

23. 1. FOR $I = 1$ TO N
    a. FOR $J = 1$ TO N
        1. SUM $(I, J) = A(I) + A(J)$
        2. OUTPUT SUM (I, J)

24. 1. FOR $I = 1$ TO N
    a. FOR $J = I$ TO N
        1. SUM $(I, J) = A(I) + A(J)$
        2. OUTPUT SUM (I, J)

Chapter 1   REVIEW

Concepts for Review

algorithm (p. 1)
SEQUENTIAL SEARCH (p. 2)

BINARY SEARCH (p. 5)
complexity (p. 6)
sorting algorithms (p. 8)
BUBBLESORT (p. 10)
Principle of Induction (p. 13)
alternate form of the Principle of Induction (p. 21)
QUICKSORT (p. 21)
greedy algorithms (p. 28)
the backpack problem (p. 28)
the traveling salesman problem (p. 30)
backtracking algorithms (p. 35)
$O(g)$ (p. 44)
polynomial complexity (p. 49)
exponential complexity (p. 49)

Review Problems

1. Describe what the following algorithm accomplishes.
   Output: S
   1. $S \leftarrow 0$
   2. FOR $I = 1$ TO N
      a. $S \leftarrow S + I^2$
   3. OUTPUT S

2. Write an algorithm that outputs POS – NEG, where POS is the number of positive entries in the list $A(1), A(2), \ldots, A(N)$, and NEG is the number of negative entries.

3. Construct a table of values of all variables that occur in Algorithm 1.2 when this algorithm is applied to the following list:
   $A(1) = AAA$ $\quad$ $A(2) = ABC$ $\quad$ $A(3) = FAR$ $\quad$ $A(4) = FUR$
   $A(5) = GUD$ $\quad$ $A(6) = HAT$ $\quad$ $A(7) = HUT$ $\quad$ $A(8) = JAR$
   $A(9) = MAR$ $\quad$ $A(10) = RAS$ $\quad$ $A(11) = TAR$
   Input: $N = RAS$

4. Follow Algorithm 1.3 through step-by-step for the list of numbers 1, 5, 10, -3, 11, 2.

5. Follow Algorithm 1.3 through step-by-step for the list of numbers -6, 4, 2, 16, -5, 1.

6. Follow Algorithm 1.4 through step-by-step for the list of numbers 1, 7, 3, -2, 6.

7. Follow Algorithm 1.4 through step-by-step for the list of numbers 3, 8, -4, -6, 5, 7.

8. Give the successive lists produced during an application of BUBBLESORT (Algorithm 1.5) to the list

   $A(1) = 5$ $\quad$ $A(2) = 6$ $\quad$ $A(3) = 4$ $\quad$ $A(4) = 7$ $\quad$ $A(5) = 1$

9. Give the successive lists produced during an application of BUBBLESORT (Algorithm 1.5) to the list

$$A(1) = 3 \qquad A(2) = -4 \qquad A(3) = 4 \qquad A(4) = 2$$
$$A(5) = 1 \qquad A(6) = 0$$

10. Write an algorithm that determines the total number of *different* entries in a list. (For instance, in the list A, C, B, B, C, D, A, there are four different entries.)

11. Use the Principle of Induction to prove that

$$3 + 5 + 7 + \cdots + (2n - 1) = n(n + 2) \qquad \text{for } n = 1, 2, \ldots$$

12. Use the Principle of Induction to prove that $x - 1$ is a factor of $x^n - 1$ for $n = 1, 2, \ldots$.

13. Give the successive lists that result from an application of QUICKSORT to the list

$$A(1) = 5 \qquad A(2) = 6 \qquad A(3) = 4 \qquad A(4) = 7 \qquad A(5) = 1$$

14. Give the successive lists that result from an application of QUICKSORT to the list

$$A(1) = 9 \qquad A(2) = 4 \qquad A(3) = -4 \qquad A(4) = 7$$
$$A(5) = 2 \qquad A(6) = 8 \qquad A(7) = -4$$

15. Use the alternate form of the Principle of Induction to prove that if

$$u_{n+2} - u_{n+1} = 2u_n \qquad \text{for all } n$$

and $u_1 = 1$ and $u_2 = 5$, then $u_n = 2^n + (-1)^n$ for all positive integers n.

16. Use the alternate form of the Principle of Induction to prove that if

$$u_{n+2} + 6u_{n+1} + 9u_n = 0 \qquad \text{for all } n$$

and if $u_1 = -6$ and $u_2 = 27$, then $u_n = (-3)^n (1 + n)$ for all positive integers n.

17. Apply Algorithms 1.7 and 1.8 to the data

Item	Weight	Value
1	$w_1 = 8$	$v_1 = 1$
2	$w_2 = 3$	$v_2 = 7$
3	$w_3 = 10$	$v_3 = 6$
4	$w_4 = 2$	$v_4 = 3$
5	$w_5 = 5$	$v_5 = 6$
6	$w_6 = 3$	$v_6 = 5$
7	$w_7 = 4$	$v_7 = 2$

Maximum weight $W = 25$

18. Apply a greedy algorithm to the following data for the traveling salesman problem. The route is to start and end with city x_1.

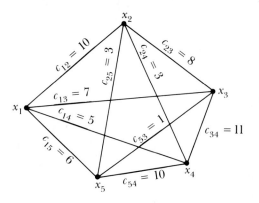

19. Explain how backtracking can be used to find a path between two given cities of a network.

20. Show that $an^3 + bn^2 \in O(n^3)$ for all constants a and b.

21. Show that $n^4 \in O(3^n)$.

22. Determine the number of times the statement

$$\text{a. SUM } (I, J, K) = A(I) + A(J) + A(K)$$

is executed in the following portion of an algorithm:

1. FOR $I = 1$ TO N
   a. FOR $J = 1$ TO N
      1. FOR $K = 1$ TO N
         a. SUM $(I, J, K) = A(I) + A(J) + A(K)$

2 Number Systems and Modular Arithmetic

We generally regard the decimal system, which uses ten digits, as the most natural number system. Computers, however, because of their "off–on" nature, use a number system with just two digits, 0 and 1. This system, called the *binary number system*, is the principal focus of this chapter.

We first develop algorithms for converting numbers from one number system to another, and then we see how to add and subtract in the binary number system. In addition, we briefly examine the concept of modular arithmetic, which plays an important role in computer operations, and then apply it to develop an "unbreakable" secret code.

2.1 NUMBERS IN OTHER BASES

We are so accustomed to working within the decimal number system that we seldom give thought to "interpreting" what a string of digits such as 3624 really means. In the decimal system this string of digits expresses in a compact way the number three thousand six hundred twenty-four: There are 3 thousands, 6 hundreds, 2 tens, and 4 ones. Written as the sum of the powers of ten, 3624 in base 10 becomes

$$3624 = 3000 + 600 + 20 + 4$$
$$= 3(10^3) + 6(10^2) + 2(10^1) + 4(10^0)$$

In general, any nonnegative integer a can be written in base 10 as

$$a = a_n(10^n) + a_{n-1}(10^{n-1}) + \cdots + a_1(10^1) + a_0(10^0) \qquad (1)$$

where $a_0, a_1, \ldots, a_n$ are integers ranging from 0 to 9. Negative powers of 10 can be used to express nonintegers. For example,

$$346.742 = 3(10^2) + 4(10^1) + 6(10^0) + 7(10^{-1}) + 4(10^{-2}) + 2(10^{-3})$$

Not all number systems are based on the integer 10. The Babylonians used the integer 60 as the base of their number system (and as a result, we still divide an hour into sixty minutes and a minute into sixty seconds). The Mayans based their number system on the integer 20 (perhaps in deference to the number of their fingers and toes). Computers often use either 2, 8, or even 16 as their number system base. In fact, any integer $b > 1$ can be a base for a number system. To express a nonnegative integer a in the base b, we first mimic (1) and write

$$a = a_n b^n + a_{n-1} b^{n-1} + \cdots + a_1 b^1 + a_0 b^0 \tag{2}$$

where the coefficients $a_0, a_1, \ldots, a_n$ may take on values $0, 1, \ldots, b - 1$. Then the number a, written in the base b, is the string of digits

$$a = (a_n a_{n-1} \cdots a_1 a_0)_b \tag{3}$$

where the subscript b indicates the particular base we are using.

Example 1   Observe that since

$$\begin{aligned} 892 &= 3(256) + 1(64) + 3(16) + 3(4) + 0(1) \\ &= 3(4^4) + 1(4^3) + 3(4^2) + 3(4^1) + 0(4^0) \end{aligned} \tag{4}$$

it follows from (2) and (3) that

$$892 = 31330_4 \qquad \square$$

In other words, the base four equivalent of the decimal number 892 is 31330.

$\square$

An algorithm that generates equations such as (4) is given in Section 2.3.

A digital computer generally uses two-state (off–on) devices. For this reason *binary numbers*—numbers written in base 2 where only two digits, 0 and 1, are used—are particularly important.

Example 2   Since

$$43 = 1(2^5) + 0(2^4) + 1(2^3) + 0(2^2) + 1(2) + 1$$

it follows from (2) and (3) that

$$43 = 101011_2$$

In other words, the binary equivalent of the decimal number 43 is 101011.

$\square$

The *octal system*, which is based on the integer 8, is also important in many computer applications.

Example 3   Since

$$216 = 3(8^2) + 3(8) + 0$$

it follows from (2) and (3) that

$$216 = 330_8$$

Thus, the octal equivalent of the decimal number 216 is 330.   □

The *hexadecimal system*, also frequently used in computer arithmetic, consists of numbers written in the base 16. The hexadecimal digits are the integers 0, 1, ..., 9 and the letters A, B, C, D, E, and F, which represent the "digits" 10, 11, 12, 13, 14, and 15, respectively (much like a deck of cards in which a Jack can represent 11, a Queen 12, and a King 13).

Example 4   Since

$$6883 = 16^3 + 10(16^2) + 14(16) + 3$$

it follows from (2) and (3) that

$$6883 = 1AE3_{16}$$

Therefore, the hexadecimal equivalent of the decimal number 6883 is $1AE3$.   □

Note that one advantage the octal and hexadecimal systems have over the binary number system is that the former require shorter strings of digits to represent a given decimal number.

Next observe that it follows from (2) and (3) that

$$(a_n a_{n-1} \cdots a_1 a_0)_b = a_n b^n + a_{n-1} b^{n-1} + \cdots + a_1 b^1 + a_0 b^0 \qquad (5)$$

In the next two examples we use (5) to convert a number written in base b to its decimal equivalent.

Example 5

(a) To convert 731_8 to its decimal equivalent, we use (5) to write

$$731_8 = 7(8^2) + 3(8^1) + 1(8^0) = 473$$

Thus, 473 is the decimal equivalent of 731_8.

(b) To convert $AB5_{16}$ to its decimal equivalent, we use (5) to write

$$AB5_{16} = 10(16^2) + 11(16) + 5 = 2741$$

Thus, 2741 is the decimal equivalent of $AB5_{16}$.   □

As in the decimal system, negative powers of base b can be used to express nonintegers. In general we have

$$(a_n \ldots a_0 a_{-1}.a_{-2} \ldots a_{-k})_b = a_n b^n + \cdots + a_0 + a_{-1} b^{-1}$$
$$+ a_{-2} b^{-2} + \cdots + a_{-k} b^{-k}$$

Example 6

(a) $101.1101_2 = 1(2^2) + 0(2^1) + 1(2^0) + 1(2^{-1}) + 1(2^{-2})$
$$+ 0(2^{-3}) + 1(2^{-4})$$

$$= 4 + 0 + 1 + \frac{1}{2} + \frac{1}{4} + 0 + \frac{1}{16}$$

$$= 5\frac{13}{16}$$

(b) $F.1CA_{16} = 15(16^0) + 1(16^{-1}) + 12(16^{-2}) + 10(16^{-3})$

$$= 15 + \frac{1}{16} + \frac{12}{16^2} + \frac{10}{16^3}$$

$$= 15\frac{458}{4096} \qquad\qquad \square$$

In the next few sections, we will consider more efficient algorithms that will convert numbers from one base to another.

2.1   Problems

1. Use the fact that $1132 = 3(7^3) + 2(7^2) + 5$ to convert 1132 to its base 7 equivalent.
2. Use that fact that $21 = 2^4 + 2^2 + 1$ to convert 21 to its binary equivalent.
3. Use the fact that $20{,}976 = 5(8^4) + 7(8^2) + 6(8)$ to convert 20,976 to its octal equivalent.
4. Use the fact that $1451 = 5(16^2) + 10(16) + 11$ to convert 1451 to its hexadecimal equivalent.
5. Use the fact that $1{,}027{,}565 = 15(16^4) + 10(16^3) + 13(16^2) + 14(16) + 13$ to convert 1,027,565 to its hexadecimal equivalent.

In Problems 6–12 convert the given numbers to decimal numbers.

6. 1001_2
7. 3562_8
8. 45.23_8
9. $3FA0_{16}$
10. 1.633_8
11. 60_{60}

12. 0_4

13. Why is it impossible to write numbers in base 1?

14. Explain why negative integers generally are not used as bases for number systems.

15. Show that if $a_0 + a_1 + \cdots + a_n$ is a multiple of 9, then so is $(a_n a_{n-1} \cdots a_1 a_0)_{10}$.

16. Show that if $a_0 + a_1 + \cdots + a_n$ is divisible by 11, then so is $(a_n a_{n-1} \cdots a_1 a_0)_{12}$.

2.2 HORNER'S ALGORITHM: CONVERTING TO THE DECIMAL SYSTEM

In the previous section we used the equation

$$(a_n a_{n-1} \cdots a_1 a_0)_b = a_n b^n + a_{n-1} b^{n-1} + \cdots + a_1 b + a_0 \qquad (1)$$

to compute the decimal equivalent of a number written in base b. Rather surprisingly, there is a more efficient method for evaluating the right-hand side of (1). This method involves nothing more than finding a more efficient means for evaluating polynomials.

A *polynomial of degree n* is a function of the form

$$f(x) = a_n x^n + a_{n-1} x^{n-1} + \cdots + a_1 x + a_0 \qquad (2)$$

where $a_n \neq 0$. Thus, for example,

$$f(x) = -2x^4 + 3x^3 - 5x + \tfrac{1}{2}$$

is a fourth-degree polynomial. In this section, we will compare two ways to evaluate a polynomial at a given number $x = b$, and then we will apply one of these methods (Horner's Algorithm) to convert nondecimal-based numbers into base 10.

The most obvious way to evaluate (2) for a given number $x = b$, is to substitute b into (2) and perform the necessary multiplications and additions to calculate $f(b)$. To determine this method's complexity, first note that we must make $n - 1$ multiplications to compute the $n - 1$ numbers $b^2, b^3, \ldots, b^n$. We must then make an additional n multiplications to calculate the products $a_1 b$, $a_2 b^2, \ldots, a_n b^n$. Finally, we must perform n additions to form the sum

$$a_n b^n + a_{n-1} b^{n-1} + \cdots + a_1 b + a_0$$

Thus, we find that an algorithm using direct substitution into (2) would require

$$(n - 1) + n + n = 3n - 1$$

operations to calculate $f(b)$. In other words, such an algorithm will have complexity $3n - 1$.

There is a way of finding $f(b)$ that requires only $2n$ operations. We first rewrite (2) as follows:

$$
\begin{aligned}
f(x) &= a_n x^n + a_{n-1} x^{n-1} + a_{n-2} x^{n-2} + \cdots + a_1 x^1 + a_0 \\
&= x^{n-1}(a_n x + a_{n-1}) + a_{n-2} x^{n-2} + a_{n-3} x^{n-3} + \cdots + a_1 x^1 + a_0 \\
&= x^{n-2}(x(a_n x + a_{n-1}) + a_{n-2}) + a_{n-3} x^{n-3} + \cdots + a_1 x^1 + a_0 \\
&= x^{n-3}(x(x(a_n x + a_{n-1}) + a_{n-2}) + a_{n-3}) + \cdots + a_1 x^1 + a_0 \\
&\ \ \vdots \\
&= \underbrace{x(x(x \ldots (x(x a_n + a_{n-1}) + a_{n-2}) + a_{n-3}) + \cdots + a_1)}_{n\ x\text{'s}} + a_0 \quad (3)
\end{aligned}
$$

Before seeing that (3) yields a more efficient method for evaluating a polynomial, we consider a specific example to illustrate the procedure for writing a polynomial in the form of equation (3).

Example 1 From (3) we find that the polynomial $f(x) = 3x^4 + 7x^3 + 2x^2 + 7x - 2$ can be written as

$$
\begin{aligned}
f(x) = 3x^4 + 7x^3 + 2x^2 + 7x - 2 &= x^3(3x + 7) + 2x^2 + 7x - 2 \\
&= x^2(x(3x + 7) + 2) + 7x - 2 \\
&= x(x(x(3x + 7) + 2) + 7) - 2
\end{aligned}
$$

$\square$

Using (3), we may evaluate $f(b)$ as follows:

$$
f(b) = \underbrace{b(b(b \ldots (b(b a_n + a_{n-1}) + a_{n-2}) + a_{n-3}) + \cdots + a_1)}_{n\ b\text{'s}} + a_0 \quad (4)
$$

To count the number of operations required to carry out (4), we first observe that the number of b's in (4) is n. To each of these b's there corresponds one multiplication, $b(\underline{\quad})$, and one addition, $b(\underline{\quad}) + \underline{\quad}$. Consequently, computing $f(b)$ using (4) requires only $n + n = 2n$ operations, whereas direct substitution requires $3n - 1$ operations. Equation (4) is the basis for *Horner's Algorithm*.

Algorithm 2.1 *Horner's Algorithm.* To compute $F(B)$ if $F(X) = A_N X^N + A_{N-1} X^{N-1} \cdots + A_1 X + A_0$.

Input: $A(N), A(N - 1), \ldots, A(0), B$ (where $A(I) = A_I$ for $I = 1, \ldots, N$)
Output: $S = F(B)$

1. $S \leftarrow A(N)$ S is initialized.

2. FOR $K = 1$ TO N Evaluation of (3) begins with the
    a. $S \leftarrow BS + A(N - K)$ innermost parenthetical expression.
                                              Successive values of S are the values
                                              of successive parenthetical
                                              expressions.

3. OUTPUT S

Example 2   To apply Horner's Algorithm to the polynomial

$$f(x) = 4x^3 + 2x^2 - x + 16$$

for $B = 2$, we input $A(3) = 4$, $A(2) = 2$, $A(1) = -1$, $A(0) = 16$, and the algorithm proceeds as follows:

$$S \leftarrow 4$$
$$K = 1, S \leftarrow 2(4) + 2 = 10$$
$$K = 2, S \leftarrow 2(10) - 1 = 19$$
$$K = 3, S \leftarrow 2(19) + 16 = 54$$

Thus, $f(2) = 54$ □

In the next two examples we use Horner's Algorithm to convert a number $(a_n a_{n-1} \cdots a_1 a_0)_b$ to its decimal equivalent. Note that by (1)

$$(a_n a_{n-1} \cdots a_1 a_0)_b = a_n b^n + a_{n-1} b^{n-1} + \cdots + a_1 b + a_0$$

and hence to find the decimal equivalent of $(a_n a_{n-1} \cdots a_1 a_0)_b$, it suffices to find $f(b)$ where

$$f(x) = a_n x^n + a_{n-1} x^{n-1} + \cdots + a_1 x + a_0$$

Example 3   To use Horner's Algorithm to convert the binary number

$$10111010_2 = 1(2^7) + 0(2^6) + 1(2^5) + 1(2^4) + 1(2^3) + 0(2^2) + 1(2) + 0$$

to its decimal equivalent, input $B = 2$, $A(7) = 1$, $A(6) = 0$, $A(5) = 1$, $A(4) = 1, A(3) = 1, A(2) = 0, A(1) = 1, A(0) = 0$. The algorithm yields the following:

$$S \leftarrow 1$$
$$K = 1, S \leftarrow 2(1) + 0 = 2$$
$$K = 2, S \leftarrow 2(2) + 1 = 5$$
$$K = 3, S \leftarrow 2(5) + 1 = 11$$
$$K = 4, S \leftarrow 2(11) + 1 = 23$$
$$K = 5, S \leftarrow 2(23) + 0 = 46$$
$$K = 6, S \leftarrow 2(46) + 1 = 93$$
$$K = 7, S \leftarrow 2(93) + 0 = 186$$

Thus $f(2) = 186$, so the decimal equivalent of 10111010_2 is 186.   □

Example 4   To use Horner's Algorithm to convert the hexadecimal number

$$ACE_{16} = 10(16^2) + 12(16) + 14$$

to its decimal equivalent, input $B = 16$, $A(2) = 10$, $A(1) = 12$, $A(0) = 14$. The algorithm yields the following:

$$S \leftarrow 10$$
$$K = 1, S \leftarrow 16(10) + 12 = 172$$
$$K = 2, S \leftarrow 16(172) + 14 = 2766$$

Thus $f(16) = 2766$, so the decimal equivalent of ACE_{16} is 2766.   □

2.2   Problems

In Problems 1–4 rewrite the given polynomial in the form used for Horner's Algorithm.

1. $f(x) = 2x^3 - 3x^2 + 4x - \frac{1}{2}$ 2. $f(x) = x^4 + 2x^2 + 6$
3. $f(x) = 3x^4 - x^3 + x$ 4. $f(x) = 2x^5 + 3x^4 - x^2 + 7$

5. Use Horner's Algorithm to find $f(4)$ for each of the polynomials given in Problems 1–4. Arrange your computations as in Example 2.
6. Use Horner's Algorithm to find $f(-2)$ for each of the polynomials given in Problems 1–4. Arrange your computations as in Example 2.

In Problems 7–11 use Horner's Algorithm to convert the given numbers to decimal numbers.

7. 10011_2 8. $3AF2_{16}$
9. 1704_8 10. 11011011_2
11. 3210_4

12. If $b^2, b^3, \ldots, b^n$ are all calculated independently of one another, how many operations are necessary to compute

$$a_n b^n + a_{n-1} b^{n-1} + \cdots + a_1 b + a_0$$

2.3   CONVERTING FROM THE DECIMAL SYSTEM TO OTHER NUMBER SYSTEMS

To convert from the decimal to other number systems, we use the following familiar result, which is known (rather inappropriately) as the Division Algorithm.

Division Algorithm   If a and b are integers and $b > 0$, then there is a unique integer q (called the quotient) and a unique integer r (called the remainder) such that

$$a = bq + r \quad \text{and} \quad 0 \le r < b$$

The following algorithm may be used to calculate the numbers q and r whose existence is assured by the Division Algorithm.

Algorithm 2.2   To find the quotient Q and remainder R of the Division Algorithm.

> Input: Integers A, B $(B > 0)$
> Output: Q, R

1. $Q \leftarrow \lfloor A/B \rfloor$
2. $R \leftarrow A - QB$
3. OUTPUT Q, R

Example 1

(a) Applying Algorithm 2.2 to $A = 268$ and $B = 5$ yields

$$Q = \left\lfloor \frac{268}{5} \right\rfloor = \lfloor 53.6 \rfloor = 53$$

and

$$R = 268 - 5(53) = 3$$

Thus, $268 = 5(53) + 3$. Note that $0 \le 3 < 5$.

(b) Applying Algorithm 2.2 to $A = -12$ and $B = 5$ yields

$$Q = \left\lfloor \frac{-12}{5} \right\rfloor = \lfloor -2.4 \rfloor = -3$$

and

$$R = -12 - (-3)(5) = 3$$

Thus, $-12 = 5(-3) + 3$. Note that $0 \le 3 < 5$.   □

As we shall see, the following algorithm, based on repeated application of the Division Algorithm, converts a decimal number to an equivalent base b number.

Algorithm 2.3   For converting a decimal number A to its equivalent base B number, $R(N + 1)R(N) \ldots R(2)R(1)R(0)$.

Input: A, B

Output: $R(0), R(1), \ldots, R(N), R(N + 1)$

1. $K \leftarrow 0$; $Q(-1) \leftarrow A$ Initialize.
2. WHILE $Q(K) \neq 0$ DO   The computer repeatedly uses the
   a. Use ALGORITHM 2.2   Division Algorithm until $Q(K) = 0$ is
      to find $Q(K)$ and   attained.
      $R(K), 0 \leq R(K) < B$
      such that
      $Q(K - 1) = BQ(K) + R(K)$
   b. $K \leftarrow K + 1$
3. FOR $K = N + 1$ TO 0
   a. OUTPUT $R(K)$

Example 2 Table 2.1 traces the values obtained when Algorithm 2.3 converts the decimal number $A = 126$ to its octal equivalent.

K	$Q(K - 1)$	$R(K - 1)$	
0	126		
1	15	6	$(126 = 8(15) + 6)$
2	1	7	$(15 = 8(1) + 7)$
3	0	1	$(1 = 8(0) + 1)$

Table 2.1

Thus the octal equivalent of the decimal number 126 is 176.   □

Example 3 Table 2.2 traces the values obtained when Algorithm 2.3 converts the decimal number $A = 4672$ to its hexadecimal equivalent.

K	$Q(K - 1)$	$R(K - 1)$	
0	4672		
1	292	0	$(4672 = 16(292) + 0)$
2	18	4	$(292 = 16(18) + 4)$
3	1	2	$(18 = 16(1) + 2)$
4	0	1	$(1 = 16(0) + 1)$

Table 2.2

Thus the hexadecimal equivalent of the decimal number 4672 is 1240.   □

To see why Algorithm 2.3 works, we observe how it computes $q_i = Q(I)$ and $r_i = R(I)$:

$$a = q_{-1}$$
$$q_{-1} = bq_0 + r_0$$
$$q_0 = bq_1 + r_1$$
$$q_1 = bq_2 + r_2$$
$$\vdots$$
$$q_{n-1} = bq_n + r_n$$
$$q_n = b(0) + r_{n+1} \tag{1}$$

The first two equations yield

$$a = bq_0 + r_0$$

Then since $q_0 = bq_1 + r_1$, we have

$$a = b(bq_1 + r_1) + r_0$$
$$= b^2 q_1 + br_1 + r_0$$

Substituting $q_1 = bq_2 + r_2$ gives us

$$a = b^2(bq_2 + r_2) + br_1 + r_0$$
$$= b^3 q_2 + b^2 r_2 + br_1 + r_0$$

If we continue in this manner, we find

$$a = b^{n+1}q_n + b^n r_n + b^{n-1} r_{n-1} + \cdots + b^2 r_2 + br_1 + r_0$$

Since by (1), $q_n = r_{n+1}$, we have

$$a = b^{n+1}r_{n+1} + b^n r_n + \cdots + b^2 r_2 + br_1 + r_0$$

so by (2) and (3) of Section 2.1, $(r_{n+1}r_n \ldots r_2 r_1 r_0)_b$ is the base b equivalent of a.

The conversions between binary and octal numbers and between binary and hexadecimal numbers are particularly simple. To convert an octal number to its binary equivalent, note that each integer 0, 1, 2, 3, 4, 5, 6, 7 used in the octal system corresponds to a "triple" in the binary system, as Table 2.3 shows. If we wish to convert from the octal to the binary system, we refer to Table 2.3 and find, for example, that

$$56421_8 = \overbrace{101}^{5}\,\overbrace{110}^{6}\,\overbrace{100}^{4}\,\overbrace{010}^{2}\,\overbrace{001}^{1} = 101110100010001_2$$

octal	binary	octal	binary
0	000	4	100
1	001	5	101
2	010	6	110
3	011	7	111

Table 2.3

Conversely, if we wish to go from the binary system to the octal system, we can arrange the digits of a given binary number into groups of three. (If necessary, we add zeros to the left-hand side of the given binary number to "come out even"):

$$1011101101_2 = \underset{1}{\underline{001}}\ \underset{3}{\underline{011}}\ \underset{5}{\underline{101}}\ \underset{5}{\underline{101}} = 1355_8$$

We can handle conversions between the binary and hexadecimal systems in much the same way, except that each hexadecimal digit corresponds to a "quadruple" in the binary system, as indicated in Table 2.4. Using Table 2.4, we convert from a hexadecimal number to a binary number as follows:

$$FACE_{16} = \overset{F}{\overline{1111}}\ \overset{A}{\overline{1010}}\ \overset{C}{\overline{1100}}\ \overset{E}{\overline{1110}} = 1111101011001110_2$$

To go from binary to hexadecimal, we arrange the digits of the given binary number into blocks of four. As before, we can add zeros on the left:

$$101010010101101011_2 = \underset{2}{\underline{0010}}\ \underset{A}{\underline{1010}}\ \underset{5}{\underline{0101}}\ \underset{6}{\underline{0110}}\ \underset{B}{\underline{1011}} = 2A56B_{16}$$

hexadecimal	binary	hexadecimal	binary
0	0000	8	1000
1	0001	9	1001
2	0010	A	1010
3	0011	B	1011
4	0100	C	1100
5	0101	D	1101
6	0110	E	1110
7	0111	F	1111

Table 2.4

2.3 Problems

In Problems 1–8 use Algorithm 2.3 to convert the decimal number a to a number in the given base b.

1. $a = 946$; $b = 8$
2. $a = 165$; $b = 2$
3. $a = 1096$; $b = 16$
4. $a = 64$; $b = 2$
5. $a = 87$; $b = 3$
6. $a = 96$; $b = 16$
7. $a = 4792$; $b = 8$
8. $a = 1000$; $b = 2$

In Problems 9–12 convert the given binary number to its octal equivalent.

9. 10110110_2
10. 1111111_2
11. 1000000_2
12. 1001101010110_2

In Problems 13–16 convert the given octal number to its binary equivalent.

13. 741_8

14. 1352_8

15. 1001_8

16. 7025_8

In Problems 17–20 convert the given binary number to its hexadecimal equivalent.

17. 1001110101_2

18. 1111111111_2

19. 1001101110010_2

20. 101_2

In Problems 21–24 convert the given hexadecimal number to its binary equivalent.

21. $9C2_{16}$

22. 1111_{16}

23. $CAFE_{16}$

24. $60D1_{16}$

In Problems 25–28 convert the given hexadecimal number to its octal equivalent.

25. $4B32_{16}$

26. $F00F_{16}$

27. $30C2_{16}$

28. $2A5C3_{16}$

In Problems 29–32 convert the given octal number to its hexadecimal equivalent.

29. 176_8

30. 4612_8

31. 6003_8

32. 1423_8

33. A method for multiplication, called the "Russian peasant algorithm," was used in the early part of this century in Russia. We describe this algorithm by means of an example. Suppose we wish to multiply 25 by 13. We write the two numbers at the head of two columns. In the first column we successively divide by two and discard any remainders. In the second column we successively double the numbers. Thus, we have

25	13
12	26
6	52
3	104
1	208

We then cross out all the even numbers in the left-hand column, along with the corresponding number in the right-hand column. The desired product is the sum of the remaining numbers in the right-hand column.

Thus, we have

25	13
12	26
6	52
3	104
1	208
	325

(a) Use the Russian peasant algorithm to compute $17(43)$ and $41(126)$.

(b) Explain why the Russian peasant algorithm works. (*Hint:* In the given example note that $25 = 1(2^4) + 1(2^3) + 0(2^2) + 0(2^1) + 1(2^0)$.)

34. A procedure for converting a fractional decimal number such as .365 to a binary number can be described as follows. (Technically, this procedure is not an algorithm because it does not necessarily terminate after a finite number of steps.)

"*Algorithm*" For converting a decimal number $.A_1A_2A_3\ldots$ to a binary number $.B_1B_2B_3\ldots$.

Input: $A_1, A_2, A_3, \ldots$
Output: $B_1, B_2, B_3, \ldots$

1. $K \leftarrow 1$
2. FOR $I = 1, 2, \ldots$
    a. $A_I^{(1)} \leftarrow A_I$
3. WHILE $2(.A_1^{(K)}A_2^{(K)} \ldots) \neq 1$ DO

    For notational convenience we have replaced $A(J, K)$ by $A_J^{(K)}$.

    a. $2(.A_1^{(K)}A_2^{(K)} \ldots) = C + .A_1^{(K+1)}A_2^{(K+1)} \ldots$
        where $C = 0$ or $C = 1$

    $C = 0$ if $2(.A_1^{(K)}A_2^{(K)} \ldots) < 1$ and
    $C = 1$ if $2(.A_1^{(K)}A_2^{(K)} \ldots) > 1$.

    b. IF $2(.A_1^{(K)}A_2^{(K)} \ldots) > 1$ THEN
        1. $B_K \leftarrow 1$
    c. IF $2(.A_1^{(K)}A_2^{(K)} \ldots) < 1$ THEN
        1. $B_K \leftarrow 0$
    d. $K \leftarrow K + 1$
4. $B_K \leftarrow 1$
5. OUTPUT $B_1, B_2, \ldots$

To illustrate this algorithm we determine the binary representation of the fractional decimal .625. First we set $A_1^{(1)} = 6$, $A_2^{(1)} = 2$, $A_3^{(1)} = 5$. Applying the algorithm yields the following values

K	$2(.A_1^{(K)}A_2^{(K)} \ldots) = C + .A_1^{(K+1)}A_2^{(K+1)} \ldots$	B_K
1	$2(.625) = 1 + .250$	1
2	$2(.250) = 0 + .500$	0
3	$2(.500) = 1 + 0$	1

(Note that the procedure stops at this point since $2(.500) = 1$.) Thus we find that the binary equivalent of .625 is .101.

For Problems (a)–(d), apply this algorithm to the given fractional decimal.

(a) .125   (b) .101

(c) .3   (d) .0025

(e) Explain why this "algorithm" works.

2.4 OPERATIONS WITH BINARY NUMBERS

Adding binary numbers is similar to adding decimal numbers: Whenever a column adds up to two or more (rather than adding up to ten or more as in the decimal system), we carry over a 1 in the column to the left. The following example illustrates the procedure.

$$
\begin{array}{r}
1 \quad 1 \quad 1\,1 \\
1\,0\,1\,0\,1\,0\,1\,1\,0 \\
+ \quad\;\; 1\,0\,1\,0\,1\,1\,1 \\
\hline
1\,1\,0\,1\,0\,1\,1\,0\,1
\end{array}
\qquad
\begin{array}{r}
1 \\
342 \\
+\; 87 \\
\hline
429
\end{array}
$$

In Chapter 6 we shall employ switching system theory to implement binary addition on the computer.

Subtracting binary numbers is similar to subtracting decimal numbers. As in the decimal system, we can "borrow" from one column to another. To do so, we use the fact that in the binary number system

$$
\begin{array}{r}
10 \\
-\; 1 \\
\hline
1
\end{array}
$$

For example, to subtract 1011110_2 from 10101101_2, we have

$$
\begin{array}{r}
1\;10\,10 \\
0\;10\,0\;\not{10}\;\not{0}\;\not{0}\;10 \\
\not{1}\,\not{0}\,\not{1}\,\not{0}\,\not{1}\,\not{1}\,\not{0}\,1 \\
-\quad\; 1\,0\,1\,1\,1\,1\,0 \\
\hline
1\,0\,0\,1\,1\,1\,1
\end{array}
\qquad
\begin{array}{r}
16\;13 \\
\not{1}\;\not{7}\;\not{3} \\
-\quad 9\;4 \\
\hline
7\;9
\end{array}
$$

One method for subtracting numbers that is particularly adaptable to computer use employs the notion of "complements" and does not require borrowing. To explain the concept of complements, we will first see how it is used in the decimal system.

We obtain the *nine's complement* of a number a by subtracting each digit of a from 9, and we obtain the *ten's complement* of a by adding 1 to the nine's complement of a.

Example 1

(a) If $a = 34,506$, then the nine's complement, a^*, of a is found from the subtractions

$$
\begin{array}{ccccc}
9 & 9 & 9 & 9 & 9 \\
-3 & -4 & -5 & -0 & -6 \\
\hline
6 & 5 & 4 & 9 & 3
\end{array}
$$

Thus

$$a^* = 65,493$$

and the ten's complement, a^{**}, of a is given by

$$a^{**} = a^* + 1 = 65,494$$

(b) When using nine's complements to subtract, we will often need to place zeros on the left of a number. For instance, we could rewrite $a = 4702$ as

$$a = 0,004,702$$

Then the nine's complement is given by

$$a^* = 9,995,297$$

and the ten's complement is given by

$$a^{**} = a^* + 1 = 9,995,298 \qquad \square$$

Using complements avoids the problem of borrowing when we subtract decimal numbers. Suppose we wish to perform the subtraction

$$
\begin{array}{r}
2,355,810 \\
-\quad 34,506 \\
\hline
\end{array}
$$

This is obviously equivalent to the problem

$$
\begin{array}{r}
2,355,810 \\
-0,034,506 \\
\hline
\end{array}
$$

Observe that the nine's complement of $a = 0,034,506$ is $a^* = 9,965,493$, and the ten's complement of a is $a^{**} = 9,965,493 + 1$. We have then

$$2,355,810 - 34,506$$
$$= 2,355,810 - 0,034,506$$

$$= 2,355,810 + (\underbrace{\overbrace{9,999,999 - 0,034,506}^{a^*} + 1}_{a^{**}}) - 10,000,000$$

$$= 2,355,810 + \underbrace{(9,965,493 + 1)}_{a^{**}} - 10,000,000$$

$$= 12,321,304 - 10,000,000 = 2,321,304$$

Thus, we have subtracted without resorting to borrowing. Moreover, we subtract $10,000,000$ by simply "ignoring" the leftmost 1 that occurs when we add the ten's complement of $0,034,506$ to $2,355,810$.

In general then, to subtract a decimal number $a = a_k a_{k-1} \ldots a_1 a_0$ from a decimal number $b = b_{n-1} \ldots b_1 b_0$, we use the equality

$$b - a = b + a^{**} - 10^n$$

where a^{**} is the ten's complement of

$$\underbrace{00 \ldots 0 a_k a_{k-1} \ldots a_1 a_0}_{n \text{ digits}}$$

and n is the number of digits appearing in b.

Example 2   If $b = 4,703,121$ and $a = 2886$, we can rewrite a as

$$a = 0,002,886$$

Then the ten's complement a^{**} of a is given by

$$a^{**} = a^* + 1 = 9,997,113 + 1$$

To calculate $4,703,121 - 2886$, we add $4,703,121$ and a^{**} and discard the leftmost 1 in this sum to obtain

$$
\begin{array}{r}
4,703,121 \leftarrow b \\
+ \ 9,997,114 \leftarrow a^{**} \\
\hline
\cancel{1}4,700,235
\end{array}
$$

Consequently, we find (without borrowing) that $4,703,121 - 2886 = 4,700,235$. □

To subtract binary numbers, we can use essentially the same procedure. We obtain the *one's complement* of a binary number by subtracting each digit of the binary number from 1; we obtain the *two's complement* of a binary number by adding 1 to the one's complement.

Example 3
   (a) If $a = 11001011110$, then the one's complement, a^*, of a is

$$a^* = 00110100001$$

and the two's complement, a^{**}, of a is

$$a^{**} = 00110100001 + 1 = 00110100010$$

(b) If $a = 00001010$, then the one's complement is

$$a* = 11110101$$

and the two's complement is

$$a** = a* + 1 = 11110110 \qquad \square$$

The following example illustrates how to subtract binary numbers without borrowing.

Example 4 To find

$$\begin{array}{r} 111011011 \\ - \quad 101100 \\ \hline \end{array}$$

we first determine $a*$, the one's complement of 000101100, and see that

$$a* = 111010011$$

The two's complement $a**$ of 000101100 is

$$a** = a* + 1 = 111010011 + 1 = 111010100$$

Next, we add $b = 111011011$ and $a**$ and discard the leftmost 1 of this sum:

$$\begin{array}{r} 111011011 \leftarrow b \\ + \ 111010100 \leftarrow a** \\ \hline \cancel{1}110101111 \end{array}$$

Therefore $111011011 - 101100 = 110101111$. $\qquad \square$

2.4 Problems

1. Find the following binary sums:

   (a) $\begin{array}{r} 10010 \\ + \ 1101 \\ \hline \end{array}$ (b) $\begin{array}{r} 10101101 \\ + 11001111 \\ \hline \end{array}$

   (c) $\begin{array}{r} 1010011 \\ + \quad 11111 \\ \hline \end{array}$ (d) $\begin{array}{r} 101.10011 \\ + \ 11.11011 \\ \hline \end{array}$

2. Perform the following binary subtractions without using two's complements.

   (a) $\begin{array}{r} 10111001 \\ - \quad 101101 \\ \hline \end{array}$ (b) $\begin{array}{r} 10000000 \\ - \qquad 1 \\ \hline \end{array}$

   (c) $\begin{array}{r} 1011001 \\ - \quad 1111 \\ \hline \end{array}$ (d) $\begin{array}{r} 101.00101 \\ - \ 11.01111 \\ \hline \end{array}$

3. Use ten's complements to perform the following subtractions.

   (a) $\begin{array}{r} 37804 \\ -\ \ 5693 \\ \hline \end{array}$ (b) $\begin{array}{r} 1004567 \\ -\ \ \ \ 5983 \\ \hline \end{array}$ (c) $\begin{array}{r} 678234 \\ -\ \ \ 9999 \\ \hline \end{array}$

4. Use two's complements to perform the following binary subtractions.

   (a) $\begin{array}{r} 1001100 \\ -\ \ \ 11010 \\ \hline \end{array}$ (b) $\begin{array}{r} 100000 \\ -\ \ \ \ \ \ 1 \\ \hline \end{array}$

   (c) $\begin{array}{r} 10010010 \\ -\ \ \ \ 11111 \\ \hline \end{array}$ (d) $\begin{array}{r} 10101101 \\ -\ 1101110 \\ \hline \end{array}$

5. Develop a method to multiply binary numbers.
6. Use your method from Problem 5 to perform the given binary multiplications, and then check your answers by converting the binary numbers to decimal numbers.

   (a) $\begin{array}{r} 101101 \\ \times\ \ \ 1101 \\ \hline \end{array}$ (b) $\begin{array}{r} 1101101 \\ \times\ \ \ \ \ 111 \\ \hline \end{array}$

   (c) $\begin{array}{r} 11.011 \\ \times\ \ \ 1011 \\ \hline \end{array}$ (d) $\begin{array}{r} 110.11 \\ \times\ 1.101 \\ \hline \end{array}$

7. Develop a method to divide binary numbers.
8. Use your method from Problem 7 to perform the following binary divisions, and then check your answers by converting the binary numbers to decimal numbers.

   (a) $101 \overline{)110110}$ (b) $11 \overline{)110001}$ (c) $110 \overline{)1011011}$

9. Write an algorithm to compute the sum of two binary numbers

$$(a_n a_{n-1} \ldots a_1 a_0)_2 \quad \text{and} \quad (b_m b_{m-1} \ldots b_1 b_0)_2$$

10. Write an algorithm to compute the difference of two binary numbers

$(a_n a_{n-1} \ldots a_1 a_0)_2$ and $(b_m b_{m-1} \ldots b_1 b_0)_2$ (*Hint:* Use two's complements.)

2.5   MODULAR ARITHMETIC

Computers are basically finite machines. They have finite storage and can only deal with numbers of some finite length. Because of such limitations, computers often use modular arithmetic.[1] Modular arithmetic also has

---

[1] Modular arithmetic relies heavily on relations, which are discussed in Section 5 of the Appendix.

applications in such problems as hashing, encryption (the development of secret codes), and, as we shall see, random number generation. Modular arithmetic utilizes the following relation R defined on the set of integers $\mathbf{Z}$.

Let m be a positive integer. Define the relation R on $\mathbf{Z}$ by

$$xRy \quad \text{if} \quad x = y + am \quad \text{for some integer } a$$

This particular relation is discussed further in Section 5 of the Appendix. As shown in the Appendix, R is an equivalence relation with equivalence classes $[0], [1], [2], \ldots, [m-1]$. In this context, it is customary to replace the symbol xRy with

$$x \equiv y \pmod{m}$$

The set of equivalence classes generated by the equivalence relation $x \equiv y \pmod{m}$ is frequently denoted by $\mathbf{Z}_m$. Thus,

$$\mathbf{Z}_m = \{[0], [1], \ldots, [m-1]\}$$

Example 1 If $m = 5$, the set $\mathbf{Z}_5$ of equivalence classes resulting from the equivalence relation $x \equiv y \pmod 5$ consists of the elements $[0], [1], [2], [3]$, and $[4]$, where

$$
\begin{aligned}
[0] &= \{\ldots -10, -5, 0, 5, 10, \ldots\} \\
[1] &= \{\ldots -9, -4, 1, 6, 11, \ldots\} \\
[2] &= \{\ldots -8, -3, 2, 7, 12, \ldots\} \\
[3] &= \{\ldots -7, -2, 3, 8, 13, \ldots\} \\
[4] &= \{\ldots -6, -1, 4, 9, 14, \ldots\}
\end{aligned}
$$
$\square$

For each positive integer m, the basic operations of addition, subtraction, and multiplication are defined on the set $\mathbf{Z}_m$ of equivalence classes by

$$[x] + [y] = [x + y] \tag{1}$$
$$[x] - [y] = [x - y] \tag{2}$$
$$[x][y] = [xy] \tag{3}$$

Thus, for instance if $m = 5$, we have

$$
\begin{aligned}
[6] + [9] &= [15] = [0] \\
[14] - [5] &= [9] = [4] \\
[-6][7] &= [-42] = [3]
\end{aligned}
$$

There is a potential problem, however, with these definitions of addition, subtraction, and multiplication. Suppose, for example, that

$$[x] = [w] \quad \text{and} \quad [y] = [z] \tag{4 and 5}$$

It would seem reasonable that

$$[x] + [y] = [w] + [z] \tag{6}$$

But by definition

$$[x] + [y] = [x + y] \quad \text{and} \quad [w] + [z] = [w + z]$$

Thus, in order for (6) to hold, we must have

$$[x + y] = [w + z] \tag{7}$$

The question then is whether (7) necessarily follows from (4) and (5).

For instance, let $m = 6$. Then $[9] = [63]$ and $[2] = [164]$. By (1) we have

$$[9] + [2] = [11] \tag{8}$$

and

$$[63] + [164] = [227] \tag{9}$$

But does it necessarily to follow from (8) and (9) that $[11] = [227]$? Certainly this must be the case if we are to conclude that

$$[9] + [2] = [63] + [164]$$

In this particular instance, we can verify directly that $11 \equiv 227 \pmod{6}$, and, hence, $[11] = [227]$. We must, however, verify that this will be true *in general*.

Theorem 2.4   Let m be a positive integer and let $\mathbf{Z}_m$ be the set of equivalence classes generated by the relation $x \equiv y \pmod{m}$. Suppose that $[x] = [w]$ and $[y] = [z]$. Then

$$[x + y] = [w + z]$$

Proof: Since $[x] = [w]$, it follows that $x \equiv w \pmod{m}$, and hence,

$$x = w + am \tag{10}$$

for some integer a. Similarly, since $[y] = [z]$, it follows that $y \equiv z \pmod{m}$ and

$$y = z + bm \tag{11}$$

for some integer b.

Adding (10) and (11) we have

$$x + y = w + z + am + bm = w + z + (a + b)m$$

and, therefore, $x + y = w + z + cm$ (where $c = a + b$). Consequently, $x + y \equiv w + z \pmod{m}$ and we have $[x + y] = [w + z]$, which is what we wished to show.   ■

Subtraction and multiplication are also well defined in modular arithmetic. That is, if $[x] = [w]$ and $[y] = [z]$, then $[x - y] = [w - z]$ and $[xy] = [wz]$ (see Problems 4 and 5).

The commutative, associative, and distributive properties also hold for modular arithmetic. In other words:

$$[x] + [y] = [y] + [x]$$
$$[x][y] = [y][x]$$
$$([x] + [y]) + [z] = [x] + ([y] + [z])$$
$$([x][y])[z] = [x]([y][z])$$
$$[x]([y] + [z]) = [x][y] + [x][z]$$

For example, to establish the distributive property

$$[x]([y] + [z]) = [x][y] + [x][z]$$

note that

$$[x]([y] + [z]) = [x][y + z] = [x(y + z)] = [xy + xz]$$
$$= [xy] + [xz] = [x][y] + [x][z]$$

The other properties may be established in a similar manner (see Problems 6 and 7).

Modular arithmetic is helpful for generating lists of random numbers. Random number generation is of considerable interest in computer science because random numbers are useful in simulation, coding, and a host of other contexts. A sequence of numbers is random if each number in the sequence is independent of the preceding numbers; there is no pattern to help us predict any number of the sequence.

When a computer generates a random sequence, it is actually producing *pseudorandom numbers*, because the computer must utilize some completely predetermined calculation to "choose" the numbers. With modular arithmetic, however, we can generate a sequence of numbers that appear to be essentially random.

In one quite simple but commonly used procedure for generating lists of pseudorandom numbers, we choose four integers, x_0, a, b, and m, so that

a. x_0 is the initial number of the sequence (frequently called the *seed* or *generator*);
b. a is the multiplier;
c. b is the increment; and
d. m is the modulus.

The sequence generated by this procedure is defined by

$$x_0$$
$$x_1 \equiv ax_0 + b \pmod{m}$$
$$x_2 \equiv ax_1 + b \pmod{m}$$
$$x_3 \equiv ax_2 + b \pmod{m}$$
$$\vdots$$
$$x_{n+1} \equiv ax_n + b \pmod{m}$$

(12)

where $0 \leq x_i < m$.

For instance, if $x_0 = 4$, $a = 6$, $b = 3$, and $m = 11$, then we obtain

$x_0 \equiv 4$

$x_1 \equiv (6 \cdot 4 + 3) \ (\text{mod } 11) \equiv 5$

$x_2 \equiv (6 \cdot 5 + 3) \ (\text{mod } 11) \equiv 0$

$x_3 \equiv (6 \cdot 0 + 3) \ (\text{mod } 11) \equiv 3$

$x_4 \equiv (6 \cdot 3 + 3) \ (\text{mod } 11) \equiv 10$

$x_5 \equiv (6 \cdot 10 + 3) \ (\text{mod } 11) \equiv 8$

$x_6 \equiv (6 \cdot 8 + 3) \ (\text{mod } 11) \equiv 7$

$x_7 \equiv (6 \cdot 7 + 3) \ (\text{mod } 11) \equiv 1$

$x_8 \equiv (6 \cdot 1 + 3) \ (\text{mod } 11) \equiv 9$

$x_9 \equiv (6 \cdot 9 + 3) \ (\text{mod } 11) \equiv 2$

$x_{10} \equiv (6 \cdot 2 + 3) \ (\text{mod } 11) \equiv 4$

$x_{11} \equiv (6 \cdot 4 + 3) \ (\text{mod } 11) \equiv 5$

$x_{12} \equiv (6 \cdot 5 + 3) \ (\text{mod } 11) \equiv 0$

$\vdots$

Note that beginning with x_{10}, this sequence begins to repeat itself. Therefore, in this case, the random sequence this procedure generates consists of the numbers $x_0, x_1, \ldots, x_9$.

Observe, however, that if we set $x_0 = 2$, $a = 5$, $b = 3$, and $m = 6$, then we obtain

$$x_0 \equiv 2$$
$$x_1 \equiv (5 \cdot 2 + 3)(\text{mod } 6) \equiv 1$$
$$x_2 \equiv (5 \cdot 1 + 3)(\text{mod } 6) \equiv 2$$

and the sequence has already begun to repeat itself. Thus what we choose as x_0, a, b, and m is crucial to the process of generating long lists of random numbers.

2.5   Problems

1. Determine whether the following are true or false.
   (a) $9 \equiv 3 (\text{mod } 7)$
   (b) $6 \equiv -15 \ (\text{mod } 7)$
   (c) $5 \equiv 5 \ (\text{mod } 19)$
   (d) $-2 \equiv 162 \ (\text{mod } 4)$

2. Determine whether the following are true or false.
   (a) $15 \equiv -15 \ (\text{mod } 15)$
   (b) $6 \equiv 14x \ (\text{mod } 4)$, for each integer x
   (c) $-7 \equiv -159 \ (\text{mod } 6)$
   (d) $1 \equiv p \ (\text{mod } 2)$ for each prime integer $p > 2$.

3. (a) List the equivalence classes that comprise $\mathbf{Z}_6$.
   (b) List the equivalence classes that comprise $\mathbf{Z}_2$.

4. Show that modular subtraction is well defined; that is, show that if $[x]$, $[y]$, $[w]$, and $[z]$ are members of $\mathbf{Z}_m$, and if $[x] = [w]$ and $[y] = [z]$, then $[x - y] = [w - z]$.

5. Show that modular multiplication is well defined; that is, show that if $[x]$, $[y]$, $[w]$, and $[z]$ are members of $\mathbf{Z}_m$, and if $[x] = [w]$ and $[y] = [z]$, then $[xy] = [wz]$.

6. Show that modular addition and multiplication are commutative. That is, show
   (a) $[x] + [y] = [y] + [x]$
   (b) $[x][y] = [y][x]$

7. Show that modular addition and multiplication are associative. That is, show
   (a) $([x] + [y]) + [z] = [x] + ([y] + [z])$
   (b) $([x][y])[z] = [x]([y][z])$

8. Investigate possible ways of defining modular division and discuss problems that arise with your definitions.

9. Compute and express your answer in terms of an equivalence class $[a]$ in $\mathbf{Z}_m$ where $0 \le a < m$.
   (a) $[17] + [34]$; $m = 8$ (b) $[15] + [-63]$; $m = 4$
   (c) $[19] - [6]$; $m = 5$ (d) $[21][-5]$; $m = 7$

10. Compute and express your answer in terms of an equivalence class $[a]$ in $\mathbf{Z}_m$ where $0 \le a < m$.
    (a) $[256] + [34]$; $m = 11$
    (b) $[14] - [38]$; $m = 5$
    (c) $[32][17]$; $m = 32$
    (d) $[15][31]$; $m = 4$

11. Show that if $a \equiv b \pmod m$ and $b \equiv c \pmod m$, then $a \equiv c \pmod m$.

12. Show that if $a \equiv b \pmod m$ and d is a positive integer that divides m, then $a \equiv b \pmod d$.

13. Suppose that $[x] \in \mathbf{Z}_m$. Then $[y] \in \mathbf{Z}_m$ is said to be an *inverse* of $[x]$ if $[x][y] = [1]$. Show that $[3]$ is an inverse of $[5]$ in $\mathbf{Z}_7$. Does $[3]$ have an inverse in $\mathbf{Z}_6$?

14. Show that if m is prime, then every equivalence class $[0], [1], \ldots, [m-1]$ has an inverse in $\mathbf{Z}_m$. (*Hint:* You may use the fact that if integers a and b have no common divisor (other than 1 and -1), then there is an integer x such that $ax \equiv 1 \pmod b$.)

15. Show that if p is a prime number, then $p \equiv 1 \pmod 4$ or $p \equiv 3 \pmod 4$.

16. Determine the pseudorandom numbers that are generated if
    (a) $x_0 = 3$, $a = 7$, $b = 4$, $m = 15$
    (b) $x_0 = 8$, $a = 3$, $b = 2$, $m = 7$
    (c) $x_0 = 4$, $a = 5$, $b = 11$, $m = 14$

17. Show that the maximum number of pseudorandom numbers that can be generated using the method described by (12) of this section is m.

*2.6 A MODULAR CODE

In this section we will develop a secret code based on modular arithmetic and certain properties of the integers. This code is quite difficult to crack, even with high-speed computers.

To describe this code we will need some basic results from number theory. We will not prove all of these results, but those of you who are interested in pursuing them in more depth are encouraged to consult any standard number theory text.

Recall that the *greatest common divisor* of two integers a and b is the largest positive integer that divides both a and b. For instance, the greatest common divisor of -60 and 32 is 4. We shall use the notation $\gcd(a, b)$ to denote the greatest common divisor of the integers a and b. (Problem 19 of Section 1.1 gives an algorithm for the computation of gcd (a, b).)

We say that two integers a and b are *relatively prime* if $\gcd(a, b) = 1$. The integers 12 and 35 are relatively prime since $\gcd(12, 35) = 1$.

The following theorem is one of the most useful results in number theory.

Theorem 2.5   If a and b are integers such that $\gcd(a, b) = 1$, then there is a positive integer x such that

$$ax \equiv 1 \ (\mathrm{mod} \ b) \tag{1}$$

An integer x that satisfies (1) may be found by repeatedly applying the Division Algorithm (see Section 2.3). We illustrate this process in the next two examples.

Example 1   We can easily verify that the integers 735 and 52 are relatively prime. Hence, by Theorem 2.5 there is a positive integer x such that

$$735x \equiv 1 \ (\mathrm{mod} \ 52)$$

To find x, we apply the Division Algorithm to the numbers 735 and 52 and then to the pairs of circled numbers in each line below. We obtain the equations:

$$735 = 14(\,\boxed{52}\,) + \boxed{7}$$
$$52 = 7(\boxed{7}) + \boxed{3}$$
$$7 = 2(3) + 1$$

We terminate this procedure when a remainder of 1 occurs. Starting with the last of these equations and working our way up with the appropriate substitutions, we now find

$$\begin{aligned}
1 &= 7 - 2(3) \\
&= 7 - 2(52 - 7(7)) \\
&= 15(7) - 2(52) \\
&= 15(735 - 14(52)) - 2(52) \\
&= 15(735) - 212(52)
\end{aligned}$$

Consequently, we have

$$735(15) - 1 = 212(52)$$

Therefore 52 divides $735(15) - 1$, so $735(15) \equiv 1 \pmod{52}$. Thus, $x = 15$ satisfies (2).   □

Example 2   Since 17 is a prime number, it is clear that the integers 17 and 840 are relatively prime. To find a positive integer x such that

$$17x \equiv 1 \pmod{840} \tag{3}$$

we first use the Division Algorithm to obtain the equations

$$840 = 49(17) + 7$$
$$17 = 2(7) + 3$$
$$7 = 2(3) + 1$$

Beginning with the last of these equations and working backwards, we find

$$\begin{aligned}
1 &= 7 - 2(3) \\
&= 7 - 2(17 - 2(7)) \\
&= 5(7) - 2(17) \\
&= 5(840 - 49(17)) - 2(17) \\
&= 5(840) - 247(17)
\end{aligned}$$

Consequently

$$17(-247) - 1 = -5(840)$$

Thus $x = -247$ satisfies (3). To find a *positive* integer x which can be used in (3), we observe that since $17(840) \equiv 0 \pmod{840}$, it follows that

$$17(-247) + 17(840) \equiv 1 \pmod{840}$$

Consequently, since $-247 + 840 = 593$, we have $17(593) \equiv 1 \pmod{840}$, and therefore $x = 593$ also satisfies (3).   □

We omit the proof of the next result.

Theorem 2.6   Let p and q be prime numbers and let $m = pq$ and $s = (p - 1)(q - 1)$. Let $[c] \in \mathbf{Z}_m$, where $\mathbf{Z}_m$ is the set of equivalence classes generated by the equivalence relation $x \equiv y \pmod{m}$, and suppose that $\gcd(c, m) = 1$. Then $c^s \equiv 1 \pmod{m}$ (or equivalently, $[c]^s = [1]$, since by (3) of Section 2.5, we have $[c]^s = [c^s]$).

Example 3   If $p = 29$ and $q = 31$, then $m = pq = 899$ and $s = (p - 1)(q - 1) = 840$. Let $c = 14$. Then $\gcd(c, m) = \gcd(14, 899) = 1$ and hence, by Theorem 2.6, $14^{840} \equiv 1 \pmod{m}$, or, equivalently, $[14]^{840} = [1]$.   □

The code that we will develop is based on the following observation. Let p and q be prime numbers and let m and s be defined as in Theorem 2.6. Choose integers c and t so that $\gcd(c, m) = 1$ and $\gcd(t, s) = 1$. Then there is a positive integer x such that

$$[c]^{tx} = [c] \qquad (4)$$

To see this, note that since $\gcd(t, s) = 1$, it follows from Theorem 2.5 that there is a positive integer x such that

$$tx \equiv 1 \pmod{s}$$

Thus, $tx = 1 + sy$ for some integer y, and we have

$$[c]^{tx} = [c]^{sy+1} = ([c]^s)^y [c] \qquad (5)$$

But by Theorem 2.6, $[c]^s = [1]$, and hence from (5) we have

$$[c]^{tx} = [1]^y [c] = [c]$$

For our purposes, the preceding observation is important because if we are given $[c]^t$, then we can follow the procedure illustrated in Examples 1 and 2 to find a positive integer x such that $tx \equiv 1 \pmod{s}$. Then we can use (4) to find $[c]$. This is the procedure we will follow to develop the secret code. The received coded message will be an equivalence class $[c]^t = [c^t]$, and to decode the message we will need to find $[c]$.

To develop this code, we associate the letters of the alphabet with numbers, as indicated below:

$$
\begin{array}{cccccccccccccc}
a & b & c & d & e & f & g & h & i & j & k & l & m & n \\
\downarrow & \downarrow & \downarrow & \downarrow & \downarrow & \downarrow & \downarrow & \downarrow & \downarrow & \downarrow & \downarrow & \downarrow & \downarrow & \downarrow \\
2 & 3 & 4 & 5 & 6 & 7 & 8 & 9 & 10 & 11 & 12 & 13 & 14 & 15
\end{array}
$$

$$
\begin{array}{cccccccccccc}
o & p & q & r & s & t & u & v & w & x & y & z \\
\downarrow & \downarrow & \downarrow & \downarrow & \downarrow & \downarrow & \downarrow & \downarrow & \downarrow & \downarrow & \downarrow & \downarrow \\
16 & 17 & 18 & 19 & 20 & 21 & 22 & 23 & 24 & 25 & 26 & 27
\end{array}
$$

There is a receiver and a sender of the code. The receiver chooses two prime numbers and gives the *product* of these prime numbers to the senders of the messages. Because only the receiver knows what the prime numbers actually are, many people will be able to send messages, but only the receiver will be able to decode them. In practice, a receiver would choose extremely large prime numbers, often of more than a hundred digits each, but for purposes of illustration we choose $p = 29$ and $q = 31$. As in Example 1, we let $m = pq = 899$, and $s = (p - 1)(q - 1) = 840$, and we let $t = 17$; t is chosen so that $\gcd(t, s) = 1$. The sender of the message knows m and t, but not the values of p and q; the receiver knows the values t, p, q (and consequently knows m and s).

Suppose we wish to transmit the letter g. This letter corresponds to 8 in the chart above, and we encode g as $[8]^{17}$, where $[8] \in \mathbf{Z}_m$. To aid in the calculation of $[8]^{17}$ for transmission, we note that $900 \equiv 1 \pmod{899}$, so for any integers k and r we have

$$[k \cdot 900 + r] = [k \cdot 900] + [r] = [k][900] + [r] = [k][1] + [r] = [k + r] \qquad (6)$$

Applying (6) when necessary, we have

$$[8]^2 = [64]$$
$$[8]^4 = [64]^2 = [4096] = [4 \cdot 900 + 496] = [500]$$
$$[8]^8 = [500]^2 = [250,000] = [277 \cdot 900 + 700] = [977]$$
$$= [1 \cdot 900 + 77] = [78]$$
$$[8]^{16} = [78]^2 = [6084] = [6 \cdot 900 + 684] = [690]$$
$$[8]^{17} = [8]^{16}[8] = [690][8] = [5520] = [6 \cdot 900 + 120] = [126]$$

Thus, the transmitted message for g is $[126]$.

At this point we have $[c^t] = [8^{17}] = [126]$. To decode the message $[126]$ the receiver must determine c. The receiver knows that $t = 17$, $p = 29$, and $q = 31$, and therefore $s = (p - 1)(q - 1) = 840$. From Example 2 we have

$$17(593) \equiv 1 \ (\text{mod } 840)$$

From (4) the receiver finds that

$$[c] = [c]^{tx} = [c^t]^x = [126]^x = [126]^{593}$$

and with the aid of a computer, determines that

$$126^{593} \equiv 8 \ (\text{mod } 899)$$

thus obtaining the original message g.

Obviously, computers are absolutely essential to both transmit and receive such coded messages. The beauty of this code is that it is so difficult to break. If p and q are very large primes, then even though the sender (or anyone else) may know the value $m = pq$, it is virtually impossible to determine p and q. The enormity of the task of breaking down large numbers into their prime factors is generally beyond the capacity of any computer. Because only the receiver knows the values of p and q, only he will be able to determine $s = (p - 1)(q - 1)$ and be able to decode the message.

2.6 Problems

1. Find
   (a) gcd(64, 116)   (b) gcd(-256, 72)   (c) gcd(120, 49)
2. Find a positive integer x such that $ax \equiv 1 \ (\text{mod } m)$ if
   (a) $a = 120; m = 7$ (b) $a = 693; m = 16$
   (c) $a = 75; m = 14$ (d) $a = 100; m = 21$
3. Use $p = 29$, $q = 31$, and $t = 17$ to send the following messages using the code developed in Section 2.6.
   (a) CAT   (b) DOG
4. Use $p = 31$, $q = 37$, and $t = 11$ to send the following messages using the code developed in Section 2.6.
   (a) HI   (b) NO   (c) YES

Chapter 2 REVIEW

Concepts for Review

number base (p. 55)
binary number system (p. 56)
octal number system (p. 57)
hexadecimal number system (p. 57)
Horner's Algorithm (p. 59)
Division Algorithm (p. 63)
addition and subtraction of binary numbers (p. 69)
nine's complement (p. 69)
ten's complement (p. 70)
one's complement (p. 71)
two's complement (p. 71)
modular arithmetic (p. 73–75)
random numbers (p. 76)
greatest common divisor (p. 79)

Review Problems

In Problems 1–4 use equation (2) of Section 2.1 to convert the given number to its decimal equivalent.

1. 11010_2
2. $FB1B_{16}$
3. 7146_8
4. 1312_5
5. Write the polynomial $f(x) = 5x^6 + 3x^3 - x^2 + 8$ in the form used for Horner's Algorithm.
6. Use Horner's Algorithm to compute $g(2)$ if $g(x) = 7x^4 - 3x^3 + 7x^2 - 5$.

In Problems 7–10 use Horner's Algorithm to convert the given number to its decimal equivalent.

7. 110011_2
8. 777_8
9. $BB5C_{16}$
10. 3201_4

In Problems 11–14 use the Division Algorithm to convert the number a to its equivalent in base b.

11. $a = 735; b = 5$
12. $a = 227; b = 2$
13. $a = 957; b = 16$
14. $a = 999; b = 8$

15. Convert the binary number 1111111111_2 to both its octal and hexadecimal equivalents.

16. Convert the hexadecimal number $FBAB_{16}$ to both its octal and binary equivalents.

17. Convert the octal number 77623_8 to both its binary and hexadecimal equivalents.

18. Compute the binary sum $110010 + 111110$.

19. Perform the following binary subtractions using two's complements.
    (a) $1000110 - 10010$
    (b) $1001011 - 11111$

20. List the equivalence classes of $\mathbf{Z}_3$.

21. Compute $[23] + [19]$ in $\mathbf{Z}_8$. Express your answer in the form $[a]$ where $0 \le a < 8$.

22. True or false?
    (a) $17 \equiv -3 \pmod 5$
    (b) $14 \equiv 22 \pmod 7$

23. Use the method discussed in Section 2.5 to generate pseudorandom numbers if $x_0 = 4$, $a = 7$, $b = 2$, and $m = 11$.

24. Compute $\gcd(252, 144)$

25. Find a positive integer x such that $327x \equiv 1 \pmod{14}$.

26. Use $p = 31$, $q = 41$, and $t = 13$ to send the following messages using the code developed in Section 2.6.
    (a) OK
    (b) WHY
    (c) WATER

3 An Introduction to Graph Theory

Graphs are remarkably simple mathematical structures, and perhaps because of this simplicity, they have widespread applications. In this chapter we will first establish some of the basic properties of graphs and then introduce several special types of graphs that can be used for the algorithmic solution of a variety of applied problems.

3.1 GRAPHS

Intuitively, a graph is nothing more than a collection of points (called *vertices*) and a collection of lines or curves (called *edges*) that connect some or all of the points. One might, for example, think of a road map as a graph: The towns are the vertices and the roads connecting the towns are edges connecting the vertices, as in Figure 3.1.

The analogy between a road map and a graph does have its limitations. Although two towns, such as G and H in Figure 3.1, may be connected by

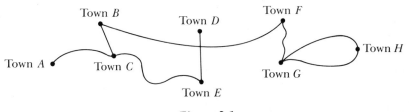

Figure 3.1

more than one road, two vertices cannot be connected by more than one edge in a graph. Thus we will permit at most one edge to connect a pair of vertices.

Because an edge is determined completely by the two vertices it connects, we can define the edge connecting vertices v and w as the set $\{v, w\}$. This observation leads us to the following definition.

Definition 3.1   A *graph G* consists of a nonempty finite set V (the *set of vertices* of G) and a set E of two-element subsets of V (the *set of edges* of G).

We will often use the notation $V(G)$ and $E(G)$, instead of V and E, to indicate the set of vertices and the set of edges, respectively, of a graph G.

If $e = \{v, w\}$ is an edge of a graph G, we say that e *connects* v and w, or that v and w are *adjacent*. The vertices v and w are called the *endpoints* of the edge e.

Example 1   The graph G with $V = \{v_1, v_2, v_3, v_4\}$ and $E = \{\{v_1, v_2\}, \{v_1, v_4\}, \{v_2, v_4\}, \{v_3, v_4\}\}$ is a graph with four vertices and four edges.   □

We can obtain a *diagram* of a graph G by using dots as the vertices and straight line segments or curves as the edges connecting the appropriate dots. The diagram in Figure 3.2 is one representation of the graph G described in Example 1.

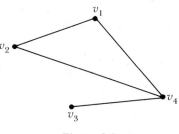

Figure 3.2

Example 2   Let G be the graph with vertex set $V = \{a, b, c, d, e\}$ and edge set $E = \{\{a, b\}, \{b, c\}, \{c, d\}, \{c, e\}, \{d, e\}, \{d, b\}\}$. Then we can represent G by the diagram in Figure 3.3.   □

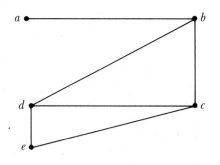

Figure 3.3

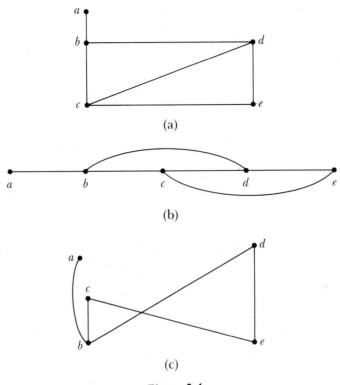

(a)

(b)

(c)

Figure 3.4

Many diagrams can represent the same graph G. Simply by rearranging the vertices, for example, we can create new diagrams to represent the graph described in Figure 3.3. Figure 3.4 shows three other options. Observe that Figure 3.4(b) has both straight line segments and curves to designate edges, while Figure 3.4(c) has two edges $\{c, e\}$ and $\{b, d\}$ which intersect at a point that is not a vertex. Neither of these facts concerns us; all three diagrams are acceptable representations of the graph G.

We will often present graphs by only showing their diagrams, and it will be up to you to supply the *set theoretic definition* (Definition 3.1) of the graph, if necessary.

Example 3   The set theoretic definition of the graph diagrammed in Figure 3.5 (p. 88), is the graph with vertex set $V = \{v_1, v_2, v_3, v_4, v_5, v_6\}$ and edge set $E = \{\{v_1, v_2\}, \{v_2, v_3\}, \{v_2, v_5\}, \{v_5, v_6\}, \{v_6, v_1\}\}$.   □

Note that Example 3 illustrates that a vertex (v_4 in this example) of a graph need not be connected to any other vertex by an edge.

At times we will consider subgraphs, which are graphs contained within other graphs.

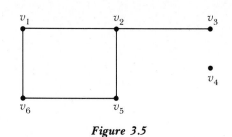

Figure 3.5

Definition 3.2   A graph H is a *subgraph* of a graph G if

    a. $V(H) \subseteq V(G)$, and

    b. $E(H) \subseteq E(G)$.

For example, the graph H defined by

$$V(H) = \{v_1, v_3, v_4, v_5, v_6\} \qquad E(H) = \{\{v_1, v_6\}, \{v_6, v_5\}\}$$

is a subgraph of the graph G given in Figure 3.5. Figure 3.6 illustrates the graphs G and H, where the vertices of H are small open circles and the edges of H are wavy lines.

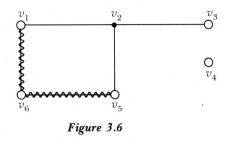

Figure 3.6

One efficient and useful way to describe a graph is by its adjacency matrix.

Definition 3.3   Let G be a graph with vertex set $\{v_1, v_2, \ldots, v_n\}$. The *adjacency matrix*[1] of G is the matrix

$$A = [a_{ij}]_{n \times n}$$

defined by

$$a_{ij} = \begin{cases} 1 & \text{if vertices } v_i \text{ and } v_j \text{ are adjacent} \\ 0 & \text{if vertices } v_i \text{ and } v_j \text{ are not adjacent} \end{cases}$$

---

[1] Matrices are discussed in Appendix Section 3.

Example 4 Suppose that

$$A = \begin{array}{c} \\ v_1 \\ v_2 \\ v_3 \\ v_4 \\ v_5 \end{array} \begin{array}{ccccc} v_1 & v_2 & v_3 & v_4 & v_5 \\ \begin{bmatrix} 0 & 1 & 1 & 1 & 0 \\ 1 & 0 & 0 & 1 & 1 \\ 1 & 0 & 0 & 0 & 0 \\ 1 & 1 & 0 & 0 & 0 \\ 0 & 1 & 0 & 0 & 0 \end{bmatrix} \end{array}$$

is the adjacency matrix of a graph G whose vertex set is $\{v_1, v_2, v_3, v_4, v_5\}$. Observe that since the entry a_{12} in matrix A equals 1, it follows that vertices v_1 and v_2 are adjacent. Similarly, since $a_{41} = 1$, the vertices v_4 and v_1 must be adjacent. The vertices v_3 and v_4 are not adjacent since $a_{34} = 0$. Continuing with this analysis we find that $E(G) = \{\{v_1, v_2\}, \{v_1, v_3\}, \{v_1, v_4\}, \{v_2, v_4\}, \{v_2, v_5\}\}$.   □

Note that the adjacency matrix of a graph G has the same entry in the (i, j)th position as it has in the (j, i)th position, that is, $a_{ij} = a_{ji}$. A matrix with this property is called a *symmetric matrix*. Note, too, that since an edge never joins a vertex to itself, a_{ii} must equal zero for all i. Finally, we observe that since an edge $\{v_i, v_j\}$ of a graph G is represented twice in the adjacency matrix of G (once in the (i, j)th entry, and once in the (j, i)th entry), it follows that precisely $2|E|$ of the entries in the adjacency matrix are 1's, where $|E|$ denotes the number of edges of the graph G. In other words, the sum of the entries of the adjacency matrix is equal to twice the number of the edges of the graph.

Example 5 We find the adjacency matrix of the graph G diagrammed in Figure 3.7. Recall that an adjacency matrix has a 1 in the (i, j)th and (j, i)th positions to indicate that $\{v_i, v_j\}$ is an edge of G.

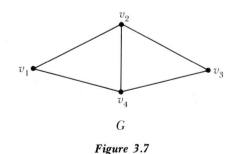

G

Figure 3.7

The edges of G are $\{v_1, v_2\}$, $\{v_2, v_3\}$, $\{v_2, v_4\}$, $\{v_3, v_4\}$, and $\{v_4, v_1\}$, so we place the number 1 in positions $(1, 2)$, $(2, 1)$, $(2, 3)$, $(3, 2)$, $(2, 4)$, $(4, 2)$, $(3, 4)$, $(4, 3)$, $(4, 1)$, and $(1, 4)$ of the adjacency matrix. We enter zeros in the remaining positions. This gives us the adjacency matrix A of the graph G:

$$A = \begin{bmatrix} 0 & 1 & 0 & 1 \\ 1 & 0 & 1 & 1 \\ 0 & 1 & 0 & 1 \\ 1 & 1 & 1 & 0 \end{bmatrix}$$

The sum of the entries in the adjacency matrix is ten since G has five edges. □

Example 6   It is also possible to diagram a graph given its adjacency matrix. For instance, let

$$A = \begin{bmatrix} 0 & 1 & 0 & 1 & 0 \\ 1 & 0 & 1 & 0 & 1 \\ 0 & 1 & 0 & 1 & 0 \\ 1 & 0 & 1 & 0 & 1 \\ 0 & 1 & 0 & 1 & 0 \end{bmatrix}$$

be the adjacency matrix of a graph G. Then G has one vertex for each row (or column) of A. Thus, the vertex set of G can be taken as $\{v_1, v_2, v_3, v_4, v_5\}$. Two vertices v_i and v_j are joined by an edge if and only if the entry a_{ij} is 1. You can verify that Figure 3.8 shows three possible diagrams of this graph. □

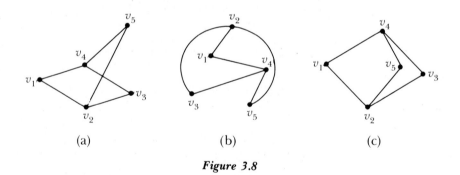

(a)   (b)   (c)

Figure 3.8

The adjacency matrix provides an efficient way to describe a graph in a digital computer. As we shall see, it is also an important theoretical tool.

Definition 3.4   Let G be a graph and let v be a vertex of G. The *degree* of v, $\delta(v)$, is the number of edges of G having v as an endpoint.

Example 7   To find the degree of a vertex, we count how many edges have v as an endpoint. For the graph diagrammed in Figure 3.5,

$$\delta(v_1) = 2 \qquad \delta(v_2) = 3 \qquad \delta(v_3) = 1$$
$$\delta(v_4) = 0 \qquad \delta(v_5) = 2 \qquad \delta(v_6) = 2 \qquad \qquad \square$$

Corresponding to each 1 in the ith row (or column) of an adjacency matrix, there is an edge which has the vertex v_i as an endpoint. It follows that the sum of the entries in the ith row (or column) of matrix A is equal to $\delta(v_i)$. We use this observation to prove the following theorem.

Theorem 3.5   If G is a graph with vertex set $V = \{v_1, v_2, \dots, v_n\}$ and edge set E, then

$$\delta(v_1) + \delta(v_2) + \cdots + \delta(v_n) = 2|E|$$

Proof: We have previously observed that the sum of all the entries in the adjacency matrix A of G is twice the number of edges, $2|E|$. One way to obtain the sum of all of the entries of A is to add the sum of the entries of the first row, $\delta(v_1)$, to the sum of the entries of the second row, $\delta(v_2)$, to the sum of the entries of the third row, $\delta(v_3)$, and so on. Thus,

$$\delta(v_1) + \delta(v_2) + \cdots + \delta(v_n) = 2|E| \qquad\qquad \blacksquare$$

Example 8   We verify Theorem 3.5 for the graph diagrammed in Figure 3.9.

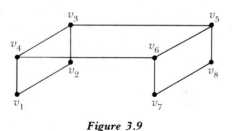

Figure 3.9

Here we have $\delta(v_1) = 2$, $\delta(v_2) = 2$, $\delta(v_3) = 3$, $\delta(v_4) = 3$, $\delta(v_5) = 3$, $\delta(v_6) = 3$, $\delta(v_7) = 2$, and $\delta(v_8) = 2$. Consequently, $\delta(v_1) + \cdots + \delta(v_8) = 20$. Since G has ten edges, we have $2|E| = 20$, which is in agreement with Theorem 3.5. $\qquad\qquad \square$

3.1   Problems

In Problems 1–8 diagram the given graph.

1. $V(G) = \{v_1, v_2, v_3, v_4\}$; $E(G) = \{\{v_1, v_2\}, \{v_3, v_4\}, \{v_1, v_4\}\}$
2. $V(G) = \{a, b, c, d, e\}$; $E(G) = \{\{a, b\}, \{b, c\}, \{c, d\}, \{d, e\}, \{e, a\}\}$
3. $V(G) = \{1, 2, 3, 4\}$; $E(G) = \{\{1, 2\}, \{1, 3\}, \{1, 4\}, \{2, 3\}, \{2, 4\}, \{3, 4\}\}$

4. $V(G) = \{u, v, w, x, y\}; E(G) = \{\{u, v\}, \{w, x\}, \{x, y\}\}$

5. $V(G) = \{a, b, c, d, e\}; E(G) = \{\{a, b\}, \{d, e\}\}$

6. $V(G) = \{v_1, v_2, v_3\}; E(G) = \varnothing$

7. $V(G) = \{1, 2, 3, 4, 5, 6, 7, 8\}; E(G) = \{\{x, y\} | xy \text{ is even}\}$

8. $V(G) = \{1, 2, 3, 4, 5\}; E(G) = \{\{x, y\} | x/y \text{ is an integer}\}$

In Problems 9–14 give the set theoretic definition of the diagrammed graph.

9.

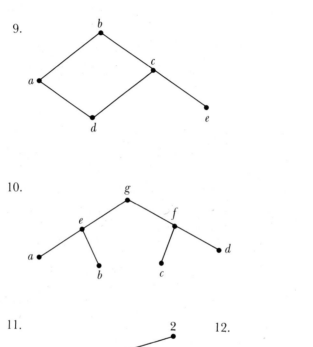

10.

11.   12.

13.   14.

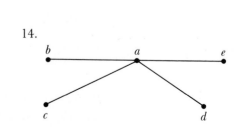

In Problems 15–20 give the adjacency matrix of the diagrammed graph.

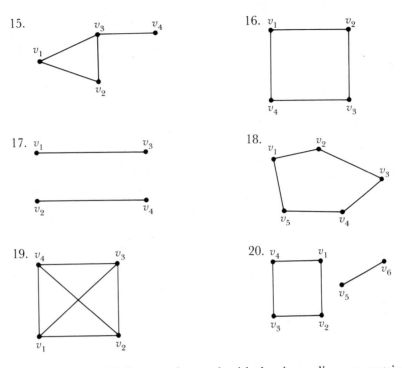

15. v_3 v_4 v_1 v_2

16. v_1 v_2 v_4 v_3

17. v_1 v_3 v_2 v_4

18. v_2 v_1 v_3 v_5 v_4

19. v_4 v_3 v_1 v_2

20. v_4 v_1 v_6 v_5 v_3 v_2

In Problems 21–26 diagram the graph with the given adjacency matrix.

21. $\begin{bmatrix} 0 & 1 & 1 & 0 \\ 1 & 0 & 1 & 1 \\ 1 & 1 & 0 & 0 \\ 0 & 1 & 0 & 0 \end{bmatrix}$

22. $\begin{bmatrix} 0 & 0 & 0 \\ 0 & 0 & 0 \\ 0 & 0 & 0 \end{bmatrix}$

23. $\begin{bmatrix} 0 & 1 & 0 & 1 \\ 1 & 0 & 1 & 0 \\ 0 & 1 & 0 & 1 \\ 1 & 0 & 1 & 0 \end{bmatrix}$

24. $\begin{bmatrix} 0 & 1 & 1 \\ 1 & 0 & 1 \\ 1 & 1 & 0 \end{bmatrix}$

25. $\begin{bmatrix} 0 & 1 & 1 & 1 \\ 1 & 0 & 1 & 1 \\ 1 & 1 & 0 & 1 \\ 1 & 1 & 1 & 0 \end{bmatrix}$

26. $\begin{bmatrix} 0 & 1 & 1 & 1 \\ 1 & 0 & 0 & 0 \\ 1 & 0 & 0 & 0 \\ 1 & 0 & 0 & 0 \end{bmatrix}$

27. Prove that there is no 4-vertex-graph G with $\delta(v_1) = 3$, $\delta(v_2) = 2$, $\delta(v_3) = 2$, $\delta(v_4) = 2$. (*Hint:* Use Theorem 3.5.)

28. Prove that there is no 6-vertex-graph G with $\delta(v_1) = 1$, $\delta(v_2) = 3$, $\delta(v_3) = 2$, $\delta(v_4) = 3$, $\delta(v_5) = 3$, $\delta(v_6) = 3$.

29. Fill in the blank in the following generalization of Problems 27 and 28.
    Theorem: There must be a(n) _____ number of vertices of odd degree in any graph.

30. Describe the relation between the adjacency matrix of a graph G and the adjacency matrix of a subgraph of G.

31. The complement $\bar{G}$ of a graph G is defined as follows: $V(\bar{G}) = V(G)$; for $u, v \in V(G)$ $\{u, v\} \in E(\bar{G})$ if and only if $\{u, v\} \notin E(G)$. Describe the relationship between the adjacency matrix of G and the adjacency matrix of $\bar{G}$.

32. A graph G has twenty-three edges and every vertex of G has degree four or larger. What is the largest possible number of vertices that G can have?

33. What is the largest possible number of vertices in a graph with thirty-six edges if all of the vertices have degree at least four?

3.2 CONNECTEDNESS

When diagrammed, some graphs appear to be broken into parts, while others are in one piece. To formalize this rather intuitive distinction, we will make use of the following idea.

Definition 3.6 A *path* joining vertices a and b of a graph G is a sequence of vertices of G, $(a = w_0), w_1, \ldots, (w_m = b)$ such that $\{w_i, w_{i+1}\} \in E$ for $i = 0, 1, \ldots, m - 1$. The *length* of the path $w_0, w_1, \ldots, w_m$ is m.

Example 1 Let G be the graph with vertex set $V = \{a, b, c, d, e\}$ and edge set $E = \{\{a, e\}, \{a, f\}, \{e, f\}, \{e, d\}, \{d, c\}, \{d, b\}\}$. Figure 3.10 provides a diagram of G.

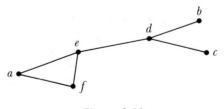

Figure 3.10

According to Definition 3.6, paths connecting vertices a and b include a, e, d, b and a, e, f, a, e, d, b. □

Definition 3.7 A graph G is *connected* if each pair of vertices of G is joined by a path. If there is a pair of vertices that are not joined by any path, the graph is said to be *disconnected*.

Example 2 The graph diagrammed in Figure 3.11(a) is connected. The graph in Figure 3.11(b) is disconnected, because, for example, the vertices u and v are not joined by a path. □

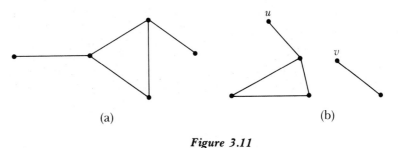

(a)   (b)

Figure 3.11

Paths that do not have repeated vertices will be especially useful in the following sections.

Definition 3.8 A *simple path* joining vertices a and b of a graph G is a path $(a = w_0)$, $w_1, \ldots, (w_m = b)$ in which $w_i \neq w_j$ if $i \neq j$.

Thus a simple path neither backtracks nor crosses itself. In Figure 3.10 the path a, e, d, b is a simple path, while the paths a, e, f, a, e, d, b and a, e, d, e, d, b are not. The next theorem states that if a path joins two vertices, then we can always find a simple path joining these vertices.

Theorem 3.9 If vertices a and b in a graph G are joined by a path, then they are also joined by a simple path.

Proof: The proof employs the alternate form of the Principle of Induction (see Section 1.4). Induction is on the number, n, of repeated vertices in any path that joins a to b.

a. $n = 0$; if no vertex is repeated, then the path is simple and the theorem clearly holds.

b. Suppose that the theorem is true for $n = 1, 2, \ldots, k$; that is, suppose that if there are k or fewer repeated vertices in a path joining a and b, then there is a simple path joining a to b. We must show that the theorem is also true for $n = k + 1$.

Suppose then that $(a = w_0)$, $w_1, \ldots,$ $(w_m = b)$ is a path with $k + 1$ repeated vertices, and let v be one of the repeated vertices. If v first appears as vertex w_i and last appears as vertex w_j, then this path can be written as

$$(a = w_0), w_1, \ldots, w_{i-1}, (w_i = v), \ldots, (w_j = v), w_{j+1}, \ldots, (w_m = b)$$

Note that the path

$$(a = w_0), w_1, \ldots, w_{i-1}, v, w_{j+1}, \ldots, (w_m = b)$$

has at least one less repeated vertex (v is no longer repeated) and still connects vertices a and b. Therefore, by the induction hypothesis, vertices a and b are joined by a simple path, and this concludes the induction argument. ∎

Corollary 3.10   Let G be a connected graph with n vertices. Then any two vertices are joined by a simple path that has length less than or equal to $n - 1$.

Proof: Any simple path in G must have n or fewer vertices. Because the length of a simple path is one less than the number of vertices forming the path, the result follows. ∎

By examining the powers of a graph's adjacency matrix A, we can determine if the graph is connected. We can also use powers of the matrix to compute the number of paths of a specified length that join any pair of vertices. Suppose that $A = [a_{ij}]$ is the $n \times n$ adjacency matrix of a graph G. Then by the definition of matrix multiplication (see Appendix Section 3), $A^2 = [b_{ij}]$, where for $1 \le i \le n$ and $1 \le j \le n$,

$$b_{ij} = a_{i1}a_{1j} + a_{i2}a_{2j} + \cdots + a_{in}a_{nj}$$

Note that for each k, $a_{ik}a_{kj}$ is either 0 or 1; moreover, $a_{ik}a_{kj} = 1$ if and only if both $a_{ik} = 1$ and $a_{kj} = 1$. It follows that $a_{ik}a_{kj} = 1$ if and only if both $\{v_i, v_k\}$ and $\{v_k, v_j\}$ are edges of G, and therefore the sum

$$a_{i1}a_{1j} + a_{i2}a_{2j} + \cdots + a_{in}a_{nj}$$

counts the number of vertices v_k such that v_i, v_k, v_j is a path from v_i to v_j.

Thus we have established that the (i, j)th entry of A^2 is the number of paths of length two joining v_i and v_j. You are asked to use induction to prove the following general result in Problem 11.

Theorem 3.11   Let A be the adjacency matrix of a graph G. Then the (i, j)th entry of the matrix A^m gives the number of paths in G of length m that join the vertices v_i and v_j.

Example 3 Let G be a graph with vertex set $V = \{v_1, v_2, v_3\}$ and edge set $E = \{\{v_1, v_2\}, \{v_2, v_3\}\}$, as in Figure 3.12. We want to find how many paths of length three join vertices v_1 and v_2.

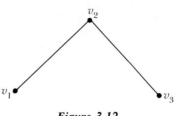

Figure 3.12

The adjacency matrix of G is

$$A = \begin{bmatrix} 0 & 1 & 0 \\ 1 & 0 & 1 \\ 0 & 1 & 0 \end{bmatrix}$$

Routine calculations show that

$$A^2 = \begin{bmatrix} 1 & 0 & 1 \\ 0 & 2 & 0 \\ 1 & 0 & 1 \end{bmatrix} \quad \text{and} \quad A^3 = \begin{bmatrix} 0 & 2 & 0 \\ 2 & 0 & 2 \\ 0 & 2 & 0 \end{bmatrix}$$

Since the $(1, 2)$ entry of A^3 is 2, it follows from Theorem 3.11 that there are two paths of length three connecting vertices v_1 and v_2. These paths are v_1, v_2, v_3, v_2 and v_1, v_2, v_1, v_2. $\qquad\square$

If a connected graph G has n vertices, then by Corollary 3.10 any two vertices are joined by a path of length less than or equal to $n - 1$. Hence, if a graph G with n vertices is connected, and if A is the adjacency matrix of G, then the (i, j) th entry in at least one of the matrices $A, A^2, \ldots, A^{n-1}$ must be greater than or equal to 1. This observation leads us to the following theorem, which you are asked to prove in Problem 12 using the definition of matrix addition (see Appendix Section 3).

Theorem 3.12   Let G be a graph with n vertices $(n > 2)$ and let A be the adjacency matrix of G. Then G is connected if and only if every entry in the matrix $A + A^2 + \cdots + A^{n-1}$ is greater than or equal to 1.

Example 4   Let G be a graph with the adjacency matrix

$$A = \begin{bmatrix} 0 & 0 & 0 & 1 & 1 \\ 0 & 0 & 1 & 0 & 0 \\ 0 & 1 & 0 & 0 & 0 \\ 1 & 0 & 0 & 0 & 1 \\ 1 & 0 & 0 & 1 & 0 \end{bmatrix}$$

To determine if G is connected we apply Theorem 3.12.

Direct computation gives

$$A^2 = \begin{bmatrix} 2 & 0 & 0 & 1 & 1 \\ 0 & 1 & 0 & 0 & 0 \\ 0 & 0 & 1 & 0 & 0 \\ 1 & 0 & 0 & 2 & 1 \\ 1 & 0 & 0 & 1 & 2 \end{bmatrix}, \qquad A^3 = \begin{bmatrix} 2 & 0 & 0 & 3 & 3 \\ 0 & 0 & 1 & 0 & 0 \\ 0 & 1 & 0 & 0 & 0 \\ 3 & 0 & 0 & 2 & 3 \\ 3 & 0 & 0 & 3 & 2 \end{bmatrix}$$

and

$$A^4 = \begin{bmatrix} 6 & 0 & 0 & 5 & 5 \\ 0 & 1 & 0 & 0 & 0 \\ 0 & 0 & 1 & 0 & 0 \\ 5 & 0 & 0 & 6 & 6 \\ 5 & 0 & 0 & 6 & 6 \end{bmatrix}$$

Since

$$A + A^2 + A^3 + A^4 = \begin{bmatrix} 10 & 0 & 0 & 10 & 10 \\ 0 & 2 & 2 & 0 & 0 \\ 0 & 2 & 2 & 0 & 0 \\ 10 & 0 & 0 & 10 & 11 \\ 10 & 0 & 0 & 11 & 10 \end{bmatrix}$$

it follows from Theorem 3.12 that the graph G is not connected. For example, vertices v_1 and v_2 are not joined by a path. $\qquad \square$

Instead of using Theorem 3.12 to find out that graph G of Example 4 is disconnected, we could have used the adjacency matrix of G to draw a diagram of the graph. From Figure 3.13, it is clear that vertices v_1 and v_2 are not joined by a path and that graph G is not connected.

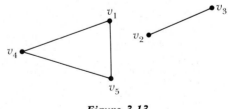

Figure 3.13

Thus we have two procedures—one algebraic and one pictorial—for determining whether or not a graph is connected. Although it is often easy to diagram a small graph, it is generally simpler to determine connectedness for large graphs by using an algorithm based on Theorem 3.12. In the next section we will consider a more efficient method to determine if a graph is connected.

3.2   Problems

In Problems 1–8 use the given adjacency matrix to determine (a) the number of paths of length four joining vertices v_1 and v_2; and (b) if the respective graph is connected.

1. $\begin{bmatrix} 0 & 1 & 1 \\ 1 & 0 & 1 \\ 1 & 1 & 0 \end{bmatrix}$

2. $\begin{bmatrix} 0 & 1 & 0 \\ 1 & 0 & 1 \\ 0 & 1 & 0 \end{bmatrix}$

3. $\begin{bmatrix} 0 & 1 & 0 \\ 1 & 0 & 0 \\ 0 & 0 & 0 \end{bmatrix}$

4. $\begin{bmatrix} 0 & 0 & 1 \\ 0 & 0 & 1 \\ 1 & 1 & 0 \end{bmatrix}$

5. $\begin{bmatrix} 0 & 1 & 1 & 1 \\ 1 & 0 & 0 & 0 \\ 1 & 0 & 0 & 0 \\ 1 & 0 & 0 & 0 \end{bmatrix}$

6. $\begin{bmatrix} 0 & 1 & 0 & 1 \\ 1 & 0 & 0 & 1 \\ 0 & 0 & 0 & 0 \\ 1 & 1 & 0 & 0 \end{bmatrix}$

7. $\begin{bmatrix} 0 & 1 & 0 & 0 \\ 1 & 0 & 1 & 0 \\ 0 & 1 & 0 & 1 \\ 0 & 0 & 1 & 0 \end{bmatrix}$

8. $\begin{bmatrix} 0 & 1 \\ 1 & 0 \end{bmatrix}$

9. Let A be the adjacency matrix of a graph G where $V(G) = \{v_1, v_2, \ldots, v_n\}$. Show that the (i, i)th entry in the matrix A^2 is equal to $\delta(v_i)$, the degree of the ith vertex.

10. Prove that if G_1 and G_2 are connected graphs and $V(G_1) \cap V(G_2) \neq \varnothing$, then the graph G, defined by $V(G) = V(G_1) \cup V(G_2)$ and $E(G) = E(G_1) \cup E(G_2)$, is connected.

11. Use induction to prove Theorem 3.11.

12. Prove Theorem 3.12.

3.3   THE COMPONENTS OF A GRAPH

Theorem 3.12 provides one method for determining whether or not a graph with a given adjacency matrix is connected. In this section, we will consider an alternate procedure that will allow us to check the connectedness of a graph. This procedure generates a list of the vertices in any component of the graph.

Definition 3.13   A subgraph H of a graph G is a *component* of G if

a. H is connected, and

b. H is not contained in any other, larger, connected subgraph of G.

Thus, for example, the graph diagrammed in Figure 3.13 has two components, H_1 and H_2, where

$$V(H_1) = \{1, 4, 5\} \qquad E(H_1) = \{\{1, 5\}, \{5, 4\}, \{4, 1\}\}$$

and

$$V(H_2) = \{2, 3\} \qquad E(H_2) = \{\{2, 3\}\}$$

Another subgraph, H_3, of G is

$$V(H_3) = \{1, 4, 5\} \qquad E(H_3) = \{\{5, 4\}, \{4, 1\}\}$$

This, however, is not a component of the graph G because it is contained in H_1, a larger connected subgraph of G, and therefore does not fulfill part **b.** of Definition 3.13.

Our next objective is to list all the vertices of a component of a graph G that includes a given vertex v. It is sufficient to create a list C that contains all vertices connected by paths to v. First we place v in C, because certainly v, if nothing else, is connected to v. Next we place all those vertices adjacent to v into C. Then we add to C all the vertices adjacent to those vertices most recently placed into C. We continue to place vertices into C in this manner until we can no longer enlarge C. Note that if all vertices of G have been placed into C by this process, then all vertices of G are connected to v by a path. It then follows that G is connected (Why?).

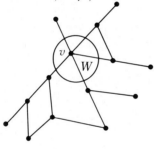

(i)

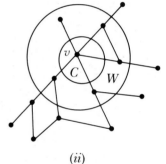

(ii)

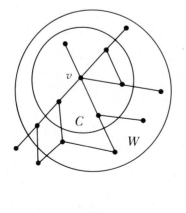

(iii)

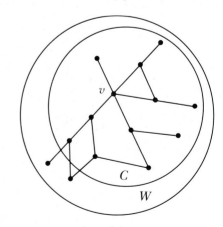

(iv)

Figure 3.14

To write this procedure more formally, we let W represent the set of vertices most recently joined to C. Then we only need to check the vertices of W for additional adjacencies, since all vertices previously checked for adjacencies have already been included in C. The set W is, in effect, the "cutting edge" of the amoeba-like growth of C. Figure 3.14 pictorially explains this process, which underlies the following algorithm.

Algorithm 3.14 For generating a list of all vertices of a graph G lying in the component of G that contains a given vertex V of G. The vertex set of G is $\{1, 2, \ldots, N\}$.

Input: $A(I, J)$, the $N \times N$ adjacency matrix of G;
V, a vertex in G

Output: C, the set of vertices in the component of G containing V

1. $W \leftarrow \{V\}; C \leftarrow \varnothing;$ Initialize.

2. WHILE $W \neq \varnothing$ DO   The algorithm will continue this loop as long as new vertices are added to C. When no new vertices are added, the algorithm terminates. *NEW* is used to keep track of the vertices adjacent to vertices of W. *NEW* will become the "new" W.

   a. *NEW* $\leftarrow \varnothing$
   b. FOR ALL $I \in W$ All vertices of W are checked for new adjacencies.

      1. IF $A(I, J) = 1$ AND $J \notin C \cup W$ THEN   If J is adjacent to I, and J has not been previously considered, then J is put in the new W.
         a. *NEW* $\leftarrow$ *NEW* $\cup \{J\}$

   c. $C \leftarrow C \cup W$ After all vertices of W are checked, C and W are updated (steps c. and d.).

   d. $W \leftarrow$ *NEW*
3. OUTPUT C

Example 1 In this example we trace an application of Algorithm 3.14 to the graph G with adjacency matrix

$$A = \begin{bmatrix} 0 & 1 & 0 & 0 & 0 \\ 1 & 0 & 1 & 1 & 0 \\ 0 & 1 & 0 & 0 & 1 \\ 0 & 1 & 0 & 0 & 0 \\ 0 & 0 & 1 & 0 & 0 \end{bmatrix}$$

Our objective is to list all the vertices in the component containing the vertex

$v = 1$. Thus we initially have $W = \{1\}$. It is convenient to organize the work as shown in Table 3.1.

NEW	C	W
	$\varnothing$	$\{1\}$
$\{2\}$	$\{1\}$	$\{2\}$
$\{3, 4\}$	$\{1, 2\}$	$\{3, 4\}$
$\{5\}$	$\{1, 2, 3, 4\}$	$\{5\}$
$\varnothing$	$\{1, 2, 3, 4, 5\}$	$\varnothing$

Table 3.1

Note that since $C = \{1, 2, 3, 4, 5\} = V(G)$, G is connected.   □

3.3   Problems

In Problems 1–10 use the given adjacency matrices and Algorithm 3.14 to generate a list of the vertices found in the component containing the vertex 1.

1. $\begin{bmatrix} 0 & 1 & 1 \\ 1 & 0 & 1 \\ 1 & 1 & 0 \end{bmatrix}$

2. $\begin{bmatrix} 0 & 1 & 0 & 1 \\ 1 & 0 & 1 & 0 \\ 0 & 1 & 0 & 1 \\ 1 & 0 & 1 & 0 \end{bmatrix}$

3. $\begin{bmatrix} 0 & 1 & 1 & 1 \\ 1 & 0 & 1 & 1 \\ 1 & 1 & 0 & 1 \\ 1 & 1 & 1 & 0 \end{bmatrix}$

4. $\begin{bmatrix} 0 & 0 & 1 \\ 0 & 0 & 0 \\ 1 & 0 & 0 \end{bmatrix}$

5. $\begin{bmatrix} 0 & 0 & 1 & 0 \\ 0 & 0 & 0 & 0 \\ 1 & 0 & 0 & 1 \\ 0 & 0 & 1 & 0 \end{bmatrix}$

6. $\begin{bmatrix} 0 & 1 & 1 & 0 \\ 1 & 0 & 1 & 0 \\ 1 & 1 & 0 & 1 \\ 0 & 0 & 1 & 0 \end{bmatrix}$

7. $\begin{bmatrix} 0 & 1 & 0 & 0 \\ 1 & 0 & 0 & 0 \\ 0 & 0 & 0 & 1 \\ 0 & 0 & 1 & 0 \end{bmatrix}$

8. $\begin{bmatrix} 0 & 0 & 1 & 0 & 0 \\ 0 & 0 & 0 & 1 & 1 \\ 1 & 0 & 0 & 0 & 0 \\ 0 & 1 & 0 & 0 & 1 \\ 0 & 1 & 0 & 1 & 0 \end{bmatrix}$

9. $\begin{bmatrix} 0 & 1 & 0 & 1 & 1 \\ 1 & 0 & 1 & 0 & 1 \\ 0 & 1 & 0 & 1 & 1 \\ 1 & 0 & 1 & 0 & 1 \\ 1 & 1 & 1 & 1 & 0 \end{bmatrix}$

10. $\begin{bmatrix} 0 & 0 & 0 & 0 & 1 \\ 0 & 0 & 1 & 1 & 0 \\ 0 & 1 & 0 & 1 & 0 \\ 0 & 1 & 1 & 0 & 0 \\ 1 & 0 & 0 & 0 & 0 \end{bmatrix}$

11. Prove that if $H_1, H_2, \ldots, H_k$ are the components of a graph G, then $V(H_1), V(H_2), \ldots, V(H_k)$ is a *partition* of $V(G)$, that is, $V(H_i) \cap V(H_j) = \emptyset$ if $i \neq j$, and $\bigcup_{i=1}^{k} V(H_i) = V(G)$.

12. Modify Algorithm 3.14 so that its output is either "CONNECTED" or "NOT CONNECTED" depending on whether or not the adjacency matrix $[a_{ij}]_{n \times n}$ is that of a connected graph.

3.4   TREES

A special type of graph, called a *tree*, holds an important place in both the theory and application of graphs. As we shall see, we can use trees to construct minimum cost transportation networks and schedule interdependent activities. In this section we will define this special graph and consider some of its properties.

Definition 3.15   A *cycle C* in a graph G is a path $w_0, w_1, \ldots, w_m$ in G such that $m \geq 3$ and only the first and last vertices of C are the same. That is, $w_0 = w_m$ but $w_i \neq w_j$ if $i \neq j$ and $\{i, j\} \neq \{0, m\}$.

Example 1   In Figure 3.15, the wavy lines indicate the cycle a, b, c, d, e, f, a.

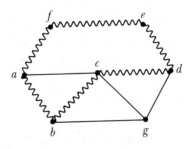

Figure 3.15

The requirement that $m \geq 3$ means, for example, that the path d, g, d in Figure 3.15 is not a cycle.   □

Definition 3.16   A *tree* is a connected graph without cycles.

Figure 3.16 illustrates the general form of all trees having five vertices. Observe that each of the trees has at least one vertex of degree one. This is true in general, and we will prove it as a corollary to the following theorem.

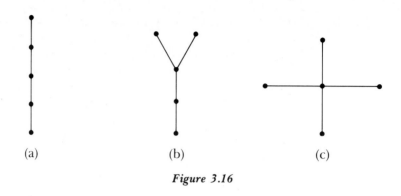

(a)   (b)   (c)

Figure 3.16

Theorem 3.17   If G is a graph and if $\delta(v) \geq 2$ for all vertices v of G, then G has a cycle.

> *Proof:*   The idea of the proof is quite straightforward: We begin with an arbitrary vertex and then move from vertex to adjacent vertex. At some point we must return to a vertex that has already been used, and when this occurs, we will have formed a cycle.
>
> More formally, let w_0 be an arbitrary vertex of G. Since $\delta(w_0) \geq 2$, there is another vertex, w_1, that is adjacent to w_0. Since $\delta(w_1) \geq 2$, there is a vertex other than w_0 that is adjacent to w_1. Call this vertex w_2.
>
> Since $\delta(w_2) \geq 2$ there is a vertex, w_3, adjacent to w_2 and distinct from the vertex w_1. It may happen that $w_3 = w_0$. If so, we have completed a cycle w_0, w_1, w_2, w_0 and thus have established the theorem. If $w_3 \neq w_0$, then, since $\delta(w_3) \geq 2$, there is a vertex w_4, not equal to w_2, that is adjacent to w_3. If w_4 is equal to either w_0 or w_1, then we have completed a cycle; if w_4 is distinct from both w_0 and w_1, then we continue this process to find a vertex, w_5, and so on. Ultimately, because the graph has only a finite number of vertices, some vertex will be repeated. When this happens, a cycle will be formed. In other words, if $w_q = v = w_{q+r}$, then
>
> $$(v = w_q), w_{q+1}, \ldots, (w_{q+r} = v)$$
>
> is the desired cycle, as Figure 3.17 indicates.   ■

Figure 3.17

Corollary 3.18 If T is a tree with at least one edge, then T has a vertex of degree one.

> ***Proof:*** Since T has at least one edge, it has at least two vertices. Thus, since T is connected, none of its vertices can have degree zero. Consequently $\delta(v) \geq 1$ for all $v \in V(T)$.
>
> If $\delta(v) \geq 2$ for all vertices, then by Theorem 3.17, T would have a cycle, which, by definition, it may not have. Hence, there must be at least one vertex, a, such that $1 \leq \delta(a) < 2$. In other words, there is at least one vertex of degree one. ∎

As we will see, finding vertices of degree one in a graph is often important in applying induction arguments to propositions concerning graphs. The following theorem gives yet another condition that assures the existence of a vertex of degree one.

Theorem 3.19 If G is a connected graph with at least two vertices and if $|E(G)| < |V(G)|$, then G has a vertex of degree one.

> ***Proof:*** Since G has at least two vertices and is connected, it has no vertices of degree zero. Thus $\delta(v) \geq 1$ for all $v \in V(G)$. If G has no vertex of degree one, then it follows that $\delta(v) \geq 2$ for each $v \in V(G)$.
>
> Suppose that $V(G) = \{v_1, v_2, \ldots, v_n\}$. Then by Theorem 3.5
>
> $$\delta(v_1) + \delta(v_2) + \cdots + \delta(v_n) = 2|E(G)| \qquad (1)$$
>
> If for each i, $\delta(v_i) \geq 2$, then since $n = |V(G)|$, we have
>
> $$\delta(v_1) + \delta(v_2) + \cdots + \delta(v_n) \geq 2n = 2|V(G)| \qquad (2)$$
>
> Combining (1) and (2) gives us
>
> $$2|V(G)| \leq 2|E(G)|$$
>
> and consequently $|V(G)| \leq |E(G)|$, which contradicts the hypothesis of the Theorem. Therefore, G must have at least one vertex of degree one. ∎

We can now prove one of the most useful theorems concerning trees.

Theorem 3.20 A connected graph G is a tree if and only if $|E(G)| = |V(G)| - 1$.

> ***Proof:*** Suppose that G is a tree. We use induction on the number of edges, $|E(G)|$, of G to prove that $|E(G)| = |V(G)| - 1$.
>
> The graph G is connected; therefore, if $|E(G)| = 0$, then G must consist of a single vertex. Consequently, $|E(G)| = |V(G)| - 1$ in this case.
>
> Now assume the theorem is true for trees with k edges, and consider a tree G with $|E(G)| = k + 1$. By Corollary 3.18, G must have a vertex v of degree

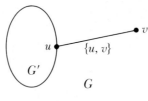

$$V(G') = V(G) \setminus \{v\}$$

$$E(G') = E(G) \setminus \{\{u, v\}\}$$

Figure 3.18

one. Thus, v is the endpoint of only one edge; denote this edge by $\{u, v\}$. Let G' be the graph formed from G by deleting the vertex v and the edge $\{u, v\}$. Figure 3.18 illustrates the relation between G and G'.

Note that G' is a tree with one less edge than G. Therefore, by the induction assumption, we have $|E(G')| = |V(G')| - 1$. Since $|E(G)| = |E(G')| + 1$ and $|V(G)| = |V(G')| + 1$, it follows that $|E(G)| = |V(G)| - 1$, as we were to prove.

Conversely, we must show that if G is a connected graph with $|E(G)| = |V(G)| - 1$, then G is a tree. That is, we must show that G has no cycles. We will again use induction on $|E(G)|$.

If $|E(G)| = 0$, then $|V(G)| - 1 = 0$, and therefore G consists of a single vertex and so is a tree.

Now suppose the result is true for graphs with k edges. Consider a connected graph G with $|E(G)| = |V(G)| - 1$ and $E(G) = k + 1$. By Theorem 3.19, G must have a vertex v of degree one. Thus v is the endpoint of only one edge; denote this edge by $\{u, v\}$. As in the first part of this proof, we will let G' be the graph given by

$$V(G') = V(G) \setminus \{v\} \quad \text{and} \quad E(G') = E(G) \setminus \{\{u, v\}\}.$$

Figure 3.18 again illustrates the relation between G and G'. Note that $|E(G')| = |E(G)| - 1 = k$. Moreover,

$$|E(G')| = |E(G)| - 1 = (|V(G)| - 1) - 1 = |V(G')| - 1$$

so by the induction assumption, G' is a tree and thus has no cycles. The edge $\{u, v\}$ cannot be part of a cycle. Consequently, G itself has no cycles, and therefore is a tree, as we were to show. ∎

3.4 Problems

1. Is it possible to construct a tree with vertices of the following degrees? If not, explain why.

$$\delta(v_1) = 1 \quad \delta(v_2) = 3 \quad \delta(v_3) = 1 \quad \delta(v_4) = 3$$
$$\delta(v_5) = 3 \quad \delta(v_6) = 3 \quad \delta(v_7) = 1 \quad \delta(v_8) = 2$$

2. Is it possible to construct a tree with vertices of the following degrees? If not, explain why.

$$\delta(v_1) = 1 \quad \delta(v_2) = 3 \quad \delta(v_3) = 1 \quad \delta(v_4) = 3$$
$$\delta(v_5) = 2 \quad \delta(v_6) = 3 \quad \delta(v_7) = 1 \quad \delta(v_8) = 2$$

3. Is it possible to construct a tree with vertices of the following degrees? If not, explain why.

$$\delta(v_1) = 1 \quad \delta(v_2) = 3 \quad \delta(v_3) = 1 \quad \delta(v_4) = 3$$
$$\delta(v_5) = 3 \quad \delta(v_6) = 3 \quad \delta(v_7) = 1 \quad \delta(v_8) = 1$$

4. Develop a method whereby one may check whether there is a tree whose vertices have degrees $\delta(v_1) = d_1, \delta(v_2) = d_2, \ldots, \delta(v_n) = d_n$.

5. Prove that if an edge is added to a tree but no new vertices are introduced, then the resulting graph has a cycle. (*Hint:* Let $T = (V, E)$ denote the original tree and let $G = (V', E')$ be the tree with an added edge. Then $|E| = |V| - 1$ and $|E'| = |E| + 1$. Show G is not a tree.)

6. Prove that a tree must have at least two vertices of degree one.

7. True or False? (If False, give a counterexample.) Suppose that G is a graph. If $|E(G)| = |V(G)| - 1$, then G is a tree.

8. Prove that if T is a tree and v is a vertex of T of degree one with $\{u, v\} \in E(T)$, then the graph G defined as $V(G) = V(T) \setminus \{v\}$, $E(G) = E(T) \setminus \{\{u, v\}\}$, is a tree.

9. A *binary tree* has one vertex of degree two (called the root of the tree), several vertices of degree three (called the internal vertices) and several vertices of degree one (called the leaves). Prove that a binary tree has two more leaves than it has interior vertices.

10. A *forest* is a graph without cycles. If a forest F has c components, v vertices and e edges, show that $v = e + c$.

3.5 SPANNING TREES

A *spanning subgraph* of a graph G is a subgraph of G that includes all the vertices (but not necessarily all of the edges) of G. Of particular interest is a spanning tree.

Definition 3.21 A subgraph T of G is a *spanning tree* of G if

    a. T is a tree, and

    b. $V(T) = V(G)$.

Spanning trees are important because they use the least possible number of edges to connect all vertices of a graph. For example, suppose that we wish to

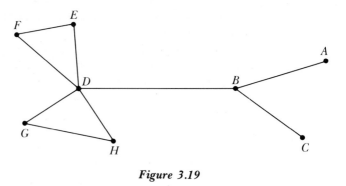

Figure 3.19

set up a communications network with a minimum number of links in the network. As in Chapter 1, we let the vertices of G represent communications centers, and we join two vertices of G with an edge to indicate that there is a direct communications link between the two corresponding centers. Figure 3.19 illustrates one such communications network.

Because maintaining links is expensive, we want to use only a minimum number of links in the network without eliminating communication (not necessarily direct) between any two centers in the network. A spanning tree such as the one indicated by wavy line segments in Figure 3.20 will work.

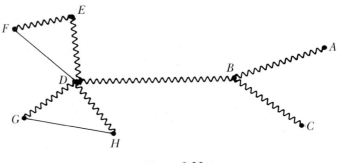

Figure 3.20

It is easy to see that a connected graph has at least one spanning tree. It is more important, however, to find an algorithm that will construct spanning trees. To develop such an algorithm, let G be a connected graph and let e be an arbitrary edge in G. Using e as the first "branch" in the spanning tree, we consider the remaining edges of G in any order, and we continue to add edges of G to the tree if doing so does not create a cycle. Once we have considered all of the edges, we will have created the desired spanning tree.

This simple idea is an excellent example of the gulf that often separates the idea behind an algorithm from its implementation. While it may be an easy matter to determine if a cycle has been completed in a small, easily drawn graph, it is quite a different matter if the graph has thousands of vertices!

The following algorithm will construct a spanning tree. It works by creating one portion of the tree at a time. Until the entire tree is complete, however, the graph under construction may not be connected. For this reason, at each stage we will need to keep track of the components of the graph being constructed. To this end, we associate a number $N(I)$ with each vertex I. Initially we set $N(I) = I$. Thereafter, we modify the numbers $N(I)$ so that if $N(I) = N(J)$, then the vertices I and J are in the same component of the graph under construction.

Algorithm 3.22 For generating a spanning tree for a connected graph G with vertex set $\{1, 2, \ldots, N\}$.

Input: $\{E(1), E(2), \ldots, E(M)\}$ the edge set of G

Output: T, the edge set of a spanning tree of G

1. FOR $I = 1$ TO N
   a. $N(I) \leftarrow I$ Initialize.
2. $T \leftarrow \varnothing$
3. $K \leftarrow 0; L \leftarrow 1$ K will count the number of edges added to the spanning tree T; when $K = N - 1$, the algorithm terminates. L is the edge index.

4. WHILE $K < N - 1$ DO
   a. IF $N(I) \neq N(J)$ Consider the edge $E(L) = \{I, J\}$. If I and J are in different components of the graph T constructed thus far, then add E to T and modify the numbers $N(R)$ to keep track of the fact that I and J are now in the same component.
      THEN
      1. $T \leftarrow T \cup \{I, J\}$
      2. FOR ALL $N(R) = $
         $\mathrm{MAX}\{N(I), N(J)\}$
         a. $N(R) \leftarrow$
            $\mathrm{MIN}\{N(I), N(J)\}$
      3. $K \leftarrow K + 1$ Since one more edge has been added to T, K must be incremented by one.

   b. $L \leftarrow L + 1$ Incrementing L prepares for the consideration of the next edge.

5. OUTPUT T

Figure 3.21 (p. 110) illustrates the values of $N(V)$ before and after $E = \{I, J\}$ is added to T. The current value of $N(V)$ is circled near each vertex.

Example 1 We use Algorithm 3.22 to construct a spanning tree T of the graph G where

$$V(G) = \{1, 2, 3, 4, 5, 6, 7\}$$
$$E(G) = \{e_1, e_2, e_3, e_4, e_5, e_6, e_7, e_8, e_9, e_{10}\}$$

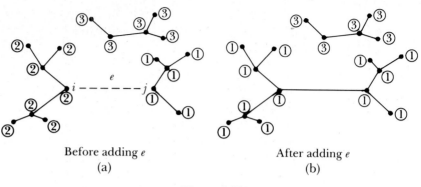

Before adding e
(a)

After adding e
(b)

Figure 3.21

and $e_1 = \{1, 2\}, \quad e_2 = \{4, 5\}, \quad e_3 = \{6, 7\}, \quad e_4 = \{5, 6\}, \quad e_5 = \{5, 7\}, \quad e_6 = \{3, 6\}, e_7 = \{2, 3\}, e_8 = \{2, 4\}, e_9 = \{3, 4\}, e_{10} = \{1, 4\}.$

It is easiest to organize our work as shown in Table 3.2. At each step we list the values of $N(I)$ and the current status of T. Figure 3.22 diagrams the graph G and the spanning tree Algorithm 3.22 generated. If the edges were to be considered in a different order, a different spanning tree might result.   □

Edge considered	$N(1)$	$N(2)$	$N(3)$	$N(4)$	$N(5)$	$N(6)$	$N(7)$	T	K
	1	2	3	4	5	6	7	$\varnothing$	0
e_1	1	1	3	4	5	6	7	$\{e_1\}$	1
e_2	1	1	3	4	4	6	7	$\{e_1, e_2\}$	2
e_3	1	1	3	4	4	6	6	$\{e_1, e_2, e_3\}$	3
e_4	1	1	3	4	4	4	4	$\{e_1, e_2, e_3, e_4\}$	4
e_5	1	1	3	4	4	4	4	$\{e_1, e_2, e_3, e_4\}$	4
e_6	1	1	3	3	3	3	3	$\{e_1, e_2, e_3, e_4, e_6\}$	5
e_7	1	1	1	1	1	1	1	$\{e_1, e_2, e_3, e_4, e_6, e_7\}$	6

Table 3.2

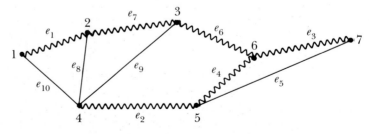

Figure 3.22

Let us show that Algorithm 3.22 does indeed create a spanning tree. It is clear that this algorithm yields a graph T with

$$|E(T)| = |V(T)| - 1$$

because the algorithm executes the loop beginning at step 4 a total of $n - 1$ times, where $n = |V(T)|$. To show that T is a tree, we prove that T is connected and apply Theorem 3.20.

Observe that there are three possible options at the loop in Algorithm 3.22: Either 0, 1, or 2 *new* vertices are added to $V(T)$. In Table 3.3 we illustrate the geometry of this application, and how each type of application changes the number of vertices, edges, and components.

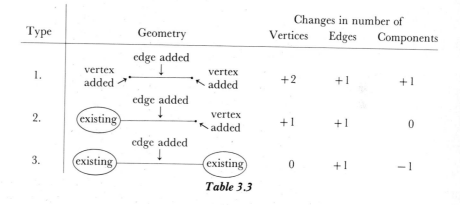

Type	Geometry	Changes in number of		
		Vertices	Edges	Components
1.	edge added / vertex added — vertex added	+2	+1	+1
2.	existing — edge added — vertex added	+1	+1	0
3.	existing — edge added — existing	0	+1	−1

Table 3.3

Suppose now that n_1 steps of Type 1 occur, n_2 steps of Type 2 occur, and n_3 steps of Type 3 occur. The number of vertices and edges of the resulting graph T are given by

$$|V(T)| = 2n_1 + n_2 \tag{1}$$

$$|E(T)| = n_1 + n_2 + n_3 \tag{2}$$

respectively. Moreover, T has $n_1 - n_3$ components.

We have previously observed that $|E(T)| = |V(T)| - 1$. Substituting from (1) and (2), we have

$$n_1 + n_2 + n_3 = 2n_1 + n_2 - 1$$

which gives us $n_1 - n_3 = 1$. Therefore, since T has $n_1 - n_3$ components, it follows that T must have just one component. In other words, T is connected and, consequently, by Theorem 3.20, T is a tree. Since T includes all vertices of G, it is a spanning tree.

Finally, note that when we apply Algorithm 3.22 in the worst case, the computer must consider *all* edges before obtaining a spanning tree. The algorithm's worst-case complexity is therefore $|E(T)|$.

3.5   Problems

In Problems 1–6 use Algorithm 3.22 to construct a spanning tree. Organize your work as shown in Example 1 and follow each construction on a diagram of the graph.

1. $V(G) = \{v_1, v_2, \ldots, v_7\}$ and $E(G) = \{e_1, e_2, \ldots, e_9\}$,
   where $e_1 = \{v_1, v_2\}$, $e_2 = \{v_1, v_3\}$, $e_3 = \{v_2, v_3\}$, $e_4 = \{v_3, v_4\}$,
   $e_5 = \{v_2, v_4\}$, $e_6 = \{v_2, v_5\}$, $e_7 = \{v_5, v_6\}$, $e_8 = \{v_6, v_7\}$, $e_9 = \{v_4, v_7\}$.

2. $V(G) = \{v_1, v_2, \ldots, v_7\}$ and $E(G) = \{e_1, e_2, \ldots, e_9\}$,
   where $e_1 = \{v_1, v_3\}$, $e_2 = \{v_3, v_4\}$, $e_3 = \{v_2, v_5\}$, $e_4 = \{v_6, v_7\}$,
   $e_5 = \{v_2, v_3\}$, $e_6 = \{v_2, v_4\}$, $e_7 = \{v_5, v_6\}$, $e_8 = \{v_4, v_7\}$, $e_9 = \{v_1, v_5\}$.

3. $V(G) = \{v_1, v_2, \ldots, v_7\}$ and $E(G) = \{e_1, e_2, \ldots, e_9\}$,
   where $e_1 = \{v_1, v_2\}$, $e_2 = \{v_2, v_5\}$, $e_3 = \{v_1, v_3\}$, $e_4 = \{v_4, v_2\}$,
   $e_5 = \{v_3, v_4\}$, $e_6 = \{v_5, v_6\}$, $e_7 = \{v_1, v_7\}$, $e_8 = \{v_4, v_7\}$, $e_9 = \{v_2, v_3\}$.

4. $V(G) = \{v_1, v_2, \ldots, v_9\}$ and $E(G) = \{e_1, e_2, \ldots, e_{11}\}$,
   where $e_1 = \{v_1, v_2\}$, $e_2 = \{v_1, v_3\}$, $e_3 = \{v_2, v_3\}$, $e_4 = \{v_2, v_4\}$,
   $e_5 = \{v_4, v_5\}$, $e_6 = \{v_3, v_5\}$, $e_7 = \{v_4, v_6\}$, $e_8 = \{v_5, v_6\}$, $e_9 = \{v_5, v_8\}$,
   $e_{10} = \{v_6, v_7\}$, $e_{11} = \{v_8, v_9\}$.

5. $V(G) = \{v_1, v_2, \ldots, v_9\}$ and $E(G) = \{e_1, e_2, \ldots, e_{11}\}$,
   where $e_1 = \{v_8, v_9\}$, $e_2 = \{v_6, v_7\}$, $e_3 = \{v_5, v_8\}$, $e_4 = \{v_5, v_6\}$,
   $e_5 = \{v_4, v_6\}$, $e_6 = \{v_3, v_5\}$, $e_7 = \{v_4, v_5\}$, $e_8 = \{v_2, v_4\}$, $e_9 = \{v_2, v_3\}$,
   $e_{10} = \{v_1, v_3\}$, $e_{11} = \{v_1, v_2\}$.

6. $V(G) = \{v_1, v_2, \ldots, v_9\}$ and $E(G) = \{e_1, e_2, \ldots, e_{11}\}$,
   where $e_1 = \{v_2, v_4\}$, $e_2 = \{v_4, v_5\}$, $e_3 = \{v_4, v_6\}$, $e_4 = \{v_5, v_6\}$,
   $e_5 = \{v_2, v_3\}$, $e_6 = \{v_6, v_7\}$, $e_7 = \{v_8, v_9\}$, $e_8 = \{v_1, v_2\}$, $e_9 = \{v_1, v_3\}$,
   $e_{10} = \{v_3, v_5\}$, $e_{11} = \{v_5, v_8\}$.

7. For Algorithm 3.22, when all values of N are equal, they are all equal to 1. Explain why.

8. It is sometimes necessary to find *all* spanning trees of a given graph. Can you create an algorithm that will generate all such trees? How many applications of Algorithm 3.22 would your algorithm require for a graph G with $V(G) = 100$ and $E(G) = 2000$?

9. Another algorithm that determines spanning trees creates them by removing one edge at a time from a connected graph G, so long as the remaining graph is still connected. The procedure stops when all edges have been considered for removal or when the number of edges remaining equals $|V(G)| - 1$.
   (a) Give a more formal description of this algorithm.
   (b) Apply this algorithm to Problems 1–6.

10. Explain why any two spanning trees of a graph G must have the same number of edges.

11. Prove that if a graph G has one cycle, then $|E(G)| = |V(G)|$.

3.6   WEIGHTED GRAPHS AND MINIMAL SPANNING TREES

In our previous work with graphs, the only property of an edge that has interested us is knowing which pair of vertices it joins. In this section, however, we examine graphs that have numerical weights assigned to their edges. This allows us to complicate the basic graph structure so that it can carry more information.

To illustrate this idea we once again consider the communications network we discussed in Section 3.5. Generally, certain communications links are more costly to maintain than others. Thus a graph that indicates the costs of these links would be more useful than one that does not. This leads us to the idea of a weighted graph. In the following definition $\mathbf{R}^+$ is the set of all nonnegative real numbers.

Definition 3.23   A *weighted graph* G is a graph—also called G—with a weight function $w : E(G) \to \mathbf{R}^+$. If $e \in E(G)$, then $w(e)$ is called the *weight* of e.

For example, in a communications network, $w(e)$ could designate the cost of maintaining link e, and we would diagram the associated graph by writing the number $w(e)$ near the edge e, as shown in Figure 3.23.

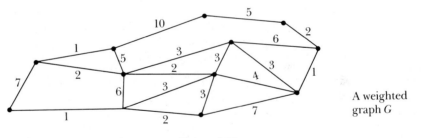

A weighted graph G

Figure 3.23

As with other graphs, we can use matrices to describe weighted graphs.

Definition 3.24   Let G be a weighted graph with vertex set $\{v_1, v_2, \ldots, v_n\}$ and weight function $w : E(G) \to \mathbf{R}^+$. The *weight matrix* of G is the matrix $W = [w_{ij}]_{n \times n}$, where

$$w_{ij} = \begin{cases} w(e) & \text{if } e = \{v_i, v_j\} \in E(G) \\ 0 & \text{if } i = j \\ \infty & \text{if } \{v_i, v_j\} \notin E(G) \end{cases}$$

The symbol ∞ indicates that the vertices v_i and v_j are not connected by an edge.

Example 1 The weight matrix of the weighted graph diagrammed in Figure 3.24 is

$$
\begin{array}{c c c c c c c}
 & v_1 & v_2 & v_3 & v_4 & v_5 & v_6 \\
v_1 & 0 & 1 & \infty & \infty & 3 & \infty \\
v_2 & 1 & 0 & 6 & \infty & \infty & 7 \\
v_3 & \infty & 6 & 0 & 1 & \infty & 2 \\
v_4 & \infty & \infty & 1 & 0 & 4 & \infty \\
v_5 & 3 & \infty & \infty & 4 & 0 & 3 \\
v_6 & \infty & 7 & 2 & \infty & 3 & 0
\end{array}
$$

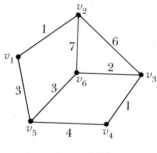

Figure 3.24

Now let us consider the communications network G with centers $v_1, v_2, \ldots, v_8$, as shown in Figure 3.25.

Suppose that the costs of the individual links (edges) of G do not differ. Because all spanning trees of G have the same number of vertices, it follows from Theorem 3.20, that they have the same number of edges. Thus, any spanning tree will provide us with a network that connects all vertices at minimal cost.

Suppose, however, that the costs of the various links do differ, as in the weighted graph in Figure 3.26.

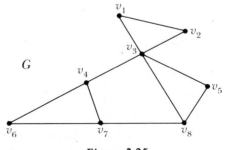

Figure 3.25

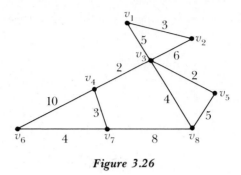

Figure 3.26

It would now make a difference which tree we selected as a spanning tree because the weights (costs) associated with different spanning trees will vary. Figure 3.27 illustrates two possibilities.

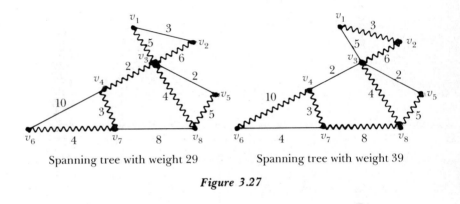

Spanning tree with weight 29   Spanning tree with weight 39

Figure 3.27

This leads us to the problem of finding minimal spanning trees.

Definition 3.25   A *minimal spanning tree* of a connected weighted graph G is a spanning tree of G with the smallest possible weight.

The task of finding a minimal spanning tree is called the *minimal connector* problem. It is of particular importance because discovering a minimal spanning tree can lead to a solution of many communications, transportation, and other problems. *Kruskal's Algorithm*, a simple variation on Algorithm 3.22, is one of several algorithms that generate minimal spanning trees. While Algorithm 3.22 examines the edges of the graph in any order, Kruskal's Algorithm considers the edges in order of increasing weight. In describing Kruskal's Algorithm we use the notation employed in describing Algorithm 3.22.

Algorithm 3.26   *Kruskal's Algorithm.* For generating a minimal spanning tree in a connected graph G with vertex set $\{1, 2, \ldots, N\}$

Input: $E(1), E(2), \ldots, E(M)$, the edge set of G ordered so that $W(E(1)) \le W(E(2)) \le \cdots \le W(E(M))$

Algorithm 1.5 (BUBBLESORT) can be used to order these weights.

Output: T, the edge set of a minimal spanning tree of G

1. FOR $I = 1$ TO N
   a. $N(I) \leftarrow I$ Initialize.
2. $T \leftarrow \varnothing$
3. $K \leftarrow 0; L \leftarrow 1$ K counts the number of edges added to the spanning tree T; when $K = N - 1$, the algorithm terminates. L is the edge index.

4. WHILE $K < N - 1$ DO
   a. IF $N(I) \ne N(J)$ Consider the edge $E(L) = \{I, J\}$.
      THEN   If I and J are in different
      1. $T \leftarrow T \cup \{I, J\}$ components of the graph T
      2. FOR ALL $N(R) =$ constructed thus far, then add E to
         $\mathrm{MAX}\{N(I), N(J)\}$ T and modify the numbers $N(R)$ to
         a. $N(R) \leftarrow$ keep track of the fact that I and J
            $\mathrm{MIN}\{N(I), N(J)\}$ are now in the same component.
      3. $K \leftarrow K + 1$ Since one more edge has been added
   b. $L \leftarrow L + 1$ to T, K is incremented by one. Incrementing L prepares for the consideration of the next edge.

5. OUTPUT T

Because Kruskal's Algorithm is a special case of Algorithm 3.22, it is clear that Kruskal's Algorithm will result in a spanning tree. The proof that this spanning tree is a *minimal* spanning tree is fairly intricate, and we will pursue it in the following section.

Example 2   We apply Kruskal's Algorithm to the weighted graph in Figure 3.28. Labeling the edges so that their weights are in increasing order, we obtain

$$e_1 = \{3, 7\} \qquad e_2 = \{5, 6\} \qquad e_3 = \{3, 4\}$$
$$e_4 = \{1, 2\} \qquad e_5 = \{1, 3\} \qquad e_6 = \{4, 7\}$$
$$e_7 = \{1, 7\} \qquad e_8 = \{5, 7\} \qquad e_9 = \{2, 3\}$$

We organize the work as indicated in Table 3.4. At each step we give the values of $N(i)$ and the current status of T.

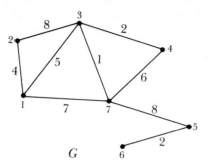

Figure 3.28

Edge considered	N(1)	N(2)	N(3)	N(4)	N(5)	N(6)	N(7)	T	Weight of T
	1	2	3	4	5	6	7	$\emptyset$	
e_1	1	2	3	4	5	6	3	$\{e_1\}$	1
e_2	1	2	3	4	5	5	3	$\{e_1, e_2\}$	3
e_3	1	2	3	3	5	5	3	$\{e_1, e_2, e_3\}$	5
e_4	1	1	3	3	5	5	3	$\{e_1, e_2, e_3, e_4\}$	9
e_5	1	1	1	1	5	5	1	$\{e_1, e_2, e_3, e_4, e_5\}$	14
e_6	1	1	1	1	5	5	1	$\{e_1, e_2, e_3, e_4, e_5\}$	14
e_7	1	1	1	1	5	5	1	$\{e_1, e_3, e_3, e_4, e_5\}$	14
e_8	1	1	1	1	1	1	1	$\{e_1, e_2, e_3, e_4, e_5, e_8\}$	22

Table 3.4

In this example Kruskal's Algorithm generates a minimal spanning tree with edge set $\{e_1, e_2, e_3, e_4, e_5, e_8\}$ and total weight 22. $\square$

3.6 Problems

Problems 1 and 2 refer to the weighted graph G shown in Figure 3.29 (p. 118).

1. Compute $W(H)$, where H is the subgraph of G given by $V(H) = \{v_1, v_2, v_3, v_4, v_5, v_{10}\}$ and $E(H) = \{\{v_1, v_2\},\ \{v_2, v_3\},\ \{v_3, v_4\}, \{v_4, v_5\}\}$.

2. Compute $W(H)$, where H is the subgraph of G given by $V(H) = \{v_5, v_6, v_7, v_8, v_9, v_{10}\}$ and $E(H) = \{\{v_5, v_{10}\},\ \{v_6, v_7\},\ \{v_8, v_7\}, \{v_7, v_9\}\}$.

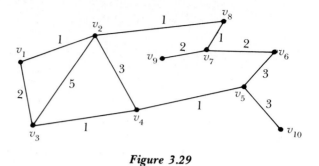

Figure 3.29

3. Give the weight matrix of the graph pictured in Figure 3.30

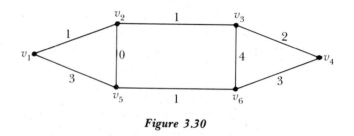

Figure 3.30

4. Give the weight matrix of the graph pictured in Figure 3.31.

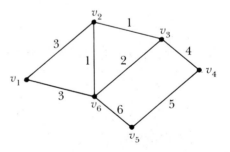

Figure 3.31

In Problems 5 and 6 draw a diagram of the graph with weight matrix.

5. $\begin{bmatrix} 0 & 1 & \infty & 1 & \infty & \infty \\ 1 & 0 & 2 & 1 & \infty & \infty \\ \infty & 2 & 0 & 2 & \infty & 3 \\ 1 & 1 & 2 & 0 & 4 & 3 \\ \infty & \infty & \infty & 4 & 0 & 3 \\ \infty & \infty & 3 & 3 & 3 & 0 \end{bmatrix}$

6. $\begin{bmatrix} 0 & 2 & \infty & 2 & \infty & \infty \\ 2 & 0 & 3 & 1 & \infty & \infty \\ \infty & 3 & 0 & 2 & 4 & \infty \\ 2 & 1 & 2 & 0 & 1 & 2 \\ \infty & \infty & 4 & 1 & 0 & 2 \\ \infty & \infty & \infty & 2 & 2 & 0 \end{bmatrix}$

In Problems 7–12 apply Kruskal's Algorithm to the graph whose edges have the indicated weights.

7. $V(G) = \{1, 2, 3, 4, 5, 6\}$; $w(\{1, 2\}) = 1$, $w(\{1, 4\}) = 1$, $w(\{2, 4\}) = 1$, $w(\{2, 3\}) = 2$, $w(\{3, 4\}) = 2$, $w(\{3, 6\}) = 3$, $w(\{4, 6\}) = 3$, $w(\{5, 6\}) = 3$, $w(\{4, 5\}) = 4$.

8. $V(G) = \{1, 2, 3, 4, 5, 6\}$; $w(\{4, 2\}) = 1$, $w(\{4, 5\}) = 1$, $w(\{1, 4\}) = 2$, $w(\{1, 2\}) = 2$, $w(\{4, 6\}) = 2$, $w(\{5, 6\}) = 2$, $w(\{4, 3\}) = 2$, $w(\{2, 3\}) = 3$, $w(\{3, 5\}) = 4$.

9. $V(G) = \{1, 2, 3, 4, 5, 6\}$; $w(\{1, 2\}) = 1$, $w(\{4, 5\}) = 3$, $w(\{3, 4\}) = 2$, $w(\{5, 6\}) = 1$, $w(\{1, 6\}) = 3$, $w(\{2, 6\}) = 2$, $w(\{2, 3\}) = 1$, $w(\{3, 5\}) = 1$.

10. $V(G) = \{1, 2, 3, 4, 5, 6\}$; $w(\{1, 2\}) = 2$, $w(\{1, 6\}) = 3$, $w(\{2, 6\}) = 1$, $w(\{2, 3\}) = 1$, $w(\{3, 4\}) = 4$, $w(\{3, 6\}) = 2$, $w(\{4, 5\}) = 5$, $w(\{5, 6\}) = 6$.

11. $V(G) = \{1, 2, 3, 4, 5, 6, 7, 8\}$; $w(\{1, 2\}) = 1$, $w(\{1, 3\}) = 1$, $w(\{1, 4\}) = 2$, $w(\{2, 3\}) = 3$, $w(\{3, 6\}) = 3$, $w(\{3, 7\}) = 4$, $w(\{4, 5\}) = 4$, $w(\{4, 6\}) = 2$, $w(\{5, 6\}) = 1$, $w(\{6, 7\}) = 5$, $w(\{7, 8\}) = 6$.

12. $V(G) = \{1, 2, 3, 4, 5, 6, 7, 8, 9\}$; $w(\{1, 2\}) = 1$, $w(\{1, 3\}) = 1$, $w(\{2, 3\}) = 3$, $w(\{2, 4\}) = 2$, $w(\{3, 5\}) = 2$, $w(\{3, 7\}) = 4$, $w(\{4, 5\}) = 5$, $w(\{4, 6\}) = 4$, $w(\{5, 6\}) = 1$, $w(\{7, 8\}) = 2$, $w(\{7, 9\}) = 1$, $w(\{8, 9\}) = 3$.

In Problem 13–18 apply Kruskal's Algorithm to the graph with the given weight matrix.

13.
$$\begin{bmatrix} 0 & 2 & \infty & 2 & 5 \\ 2 & 0 & 2 & \infty & \infty \\ \infty & 2 & 0 & 2 & \infty \\ 2 & \infty & 2 & 0 & 5 \\ 5 & \infty & \infty & 5 & 0 \end{bmatrix}$$

14.
$$\begin{bmatrix} 0 & 2 & \infty & \infty & 1 \\ 2 & 0 & 2 & 3 & \infty \\ \infty & 2 & 0 & 1 & \infty \\ \infty & 3 & 1 & 0 & 3 \\ 1 & \infty & \infty & 3 & 0 \end{bmatrix}$$

15.
$$\begin{bmatrix} 0 & 1 & \infty & \infty & 1 \\ 1 & 0 & 2 & 3 & \infty \\ \infty & 2 & 0 & 4 & 2 \\ \infty & 3 & 4 & 0 & 3 \\ 1 & \infty & 2 & 3 & 0 \end{bmatrix}$$

16.
$$\begin{bmatrix} 0 & 1 & \infty & 1 & \infty & 2 \\ 1 & 0 & 2 & \infty & \infty & \infty \\ \infty & 2 & 0 & 3 & \infty & 2 \\ 1 & \infty & 3 & 0 & 2 & \infty \\ \infty & \infty & \infty & 2 & 0 & 1 \\ 2 & \infty & 2 & \infty & 1 & \infty \end{bmatrix}$$

17.
$$\begin{bmatrix} 0 & 1 & \infty & \infty & \infty & 2 \\ 1 & 0 & 2 & \infty & \infty & 1 \\ \infty & 2 & 0 & 3 & 2 & \infty \\ \infty & \infty & 3 & 0 & 3 & \infty \\ \infty & \infty & 2 & 3 & 0 & 1 \\ 2 & 1 & \infty & \infty & 1 & 0 \end{bmatrix}$$

18.
$$\begin{bmatrix} 0 & 1 & \infty & \infty & \infty & 1 \\ 1 & 0 & 3 & \infty & \infty & 1 \\ \infty & 3 & 0 & 1 & \infty & \infty \\ \infty & \infty & 1 & 0 & 2 & \infty \\ \infty & \infty & \infty & 2 & 0 & 3 \\ 1 & 1 & \infty & \infty & 3 & 0 \end{bmatrix}$$

19. A greedy algorithm for finding a minimal spanning tree of a graph G is based on the following idea. Begin a spanning tree T with a single vertex v_1. Check all edges not yet in T that have a vertex in common with some vertex in T. Let e be the edge with the least weight such that if e is added to T, no cycle will be created. Add e to T to form a new tree, and continue this process until T has $n - 1$ edges.

    (a) Why is this algorithm (known as Prim's Algorithm) a greedy algorithm?

    (b) Write a more formal description of this algorithm.

    (c) Apply this algorithm to the graphs described in Problems 7–12.

*3.7   A PROOF OF KRUSKAL'S ALGORITHM

In the previous section we indicated that Kruskal's Algorithm is a special case of Algorithm 3.22, and therefore it generates a spanning tree. We now show that Kruskal's Algorithm generates a *minimal* spanning tree.

    The proof is difficult, but the clever reasoning should more than compensate for the effort required to read it through in detail.

Theorem 3.27   If G is a connected weighted graph, then the tree generated by Kruskal's Algorithm when applied to G is a minimal spanning tree.

    Proof: Let K denote the tree Kruskal's Algorithm generates when it is applied to the weighted graph G. From the set of all minimal spanning trees in G, select the tree, T, that differs from K in the fewest number of edges. Our goal is to show that K and T have exactly the same edges. To show this, we assume to the contrary—that K and T *do not* have the same edge set—and we eventually reach a contradiction.

    Since K and T are both spanning trees, it follows from Theorem 3.20 that they have the same number of edges. Thus, if $E(T) \neq E(K)$, then $E(K) \setminus E(T) \neq \varnothing$. Let e be the edge of smallest weight in $E(K) \setminus E(T)$. Consider the graph formed by adjoining e to T. Since e is not an edge of T, adding e to T will create a cycle C, as Figure 3.32(a) shows.

    Observe now that since K contains no cycle (by the definition of a tree), the cycle C must have at least one edge, f, not in K. Of course, since e is an edge of K, f is not e.

    Let T' denote the spanning tree obtained when we add e to T and delete f; T' is shown in Figure 3.32(b). Note that

$$W(T') = W(T) + w(e) - w(f) \tag{1}$$

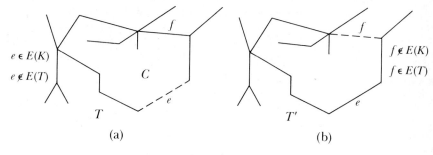

$e \in E(K)$

$e \notin E(T)$

$f \notin E(K)$

$f \in E(T)$

Figure 3.32

Because T is a minimal tree, we have that $W(T) \leq W(T')$, which gives us $W(T') - W(T) \geq 0$. By rewriting (1) as

$$W(T') - W(T) = w(e) - w(f)$$

we see that $w(e) - w(f) \geq 0$, or

$$w(e) \geq w(f) \qquad (2)$$

Suppose now that $w(e) = w(f)$. Then $W(T) = W(T')$, so T' is also a minimal tree. However, T' has one more edge, namely e, in common with K than does T. These conclusions (T' is a minimal tree and has more edges in common with K than does T) contradict our choice of T. Consequently, we cannot have $w(e) = w(f)$, and in view of (2) we know that

$$w(f) < w(e) \qquad (3)$$

We now follow much the same procedure by adjoining the edge f to K, which creates a cycle C', as Figure 3.33 shows. As before, since T contains no cycles, there is at least one edge, call it e', of C' that is in $E(K)$ but not in $E(T)$.

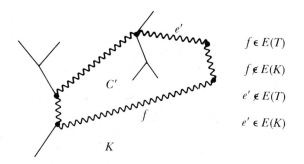

$f \in E(T)$

$f \notin E(K)$

$e' \notin E(T)$

$e' \in E(K)$

Figure 3.33

By Kruskal's Algorithm we must have $w(e') \leq w(f)$. To see this, note that if $w(f) < w(e')$, then we would have considered adding f to K prior to considering adding e'. Thus by the very process described in Kruskal's Algorithm we would have $f \in E(K)$ and $e' \notin E(K)$, which contradicts the fact that $f \notin E(K)$ and $e' \in E(K)$. Consequently,

$$w(e') \leq w(f) \tag{4}$$

Finally, combining (3) and (4) we have

$$w(e') < w(e)$$

This contradicts our choice of e as the smallest weight edge of K that is not an edge of T, and so completes the proof!   ∎

3.8   MINIMUM WEIGHT SPANNING CYCLES

Consider the following situation. We must program an automated drill press to drill a certain number of holes in a metal plate. The drill head is to follow the shortest possible path to accomplish this task, and then return to its starting point.

As a specific example, suppose that we would like to drill holes at the points indicated on the metal plate depicted in Figure 3.34.

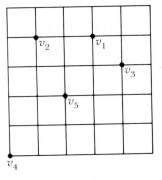

Figure 3.34

The drill head can move only vertically and horizontally, and each side of a small square on the superimposed grid represents one unit of distance. (For example, the total distance from v_3 to v_5 is 3 because the press must go down 1 unit and go across 2 units.) Calculating the distance between each pair of holes, we come up with the weighted graph G in Figure 3.35, where the weight represents the distance between the respective holes.

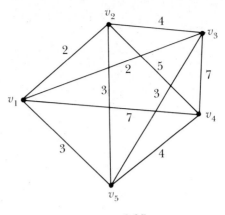

Figure 3.35

The weight (distance) matrix A associated with the graph G is defined by

$$A = \begin{bmatrix} 0 & 2 & 2 & 7 & 3 \\ 2 & 0 & 4 & 5 & 3 \\ 2 & 4 & 0 & 7 & 3 \\ 7 & 5 & 7 & 0 & 4 \\ 3 & 3 & 3 & 4 & 0 \end{bmatrix} \tag{1}$$

Because the drill head must return to its original position, our task is to determine the shortest *spanning cycle* in the weighted graph G. A spanning cycle in a graph G is a cycle in G that contains all of the vertices of G.

Determining a minimum weight spanning cycle is frequently referred to as the "traveling salesman problem," which we discussed in Section 1.5. In this context, the vertices of the graph represent cities, and the weight of an edge represents either the cost of travel between cities or the distance between cities.

Not all graphs have spanning cycles. Clearly, if a graph is not connected then it cannot have a spanning cycle. Some connected graphs, however, can also fail to contain spanning cycles (see Problem 24). A class of graphs for which spanning cycles do exist is the class of complete graphs.

Definition 3.28   A *complete graph with n vertices* is a graph with n vertices and an edge set that consists of all possible edges connecting these vertices.

The symbol K_n denotes the complete graph with n vertices. Figure 3.36 (p. 124) illustrates the complete graphs K_1, K_2, K_3, K_4, and K_5.

We would like to develop an algorithm that will determine a minimal weight spanning cycle in a weighted complete graph. To find such a cycle we

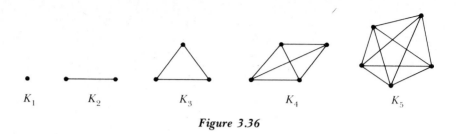

Figure 3.36

could list all possible cycles and see which has the least weight. However, for each n, K_n has

$$(n - 1)! = (n - 1)(n - 2) \ldots 3 \cdot 2 \cdot 1$$

spanning cycles (Can you show this?), and as you can easily verify, $(n - 1)!$ is an extremely large number, even for relatively small values of n (for example, $20! \cong 2.4 \times 10^{18}$). Therefore, even with a computer's help it would be impossible to calculate the weight of every possible spanning cycle.

Unfortunately, there isn't any efficient solution to this problem; that is, there is no known algorithm for computing minimal weight spanning cycles in a reasonable time span. We must therefore be content to find a fairly efficient algorithm that yields approximate solutions to the problem. One such algorithm is based on the following idea.

A cycle C is "grown" in a graph G by starting with an arbitrary vertex as the initial C, and then selecting from all vertices not in C a vertex u that is closest to some vertex v in C. As Figure 3.37 indicates, the vertex u is then included in C just prior to the vertex v. This results in a new and longer cycle C. We repeat the process until C includes all of the vertices of G.

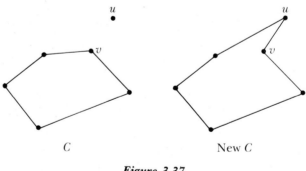

C New C

Figure 3.37

Example 1 To approximate the minimal weight spanning cycle in the graph given in Figure 3.35, we proceed as follows.

Let C_1 be the vertex v_1. From the distance (weight) matrix in (1) we see that both v_2 and v_3 are a distance of 2 from v_1. We arbitrarily select v_3 to obtain

$C_2 = v_1, v_3, v_1$. Since v_2 is the vertex closest to C_2 (it is at a distance of 2 from v_1), we insert v_2 prior to v_1 and obtain the cycle $C_3 = v_1, v_3, v_2, v_1$. Observe now that v_5 is the vertex closest to C_3 and, in fact, is at an equal distance of 3 from v_1, v_2, and v_3. We insert v_5 immediately prior to v_2 and obtain the cycle $C_4 = v_1, v_3, v_5, v_2, v_1$. Finally, since v_4 is at a distance of 4 from v_5, we have

$$C_5 = v_1, v_3, v_4, v_5, v_2, v_1$$

Note that the weight of C_5 is

$$\begin{aligned} w(C_5) &= w(\{v_1, v_3\}) + w(\{v_3, v_4\}) + w(\{v_4, v_5\}) \\ &\quad + w(\{v_5, v_2\}) + w(\{v_2, v_1\}) \\ &= 2 + 7 + 4 + 3 + 2 = 18 \end{aligned} \qquad \square$$

The following algorithm formalizes this procedure.

Algorithm 3.29   To find an approximate minimum weight spanning cycle in K_N.

Input: $[W(I, J)]$, the weight matrix of K_N
The vertex set of K_N is $VER = \{1, 2, \ldots, N\}$

Output: The set VER printed in an order that approximates a minimum weight spanning cycle.

1. $C \leftarrow \{1\}$
2. $K \leftarrow 1$
3. $P(1) \leftarrow 1$

In steps 1–4 the variables are initialized. C is the cycle being $\{1, 2, \ldots, N\}$; C is the cycle being formed; K designates the number of vertices included in C; $P(R)$ is the predecessor of R in C.

4. WHILE $K < N$ DO
  a. MIN $\leftarrow$ MAX

  b. FOR ALL
    $J \in VER \backslash C$
    1. FOR ALL $I \in C$
      a. IF $W(I, J) <$ MIN
        THEN
        1. MIN $\leftarrow W(I, J)$
        2. $V \leftarrow I$
        3. $U \leftarrow J$
  c. $P(U) \leftarrow P(V)$
  d. $P(V) \leftarrow U$
  e. $K \leftarrow K + 1$

The following steps are executed as long as there are vertices not yet included in C. MAX is the largest entry of $[W(I, J)]$.

In step b, the computer determines the vertices U and V discussed above.

The vertex U is inserted into C just prior to V. In other words, U becomes the predecessor of V, and the old predecessor of V is the predecessor of U.

5. $R \leftarrow 1$ Initialize the variables for output.
6. $K \leftarrow 1$
7. WHILE $K \leq N$ DO   Output the vertices of C in order.
  a. OUTPUT $P(R)$
  b. $R \leftarrow P(R)$
  c. $K \leftarrow K + 1$

We emphasize that Algorithm 3.29 will not necessarily yield a *minimum* weight spanning cycle. For instance, in Example 1 the algorithm generated a cycle with weight 18. A quick check shows that the spanning cycle

$$C = v_1, v_2, v_4, v_5, v_3, v_1$$

has weight 16. This raises the question of how much the weight of a graph's actual minimal spanning cycle differs from the weight of the cycle Algorithm 3.29 obtains. To deal with this question we must make an additional assumption about the weights assigned to the edges. For all $i, j,$ and k we shall assume that

$$w(v_i, v_j) + w(v_j, v_k) \geq w(v_i, v_k) \qquad (2)$$

This condition is called the *triangle inequality* because it relates the weights of three sides of a triangle as shown in Figure 3.38.

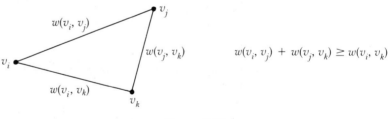

Figure 3.38

The triangle inequality (2) always holds if the weights represent actual distances; it also holds in many other applications as well. If this inequality holds, we can use the following theorem to determine bounds on the weight of the spanning cycle generated by Algorithm 3.29. (The proof of this theorem is rather tedious and is omitted.)

Theorem 3.30   Let w_m denote the weight of a minimal spanning cycle in a complete graph G, and let w_a denote the weight of the spanning cycle generated by Algorithm 3.29. If the weights assigned to the edges of G satisfy the triangle inequality, then $w_m \leq w_a \leq 2w_m$.

Applying Theorem 3.30 to Example 1, we can conclude that $w_m \leq 18 \leq 2w_m$ since $w_a = 18$. Note that it follows from this inequality that $9 \leq w_m \leq 18$.

It is interesting to compare Algorithm 3.29 with the corresponding greedy algorithm introduced in Section 1.5. The greedy algorithm starts with a path P_1 consisting of a single vertex. At each succeeding step k of this algorithm, that vertex closest to the vertex most recently adjoined to path P_k is added to P_k, producing a longer path P_{k+1}. The cycle is formed by joining the first and last vertices of P_n. We omit a formal statement of this algorithm because it is similar to the statement of Algorithm 3.29.

Example 2 To compare the results of Algorithm 3.29 and the greedy algorithm, we apply the greedy algorithm to approximate the minimal weight spanning cycle of the graph in Figure 3.35 (p. 123).

Let $P_1 = v_1$. Note that the vertex v_3 is closest to v_1, so $P_2 = v_1, v_3$. Since v_5 is the vertex (not in P_2) that is closest to the vertex v_3, we have $P_3 = v_1, v_3, v_5$. Similarly, the vertex v_2 is closest to v_5, and $P_4 = v_1, v_3, v_5, v_2$. Finally, we have $P_5 = v_1, v_3, v_5, v_2, v_4$, so

$$C = v_1, v_3, v_5, v_2, v_4, v_1$$

is the desired cycle. $\qquad\square$

Note that the greedy algorithm produced a cycle C of weight $2 + 3 + 3 + 5 + 7 = 20$, whereas Algorithm 3.29 provided a spanning cycle of weight 18. The following theorem can be used to determine a bound on the error when the greedy algorithm is used to obtain an approximation to a minimal weight spanning cycle.

Theorem 3.31 Let w_m denote the weight of a minimal spanning cycle in a complete graph G, and let w_g denote the weight of the spanning cycle generated by the greedy algorithm. If the weights assigned to the edges of G satisfy the triangle inequality, then

$$w_m \leq w_g \leq \lceil \log_2 n \rceil w_m + \tfrac{1}{2}$$

where $n = |V(G)|$.

From this theorem we can conclude that in Example 2

$$w_m \leq 20 \leq \lceil \log_2 5 \rceil w_m + \tfrac{1}{2}$$

Since $\lceil \log_2 5 \rceil = 3$, we have

$$w_m \leq 20 \leq 3w_m + \tfrac{1}{2}$$

Note that it follows from this inequality that

$$\tfrac{39}{6} \leq w_m \leq 20$$

Generally, and as we have seen here, the bound given by Theorem 3.31 for the greedy algorithm is not as narrow as that given by Theorem 3.30 for Algorithm 3.29. The greedy algorithm, however, is somewhat simpler to apply, because at each step we need only find the vertex which is closest to the last vertex of the path already obtained.

To compare the efficiencies of Algorithm 3.29 and the greedy algorithm, we will count the total number of comparisons each of these algorithms must make when applied to K_n.

The greedy algorithm must make $n - 1$ comparisons to select the second vertex. At the next step, it must compare $n - 2$ weights. In general, to select the kth vertex, the algorithm must compare the remaining $n - k + 1$ weights. Consequently, when applied to K_n the greedy algorithm makes

$$(n-1) + (n-2) + \cdots + (n-k+1) + \cdots + 2 = \frac{n(n-1)}{2} - 1 \in O(n^2)$$

comparisons.

Now suppose that we have selected k vertices using Algorithm 3.29. At this stage, the algorithm must compare each of these k vertices to each of the remaining $n - k$ vertices. Consequently, for each $k = 1, 2, \ldots, (n-2)$, Algorithm 3.29 requires $k(n-k)$ weight comparisons, and therefore, it makes a total of

$$\sum_{k=1}^{n-2} k(n-k)$$

comparisons. Note that by Example 1 of Section 1.3 and by Problem 7 of the same section, we have

$$\sum_{k=1}^{n-2} k(n-k) = n\sum_{k=1}^{n-2} k - \sum_{k=1}^{n-2} k^2$$

$$= n\frac{(n-1)(n-2)}{2} - \frac{(n-2)(n-1)(2n-3)}{6}$$

$$= (n-1)(n-2)\left(\frac{n}{2} - \frac{n}{3} + \frac{1}{2}\right) \in O(n^3).$$

Thus, the price of the increased accuracy of Algorithm 3.29 is clear: Algorithm 3.29 is of order n^3, while the greedy algorithm is of order n^2. The greedy algorithm is the more efficient—though less accurate—algorithm.

3.8　Problems

In Problems 1–4 apply Algorithm 3.29 to the pictured weighted complete graphs. In each case use v_1 as the initial vertex.

1.
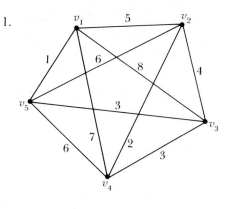

2.
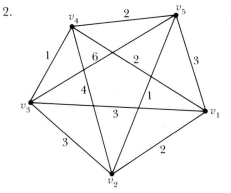

3.
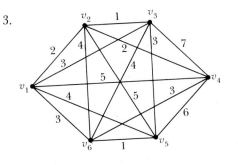

4.
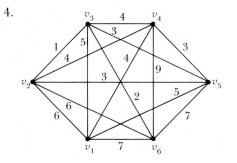

5. Repeat Problems 1–4 using the greedy algorithm and using v_1 as the initial vertex.

6. Do the weights given in Problem 1 satisfy the triangle inequality?

7. Do the weights given in Problem 2 satisfy the triangle inequality?

In Problems 8–13 apply Algorithm 3.29 to the weighted complete graph with the given weight matrix. In each case use v_1 as the initial vertex.

8. $\begin{bmatrix} 0 & 3 & 6 & 3 & 3 & 6 \\ 3 & 0 & 7 & 4 & 4 & 3 \\ 6 & 7 & 0 & 3 & 3 & 4 \\ 3 & 4 & 3 & 0 & 2 & 5 \\ 3 & 4 & 3 & 2 & 0 & 3 \\ 6 & 3 & 4 & 5 & 3 & 0 \end{bmatrix}$ 9. $\begin{bmatrix} 0 & 2 & 2 & 3 & 4 & 6 \\ 2 & 0 & 4 & 3 & 2 & 4 \\ 2 & 4 & 0 & 5 & 4 & 8 \\ 3 & 3 & 5 & 0 & 5 & 3 \\ 4 & 2 & 4 & 5 & 0 & 4 \\ 6 & 4 & 8 & 3 & 4 & 0 \end{bmatrix}$

10. $\begin{bmatrix} 0 & 1 & 2 & 1 & 4 & 3 \\ 1 & 0 & 3 & 5 & 1 & 2 \\ 2 & 3 & 0 & 1 & 2 & 1 \\ 1 & 5 & 1 & 0 & 3 & 4 \\ 4 & 1 & 2 & 3 & 0 & 5 \\ 3 & 2 & 1 & 4 & 5 & 0 \end{bmatrix}$ 11. $\begin{bmatrix} 0 & 2 & 6 & 4 & 4 & 4 \\ 2 & 0 & 4 & 2 & 2 & 4 \\ 6 & 4 & 0 & 2 & 4 & 6 \\ 4 & 2 & 2 & 0 & 2 & 4 \\ 4 & 2 & 4 & 2 & 0 & 2 \\ 4 & 4 & 6 & 4 & 2 & 0 \end{bmatrix}$

12. $\begin{bmatrix} 0 & 2 & 1 & 4 & 1 & 3 \\ 2 & 0 & 3 & 5 & 2 & 2 \\ 1 & 3 & 0 & 6 & 1 & 1 \\ 4 & 5 & 6 & 0 & 4 & 3 \\ 1 & 2 & 1 & 4 & 0 & 3 \\ 3 & 2 & 1 & 3 & 3 & 0 \end{bmatrix}$ 13. $\begin{bmatrix} 0 & 2 & 2 & 6 & 5 & 3 \\ 2 & 0 & 2 & 4 & 3 & 1 \\ 2 & 2 & 0 & 4 & 3 & 3 \\ 6 & 4 & 4 & 0 & 1 & 3 \\ 5 & 3 & 3 & 1 & 0 & 2 \\ 3 & 1 & 3 & 3 & 2 & 0 \end{bmatrix}$

14. Repeat Problems 8–13 using the greedy algorithm and using v_1 as the initial vertex.

15. Show that the weights given in the matrix in Problem 10 do not satisfy the triangle inequality.

16. Show that the weights given in the matrix in Problem 12 do not satisfy the triangle inequality.

17. Assume that the weights given in the matrix in Problem 8 do satisfy the triangle inequality. If we were to apply Algorithm 3.29 to the corresponding graph, what would be the bounds on w_a and w_m?

18. Assume that the weights given in the matrix in Problem 9 do satisfy the triangle inequality. If we were to apply Algorithm 3.29 to the corresponding graph, what would be the bounds on w_a and w_m?

19. Repeat Problems 17 and 18 using the greedy algorithm instead of Algorithm 3.29.

20. How many edges are there in the complete graph of n vertices, K_n? (*Hint:* Each vertex has degree $n - 1$.)

21. Prove that Algorithm 3.29 must terminate.

22. Write a formal presentation of the greedy algorithm for approximating a minimal weight spanning cycle.

23. Prove that the greedy algorithm you described in Problem 22 must terminate.

24. A graph is called *Hamiltonian* if it has a spanning cycle. Find a connected graph that is not Hamiltonian.

25. Prove that K_n is Hamiltonian for each $n \geq 3$.

26. Suppose the origin of a coordinate system corresponds to the lower left corner of a metal plate. A drill head capable of only vertical and horizontal motion begins at the origin and drills holes at the locations: $(1, 1), (2, 1), (3, 1), (2, 4), (2, 2), (5, 1)$. The drill head then returns to the origin.
    (a) Apply Algorithm 3.29 to approximate the minimum distance the drill head must travel.
    (b) Apply the greedy algorithm to approximate the minimum distance the drill must travel.
    (c) Find the best bounds possible on the actual minimum distance the drill head must travel.

27. Repeat Problem 26 for the locations: $(1, 1)$ $(1, 2), (1, 4), (2, 3), (2, 4)$ $(2, 6)$.

3.9 DIRECTED GRAPHS AND RELATIONS

In this section we will continue our study of more specialized graph theoretic structures. We introduce the notion of a directed graph, and then we examine a fairly natural connection between these graphs and relations.

In a directed graph, each edge has direction, indicated by an arrow, from one vertex to another. Figure 3.39 illustrates the diagram of a typical directed graph. Directed graphs have obvious applications to one-way transportation systems, pipeline networks, and so on. They can also be applied to solve the problem of scheduling interdependent activities and even to solve a variety of puzzles such as the Rubik's cube.

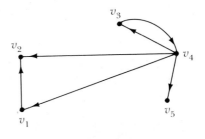

Figure 3.39

In previous sections we have formally defined an edge that connects two vertices u and v as the set $\{u, v\}$. To indicate that an edge is directed from vertex u to vertex v, we describe the edge as the *ordered pair* (u, v). The ordered pair (v, u) describes the edge that is directed from the vertex v to the vertex u. For directed graphs it will be convenient at times to allow loops. A *loop* is an edge that connects a vertex u to itself, and is denoted by (u, u).

Definition 3.32 A *directed graph* (also called a *digraph*), D, consists of a nonempty finite set, $V(D)$, of vertices and a set, $E(D)$, of ordered pairs (u, v) of vertices with $u \neq v$. The set of ordered pairs $E(D)$ is called the *edge set* of D. If loops are allowed, then we omit the restriction $u \neq v$.

Example 1

(a) Figure 3.40 indicates the diagram of the digraph D where $V(D) = \{v_1, v_2, v_3, v_4\}$ and $E(D) = \{(v_1, v_2), (v_3, v_4), (v_2, v_3), (v_2, v_1)\}$.

Figure 3.40

(b) Figure 3.41 illustrates the diagram of a directed graph D with loops. In this case $V(D) = \{v_1, v_2, v_3, v_4, v_5\}$ and $E(D) = \{(v_1, v_3), (v_1, v_4), (v_3, v_2), (v_3, v_3), (v_4, v_1), (v_2, v_5), (v_5, v_5)\}$. $\qquad\square$

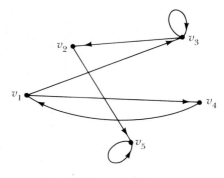

Figure 3.41

As with other graphs, we can utilize matrices to describe directed graphs.

Definition 3.33   Let D be a directed graph with vertex set $\{v_1, v_2, \ldots, v_n\}$. The *adjacency matrix* of D is an $n \times n$ matrix $[a_{ij}]_{n \times n}$ where

$$a_{ij} = \begin{cases} 1 & \text{if } (v_i, v_j) \in E(D) \\ 0 & \text{if } (v_i, v_j) \notin E(D) \end{cases}$$

The adjacency matrix of a directed graph is not necessarily symmetric. For example, the digraph described in Figure 3.40 has the adjacency matrix

$$A = \begin{bmatrix} 0 & 1 & 0 & 0 \\ 1 & 0 & 1 & 0 \\ 0 & 0 & 0 & 1 \\ 0 & 0 & 0 & 0 \end{bmatrix}$$

The previous definitions of a subgraph, and a path, carry over to digraphs with only minor notational changes.

Definition 3.34   A directed graph H is a *subgraph* of the directed graph G if $V(H) \subseteq V(G)$ and $E(H) \subseteq E(G)$.

Definition 3.35   A *directed path* from vertex a to vertex b of a digraph D is a sequence of vertices $(a = v_0), v_1, \ldots, (v_m = b)$ such that $(v_i, v_{i+1}) \in E(D)$ for $i = 0, 1, \ldots, m - 1$. The *length* of the path $v_0, v_1, \ldots, v_m$ is m.

Example 2   Figure 3.42 provides a diagram of a digraph G. A sample directed path in this digraph—from v_2 to v_6—is v_2, v_5, v_3, v_4, v_6. This directed path's length is four.   □

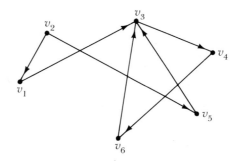

Figure 3.42

Several algorithms depend on the existence of a directed path from one vertex to another. The following definition formalizes the concept of accessibility.

Definition 3.36   A vertex b is *accessible* from a vertex a in a digraph G if there is a directed path in G from a to b.

In Problem 15 you are asked to prove the following theorem.

Theorem 3.37   Let A be the adjacency matrix of a directed graph G. Then the (i, j)th entry of A^m is the number of directed paths of length m from the vertex v_i to the vertex v_j.

Example 3   The directed graph whose adjacency matrix is

$$A = \begin{bmatrix} 0 & 1 & 1 & 0 \\ 0 & 0 & 1 & 0 \\ 0 & 0 & 0 & 1 \\ 1 & 1 & 0 & 0 \end{bmatrix}$$

has two directed paths of length four from v_1 to v_2 since

$$A^4 = \begin{bmatrix} 1 & 2 & 2 & 0 \\ 0 & 1 & 2 & 0 \\ 0 & 0 & 1 & 2 \\ 2 & 2 & 0 & 1 \end{bmatrix} \qquad \square$$

The following corollary stems directly from Theorem 3.37.

Corollary 3.38   Let G be a directed graph with n vertices. Then the vertex v_j is accessible from the vertex v_i if and only if the (i, j)th entry of $I + A + A^2 + A^3 + \cdots + A^{n-1}$ is nonzero.

Example 4   Consider the directed graph with adjacency matrix

$$A = \begin{bmatrix} 0 & 1 & 1 & 0 \\ 0 & 0 & 1 & 1 \\ 0 & 0 & 0 & 1 \\ 1 & 1 & 0 & 0 \end{bmatrix}$$

In this case

$$
I + A + A^2 + A^3 =
\begin{bmatrix}
1 & 0 & 0 & 0 \\
0 & 1 & 0 & 0 \\
0 & 0 & 1 & 0 \\
0 & 0 & 0 & 1
\end{bmatrix}
+
\begin{bmatrix}
0 & 1 & 1 & 0 \\
0 & 0 & 1 & 1 \\
0 & 0 & 0 & 1 \\
1 & 1 & 0 & 0
\end{bmatrix}
$$

$$
+
\begin{bmatrix}
0 & 0 & 1 & 2 \\
1 & 1 & 0 & 1 \\
1 & 1 & 0 & 0 \\
0 & 1 & 2 & 1
\end{bmatrix}
+
\begin{bmatrix}
2 & 2 & 0 & 1 \\
1 & 2 & 2 & 1 \\
0 & 1 & 2 & 1 \\
1 & 1 & 1 & 3
\end{bmatrix}
$$

$$
=
\begin{bmatrix}
3 & 3 & 2 & 3 \\
2 & 4 & 3 & 3 \\
1 & 2 & 3 & 2 \\
2 & 3 & 3 & 5
\end{bmatrix}
$$

and it follows that every vertex is accessible from every other vertex. Such a graph is said to be *strongly connected*.   □

As with other graphs, a digraph may be weighted to indicate relative lengths, time spans, costs, and so on. The weighted digraph is perhaps the most widely used graph theoretic structure. In the next chapter we consider a number of practical applications of weighted digraphs.

Definition 3.39 A *weighted digraph* is a digraph D with a function $w:E(D) \to \mathbf{R}^+$. If $e \in E(D)$, then $w(e)$ is called the *weight* of e.

Slightly modifying the adjacency matrix provides a way to describe a weighted digraph.

Definition 3.40 Let G be a weighted digraph with n vertices $v_1, v_2, \ldots, v_n$. The *weight matrix*, W, of G is the $n \times n$ matrix $[w_{ij}]_{n \times n}$, where

$$
w_{ij} = \begin{cases}
w(v_i, v_j) & \text{if } (v_i, v_j) \in E(G) \\
0 & \text{if } i = j \\
\infty & \text{if } (v_i, v_j) \notin E(G)
\end{cases}
$$

We use ∞ to show that two vertices v_i and v_j are not connected by an edge directed from v_i to v_j.

Example 5   The following weight matrix describes the weighted digraph illustrated in Figure 3.43.

$$\begin{bmatrix} 0 & 10 & 6 & 8 \\ \infty & 0 & \infty & \infty \\ \infty & 5 & 0 & 7 \\ \infty & \infty & \infty & 0 \end{bmatrix}$$

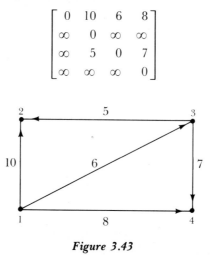

Figure 3.43   □

As we previously indicated, there is a natural connection between relations and directed graphs with loops. This connection provides geometric insight into relations. (Before proceeding you may wish to review Section 5 of the Appendix.)

Suppose that R is a relation on a set X. Then, by definition, R is a subset of $X \times X$. We define D, the directed graph with loops that is associated with R, by $V(D) = X$ and $E(D) = R$. Thus the graph D of a relation R on X has X as its vertex set, and an edge in D is directed from a vertex a to a vertex b if and only if $a R b$.

Example 6   Let

$$X = \{1, 2, 3, 4, 5\}$$

and let R be the relation defined on X by

$$R = \{(1, 2), (1, 1), (2, 3), (4, 5), (3, 4), (4, 3)\}.$$

Figure 3.44 indicates the directed graph D associated with R.   □

Example 7   Let

$$X = \mathscr{P}(\{1, 2, 3\}) = \{\varnothing, \{1\}, \{2\}, \{3\}, \{1, 2\}, \{1, 3\}, \{2, 3\}, \{1, 2, 3\}\}$$

and let R be the relation defined on X by

$$R = \{(x, y) \mid x, y \in \mathscr{P}(\{1, 2, 3\}) \text{ and } x \subseteq y\}.$$

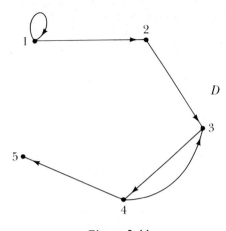

Figure 3.44

Then *R* may be viewed as the "contained in" relation applied to the subsets of the set $\{1, 2, 3\}$. Figure 3.45 shows the directed graph associated with this relation.

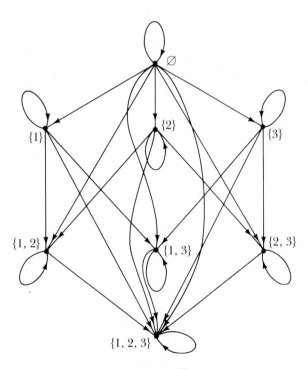

Figure 3.45

Using the connection between relations and directed graphs with loops, we can translate certain properties of relations into graphical terms. We first illustrate this with a reflexive relation. A relation on a set X is *reflexive* if whenever $a \in X$, then $(a, a) \in R$. It follows that a directed graph with loops at each vertex is the graph of a reflexive relation. Thus the graph illustrated in Figure 3.45 is the graph of a reflexive relation whereas the graph pictured in Figure 3.44 is not.

A relation R on a set X is *antisymmetric* if whenever $(a, b) \in R$ and $(b, a) \in R$, then $a = b$. This means that for a digraph to represent an antisymmetric relation, any two of its vertices can be joined by at most one directed edge. In other words, if $a \neq b$ and $(a, b) \in E(D)$, then $(b, a) \notin E(D)$. Therefore the graph in Figure 3.45 is that of an antisymmetric relation while the graph in Figure 3.44 is not.

A relation R is *transitive* if whenever $(a, b) \in R$ and $(b, c) \in R$, then $(a, c) \in R$. It follows that a directed graph D represents a transitive relation if whenever $(a, b) \in E(D)$ and $(b, c) \in E(D)$, then $(a, c) \in E(D)$. Figure 3.46 shows this concept geometrically: If two edges form a directed path from a to c in D, then there is a directed edge from a to c.

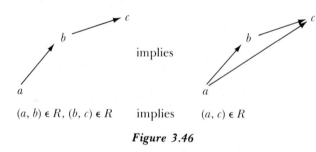

$(a, b) \in R, (b, c) \in R$ implies $(a, c) \in R$

Figure 3.46

Because of the many applications of partial orders, their graphical representations are particularly important. A relation is a *partial order* if it is reflexive, antisymmetric, and transitive. Thus a directed graph D with loops is the graph of a partial order if:

a. D has a loop at each vertex;

b. any two vertices are joined by at most one edge; and

c. for each directed path a, b, c of length two in D, there is a directed edge (a, c) in $E(D)$.

Suppose that R is partial order on a set X and D is the graph of this relation. Since D satisfies properties **a.**, **b.**, and **c.**, we can simplify D to a significant extent. For example, since by **a.** D must have a loop at each vertex, we need not diagram these loops; their presence is assumed.

In addition, since by **b.** no pair of vertices is joined by two edges, we can direct all edges in the diagram of D downward. With these two simplifications,

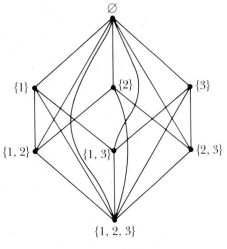

Figure 3.47

the graph of the partial order from Example 7 (Figure 3.45) becomes as shown in Figure 3.47.

By utilizing **c.**, we can make one further simplification, eventually creating what is known as a *Hasse Diagram* of the partial order. We must first, however, introduce the following idea.

Definition 3.41   Let $\preceq$ be a partial order relation on a set X and let $a \in X$ and $b \in X$. Then a is said to be the *immediate predecessor* of b if $a \preceq b$ and there does not exist a $c \in X$ such that $c \neq a$, $c \neq b$, and $a \preceq c \preceq b$.

In other words, a is an immediate predecessor of b if a precedes b and no other element falls "between" a and b.

Looking at the diagram of the relation in Example 7 given in Figure 3.47, we can see that the empty set, $\varnothing$, is an immediate predecessor of $\{1\}$, and that $\{1\}$ and $\{3\}$ are both immediate predecessors of $\{1, 3\}$. However, $\{1\}$ is not an immediate predecessor of $\{1, 2, 3\}$.

Drawing upon Definition 3.41, we can now utilize **c.** to simplify the graphical representation of a partial order. By connecting vertices a and b only if a is an immediate predecessor of b, we "understand" the presence of the remaining undrawn edges. This is because any partial order is transitive and therefore satisfies property **c.** For example, in Figure 3.48, a diagram of a partial order relation, the presence of the edges (a, c) and (d, c) is understood. This figure is a Hasse Diagram.

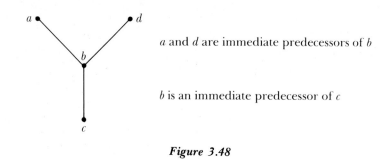

a and d are immediate predecessors of b

b is an immediate predecessor of c

Figure 3.48

Definition 3.42   A *Hasse Diagram* is the diagram of a partial order R that has been simplified by

    a. omitting all loops,

    b. directing all edges downward, and

    c. connecting two vertices only if one of the vertices is an immediate predecessor of the other.

Example 8   The Hasse Diagram of the partial order from Example 7 (Figures 3.45 and 3.47) is pictured in Figure 3.49.

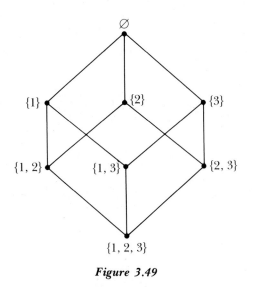

Figure 3.49

Observe that this graph is a cube. The relationship between the Hasse Diagram of the partial order $\subseteq$ on the set $\mathscr{P}(X)$ and cubes is taken up in Problems 26–28.   □

3.9   Problems

In Problems 1–4 draw a diagram of the given directed graph.

1. $V(D) = \{a, b, c, d\}; E(D) = \{(a, b), (b, c), (c, d), (a, d)\}$
2. $V(D) = \{1, 2, 3, 4, 5\}; E(D) = \{(2, 1), (5, 1), (2, 4), (4, 5), (2, 3), (3, 4)\}$
3. $V(D) = \{a, b, c, d\}, E(D) = \{(b, a), (b, c), (c, d), (c, b), (d, a)\}$
4. $V(D) = \{v_1, v_2, v_3, v_4, v_5\}, E(D) = \{(v_1, v_2), (v_4, v_1), (v_3, v_4), (v_3, v_5), (v_5, v_3), (v_2, v_1)\}$

In Problems 5–8 determine $V(D)$ and $E(D)$ for each of the graphs illustrated.

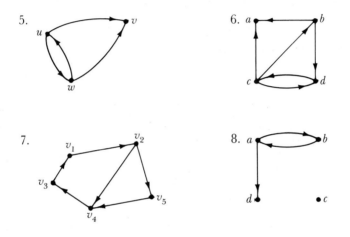

In Problems 9–12 determine the adjacency matrix of the graph illustrated and use the adjacency matrix to determine the number of directed paths of length 3 from v_1 to v_3.

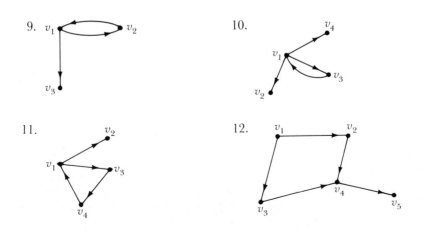

13. If a_{ij} is the (i, j)th entry of the adjacency matrix of the digraph D with k vertices, interpret:

(a) $\displaystyle\sum_{i=1}^{k} a_{ij}$ (b) $\displaystyle\sum_{j=1}^{k} a_{ij}$

14. Let A be the adjacency matrix of the digraph D. What does the (i, j)th entry of A^2 denote? Prove your conjecture.

15. Prove Theorem 3.37.

16. Prove Corollary 3.38.

In Problems 17–20 determine the weight matrix of the graph illustrated.

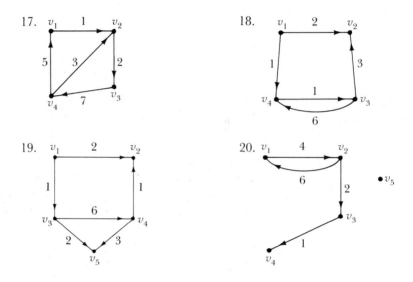

21. Diagram the directed graphs associated with each of the following relations:
(a) $X = \{a, b, c, d, e\}$; $R = \{(a, c), (d, e), (b, b), (c, c), (e, d)\}$
(b) $X = \{1, 2, 3, 4, 5, 6, 7, 8, 9\}$; $R = \{(x, y) \mid x \text{ divides } y, \text{ and } x \neq y\}$

22. Diagram the directed graphs associated with each of the following relations:
(a) $X = \{1, 2, 3, 4, 5, 6\}$; $R = \{(x, y) \mid x \leq y \text{ or } xy \text{ is an even integer}\}$
(b) $X = \{1, 2, 3, 4, 5, 6, 7, 8, 9\}$; $R = \{(x, y) \mid x + y \text{ is an odd integer}\}$

23. Give an example and diagram the associated digraph of each of the following:
(a) A relation that is transitive, but not reflexive or symmetric.
(b) A relation that is reflexive and symmetric, but not transitive.
(c) A relation that is antisymmetric, but not transitive.
(d) A relation that is reflexive and transitive, but not symmetric.

24. Give the Hasse Diagram of each of the following partial orders.
    (a) $X = \{1, 2, 3, 4, 6, 7, 8, 9, 10\}$; $R = \{x, y) \,|\, x \text{ divides } y\}$
    (b) $X = \{1, 2, 3, 4, 5, 6, 7, 8, 9, 10\}$; $R = \{(x, y) \,|\, x \leq y\}$

25. Give the Hasse Diagrams of each of the following partial orders:
    (a) $X = \mathscr{P}(\{1\})$; $R = \{(x, y) \,|\, x, y \in \mathscr{P}(\{1\}) \text{ and } x \subseteq y\}$
    (b) $X = \mathscr{P}(\{1, 2\})$; $R = \{(x, y) \,|\, x, y \in \mathscr{P}(\{1, 2\}) \text{ and } x \subseteq y\}$
    (c) $X = \mathscr{P}(\{1, 2, 3, 4\})$; $R = \{(x, y) \,|\, x, y \in \mathscr{P}(\{1, 2, 3, 4\}) \text{ and } x \subseteq y\}$

26. Let $X = \mathscr{P}(\{1, 2, \ldots, n\})$ and let R be the "contained in" relation on X. Show that $A \in \mathscr{P}(X)$ is an immediate predecessor of $B \in \mathscr{P}(X)$ if and only if $|A| < |B|$ and the characteristic functions of A and B differ in exactly one place.

27. The n-cube, Q_n, is the graph whose vertex set is

$$V = \{(x_1, x_2, \ldots, x_n) \,|\, x_i = 0 \text{ or } x_i = 1\}.$$

Two vertices are joined by an edge if and only if the two vertices differ in exactly one coordinate. Diagram the cubes Q_1, Q_2, and Q_3.

28. Use the results from Problems 26 and 27 to show that the Hasse Diagram of the relation given in Problem 26 is the cube Q_n.

29. A directed graph D is *unilaterally connected* if for all $\{u, v\} \subseteq V(D)$, u is accessible from v or v is accessible from u.
    (a) Give an example of a unilaterally connected graph.
    (b) Give an example of a unilaterally connected graph that is not strongly connected.

Chapter 3   REVIEW

Concepts for Review

graph (p. 86)
subgraph (p. 88)
adjacency matrix (p. 88)
symmetric matrix (p. 89)
degree of a vertex (p. 90)
path (p. 94)
connected graph (p. 95)
simple path (p. 95)
components of a graph (p. 99)
cycle (p. 103)
tree (p. 103)
spanning subgraph (p. 107)

spanning tree (p. 107)
weighted graph (p. 113)
weight matrix (p. 113)
minimal spanning tree (p. 115)
Kruskal's Algorithm (p. 116)
complete graph (p. 123)
triangle inequality (p. 126)
directed graph (p. 132)
accessibility (p. 134)
strongly connected graph (p. 135)
Hasse Diagram (p. 140)

Review Problems

1. Diagram the graph G where $V(G) = \{a, b, c, d, e\}$ and $E(G) = \{\{a, c\}, \{a, d\}, \{d, e\}, \{e, c\}\}$.

2. Give the adjacency matrix of the graph given in Problem 1.

3. A graph has seven vertices all having degree four. How many edges does this graph have?

4. The adjacency matrix of the complete graph with four points, K_4, is

$$\begin{bmatrix} 0 & 1 & 1 & 1 \\ 1 & 0 & 1 & 1 \\ 1 & 1 & 0 & 1 \\ 1 & 1 & 1 & 0 \end{bmatrix}$$

How many paths of length three join the pair of vertices v_2, v_3?

5. Use Algorithm 3.14 to generate a list of all vertices of G found in the component of v_1. The adjacency matrix of G is

$$\begin{bmatrix} 0 & 0 & 0 & 0 & 0 & 1 & 0 \\ 0 & 0 & 0 & 0 & 0 & 1 & 0 \\ 0 & 0 & 0 & 1 & 1 & 0 & 0 \\ 0 & 0 & 1 & 0 & 1 & 0 & 0 \\ 0 & 0 & 1 & 1 & 0 & 0 & 0 \\ 1 & 1 & 0 & 0 & 0 & 0 & 1 \\ 0 & 0 & 0 & 0 & 0 & 1 & 0 \end{bmatrix}$$

6. Is it possible to construct a tree having vertices of the following degrees?

$$\delta(v_1) = 2, \qquad \delta(v_2) = 3, \qquad \delta(v_3) = 4,$$
$$\delta(v_4) = 1, \qquad \delta(v_5) = 1, \qquad \delta(v_6) = 1$$

If not, explain why.

7. Use Algorithm 3.22 to construct a spanning tree in G defined by $V(G) = \{1, 2, 3, 4, 5, 6\}$, $E(G) = \{e_1, e_2, \ldots, e_{10}\}$ where

$$e_1 = \{1, 6\}, \qquad e_2 = \{2, 3\}, \qquad e_3 = \{4, 5\},$$
$$e_4 = \{6, 4\}, \qquad e_5 = \{5, 6\}, \qquad e_6 = \{2, 6\},$$
$$e_7 = \{6, 3\}, \qquad e_8 = \{1, 2\}, \qquad e_9 = \{5, 1\},$$
$$e_{10} = \{3, 4\}$$

8. Give the weight matrix of the graph defined by $V(G) = \{1, 2, 3, 4\}$ and $E(G) = \{\{1, 2\}, \{2, 3\}, \{2, 4\}, \{1, 4\}\}$ with

$$w(\{1, 2\}) = 1, \qquad w(\{2, 3\}) = 7, \qquad w(\{2, 4\}) = 5, \qquad w(\{1, 4\}) = 2$$

9. Draw a diagram of a graph with the weight matrix

$$\begin{bmatrix} 0 & 1 & \infty & 2 & \infty & \infty \\ 1 & 0 & 3 & 5 & \infty & 7 \\ \infty & 3 & 0 & \infty & 1 & 6 \\ 2 & 5 & \infty & 0 & 2 & \infty \\ \infty & \infty & 1 & 2 & 0 & 1 \\ \infty & 7 & 6 & \infty & 1 & 0 \end{bmatrix}$$

10. Apply Kruskal's Algorithm to the graph G where $V(G) = \{1, 2, 3, 4, 5, 6\}$ and $E(G) = \{\{2, 3\}, \{5, 1\}, \{6, 1\}, \{5, 4\}, \{3, 4\}, \{4, 6\}, \{4, 1\}, \{3, 4\}, \{5, 3\}\}$ with

$$\begin{array}{lll} w(\{2, 3\}) = 3, & w(\{5, 1\}) = 2, & w(\{6, 1\}) = 2, \\ w(\{5, 4\}) = 2, & w(\{3, 4\}) = 3, & w(\{4, 6\}) = 4, \\ w(\{4, 1\}) = 1, & w(\{3, 4\}) = 1, & w(\{5, 3\}) = 2 \end{array}$$

11. Apply Kruskal's Algorithm to a graph whose weight matrix is

$$\begin{bmatrix} 0 & 1 & 5 & 7 & \infty & 2 \\ 1 & 0 & 3 & \infty & 2 & 6 \\ 5 & 3 & 0 & 1 & 4 & 8 \\ 7 & \infty & 1 & 0 & \infty & 2 \\ \infty & 2 & 4 & \infty & 0 & \infty \\ 2 & 6 & 8 & 2 & \infty & 0 \end{bmatrix}$$

12. Give the adjacency matrix of a complete graph on five vertices.

13. Apply Algorithm 3.29 to determine an approximation to a minimum weight spanning cycle for the weighted complete graph with weight matrix

$$\begin{bmatrix} 0 & 2 & 5 & 4 & 4 & 1 \\ 2 & 0 & 3 & 4 & 4 & 3 \\ 5 & 3 & 0 & 3 & 3 & 6 \\ 4 & 4 & 3 & 0 & 2 & 3 \\ 4 & 4 & 3 & 2 & 0 & 5 \\ 1 & 3 & 6 & 3 & 5 & 0 \end{bmatrix}$$

14. Do the weights given in the following weight matrix satisfy the triangle inequality?

$$\begin{bmatrix} 0 & 1 & 3 & 4 \\ 1 & 0 & 2 & 0 \\ 3 & 2 & 0 & 1 \\ 4 & 0 & 1 & 0 \end{bmatrix}$$

15. Explain why distances satisfy the triangle inequality.

16. Apply a greedy algorithm to determine an approximation to a minimum weight spanning cycle for the graph whose weight matrix is given in Problem 13.

17. The weights given in Problem 13 satisfy the triangle inequality. Use the result of Problem 16 to compute bounds on the weight of a minimal weight cycle.

18. Draw a diagram of the directed graph D where $V(D) = \{v_1, v_2, v_3, v_4\}$ and $E(D) = \{(v_1, v_2), (v_2, v_1), (v_3, v_4), (v_4, v_2), (v_4, v_4)\}$.

19. Draw a diagram of the directed graph associated with the relation R on X where $X = \{1, 2, 3, 4, 5, 6, 7, 8, 9\}$ and $R = \{(a, b)|a, b \in X, \text{and } a - b$ is nonnegative and divisible by 3$\}$.

20. Draw the Hasse Diagram of the relation given in Problem 19.

4 Applications of Graph Theory

In this chapter we consider a number of applications of elementary graph theory. We begin by using weighted digraphs to find minimal cost distribution networks. Such networks are important to the design of transportation and communication routes.

Next we consider the *critical path method*, a procedure used in dealing with certain scheduling problems. The efficient scheduling of the component activities of a project can be difficult if some of these activities must be completed before starting others. The critical path method allows us to identify those activities which, if delayed, would also delay the completion of the entire project.

Euler paths and tours are the subjects of Section 4.3. These concepts date back to the eighteenth century, but are nevertheless quite relevant in a variety of modern-day contexts, including cryptography and *de Bruijn sequences*. The latter are discussed in Section 4.4.

We also briefly examine *Gray codes*. These codes are used to read the position of a rotating disk and eliminate errors inherent in analog to digital conversion.

4.1 MINIMAL PATH TREES

The weighted digraph in Figure 4.1 represents a distribution network. The assigned weights indicate the transportation costs between various locations; the vertex v_0 represents a production center; and the other vertices represent

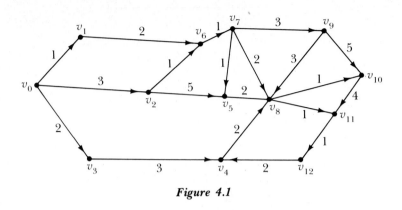

Figure 4.1

distribution centers. We would like to find minimum-cost paths from v_0 to each of the other vertices. Such paths represent the lowest cost transportation routes and thus constitute a *minimal-cost distribution network*.

Note that if a vertex v is accessible from v_0, then it suffices to select a single minimal-cost path from v_0 to v. There is no point in including several minimal-cost paths linking a vertex v to v_0, for additional paths would increase costs without adding accessibility. A minimal-cost distribution network therefore is a tree consisting of paths directed from v_0 to each of the other accessible vertices. Trees of this type are said to be *rooted*.

Definition 4.1   A tree *rooted* at v_0 is a directed graph, D, such that

      a. the underlying graph of D is a tree, and

      b. all the vertices of D are accessible from v_0.

Example 1   The directed graph in Figure 4.2(a) is a tree rooted at v_0. The directed graph in Figure 4.2(b) is not a tree rooted at v_0 since one of its vertices (the vertex v) is not accessible from v_0.   □

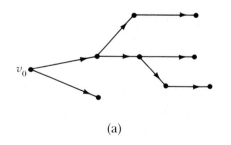

(a)

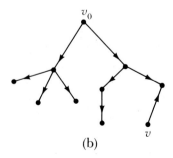

(b)

Figure 4.2

To define precisely a minimal-cost distribution network, we use the following notion. If $P = v_0, v_1, \ldots, v_k$ is a path in a weighted digraph, then the *weight* of P, $W(P) = W(v_0, v_1, \ldots, v_k)$, is the sum of the weights of the edges of P; thus,

$$W(P) = W(v_0, v_1, \ldots, v_k) = w(v_0, v_1) + w(v_1, v_2) + \cdots + w(v_{k-1}, v_k)$$

The solution of the minimal-cost distribution network problem is called a *minimal path tree*.

Definition 4.2   Let G be a weighted digraph. A subgraph T of G is called a *minimal path tree rooted at v_0* provided that

    a. T is a tree rooted at v_0,

    b. the vertex set of T consists of all vertices of G that are accessible from v_0, and

    c. if Q_v is the unique path in T from v_0 to a vertex v, then $W(Q_v) \leq W(P)$ for each path P in G from v_0 to v.

The following theorem allows us to develop an algorithm for finding minimal path trees.

Theorem 4.3   Suppose that G is a weighted digraph and v_0 is a vertex of G. Then a subgraph T of G is a minimal path tree rooted at v_0 if and only if

    a. T is a tree rooted at v_0,

    b. the vertex set of T consists of all vertices of G accessible from v_0, and

    c. $W(Q_v) \leq W(Q_u) + w(u, v)$ for each edge $(u, v) \in E(G)$, where Q_v and Q_u are the unique paths in T from v_0 to v and from v_0 to u, respectively.

Proof: First observe that if T is a minimal path tree rooted at v_0, then **a.** and **b.** are satisfied by definition. We prove **c.** by contradiction. Suppose that **c.** is not satisfied for some edge (u, v). Then we have

$$W(Q_v) > W(Q_u) + w(u, v)$$

Figure 4.3(a) portrays this situation schematically. Each of the edges of Q_u and Q_v is in $E(T)$, but (u, v) is not an edge of T.

Now consider the path Q'_v formed joining the edge (u, v) to the path Q_u as shown in Figure 4.3(b). Since

$$W(Q'_v) = W(Q_u) + w(u, v) < W(Q_v)$$

we have constructed a path of lesser weight from v_0 to v. This contradicts the fact that T is a minimal path tree. Consequently, we must have

$$W(Q_v) \leq W(Q_u) + w(u, v)$$

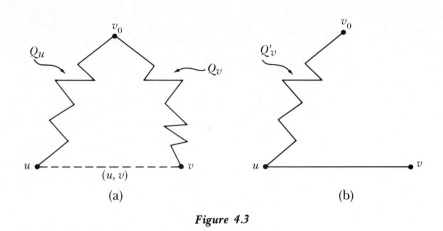

Figure 4.3

Conversely, let us suppose that T does satisfy **a.**, **b.**, and **c.** of the theorem. We must show that T satisfies Definition 4.2. Parts **a.** and **b.** of this definition are satisfied because we are assuming that **a.** and **b.** of the theorem are true. Thus we need only establish **c.**, that the paths in T from v_0 to each of the vertices are paths of minimal weight. Suppose this is not the case. Then there is at least one vertex v of G and a path $v_0, v_1, \ldots, (v_k = v)$ in G that has less weight than the path Q_v in T from v_0 to v; thus, we have

$$W(v_0, v_1, \ldots, v_k) < W(Q_v)$$

In the sequence of vertices $v_0, v_1, \ldots, v_k$ there must be a *first* vertex, v_j, (perhaps $v_j = v_k$) with the property that

$$W(v_0, v_1, \ldots, v_j) < W(Q_{v_j}) \tag{1}$$

Since v_j is the first vertex satisfying (1) we have

$$W(v_0, v_1, \ldots, v_{j-1}) \geq W(Q_{v_{j-1}}) \tag{2}$$

Also note that

$$W(v_0, v_1, \ldots, v_j) = W(v_0, v_1, \ldots, v_{j-1}) + w(v_{j-1}, v_j) \tag{3}$$

Combining (1), (2), and (3) gives

$$W(Q_{v_j}) > W(v_0, v_1, \ldots, v_j) = W(v_0, v_1, \ldots, v_{j-1}) \\ + w(v_{j-1}, v_j) \geq W(Q_{v_{j-1}}) + w(v_{j-1}, v_j)$$

or

$$W(Q_{v_j}) > W(Q_{v_{j-1}}) + w(v_{j-1}, v_j)$$

which contradicts **c.** of Theorem 4.3. Thus, T must be a minimal path tree rooted at v_0, and this completes the proof. ∎

Theorem 4.3 suggests an algorithm that will yield a minimal path tree. Let T be any tree rooted at v_0 that contains all vertices accessible from v_0. If this tree satisfies

$$W(Q_v) \leq W(Q_u) + w(u, v)$$

for all vertices u and v that are joined by an edge, then T is the desired minimal path tree. If for some vertices u and v there is an edge (u, v) such that

$$W(Q_v) > W(Q_u) + w(u, v)$$

then we form a new tree T_1, as shown in Figure 4.4. To form T_1, delete the last edge of path Q_v (the edge that goes to vertex v) and replace it with edge (u, v).

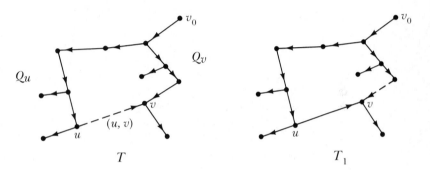

Figure 4.4

Note that the path from v_0 to v in T_1 has weight $W(Q_u) + w(u, v)$. If this weight is indeed less than the weight of the path, Q_v, from v_0 to v in T, then the rooted tree T_1 is an improvement over the original rooted tree T. We can then repeat this process with new vertices u and v to determine if there is an even "better" rooted tree than T_1. Since G is a finite graph, this procedure will eventually terminate with a minimal path tree. We will trace this procedure through for the weighted digraph G shown in Figure 4.5.

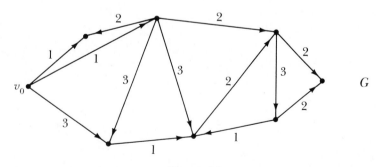

Figure 4.5

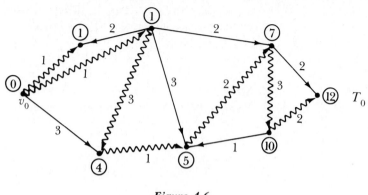

Figure 4.6

We begin with an arbitrary tree T_0 rooted at v_0 as indicated in Figure 4.6. The circled number near each vertex v is $W(Q_v)$, the weight of the path in T_0 from v_0 to v.

We associate the number

$$s(u, v) = W(Q_v) - W(Q_u) - w(u, v)$$

with each edge (u, v) in $E(G)$. This number is called the *slack* of the edge (u, v). In Figure 4.7 a boxed number near an edge indicates the slack of the edge. Since each edge already in the tree T_0 has a slack of 0 we do not show these slacks. If the slack of an edge (u, v) of G is positive, then

$$W(Q_v) > W(Q_u) + w(u, v)$$

and it follows from part **c.** of Theorem 4.3 that T_0 is not a minimal path tree. To create a minimal path tree, then, we must "take up the slack" so that all

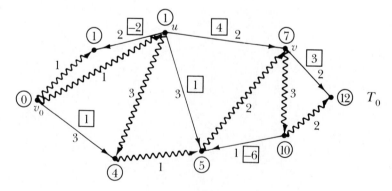

Figure 4.7

of the edges of G have either negative or zero slack. We first examine the edges that are not yet part of the tree and identify an edge that has the maximal positive slack. We label this edge as (u, v). (In Figure 4.7 the indicated edge (u, v) has the maximal slack of 4.) We then add this edge to the tree T_0 and we delete from T_0 the last edge in the path Q_v. Thus we obtain a new spanning tree T_1, indicated by the wavy line segments in Figure 4.8. The figure also shows the newly adjusted weights and the new slack values associated with the tree T_1; an edge with the largest slack is labelled (u, v).

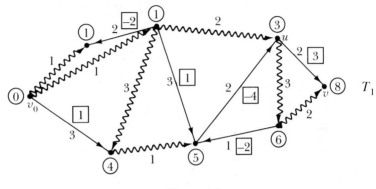

Figure 4.8

Because the slack of the edge (u, v) is positive, we automatically know that T_1 does not satisfy **c.** of Theorem 4.3, so once again we add a new edge (u, v) to T_1 and delete the last edge of path Q_v. The resulting tree, T_2, along with the new weights and slack values computed from T_2, are shown in Figure 4.9.

At this point there are two edges with equal maximal slack. By arbitrarily choosing one of these edges to add to the tree, we obtain the tree shown in Figure 4.10.

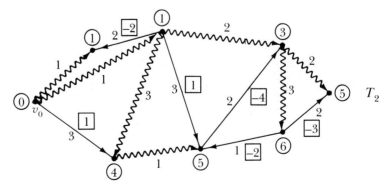

Figure 4.9

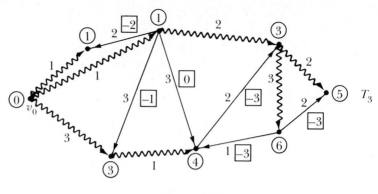

Figure 4.10

Now we have a tree that has no positive slack values. Consequently, for each $(u, v) \in E(G)$, we have

$$W(Q_v) - W(Q_u) - w(u, v) \leq 0$$

or, equivalently

$$W(Q_v) \leq W(Q_u) + w(u, v)$$

and by Theorem 4.3, T_3 is a minimal path tree.

More formally, we can state this algorithm as follows.

Algorithm 4.4   For determining the edge set of a minimal path tree in a graph G that is rooted at $V(0)$ and contains all vertices of G accessible from $V(0)$.

> Input: *EG*, the edge set of G;
>   *ET*, the edge set of a tree T rooted at $V(0)$ that contains all vertices of G accessible from $V(0)$;
>   *VERT*, the vertex set of T
>
> Output: *ET*, the edge set of a minimal path tree rooted at $V(0)$

1. $S \leftarrow \varnothing$	S designates the set of vertices of T for which $WQ(V)$ has been calculated.
2. WHILE $S(U, V) > 0$ FOR SOME $(U, V) \in EG$ DO	The following steps are performed as long as there is some edge with positive slack.

a. $WQ(V(0)) \leftarrow 0$
  b. $S \leftarrow V(0)$
  c. WHILE $S \neq VERT$
    1. IF $(U, V) \in EG$ THEN
	In steps a, b, and c, $WQ(V)$—the weight of the unique path in T from $V(0)$ to V—is calculated for all $V \in VERT$. This includes all vertices of G accessible from $V(0)$.

a. IF $U \in S$ AND
$V \in VERT \setminus S$
THEN
1. $S \leftarrow S \cup \{V\}$
2. $WQ(V) \leftarrow$
$WQ(U) + W(U, V)$
d. $M \leftarrow 0$
e. FOR ALL
$(U, V) \in EG \setminus ET$
1. IF $S(U, V) > M$ THEN
  a. $M \leftarrow S(U, V)$
  b. $UM \leftarrow U$
  c. $VM \leftarrow V$
f. $ET \leftarrow (ET \setminus \{(W, VM)\}) \cup$
$\{(UM, VM)\}$

In steps d and e calculate the greatest positive slack M; (UM, VM) is an edge with the greatest positive slack.

An edge with greatest slack is adjoined to T, and the edge of T previously ending at VM is removed.

3. OUTPUT ET

4.1   Problems

In Problems 1–6 apply Algorithm 4.4 to obtain a minimal path tree in the given graph. Begin the algorithm with the initial rooted tree, T, shown by wavy edges.

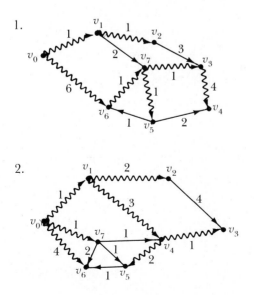

1.

2.

3.

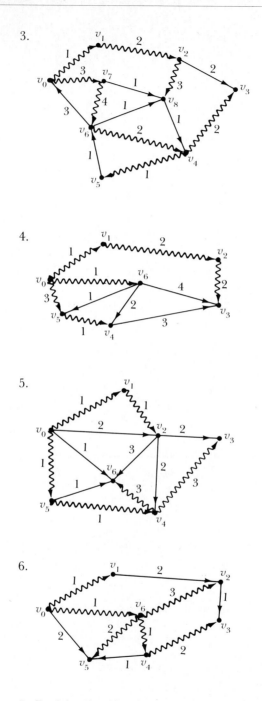

4.

5.

6.

7. Explain why Algorithm 4.4 must terminate.

4.2   THE CRITICAL PATH METHOD

Developing and manufacturing a product frequently involves many interrelated activities. It is often the case that some of these activities cannot be started until other activities are completed. As a result, scheduling the separate activities of a project is often a nontrivial problem.

The *critical path method (CPM)* is an important management tool for identifying those critical activities whose delay would postpone the completion of the entire project. Once project managers have identified these critical activities, they can usually redirect additional resources to these activities.

To apply the CPM, we will use a weighted digraph to describe the interrelationships among the various activities of a project. A weighted edge represents each activity and depicts the time necessary to complete the task. If activity *A* must be completed immediately before activity *B* begins, we require that the edges *A* and *B* be *incident*, as indicated in Figure 4.11. In this figure, activity *A* requires 3 time units, while *B* requires 6.

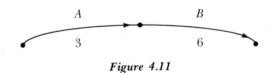

Figure 4.11

The following Example examines other scheduling possibilities.

Example 1

(a) Suppose that activities *A* and *B* must be completed immediately before activities *C* or *D* begin. Figure 4.12 graphs this situation.

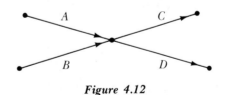

Figure 4.12

(b) Suppose that activity *A* must be completed immediately before activities *C* and *F*, and that activities *B* and *F* must be completed immediately before activity *D*. Then we have a directed graph as in Figure 4.13 (p. 158).

(c) Some scheduling situations are impossible to represent without introducing fake activities having zero completion time. For instance, suppose that activity *A* must be completed immediately before activities *C* and *D* begin.

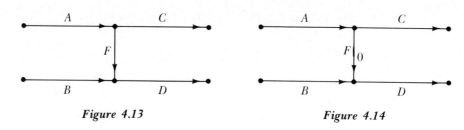

Figure 4.13   **Figure 4.14**

Suppose further that activity B must be completed before beginning activity D. In this case we introduce a fake activity F with a zero completion time as Figure 4.14 indicates.   □

In what follows we will label two special vertices, i and t, to represent the initiation and termination of the entire project.

Let us now construct a weighted digraph for a specific project. We must buy two machines, M_1 and M_2, that are to produce two objects, a and b. Machine M_1 will produce a by itself, but both machines must work together to produce b. We will then need to assemble a and b and ship them to our customers.

Table 4.1 lists the various activities involved in the project, and gives a brief description, the expected duration, and the immediate predecessors (if any) of each activity.

Activity	Description	Estimated duration (days)	Immediate predecessors
A	Purchase and install M_1	8	none
B	Purchase and install M_2	6	none
C	Test M_1	1	A
D	Test M_2	2	B
E	Produce a	3	C
F	Produce b	1	C, D
G	Assemble a and b	2	E, F
H	Ship product	1	G

Table 4.1

We are now ready to construct a weighted digraph that describes this situation. From Table 4.1 we see that activities A and B have no predecessors, so these activities may initiate the project. Thus our weighted digraph starts as in Figure 4.15.

C and D depend only on the completion of A and B, respectively, so our drawing becomes as in Figure 4.16.

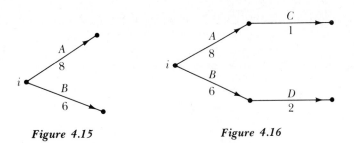

Figure 4.15   Figure 4.16

Activity E depends on the completion of C. However, since activity F requires that both C and D be completed, we must introduce a fake activity (call it I) with zero duration into our graph. Thus we obtain Figure 4.17.

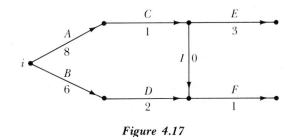

Figure 4.17

Activity G depends on the completion of both E and F; thus it is convenient to draw the edges E and F so that their common terminal vertex serves as the beginning point of the edge representing G. Finally, adding the edge H and the terminal vertex t, we obtain the desired weighted digraph in Figure 4.18.

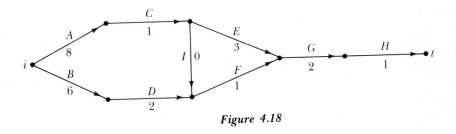

Figure 4.18

Now that we have seen how to describe a scheduled project with a weighted digraph, we will examine how to identify the project's critical activities. A *critical activity* is any task that, if delayed, will also delay the entire project. We determine these activities by finding a path of *maximal* weight from the initial vertex i to the terminal vertex t in the weighted digraph of the given project. Such a path is called a *critical path*, and its edges represent the project's critical activities. We make two important observations about critical paths:

a. The length (weight) of the critical path is the *minimal* length of time required to complete the project.

b. Increasing the time required for any critical activity will also increase the time necessary to complete the project.

To find a critical path, we will construct a *maximal path tree* rooted at the initial vertex i. We do this by slightly modifying the minimal path tree algorithm from Section 4.1. We change Algorithm 4.4 so that an edge with minimum negative slack (rather than maximum positive slack) is added. This new algorithm will terminate when there are no edges with negative slack. In Problem 18 you are asked to alter Algorithm 4.4 to effect these changes.

Before we apply this new algorithm, we observe that the existence of a path from vertex u to vertex v does not necessarily imply that there is a maximal weight path from u to v (although as we have seen previously, there must be a minimal weight path from u to v). Take, for example, the graph containing a directed cycle of positive weight shown in Figure 4.19. The vertex v is accessible from the vertex u, but by going an arbitrary number of times around the cycle, we can create a path of arbitrarily large weight. To avoid this problem, we shall only apply the maximal path tree algorithm to weighted digraphs that do not contain any directed cycles.

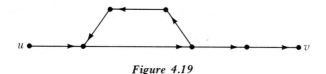

Figure 4.19

Note that on termination of the maximal path tree algorithm a maximal path tree has been generated and the slack of each edge (or activity) is either zero (for those activities included in the maximal path tree), or positive. Any positive slack of an activity $A = (u, v)$ means there is extra time available to spend on A without delaying the event v or any future event. This is because:

a. $W(Q_u)$ is the minimal time needed until activity A begins, and

b. $w(u, v)$ is the time required to complete activity A.

Consequently, the sum $W(Q_u) + w(u, v)$ is the minimal amount of time required to prepare for and complete activity A. Since $W(Q_v)$ is the minimal time required to initiate the activities that follow A, the difference

$$s(u, v) = W(Q_v) - (W(Q_u) + w(u, v))$$

is the extra or slack time that can be devoted to activity A without delaying event v.

We apply this procedure to determine the critical activities and slack times of the graph in Figure 4.20. For easy reference, we have labeled each edge with an alpha-numeric pair: the letter indicates the activity, and the number the units of time the activity requires.

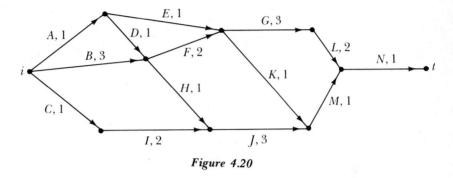

Figure 4.20

We begin with a tree rooted at i. Figure 4.21 illustrates our selection. The circled number near each vertex v is the value $W(Q_v)$.

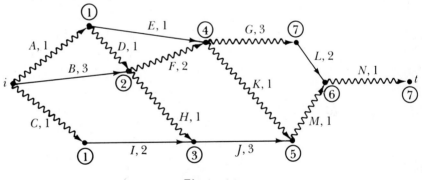

Figure 4.21

Next we identify the slack associated with each edge, as shown in the boxed numbers in Figure 4.22. Examining Figure 4.22, we see that the edge for task L has the smallest minimal negative slack, -3, so we insert that edge into the spanning tree and take away the edge that previously went to the end of edge L. Figure 4.23 illustrates the new spanning tree and the adjusted $W(Q_v)$ weights and slack times.

At this stage, we find that there are two edges with slack -1, and we may choose either one to add to our tree. If we choose to insert the edge representing task B, we obtain the tree illustrated in Figure 4.24, with the new values $W(Q_v)$ and the new slack times for each edge not in the tree.

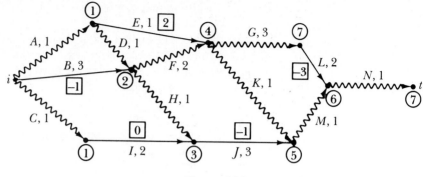

Figure 4.22

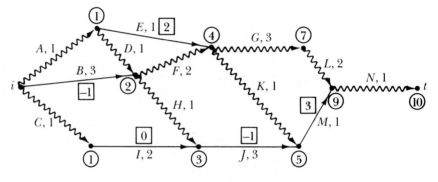

Figure 4.23

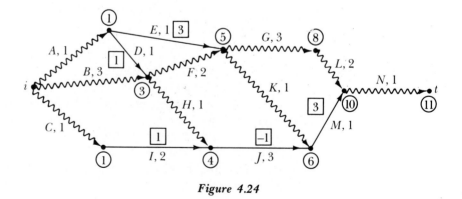

Figure 4.24

Next we insert the remaining edge that has slack −1, and we obtain the tree shown in Figure 4.25. Since all slacks are now nonnegative, this tree is a maximal path tree.

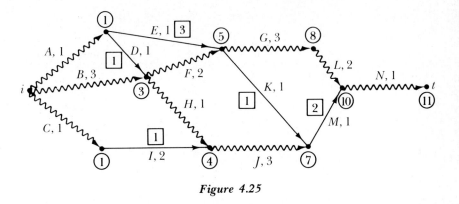

Figure 4.25

The critical path is a maximal weight (in units of time) path from i to t, and is composed of the critical activities B, F, G, L, and N. Table 4.2 lists the slack time that can be given to an activity without delaying the entire project.

Activity	Allowed slack time	Activity	Allowed slack time
A	0	H	0
B	0	I	1
C	0	J	0
D	1	K	1
E	3	L	0
F	0	M	2
G	0	N	0

Table 4.2

4.2   Problems

In Problems 1–4 draw the digraph representing each of the projects described.

1. Project: Repair an engine crankshaft

Activity	Description	Immediate predecessors
A	Remove engine	none
B	Remove crankshaft	A
C	Order necessary parts	B
D	Repair crankshaft	C
E	Replace crankshaft	D
F	Reinstall engine	E

2. Project: Repair a dented auto door

Activity	Description	Immediate predecessors
A	Remove door	none
B	Order window	none
C	Remove window	A
D	Pound out dent	C
E	Repaint	D
F	Install new window	B, D
G	Install door	F

3. Project: Pour footers for a building addition

Activity	Description	Immediate predecessors
A	Excavate area to grade	none
B	Dig trenches	A
C	Construct forms	B
D	Arrange delivery of concrete	none
E	Buy reinforcing bar	none
F	Put reinforcing bar in place	C, E
G	Put rough plumbing in place	F
H	Pour concrete	D, G,

4. Project: Produce and assemble two new products

Activity	Description	Immediate predecessors
A	Order machine A	none
B	Order machine B	none
C	Install machine A	A
D	Install machine B	B
E	Order machine J	none
F	Install machine J	E
G	Produce part a	C
H	Produce part b	D
I	Use machine J to assemble parts a and b	F, G, H

In Problems 5–8 draw the weighted digraph associated with each project

5. Activity	Immediate predecessors	Duration (time units)
A	none	3
B	A	1
C	A	2
D	B	1
E	C	4
F	D, E	3
G	C	1

6. Activity	Immediate predecessors	Duration (time units)
A	none	1
B	none	2
C	A	1
D	B	5
E	A	3
F	B	4
G	C, F	3
H	E, D	7

7. Activity	Immediate predecessors	Duration (time units)
A	none	3
B	none	1
C	none	2
D	A, B	7
E	B	5
F	C	1
G	D	3
H	A	4

8. Activity	Immediate predecessors	Duration (time units)
A	none	3
B	none	6
C	A, B	6
D	B	5
E	C, D	7
F	D	1
G	E	3
H	F	3
I	B	2

In Problems 9–12 apply the critical path method to:

(a) compute a maximal path tree beginning with the indicated tree,
(b) identify the corresponding critical activities, and
(c) record the slack time for each activity.

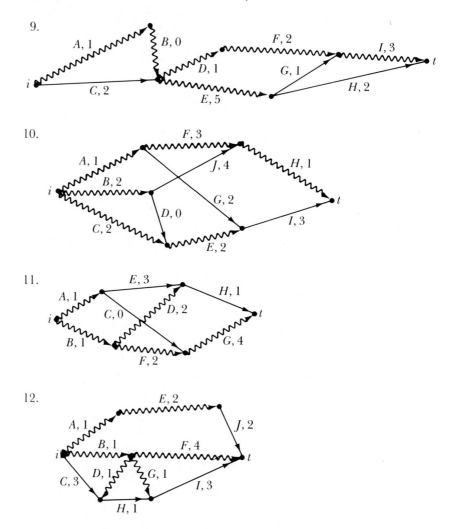

9.

10.

11.

12.

13. Three machines, A, B, and C, are to make parts a, b, and c, respectively. These machines must be purchased, installed, and tested. The test equipment, T, must also be purchased and installed. A fourth machine, D, is already in place and will assemble parts a, b, and c. Machine D, however, does require some minor modification. After parts a, b, and c are assembled, the final product must be tested prior to shipment. The follow-

ing chart gives the expected duration of each activity:

purchase machine A	7 days
purchase machine B	7 days
purchase machine C	7 days
install machine A	3 days
install machine B	1 day
install machine C	8 days
purchase test equipment	7 days
install test equipment	3 days
modify machine D	1 day
test machine A	1 day
test machine B	2 days
test machine C	1 day
produce parts a, b, and c	10 days
assemble parts a, b, and c	3 days
test final product	1 day
ship	1 day

Find a critical path and calculate the slack time for each activity.

14. Show that in a maximal path tree all slacks must be nonnegative.

15. Explain why the slack of each activity in a maximal path tree must be zero.

16. Let D' denote the digraph obtained from the diagram D associated with a project by reversing the direction of each edge. Let T' denote a maximal path tree rooted at t in D', and let Q_v' be the path in T' from t to v. As before, Q_v denotes the length of the longest path from i to v.

   (a) Interpret $W(Q_v')$.
   (b) Interpret $W(Q_t) - W(Q_v')$
   (c) Let $A = (u, v)$ be an activity. Interpret
   $$W(Q_t) - W(Q_v') - W(Q_u) - w(u, v)$$

17. Explain why a directed graph representing a project will not contain a directed cycle.

18. Write an algorithm, based on Algorithm 4.4, that will generate a maximal path tree.

4.3 EULER TOURS AND PATHS

In 1736, the Swiss mathematician Leonhard Euler (pronounced *Oi*-ler) solved the famous Königsberg bridge problem and in doing so, by many accounts, originated the study of graph theory. Beyond their historical interest, Euler's studies have applications to modern-day computer science. In the next section,

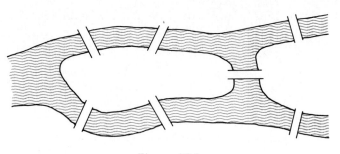

Figure 4.26

for example, we will use Euler's Theorem to study maximal-length shift register sequences.

In this section, however, we begin with Euler's Königsberg bridge problem. The eighteenth-century city of Königsberg (now Kaliningrad) was situated at the confluence of two rivers. The banks of these rivers and an island were connected by seven bridges as shown in Figure 4.26. The Königsberg bridge problem is whether or not it is possible to find a path that crosses each bridge once and only once.

Translating this problem to more modern graph theoretic terms, we obtain Figure 4.27, which associates a graph with the land mass and bridge network of Königsberg.

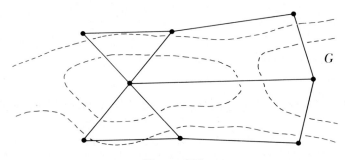

Figure 4.27

In view of the graph G in Figure 4.27, the Königsberg bridge problem becomes that of determining whether it is possible to find a path that passes over each edge exactly once in the graph G. To solve this problem and its more modern extensions, we first introduce the following definitions.

Definition 4.5   A path $v_0, v_1, v_2, \ldots, v_n$ in a graph G is an *Euler path* if every edge of G appears once and only once in the path. That is,

$$E(G) = \{\{v_i, v_{i+1}\} \mid i = 0, 1, 2, \ldots, n - 1\}$$

and

$$\{v_i, v_{i+1}\} \neq \{v_j, v_{j+1}\} \quad \text{if} \quad i \neq j$$

Definition 4.6   A *tour* in a graph G is a closed path $v_0, v_1, \ldots, (v_n = v_0)$ with no repeated edge (that is, $\{v_i, v_{i+1}\} \neq \{v_j, v_{j+1}\}$ if $i \neq j$).

   An *Euler tour* in a graph G is a tour in G that covers all the edges of G (that is, $E(G) = \{\{v_i, v_{i+1}\} \mid i = 0, \ldots, n - 1\}$).

   The graphs G and H illustrated in Figure 4.28 both have Euler paths, but, as we shall see, only H has an Euler tour. In G the path $v_2, v_3, v_1, v_4, v_3, v_0, v_2, v_4$ is an Euler path, and in H the tour $v_0, v_1, v_2, v_4, v_1, v_3, v_4, v_0$ is an Euler tour.

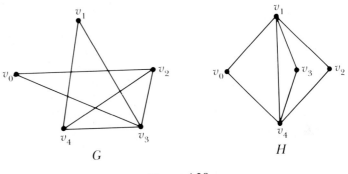

Figure 4.28

   Before taking up the problem of whether a graph has an Euler path, we consider the existence of an Euler tour.

Theorem 4.7   *Euler's Theorem.* A connected graph G has an Euler tour (is *Eulerian*) if and only if the degree of every vertex of G is even.

   Proof: First assume that G is Eulerian and let C be an Euler tour in G. Note that each time a vertex v of G appears during a traversal of C, the two edges adjacent to v are accounted for (one edge going toward v, the other edge leaving v). By definition of an Euler tour, C includes all edges of G, so it follows that the degree of each vertex in G is even.

   Now we prove the converse, that if all vertices of G are of even degree, then G is an Euler graph. If the converse is false, then there is a graph that does not contain an Euler tour, but that does have each vertex of even degree. Of all such graphs, choose one with the least number of edges (such a selection is called a *minimal counterexample*), and label it G. As you are asked to show in Problem 5, G must contain a tour. Choose the longest tour C in G. Since G is not Eulerian, C will not include all of the edges of G, and hence $E(G) \setminus E(C) \neq \emptyset$.

   Let G' be a component of the graph formed by removing the edges (*not* the vertices) of C from G. Since $E(G) \setminus E(C) \neq \emptyset$, it follows that $E(G') \neq \emptyset$. Moreover, since C itself is Eulerian, it must be the case that each vertex of C has even degree, from which it follows that every vertex of G' also has even

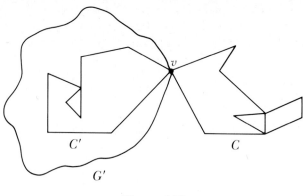

Figure 4.29

degree. Recall that G was selected as the minimal counterexample. Thus since $|E(G')| < |E(G)|$, and since every vertex of G' has even degree and G' is connected, we conclude that G' has an Euler tour C'. Since G is connected, C and C' must have a vertex v in common. We now have the situation illustrated in Figure 4.29.

The tour formed by starting at v, traversing C and then traversing C' is a tour that is longer than C. However, C was chosen as the tour of maximal length in G. This contradiction implies that G is Eulerian. ∎

Example 1 To determine if a graph is Eulerian, simply check to be sure all of the vertices are of even degree. For example, the graph illustrated in Figure 4.30(a) has only vertices of even degree, so it is Eulerian (can you find the Euler tour?). The graph in Figure 4.30(b) has two vertices of odd degree, so it is not Eulerian. □

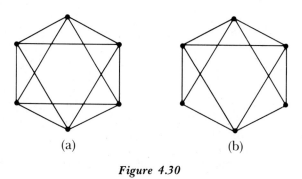

(a)   (b)

Figure 4.30

The next theorem characterizes those graphs that have an Euler path that is not an Euler tour.

Theorem 4.8 A connected graph G has an Euler path that is not an Euler tour if and only if G has exactly two vertices of odd degree.

Proof: If G has an Euler path that is not an Euler tour, then a proof similar to the proof of the previous theorem shows that each vertex except the initial and final vertices of the path has even degree. Only two vertices, the initial and final, have odd degree.

Conversely, assume that G does have only two vertices of odd degree, u and v. We must prove it has an Euler path. Form a new graph G' where

$$V(G') = V(G) \cup \{w\} \qquad w \notin V(G)$$
$$E(G') = E(G) \cup \{\{w, u\}, \{w, v\}\}$$

Figure 4.31 illustrates G'.

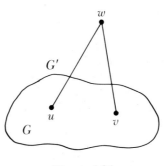

Figure 4.31

Observe now that all vertices of G' have even degree, because adding a new edge to u and to v makes u and v each have an even degree instead of an odd degree. Consequently, by Theorem 4.7 there is an Euler tour C in G'. Therefore, by removing vertex w and edges $\{u, w\}$ and $\{v, w\}$, the Euler tour C of G' becomes an Euler path of G. In other words, the path $P = C \setminus \{\{u, w\}, \{u, v\}\}$ is an Euler path in G, and this completes the proof. ∎

Using Theorem 4.8, we can quickly determine if a graph has an Euler path by counting the number of vertices of odd degree. For example, the graphical representation of the Königsberg bridge (see Figure 4.27) has no Euler path.

Certain computer science applications require that we modify Euler's Theorem so that we can apply it to directed graphs. To accomplish this modification we need some additional definitions.

Definition 4.9 Let D be a directed graph (possibly with loops) and let $v \in V(D)$.

a. The *indegree* of v, $d_i(v)$, is the number of edges of D that have v as their terminal vertex. That is,

$$d_i(v) = |\{u \mid (u, v) \in E(D)\}|$$

b. The *outdegree* of v, $d_o(v)$ is the number of edges of D that have v as their initial vertex. That is,

$$d_o(v) = |\{u \mid (v, u) \in E(D)\}|$$

Example 2   For the directed graph with loops illustrated in Figure 4.32, we have $d_i(v_0) = 2$, $d_o(v_0) = 1$, $d_i(v_1) = 0$, $d_o(v_1) = 3$, $d_i(v_2) = 1$, $d_o(v_2) = 1$, $d_i(v_3) = 2$, $d_o(v_3) = 0$.   ☐

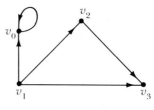

Figure 4.32

The next definition formalizes a concept introduced in Chapter 3.

Definition 4.10   A directed graph (possibly with loops) G is *strongly connected* if for every pair of vertices u and v

    a. there is a directed path in G from u to v, and

    b. there is a directed path in G from v to u.

The directed graph in Figure 4.33(a) is not strongly connected, but the one in Figure 4.33(b) is strongly connected.

We can now describe those graphs that have a directed Euler tour. This theorem's proof is so similar to that of the corresponding result, Theorem 4.7, for graphs that it is omitted.

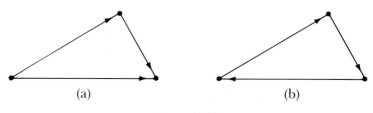

(a)   (b)

Figure 4.33

Theorem 4.11 A strongly connected directed graph (possibly with loops) D has a directed Euler tour (is Eulerian) if and only if $d_i(v) = d_o(v)$ for all $v \in V(D)$.

In the next section we will see how we can apply Theorem 4.11 to the problem of finding maximal shift register sequences.

4.3 Problems

1. Which of the following graphs

   a. has an Euler path?

   b. has an Euler tour?

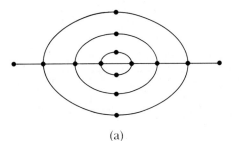

(a)

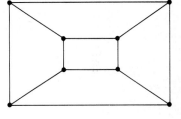

(b)

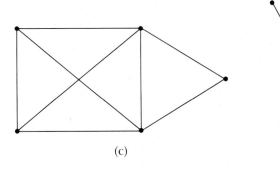

(c)

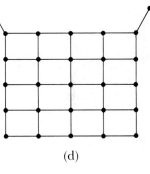

(d)

2. Which of the following directed graphs are strongly connected?

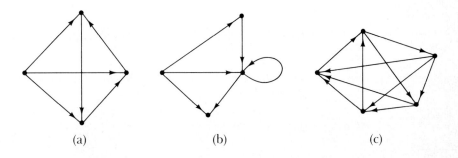

(a)   (b)   (c)

3. Explain why an Eulerian graph is strongly connected.

4. Which of the following digraphs is Eulerian?

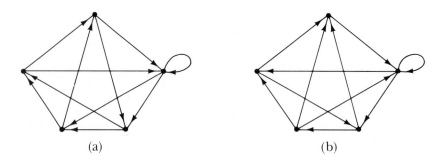

(a)   (b)

5. Show that if every vertex of a graph G has degree two or greater, then G contains a tour.

6. Show that if D is a digraph, then

$$\sum_{v \in V(D)} d_o(v) = \sum_{v \in V(D)} d_i(v) = |E(D)|$$

7. Prove that a directed graph D is not strongly connected if for some $v \in V(D)$, $d_i(v)d_o(v) = 0$.

8. Show that if D is a digraph and if $d_i(u) = d_i(v)$ and $d_o(u) = d_o(v)$ for all u and $v \in V(D)$, then $d_o(v) = d_i(v)$ for each $v \in V(D)$.

9. Show that if every vertex of a graph G has even degree, then there are tours $C_1, C_2, \ldots, C_m$ such that

   a. $E(C_i) \cap E(C_j) = \varnothing$ if $i \neq j$, and

   b. $E(G) = E(C_1) \cup E(C_2) \cup \cdots \cup E(C_m)$.

10. Show that a graph must have an even number of vertices of odd degree.

11. Show that if a connected graph G has $2m$ vertices of odd degree, then there are m paths $P_1, \ldots, P_m$ such that

   a. $E(P_i) \cap E(P_j) = \varnothing$ if $i \neq j$, and

   b. $E(G) = E(P_1) \cup E(P_2) \cup \cdots \cup E(P_m)$.

4.4   MAXIMAL SHIFT REGISTER SEQUENCES

Euler tours in a specific class of directed graphs lead to maximal shift register sequences, or de Bruijn sequences. These sequences are found in such diverse areas as coding theory, tracking systems, and deterministic simulation of random processes. We begin by considering the following problem.

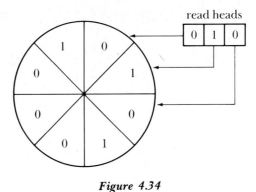

Figure 4.34

We must determine the position of a rotating disk from the output of a number of "read heads" that detect a 0 or a 1. The disk in Figure 4.34 is divided into eight sectors, and each sector is labeled with a 0 or a 1. The boxed numbers indicate the position of the disk at any given time; in this particular case, the output is 010. Rotating the disk clockwise one unit would yield an output of 101. Rotating the disk another unit clockwise would again give an output of 010. Since in this example two distinct positions yield the same output, this particular arrangement of zeros and ones is useless for determining the disk's position. This leads us to question if it is indeed possible to arrange the zeros and ones so that each position of the disk produces a unique reading. Rather than answer this question specifically, we will take a more general—and more productive—approach.

Definition 4.12   A *de Bruijn sequence* (also called a *maximal shift register sequence*) is a circular arrangement $a_1 a_2 a_3 \cdots a_{2^n}$ of zeros and ones such that every sequence of zeros and ones of length n appears exactly once. That is, given a sequence w of n zeros and ones, there is a unique i such that

$$a_i a_{i+1} \cdots a_{i+n-1} = w$$

where the indices are computed modulo 2^n.

Example 1   The sequence 11100010 is a de Bruijn sequence of length $2^3 = 8$. Every 3-bit ($n = 3$) sequence of zeros and ones appears exactly once. For example, the sequence 111 ($111 = a_1 a_2 a_3$) only occurs once. Similarly, $101 = a_7 a_8 a_1$ is the only occurrence of the sequence 101.   □

Note that in view of Example 1, it *is* possible to arrange the zeros and ones on a disk so that each position produces a unique reading. We simply label the sectors as indicated by a de Bruijn sequence to produce the desired labeling indicated in Figure 4.35.

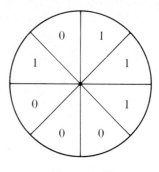

Figure 4.35

Because we need longer de Bruijn sequences to obtain a more accurate reading of a disk's position, we take up the more general problem of determining if a de Bruijn sequence exists for each positive integer n.

To solve this problem, we first introduce a special type of directed graph with loops, known as a *Good diagram*. For $n = 1, 2, \ldots$, the Good diagram D_n has the following vertex set:

$$\{v \mid v \text{ is a sequence of 0's and 1's of length } n - 1\}$$

An ordered pair $e = (a_1 a_2 \cdots a_{n-1}, b_1 b_2 \cdots b_{n-1})$ is an edge of D_n if and only if $b_1 = a_2, b_2 = a_3, \ldots, b_{b-2} = a_{n-1}$. In other words, an edge e is directed from a vertex $a_1 a_2 \cdots a_{n-1}$ to a vertex $b_1 b_2 \cdots b_{n-1}$ if and only if we can form the sequence $b_1 b_2 \cdots b_{n-1}$ from the sequence $a_1 a_2 \cdots a_{n-1}$ by eliminating the initial a_1 and adding one more entry, b_{n-1}, to the end. Consequently, all edges of D_n have the form $e = (c_1 c_2 \cdots c_{n-2} c_{n-1}, c_2 c_3 \cdots c_{n-1} c_n)$. It is helpful to label such an edge e as the sequence $c_1 c_2 \cdots c_{n-1} c_n$.

Example 2 The following examples illustrate the labeling of a Good diagram.
- (a) $(001, 011) \in E(D_4)$; the label attached to edge $(001, 011)$ is 0011.
- (b) $(001, 010) \in E(D_4)$; the label attached to edge $(001, 010)$ is 0010.
- (c) $(000, 000) \in E(D_4)$; the label attached to edge $(000, 000)$ is 0000.

$\square$

Figures 4.36(a) and (b) illustrate the Good diagrams for $n = 3$ and $n = 4$, respectively.

Observe that in the Good diagram D_n, a given sequence $c_1 c_2 \cdots c_{n-1} c_n$ appears as the label of an edge exactly once. Moreover, the labels of two successive edges of a path in D_n have the form $c_1 c_2 \cdots c_n$ and $c_2 c_3 \cdots c_{n+1}$. It follows that we can use an Euler tour in D_n to generate a de Bruijn sequence of length 2^n by using the first digit of each edge label in the tour to create the sequence. The following example illustrates this idea.

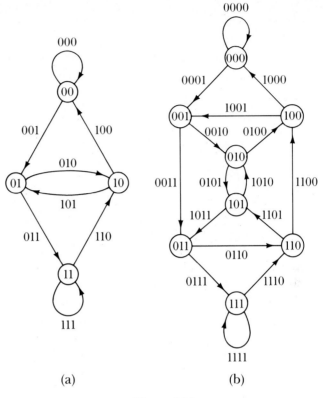

Figure 4.36

Example 3   Using the Good diagram in Figure 4.36(a), we will construct an Euler tour and consequently a de Bruijn sequence. From the figure we can see that the sequence of edges 111, 110, 100, 000, 001, 010, 101, 011 is an Euler tour of D_3. Taking the first digit of each edge of this tour, we obtain the de Bruijn sequence 11100010. □

We can now extend this example and prove that there is a de Bruijn sequence for each integer n. For a Good diagram, observe that if $v = a_1 a_2 \cdots a_{n-1} \in V(D_n)$, then there are exactly two edges that end at vertex v:

$$0a_1 a_2 \cdots a_{n-1} \quad \text{and} \quad 1a_1 a_2 \cdots a_{n-1}$$

Similarly, there are exactly two edges that begin at vertex v:

$$a_1 a_2 \cdots a_{n-1}0 \quad \text{and} \quad a_1 a_2 \cdots a_{n-1}1$$

Thus for all $v \in V(D_n)$, we have $d_i(v) = 2 = d_o(v)$, so by Theorem 4.7 every Good diagram D_n has an Euler tour. Since an Euler tour in D_n yields a de Bruijn sequence, we conclude that there are de Bruijn sequences of length 2^n for all n.

4.4   Problems

1. Find a de Bruijn sequence of length sixteen.
2. How many vertices are there in the Good diagram D_5?
3. How many edges are there in the Good diagram D_5?
4. Draw the Good diagram D_5.
5. Find a de Bruijn sequence of length thirty-two.
6. Suggest an algorithm to generate a de Bruijn sequence of length 2^n.

4.5   GRAY CODES

In the previous section we used Euler tours to solve the problem of determining the position of a rotating disk. In this section we take up a similar problem, one that is of interest in dealing with errors inherent in analog to digital conversion.

Suppose a disk rotates erratically about its center. We would like to read the disk's position into a digital computer at certain intervals. One solution is to divide the disk into 2^n sectors and label each of the sectors with an n-digit binary number. Figure 4.37 shows the disk's labeling for $n = 3$ and $n = 4$.

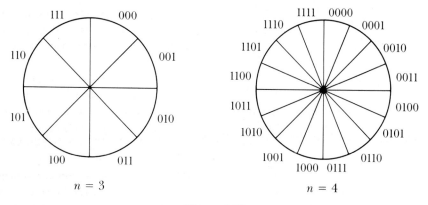

$$n = 3 \qquad\qquad\qquad n = 4$$

Figure 4.37

To read the digits associated with the sectors, we extend n photoelectric cells beneath the disk, lined up from the center to the edge. Then we place lights above the disk, in line with the photoelectric cells. Next we replace appropriate portions of each sector with translucent material. This will allow light to pass through portions of the disk and trigger the photoelectric cell below to conduct current. If light is blocked by a solid portion of the disk, the cell will not send out an impulse. When the computer receives an electrical impulse from a cell, it records a one for that cell; when it does not receive an impulse, it records a zero. Figure 4.38 illustrates the disk with eight sectors set up as just described. The white sections of the sectors represent translucent material, which will allow light through the disk.

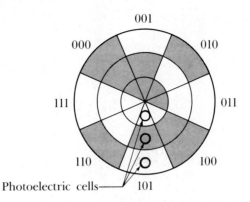

Since the inner and outer
rings of this sector are
translucent, the cells read 101.

Figure 4.38

At first glance, this solution to the problem of reading the position of the disk appears satisfactory. Consider, however, what happens if the position of the disk is read when the photoelectric cells line up along the line separating one sector from another. In this case a photoelectric cell will take its reading from either one of the two adjacent sectors. Although along some radii this presents little problem, along others the readings become entirely arbitrary. Figure 4.39 illustrates these two situations.

This partial alignment of the photoelectric cells is the source of the system's unreliability since we can never be sure when a particular reading results from an alignment such as that illustrated in Figure 4.39(b).

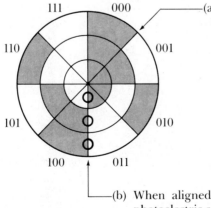

(a) When aligned with this radius, the photoelectric cells will read either 000 or 001; they will read the position of the disk correctly to within ¼ revolution.

(b) When aligned with this radius, the photoelectric cells will give an entirely arbitrary reading since each cell can read either a 0 or a 1.

Figure 4.39

Nevertheless, there is a way to ensure that *all* readings are accurate to within one-quarter of a revolution, as in 4.39(a). Before we pursue this idea, we need a little more graph theory.

Definition 4.13   The *n-cube*, Q_n, is the graph whose vertex set is

$$V = \{(x_1, x_2, \ldots, x_n) | x_i = 0 \text{ or } x_i = 1, i = 1, 2, \ldots, n\}$$

An edge joins two vertices if and only if the two vertices differ in exactly one coordinate.

In Figure 4.40 we illustrate the cubes Q_1, Q_2, and Q_3, labeled according to Definition 4.13.

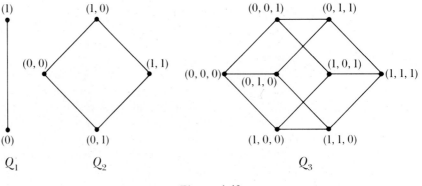

Figure 4.40

Definition 4.14   A graph is *Hamiltonian* if it has a cycle (called a *Hamiltonian cycle*) that passes through each vertex once and only once.

Example 1   The cycle v_0, v_1, v_2, v_3, v_4, v_0 in the graph illustrated in Figure 4.41(a) is a Hamiltonian cycle. You should convince yourself that the graph in Figure 4.41(b) is not Hamiltonian.   □

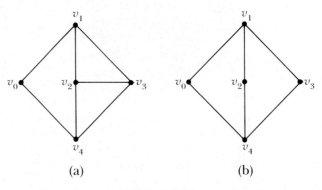

(a)   (b)

Figure 4.41

We will use the following result, which you are asked to prove in Problem 13, in connection with the rotating disk problem.

Theorem 4.15   The cube Q_n is Hamiltonian for all $n \geq 2$.

To solve the rotating disk problem, recall that if the binary numbers of two adjacent sectors differ in only one bit, then a straddled reading is correct to within two sectors. To ensure this degree of accuracy, therefore, we only need to arrange the numbering of the sectors so that adjacent sectors always differ in only one bit. A Gray code is a way of accomplishing this objective.

Definition 4.16   A *Gray code* is a Hamiltonian cycle in an *n*-cube.

In Figure 4.42, the wavy lines denote a Gray code in the cube Q_3.

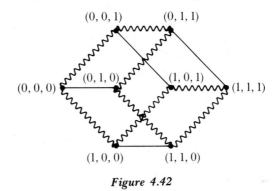

Figure 4.42

Note that a Gray code passes through all of the vertices of the cube, and that any two of its adjacent vertices differ in only one bit. Consequently, a Gray code in Q_3 provides a numbering of the sectors of a disk so that all readings are correct to within one-quarter revolution. (In the exercises you are asked to show how to increase this accuracy to one-eighth of a revolution.)

Suppose that one photoelectric cell malfunctions. This will result in an incorrect reading of the disk's position. We would like to know if the incorrect reading indicates something other than a sector adjacent to the correct sector. That a severe error could occur is seen in Figure 4.43, where a single malfunction results in a reading far removed from the correct one.

In particular, if, as illustrated in Figure 4.43, 011 is read as 111, then the position of the disk is badly misread. Such an error results because the vertices $(0, 1, 1)$ and $(1, 1, 1)$ of the cube are adjacent, although the sectors designated by these vertices are not adjacent on the disk.

As before, we also want to avoid the problem that arises when the photoelectric cells are aligned with a radius separating adjacent sectors. Thus, whatever code we do select should result from a cycle in a 3-cube. But

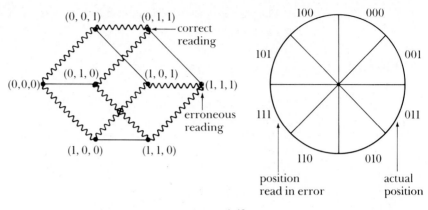

Figure 4.43

now in addition, we require that no single malfunction of a photo cell should result in a reading that indicates something other than an adjacent sector.

In general, to avoid the problems associated with a single malfunction of a photoelectric cell, we must avoid using vertices that are adjacent in the cube but are not adjacent on the cycle that determines the labeling of the disk. That is, if u and v are two vertices of the cycle that are adjacent in the cube, then they should be also adjacent on the cycle. Note in Figure 4.43 that the vertices $(0, 1, 1)$ and $(1, 1, 1)$ do not meet this requirement. A cycle all of whose vertices satisfy this requirement is called *a cycle of spread 2* (the vertices of the cycle are spread at least a distance two edges apart if they are not adjacent on the cycle). In Figure 4.44 we illustrate a cycle of spread 2 in the 3-cube, Q_3, and the associated coding of the disk.

With this coding any single error will result either in a reading corresponding to an adjacent sector or a reading that does not correspond to any sector. Thus, the consequences of any single error are minimized. However, we have paid a price for this error-detection capability: We can now use only six of the eight vertices of the 3-cube, so we can divide the disk into only six sectors.

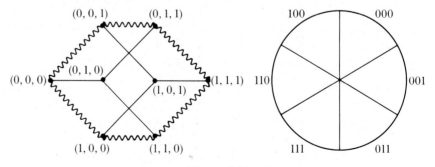

Figure 4.44

In order to divide the disk into a larger number of sectors and yet retain the error-detection capabilities, we must use a higher dimensional cube. The length of the longest cycle of spread 2 in the n-cube, Q_n, is not known.

4.5   Problems

1. Find a Hamiltonian cycle in each of the following graphs.

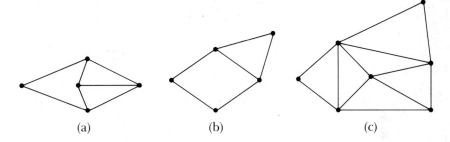

(a)   (b)   (c)

2. Explain why the following graph is not Hamiltonian.

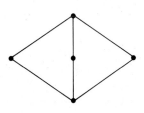

3. Show that the complete graph of n points, K_n, is Hamiltonian.
4. Draw the 4-cube Q_4.
5. Find a Gray code in Q_4.
6. Use the solution of Problem 5 to label the sectors of a disk so that all readings are correct to within one-eighth of a revolution.
7. How many photoelectric cells are necessary to obtain readings that are correct to within $1/32$ of a revolution?
8. Find a cycle of spread 2 in Q_3 other than the cycle given in Figure 4.44.
9. Find a cycle of spread 2 and length 8 in Q_4.
10. Determine the number of vertices in the n-cube Q_n.
    We can use the following steps to generate a Gray code in Q_n from a Gray code in Q_{n-1}.

    Step 1. Write a list L_1 of the vertices of Q_{n-1} in the order of their appearance in a Gray code.

Step 2. Form a new list L_2 by reversing the order of list L_1.

Step 3. Form list L_3 by appending list L_2 to the end of list L_1.

Step 4. Add a final 0 to all entries in the first half of list L_3 and a final 1 to all entries in the second half of list L_3. The resulting list, L_4, is a Gray code in Q_n.

As an example, we apply these steps to $L_1 = 00, 01, 11, 10$, which is a Gray code in Q_2:

$$L_1 = 00, 01, 11, 10$$
$$L_2 = 10, 11, 01, 00$$
$$L_3 = 00, 01, 11, 10, 10, 11, 01, 00$$
$$L_4 = 000, 010, 110, 100, 101, 111, 011, 001$$

The list L_4 is a Gray code in Q_3.

11. Use the above procedure to generate a Gray code in Q_5.

12. Use induction to prove that the above procedure generates a Gray code.

13. Use induction to prove Theorem 4.15, that Q_n is Hamiltonian for each $n \geq 2$.

Chapter 4 REVIEW

Concepts for Review

rooted tree (p. 148)
minimal path tree (p. 149)
slack value (p. 152)
critical path method (p. 157)
critical path (p. 160)
Euler path (p. 168)
tour (p. 169)
Euler tour (p. 169)
Eulerian graph (p. 169)
maximum shift register sequence (de Bruijn sequence) (p. 175)
Good diagram (p. 176)
n-cube, Q_n (p. 180)
Hamiltonian cycle (p. 180)
Hamiltonian graph (p. 180)
Gray code (p. 181)
cycle of spread 2 in Q_n (p. 182)

Review Problems

1. Apply Algorithm 4.4 to obtain a minimal path tree in the given graph. Begin the algorithm with the initial rooted tree, T, shown by wavy edges.

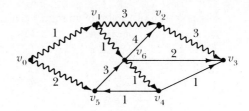

2. Draw the weighted digraph that represents the following project.

Activity	Immediate predecessors	Duration (time-units)
A	none	3
B	none	2
C	A	4
D	A, B	1
E	C	2
F	D	1
G	C, D	5

3. In the given graph:
   (a) Compute a maximal path tree beginning with the indicated tree.
   (b) Identify the corresponding critical activities.
   (c) Record the slack time for each activity.

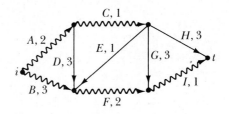

4. Indicate if the given directed graph is strongly connected.

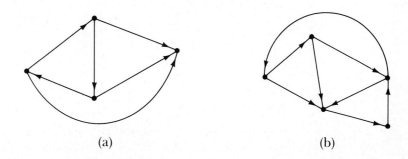

(a)   (b)

5. Which of the following graphs has
   (a) an Euler path?
   (b) an Euler tour?

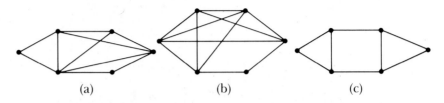

(a)   (b)   (c)

6. Determine if the given digraph is Eulerian.

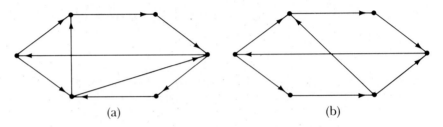

(a)   (b)

7. Use the Good diagram D_4 to find a de Bruijn sequence of length 16.

8. Find a Hamiltonian path in the given graph.

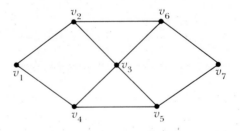

9. Explain why the given graph is not Hamiltonian.

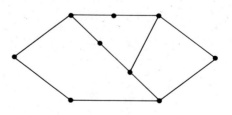

10. Draw the cube Q_4 and label its vertices.

5 Boolean Algebra and Switching Systems

This chapter focuses mainly on certain mathematical concepts that underlie the design of switching systems, used in connection with telephone networks, digital computers, and so on. We can describe these systems by a mathematical structure called Boolean algebra, named after George Boole, a noted English logician.

Boolean algebra differs considerably from the algebra you studied in high school. (For example, Boolean addition and multiplication differ from the usual addition and multiplication.) There are many examples of Boolean algebras. The Boolean algebra we use most in this chapter is especially simple but it is particularly useful for studying switching systems.

Much of this chapter is devoted to simplifying switching systems, and we examine two common procedures for this purpose: Karnaugh maps and the Quine–McCluskey Algorithm. The Quine–McCluskey Algorithm is especially adaptable to computer use, and we will study it in some detail.

Some set theory is necessary for this chapter. Sufficient background in set theory for our purposes can be found in Section 1 of the Appendix.

5.1 FROM SWITCHING SYSTEMS TO BOOLEAN ALGEBRA

A *switch* is a two-state device that is either "open" or "closed." When closed, a switch conducts electricity; when open, it does not conduct.

For our purposes, a *switching system* will consist of an energy source (for instance, a battery), an output (for instance, a light), and a number of switches designated by x, y, z and so on.

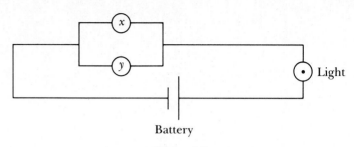

Battery

Figure 5.1

There are two basic ways to connect two switches: by a parallel connection and by a series connection. When we connect two switches x and y in parallel, we have the situation shown in Figure 5.1. If either switch x or y is closed (or if both are closed), then the entire system is closed and the light will go on. Table 5.1 summarizes the situation of two switches connected in parallel.

Parallel Connection

Switch		State of system
x	y	
open	open	open
open	closed	closed
closed	open	closed
closed	closed	closed

Table 5.1

If we connect two switches x and y in series, then we have the situation illustrated in Figure 5.2. In this case, for the light to be on—that is, for the system to be closed—both switches x and y must be closed. Table 5.2 summarizes the situation of two switches connected in series.

One other configuration is often used. Corresponding to a given switch x is

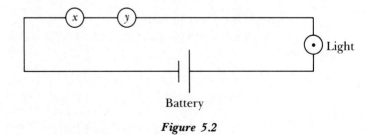

Battery

Figure 5.2

Series Connection

Switch		State of system
x	y	
open	open	open
open	closed	open
closed	open	open
closed	closed	closed

Table 5.2

Complementary Switches

x	x'
open	closed
closed	open

Table 5.3

a *complementary switch*, x', that is always in the state opposite that of x. Table 5.3 summarizes the action of the complementary switch x'.

In what follows it will be convenient to adopt the following simplified notation.

a. We substitute 0 for "open."

b. We substitute 1 for "closed."

c. We let $x + y$ represent the state of the system consisting of two switches x and y connected in parallel.

d. We let $x \cdot y$ represent the state of the system consisting of two switches x and y connected in series.

e. We let x' represent the state of the complement of x.

With this notation, we can rewrite Tables 5.1, 5.2, and 5.3 as Tables 5.4, 5.5, and 5.6, respectively. Note that the operation of addition ($+$) corresponds to a parallel connection, the operation of multiplication ($\cdot$) corresponds to the series system, and the operation of complementation ($'$) corresponds to the complement switch.

We now have three operations, addition ($+$), multiplication ($\cdot$), and complement ($'$), on the set $S = \{0, 1\}$. The following theorem shows that these operations, when applied to S as indicated in Tables 5.4, 5.5, and 5.6, satisfy many (but not all) of the basic properties of real numbers.

x	y	$x + y$
0	0	0
0	1	1
1	0	1
1	1	1

Table 5.4

x	y	$x \cdot y$
0	0	0
0	1	0
1	0	0
1	1	1

Table 5.5

x	x'
0	1
1	0

Table 5.6

Theorem 5.1   Let $S = \{0, 1\}$, and suppose that the operations $+$, $\cdot$, and $'$ are given by Tables 5.4, 5.5, and 5.6. If a, b, and c are elements of S (in other words, $a = 0$ or 1, $b = 0$ or 1, $c = 0$ or 1), then

(a) $a + b = b + a$
(b) $a \cdot b = b \cdot a$ (Commutativity)

(c) $(a + b) + c = a + (b + c)$
(d) $(a \cdot b) \cdot c = a \cdot (b \cdot c)$ (Associativity)

(e) $a + (b \cdot c) = (a + b) \cdot (a + c)$
(f) $a \cdot (b + c) = (a \cdot b) + (a \cdot c)$ (Distributivity)

(g) $a + 0 = a$
(h) $a \cdot 1 = a$ (Identity relation)

(i) $a + a' = 1$
(j) $a \cdot a' = 0$ (Complementation)

To prove this theorem, we must check the validity of properties (a)–(j) for all possible choices of a, b, c. We will verify parts (a) and (e), and you are asked to verify the remaining properties in Problems 1 and 2.

Proof: (a) Table 5.7 lists the various possibilities for a and b. Since the third and fourth columns of the table are identical, it follows that $a + b = b + a$.

a	b	$a + b$	$b + a$
0	0	0	0
0	1	1	1
1	0	1	1
1	1	1	1

Table 5.7

(e) Again we use a table to prove this property. Table 5.8 lists all possible combinations of a, b, and c. Since the fifth and eighth columns of Table 5.8 are identical, it follows that $a + (b \cdot c) = (a + b) \cdot (a + c)$.   ∎

For the rest of this chapter we will adopt the following conventions, which we illustrate in Example 1:

a. We replace $x \cdot y$ by xy.

b. We assume that multiplication takes precedence over addition in the order of operations.

a	b	c	$b \cdot c$	$a + (b \cdot c)$	$a + b$	$a + c$	$(a + b) \cdot (a + c)$
0	0	0	0	0	0	0	0
0	0	1	0	0	0	1	0
0	1	0	0	0	1	0	0
0	1	1	1	1	1	1	1
1	0	0	0	1	1	1	1
1	0	1	0	1	1	1	1
1	1	0	0	1	1	1	1
1	1	1	1	1	1	1	1

Table 5.8

c. We assume that complementation takes precedence over addition and multiplication in the order of operations.

Example 1   Under conventions **a.**, **b.**, and **c.** the expression

$$zx + x(x + y')$$

is shorthand notation for

$$(z \cdot x) + (x \cdot (x + (y')))$$ □

As we will see, the set $S = \{0, 1\}$, together with the operations $+$, $\cdot$, and $'$, provides us with a particular example of a Boolean algebra. To state the general definition of a Boolean algebra we introduce the concept of binary and unary operations. A *binary operation* on a set T is a function $f : T \times T \to T$. Thus, a binary operation on T assigns an element of T to each pair of elements of T. The operations $+$ and $\cdot$ are examples of binary operations. A *unary operation* on a set T is any function $g : T \to T$; the complement operation $'$ is a unary operation.

Definition 5.2   The sextuple $(B, \oplus, *, ', \mathbf{0}, \mathbf{1})$ is a *Boolean algebra* if B is a set; $\oplus$ and $*$ are binary operations on B; $'$ is a unary operation on B; $\mathbf{0}$ and $\mathbf{1}$ are elements of B with $\mathbf{0} \neq \mathbf{1}$; and the following ten properties are satisfied for all elements a, b, and c of B.

(a) $a \oplus b = b \oplus a$ (Commutativity)
(b) $a * b = b * a$

(c) $(a \oplus b) \oplus c = a \oplus (b \oplus c)$ (Associativity)
(d) $(a * b) * c = a * (b * c)$

(e) $a \oplus (b * c) = (a \oplus b) * (a \oplus c)$ (Distributivity)
(f) $a * (b \oplus c) = (a * b) \oplus (a * c)$

$$\left.\begin{array}{ll} \text{(g)} & a \oplus \mathbf{0} = a \\ \text{(h)} & a * \mathbf{1} = a \end{array}\right\} \quad \text{(Identity relation)}$$

$$\left.\begin{array}{ll} \text{(i)} & a \oplus a' = \mathbf{1} \\ \text{(j)} & a * a' = \mathbf{0} \end{array}\right\} \quad \text{(Complementation)}$$

It follows immediately from Theorem 5.1 that the set $S = \{0, 1\}$ with the operations $+$, $\cdot$, and $'$ is a Boolean algebra. The following is quite a different example of a Boolean algebra.

Example 2   Let X be any nonempty set and let $B = \mathscr{P}(X)$ be the power set of X. For each F and G in B, define $F \oplus G = F \cup G$, $F * G = F \cap G$, and $F' = X \backslash F$. Let $\mathbf{0} = \varnothing$ and $\mathbf{1} = X$. Then the sextuple $(\mathscr{P}(X), \cup, \cap, ', \varnothing, X)$ is a Boolean algebra. To verify this, we would have to show that each of the properties (a)–(j) of Definition 5.2 holds. We may do this by using the basic properties of set unions, intersections, and complements (see Problems 4 and 5).

$\square$

Thus far we have examined some of the basic ideas underlying Boolean algebra and have seen how these concepts are related to elementary switching systems. In the next section we introduce the notions of Boolean functions and expressions, and we use these ideas to design more complex switching systems.

5.1   Problems

1. Verify properties (b), (c), and (d) of Theorem 5.1.
2. Verify properties (f)–(j) of Theorem 5.1.
3. Which properties of Theorem 5.1 are not satisfied by the real numbers?
4. Verify properties (a)–(e) of Definition 5.2 for the sextuple $(\mathscr{P}(X), \cup, \cap, ', \varnothing, X)$ given in Example 2.
5. Verify properties (f)–(j) of Definition 5.2 for the sextuple $(\mathscr{P}(X), \cup, \cap, ', \varnothing, X)$ given in Example 2.
6. Let $B = \{1, 2, 3, 6, 7, 14, 21, 42\}$ and for $a, b \in B$ let

$$a \oplus b = \text{least common multiple of } a \text{ and } b$$

and

$$a * b = \text{greatest common divisor of } a \text{ and } b$$

For each $a \in B$, let $a' = 42/a$. Show that $(B, \oplus, *, ', 1, 42)$ is a Boolean algebra.

7. Use tables to verify that for any $a, b, c \in S$,
   (a) $a = ab + ab'$
   (b) $ac + a'b = (a + b)(a' + c)(b + c)$
   (c) $aa = a$
8. Use tables to verify that for any $a, b, c \in S$
   (a) $ab(a + b) = ab$
   (b) $a'c' + bc + ab' = b'c' + a'b + ac$
   (c) $a + bc = a + ab + ac + bc$

5.2 SWITCHING SYSTEMS, BOOLEAN FUNCTIONS, AND BOOLEAN EXPRESSIONS

In the last section we saw how the operations of addition, multiplication, and complementation on the set $S = \{0, 1\}$ could describe parallel and series connections of two switches. In this section we will combine these basic operations on S to describe more complicated switching systems.

Note. Although many of the definitions and results in this chapter are equally valid for all Boolean algebras, throughout the remainder of this chapter we shall deal exclusively with the Boolean algebra

$$(\{0, 1\}, +, \cdot, ', 0, 1) \tag{1}$$

defined in the preceding section. Unless otherwise stated, S denotes the set $\{0, 1\}$ as well as the Boolean algebra (1). It will always be clear from the context how S is to be interpreted.

Suppose we want to design a switching system that has a specified output corresponding to each possible input. Consider, for example, the problem of designing an electronic voting machine for a committee of four people. The voting rights are weighted so that one person has a block of five votes, one person has a block of four votes, and the remaining two people have blocks of three votes each.

Each committee member has a switch to use to record votes. Closing the switch registers a "yes" vote. The system is designed so that a light goes on if a particular measure gains over half of the votes.

We will designate the committee members and their associated switches as $w, x, y,$ and z so that:

w has a block of 5 votes,

x has a block of 4 votes,

y has a block of 3 votes, and

w	x	y	z	Votes for	Votes against	Circuit condition
0	0	0	0	0	15	0
0	0	0	1	3	12	0
0	0	1	0	3	12	0
0	0	1	1	6	9	0
0	1	0	0	4	11	0
0	1	0	1	7	8	0
0	1	1	0	7	8	0
0	1	1	1	10	5	1
1	0	0	0	5	10	0
1	0	0	1	8	7	1
1	0	1	0	8	7	1
1	0	1	1	11	4	1
1	1	0	0	9	6	1
1	1	0	1	12	3	1
1	1	1	0	12	3	1
1	1	1	1	15	0	1

⟵ The circuit is closed because the measure passes. It received a majority of favorable votes (10) from x, y, and z.

Table 5.9

Table 5.9 lists the desired state of the circuit for all possible situations.

Our objective is to build a circuit that will meet the requirements given in Table 5.9. First, however, we must introduce the following idea.

Definition 5.3   A *Boolean function* on the Boolean algebra S is a function

$$f : S \times S \times S \times \cdots \times S \to S$$

A Boolean function is completely specified if we know its output values for each possible input. Columns 1, 2, 3, 4, and 7 of Table 5.9 specify a Boolean function $f : S \times S \times S \times S \to S$, where, for example, $f(0, 0, 0, 0) = 0$, $f(0, 1, 1, 1) = 1$, and $f(1, 1, 0, 1) = 1$.

To express a Boolean function algebraically, we first need the concept of a Boolean expression. For this, we use a *recursive definition*, in which we first define certain objects and then give a rule for successively defining more objects in terms of those already defined.

Definition 5.4   A *Boolean expression* on the Boolean algebra S in the variables $x_1, x_2, \ldots, x_n$ is a combination of these variables determined by the following rules:

a. Each of the variables x_i is a Boolean expression.

b. If E and F are Boolean expressions, then so are $E + F$, EF, and E'.

Example 1   All of the following are examples of Boolean expressions in the variables x, y, and z.

x	$x + y$	$(x + y)'$	$(x + y)'yz$
y	$x + z$	$(x + z)'$	$x'z(y' + z)$
z	$y + z$	$(y + z)'$	$xyz + yz' + z$
x'	xy	$(xy)'$	$(x + x')y + xz$
y'	xz	$(xz)'$	$yz + y' + z'$
z'	xy	$(yz)'$	$xy'z + x + y'z'$ $\square$

There is a close relationship between Boolean functions and Boolean expressions. For instance, with a Boolean expression such as $xyz' + x'y$, we can associate the Boolean function $f: S \times S \times S \to S$ defined by $f(x, y, z) = xyz' + x'y$. More importantly, however, we can associate a Boolean expression with each Boolean function. In fact, as we shall see, we can associate many Boolean expressions with a given Boolean function. Our principal goal in this chapter will be to find the "simplest" such expression.

To find a Boolean expression associated with a Boolean function given in tabular form (such as in Table 5.10) we can proceed as follows. Each row that has a 1 in the rightmost column corresponds to a product term in the Boolean expression. We form this product term by taking the product of x (or x'), y (or y'), and z (or z') as follows: We use x if a 1 occurs in the x column, and we use x' if a 0 appears in this column. The same procedure is used for y and z. Finally, we add all of these product terms to obtain a Boolean expression corresponding to the Boolean function defined by the table. For example, in Table 5.10, 1's occur in the right-hand column of rows 3, 5, 7, and 8. From row 3 we obtain the product term $x'yz'$, and from rows 5, 7, and 8 we obtain the

x	y	z	$f(x, y, z)$
0	0	0	0
0	0	1	0
0	1	0	1
0	1	1	0
1	0	0	1
1	0	1	0
1	1	0	1
1	1	1	1

Table 5.10

product terms $xy'z'$, xyz', and xyz, respectively. Therefore we can write the Boolean function described by Table 5.10 as

$$f(x, y, z) = x'yz' + xy'z' + xyz' + xyz$$

and a Boolean expression associated with f is

$$x'yz' + xy'z' + xyz' + xyz$$

Example 2 Applying the procedure just described to the Boolean function defined in Table 5.9 (p. 194), we obtain the corresponding Boolean expression

$$w'xyz + wx'y'z + wx'yz' + wx'yz + wxy'z' + wxy'z + wxyz' + wxyz \quad \square$$

It is easy to translate a Boolean expression written as a sum of product terms into a switching system. For each product term we connect the corresponding switches in series, and then we connect these series connections in parallel. The next example illustrates this procedure.

Example 3

(a) Figure 5.3 shows the switching system that corresponds to the Boolean expression

$$x'yz' + xy'z' + xyz' + xyz$$

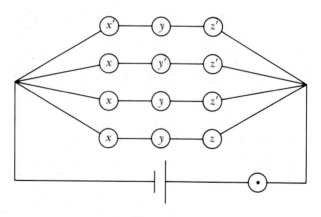

Figure 5.3

(b) Figure 5.4 shows the switching system that corresponds to the Boolean expression

$$wx'yz + w'x'y'z + wxy'z \quad \square$$

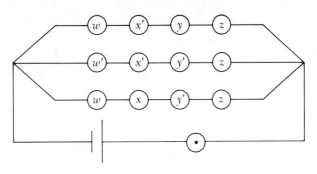

Figure 5.4

To conclude this section we introduce a way to simplify diagrams such as those in Figures 5.3 and 5.4. In a system such as the one illustrated in Figure 5.5, we assume that the light will be activated if current can flow between A and B. Otherwise, the light will remain off. In other words, the switching system is closed if current flows between A and B, and the system is open otherwise. Therefore the diagram in Figure 5.5 is equivalent to the diagram in Figure 5.3.

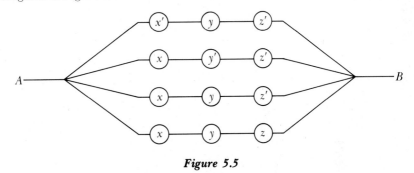

Figure 5.5

Similarly, the diagram in Figure 5.6 corresponds to the switching system in Figure 5.4.

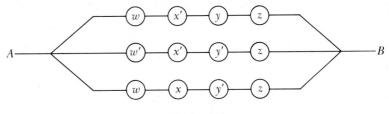

Figure 5.6

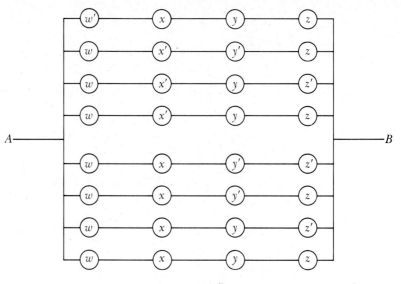

Figure 5.7

Finally, we observe that since the Boolean function defined by Table 5.9 (p. 194) can be expressed as

$$w'xyz + wx'y'z + wx'yz' + wx'yz + wxy'z' + wxy'z + wxyz' + wxyz$$

it follows that we can use the system in Figure 5.7 as a design for the electronic voting machine described earlier in this section.

As you might guess, the system drawn in Figure 5.7 is not the simplest possible circuit corresponding to the given function. The simplicity of a system often measures its cost and realiability. We will spend the remainder of this chapter determining how to make switching systems simpler.

5.2   Problems

In Problems 1–4 use the method of this section to obtain a Boolean expression that corresponds to the given Boolean function.

1.

x	y	$f(x, y)$
0	0	0
0	1	1
1	0	0
1	1	1

2.

x	y	$f(x, y)$
0	0	1
0	1	0
1	0	0
1	1	1

3. x	y	z	$f(x, y, z)$
0	0	0	0
0	0	1	1
0	1	0	1
1	0	0	0
0	1	1	1
1	0	1	0
1	1	0	1
1	1	1	0

4. x	y	z	$f(x, y, z)$
0	0	0	0
0	0	1	1
0	1	0	0
1	0	0	0
0	1	1	0
1	0	1	1
1	1	0	0
1	1	1	1

In Problems 5–8 diagram a switching system that corresponds to the given Boolean function.

5. x	y	$f(x, y)$
0	0	1
0	1	1
1	0	0
1	1	1

6. x	y	$f(x, y)$
0	0	1
0	1	0
1	0	0
1	1	1

7. x	y	z	$f(x, y, z)$
0	0	0	1
0	0	1	1
0	1	0	0
1	0	0	1
0	1	1	1
1	0	1	0
1	1	0	0
1	1	1	0

8. x	y	z	$f(x, y, z)$
0	0	0	0
0	0	1	1
0	1	0	0
1	0	0	0
0	1	1	1
1	0	1	1
1	1	0	0
1	1	1	1

In Problems 9–11 diagram a switching system that corresponds to the given Boolean expression.

9. $xy + y' + z$

10. $x'yz + xy'z + xyz'$

11. $xy + yz + xz$

In Problems 12–15, use Theorem 5.1 to write the given Boolean expression as a sum of product terms and then diagram the corresponding switching system.

12. $x(y'z + z)$

13. $(x + y')(y + z)$

14. $(x' + xy)(x + z) + (x + z)(y' + xz)$ (*Hint*: Note that $aa = a$)

15. $(x'y + z)(x + yz) + (x' + y)z'$ (*Hint*: Note that $aa = a$)

Thus far, given a Boolean function or expression, we have seen how to describe a corresponding switching system. It is also useful to be able to describe a given switching system by a Boolean expression. We can do this by identifying which switches are connected in series (these result in a product term) and which switches are connected in parallel (these result in a sum). For instance, the switching system below corresponds to the Boolean expression $(x' + y)z$.

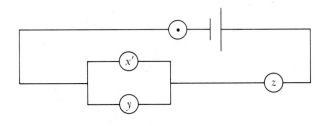

In Problems 16–20 find a Boolean expression that describes the given switching system.

16.

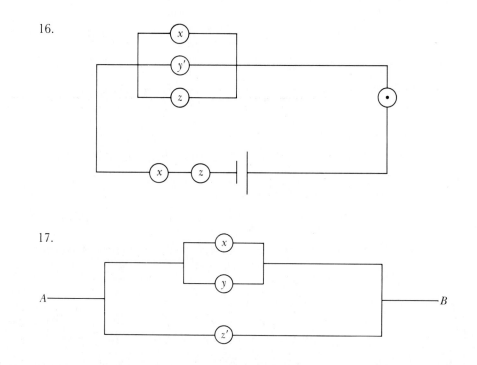

17.

18.

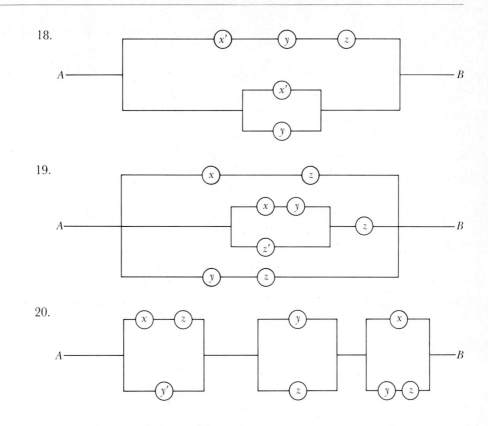

19.

20.

AN INTRODUCTION
5.3 TO SIMPLIFYING BOOLEAN EXPRESSIONS

In the preceding section we showed that Figure 5.8 (a duplicate of Figure 5.5) describes a switching system that corresponds to the Boolean expression

$$x'yz' + xy'z' + xyz' + xyz \qquad (1)$$

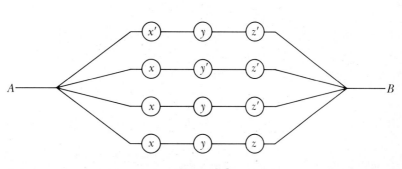

Figure 5.8

We obtained this expression from the Boolean function f defined by Table 5.11 below.

x	y	z	$f(x, y, z)$
0	0	0	0
0	0	1	0
0	1	0	1
0	1	1	0
1	0	0	1
1	0	1	0
1	1	0	1
1	1	1	1

Table 5.11

Expression (1) is not the simplest rendition of the Boolean function f, however. By applying Theorem 5.1 to (1), we can simplify this expression and therefore simplify the switching system:

$$x'yz' + xy'z' + xyz' + xyz$$
$$= x'yz' + xy'z' + xy(z' + z) \qquad \text{(part (f) of Theorem 5.1)}$$
$$= x'yz' + xy'z + xy(1) \qquad \text{(part (i) of Theorem 5.1)}$$
$$= x'yz' + xy'z + xy \qquad \text{(part (h) of Theorem 5.1)}$$

Figure 5.9 describes the system that corresponds to the last expression

$$x'yz' + xy'z + xy$$

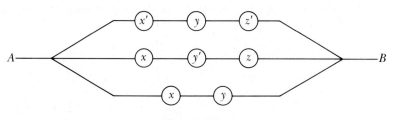

Figure 5.9

Note that this system is simpler than the one given in Figure 5.8. Since we have shown that

$$x'yz' + xy'z' + xyz' + xyz = x'yz' + xy'z + xy$$

it follows that both of these systems describe the Boolean function defined by Table 5.11.

We say that a Boolean expression E in the variables $x_1, x_2, \ldots, x_n$ *represents* a Boolean function f in $x_1, x_2, \ldots, x_n$ if whenever values are assigned to the variables so that $E = 0$, then $f(x_1, x_2, \ldots, x_n) = 0$, and if values are assigned to the variables so that $E = 1$, then $f(x_1, x_2, \ldots, x_n) = 1$. From the previous discussion we see that many Boolean expressions can represent the same Boolean function. This observation leads us to the idea of equivalent Boolean expressions.

Definition 5.5 Two Boolean expressions are *equivalent* if they represent the same Boolean function.

When two Boolean expressions E and F represent the same Boolean function, we shall designate their equivalence by $E = F$.

Example 1 The following steps (which you should verify using Theorem 5.1 and the fact that $aa = a$) show that the Boolean expressions

$$xwx'y + wxy'z + wxyz + w'y'z + wy'z + (x + y)(x + z)$$

and

$$wxz + z + x$$

are equivalent:

$$xwx'y + wxy'z + wxyz + w'y'z + wy'z + (x + y)(x + z)$$
$$= 0 + wxz(y' + y) + (w' + w)y'z + x + yz$$
$$= wxz + y'z + x + yz$$
$$= wxz + (y' + y)z + x$$
$$= wxz + z + x \qquad \square$$

In the next example we use a table to determine whether two Boolean expressions are equivalent.

Example 2 We use Table 5.12 to show that the Boolean expressions $x + y$ and $(x + y)(x' + y) + x$ are equivalent. (You may use Tables 5.4, 5.5, and 5.6 to verify the entries in Table 5.12.) Since the third and fourth columns coincide, it follows that the expressions $x + y$ and $(x + y)(x' + y) + x$ are equivalent. $\qquad \square$

x	y	$x + y$	$(x + y)(x' + y) + x$
0	0	0	0
1	0	1	1
0	1	1	1
1	1	1	1

Table 5.12

Ideally, we would like to determine the Boolean expression that is "simplest" in terms of the switching system it represents. This, however, is not possible because there is no single way to define "simplest"—simplicity often depends on the physical constraints of the switching system itself.

In Definition 5.7 we give a commonly used criterion for resolving this "minimization" problem. First, however, we introduce the following notion.

Definition 5.6   Let E be a Boolean expression on the Boolean algebra S. A *literal* of E is any variable or its complement that appears in E.

Example 3   The Boolean expression

$$xy + x'yz + x'$$

includes literals x, y, x', and z.   □

Definition 5.7   Let f be a Boolean function on the Boolean algebra S. A *minimal representation* of f is a Boolean expression E that represents f and satisfies the following conditions:

    a. E is a sum of product terms.

    b. If F is any other sum of product terms in S that represents f, then the number of product terms occurring in F is greater than or equal to the number of product terms that occur in E.

    c. If F is any other sum of product terms in S that represents f, and if the number of product terms of F is equal to the number of product terms of E, then the total number of literals found in F is greater than or equal to the total number of literals found in E (each literal is counted the number of times it appears).

Thus, to obtain a minimal representation of f, we must first minimize the number of product terms and then minimize the number of literals used.

Example 4   The following sum of product terms in S,

$$E = wy' + wz + w'xz' + w'xy$$

is a minimal representation of the Boolean function

$$f(w, x, y, z) = wx'y'z' + w'xy'z' + wx'y'z + w'xyz' + wxy'z'$$
$$+ wx'yz + wxy'z + w'xyz + wxyz$$

(We will show this is true in Section 5.8.) Because it is the minimal representation, by **b.** of Definition 5.7 any other "sum of products" representation of f must contain at least four product terms. If any "sum of products" representation of f has four product terms, then by **c.** of the same definition it must have at least ten literals.   □

Example 5   It can be shown that the Boolean expression
$$F = y'z' + xy'z$$
is a minimal representation of the Boolean function
$$f(x, y, z) = x'y'z' + xy'z' + xy'z$$
In Problem 12 you are asked to verify that the Boolean expression
$$G = y'(z' + xz)$$
is equivalent to F. Although G contains fewer literals than does F, it is not a candidate for a minimal representation of f: It is not a sum of product terms, so it does not fulfill **a.** of Definition 5.7.

For the remainder of this chapter we will develop techniques to determine a minimal representation of a given Boolean function. The following theorem gives some elementary results used to develop these techniques.

Theorem 5.8   If a and b are elements of the Boolean algebra S, then

(a) $a + a = a$

(b) $aa = a$

(c) $ab + ab' = a$

(d) $aa' = 0$

Proof: (a) Since $S = \{0, 1\}$ we need only consider the cases $a = 0$ and $a = 1$. By the definition of addition in S, we have $0 + 0 = 0$ and $1 + 1 = 1$, and this establishes part (a).

We can establish parts (b) and (c) in a similar manner (see Problems 6 and 7). Part (d) simply restates part (j) of Theorem 5.1.   ■

Observe that parts (a) and (b) of the above theorem indicate that no particular product term need be repeated in a sum, and that no literal need be repeated in a product term. This allows us to immediately simplify many expressions.

Example 6   Using Theorem 5.8 we see that the following Boolean expressions on S are all equivalent:

(a) $xy + xy + zzz + xx' + xy' + xy'$

(b) $xy + xy' + zz$ Use parts (a), (b), and (d)

(c) $x + zz$ Use part (c)

(d) $x + z$ Use part (b)

Expressions (c) and (d) both contain two product terms; however, expression (d) has only two literals while expression (c) has three. It can be shown that expression (d) is a minimal representation of the Boolean function

$$f(x, y, z) = xy + xy + zzz + xx' + xy' + xy' \qquad \Box$$

Theorem 5.8 holds for all Boolean algebras, not just for the Boolean algebra S. To establish (a) for the general case, we can use Definition 5.2 to obtain the following equalities:

$$x \oplus x = 1 * (x \oplus x) = (x \oplus x') * (x \oplus x) = x \oplus (x * x') = x \oplus 0 = x$$

Parts (b) and (c) may be proven in a similar manner (see Problems 13 and 14).

5.3   Problems

1. List the variables and literals found in the following Boolean expressions.

   (a) $xy'x + xz' + y$ (b) $wx'(x + y) + wz + z$

   In Problems 2–5 construct tables to show that the given pair of Boolean expressions on the Boolean algebra S are equivalent.

2. $x; xy + xy'$ 3. $(x + y)(x + z); x + yz$

4. $x(y + z); xy + xz$ 5. $xx' + yz; yz$

6. Prove part (b) of Theorem 5.8.

7. Prove part (c) of Theorem 5.8.

   In Problems 8–11 use Theorems 5.1 and 5.8 to show that the following Boolean expressions on S are equivalent.

8. $x; xy + xy'$ 9. $x(y + z); xy + xz$

10. $(x + y)(x + z); x + yz$ 11. $xx' + yz; yz$

12. Show that expressions F and G in Example 5 are equivalent.

13. Prove part (b) of Theorem 5.8 for any Boolean algebra by using Definition 5.2 to verify the following equalities:

$$x * x = 0 \oplus (x * x) = (x * x') \oplus (x * x) = x * (x' \oplus x) = x * 1 = x$$

14. Prove part (c) of Theorem 5.8 for any Boolean algebra.

5.4   KARNAUGH MAPS

There are several methods for minimizing Boolean expressions. In Sections 5.6–5.8 we will examine the Quine–McCluskey procedure, which provides a computer-implementable approach to this problem. In this section, however, we briefly examine another approach, the Karnaugh map, which is a pictorial method that is relatively easy to carry out without a computer. We can use this method to minimize Boolean expressions that contain no more than six or seven variables. For larger problems, a procedure such as the Quine–McCluskey method is more effective.

The Karnaugh map method is based largely on the identity

$$Aq + Aq' = A \tag{1}$$

which follows immediately from part (c) of Theorem 5.8. In (1) q is a literal and A represents a product term. For example, we have

$$w'xyz + w'xy'z = w'xz$$

where, in this case, $A = w'xz$ and $q = y$.

The concept of adjacency is important for Karnaugh maps.

Definition 5.9   If two product terms differ in exactly one literal, and if this literal appears in one term and its complement appears in the other, then the two product terms are said to be *adjacent*.

Example 1   The product terms $w'xyz'$ and $w'x'yz'$ are adjacent, as are the product terms wxy and wxy'. The product terms $w'xy'z$ and $wxyz$ are not adjacent because they differ in more than one literal.   □

It follows from (1) and the definition of adjacency that we can "reduce" a sum of adjacent product terms to a single product term containing one less literal. For instance, the product terms $w'xyz'$ and $w'x'yz'$ are adjacent, and by (1) we have

$$w'xyz' + w'x'yz' = w'yz'$$

Note that the term $w'yz'$ consists precisely of the literals common to both of the product terms $w'xyz'$ and $w'x'yz'$.

We are now ready to describe Karnaugh maps for certain Boolean expressions. We shall describe these maps for sums of product terms in four variables and introduce the necessary modifications for dealing with other Boolean expressions in the Problem set.

To obtain a Karnaugh map, we use a 4×4 table divided into sixteen boxes. Each box corresponds to a product term of four literals. In Figure 5.10 we indicate the boxes that correspond to the product terms $wx'yz$ and $w'x'y'z'$.

The most important observation to make about such a table is that adjacent boxes (either vertically or horizontally) represent adjacent product terms as defined in Definition 5.9. Moreover, a box in the top row of the table is adjacent (in the sense of Definition 5.9) to the box in the bottom row of the same column, and similarly a box in the left-hand column is adjacent to the box

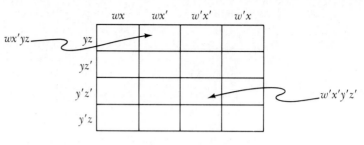

Figure 5.10

in the right-hand column of the same row. For example, the product terms corresponding to the boxes indicated in Figure 5.11(a) are adjacent, as are the product terms corresponding to the boxes indicated in Figure 5.11(b).

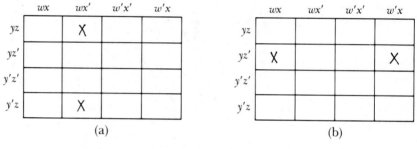

Figure 5.11

To obtain the Karnaugh map for the Boolean expression

$$E = w'x'yz + wxyz' + w'xyz' + wx'y'z' + wxy'z + wx'y'z \qquad (2)$$

we fill in the table as follows. We represent each product term P in E by placing a 1 in the box that corresponds to P. Figure 5.12 is the Karnaugh map for the Boolean expression E.

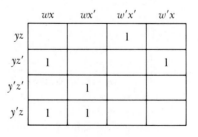

Figure 5.12

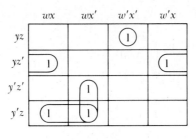

Figure 5.13

Next we see how we can use this Karnaugh map to simplify E. First we circle adjacent 1's in the Karnaugh map, as in Figure 5.13. We also circle the "isolated" 1 that corresponds to the product term $w'x'yz$.

Each circled *pair* of 1's represents two product terms that differ in exactly one literal. By identity (1), therefore, we can simplify each such pair of product terms to a single product term. Any unpaired 1 represents a term that will remain unchanged in the final simplified expression.

Applying (1), we can reduce the adjacent product terms in Figure 5.13 as follows: $wxyz'$ and $w'xyz'$ become xyz'; $wxy'z$ and $wx'y'z$ become $wy'z$; and $wx'y'z'$ and $wx'y'z$ become $wx'y'$. Thus we obtain the Boolean expression

$$w'x'yz + xyz' + wy'z + wx'y' \tag{3}$$

which is simpler than, but equivalent to, (2). In fact, it can be verified that (3) is a minimal representation of (2).

Thus by covering all of the 1's in the Karnaugh map of expression (2) we have obtained a minimal representation of (2). Generally, however, covering all of the 1's may be complicated by multiple adjacencies. In the following example, an adjacency circle made of four 1's in a Karnaugh map leads to a single product term that has two fewer literals than each of the terms represented by the 1's.

Example 2   The circled 1's in the Karnaugh map in Figure 5.14 correspond to the sum of product terms

$$w'x'yz' + w'xyz' + w'x'y'z' + w'xy'z' \tag{4}$$

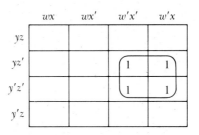

Figure 5.14

Applying identity (1) twice to this sum, we obtain

$$w'x'yz' + w'xyz' + w'x'y'z' + w'xy'z' = w'yz' + w'y'z' = w'z' \qquad (5)$$

Note that w' and z' are precisely the two variables common to all four product terms in (4). $\qquad\square$

A similar analysis shows that the circled 1's in Figure 5.15(a) correspond to four product terms that we can reduce to a single product term wx'. Similarly, we can "reduce" the circled 1's in Figure 5.15(b) to obtain the product term xy'.

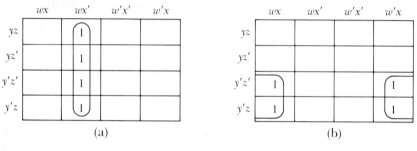

(a)   (b)

Figure 5.15

With a little experimenting you will realize that adjacency circles around 1's in three contiguous boxes do not lead to a *single* product term in fewer variables. In fact, only adjacency circles made up of 2, 4, or 8 boxes having one of the rectangular configurations illustrated in Figure 5.16 will lead to such reductions. "Rectangular" configurations such as those found in Figure 5.17 also lead to single product terms.

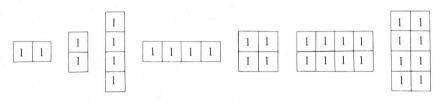

Figure 5.16

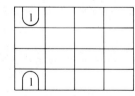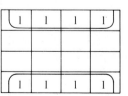

Figure 5.17

In general, when we use Karnaugh maps, we will want to:

a. Use the least number of adjacency circles to cover all of the 1's; since each adjacency circle results in a product term, this will minimize the number of product terms.

b. Include each 1 in the largest possible adjacency circle; since larger groupings result in the use of fewer literals, this will minimize the total number of literals finally used.

Condition **a.** takes precedence over condition **b.** In other words, we will first satisfy condition **a.** and then find an appropriate grouping to best satisfy condition **b.** The next example illustrates this order of operations.

Example 3 The Karnaugh map in Figure 5.18 represents the Boolean expression

$$w'x'yz + wxyz' + wx'yz' + w'x'yz' + wx'y'z' + w'x'y'z' + w'xy'z' + wx'y'z$$

	wx	wx'	$w'x'$	$w'x$
yz			1	
yz'	1	1	1	
$y'z'$		1	1	1
$y'z$		1		

Figure 5.18

Figure 5.19 shows two ways to cover all of the 1's in this map. In Figure 5.19(a) five adjacency circles result in the Boolean expression

$$A = x'z' + wxyz' + wx'y'z + w'x'yz + w'xy'z'$$

which has five product terms.

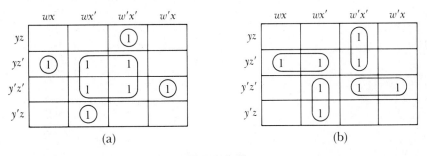

(a)   (b)

Figure 5.19

The selection given in Figure 5.19(b), on the other hand, uses four adjacency circles and results in the Boolean expression

$$B = w'x'y + wyz' + w'y'z' + wx'y'$$

which has four product terms. From the standpoint of minimization, expression B is simpler than expression A because B has one less product term.

□

In the next example we again see that a single Karnaugh map can lead to different simplifications, depending on how we group adjacent 1's.

Example 4 The Boolean expression

$$w'xyz + wx'y'z' + wxy'z + wx'y'z + w'xy'z \tag{6}$$

has the Karnaugh map shown in Figure 5.20.

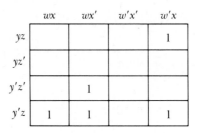

Figure 5.20

We can select adjacent 1's as indicated in Figure 5.21(a) and (b).

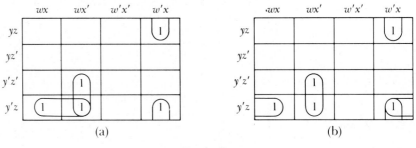

Figure 5.21

From Figure 5.21(a) we simplify (6) as

$$w'xz + wx'y' + wy'z \tag{7}$$

and from Figure 5.21(b) we simplify it as

$$w'xz + xy'z + wx'y' \tag{8}$$

Either (7) or (8) provides a minimization of the Boolean expression (6), so, in this case, it does not matter how we group the adjacent 1's.   □

The following steps indicate how to cover all of the 1's with a minimal number of adjacency circles and, using this minimal number of adjacency circles, how to include each 1 in the largest possible adjacency circle. When we apply these steps to a Karnaugh map of a Boolean expression E, we will obtain a minimal representation of E.

a. Circle all isolated 1's appearing in the Karnaugh map; these 1's represent product terms not adjacent to any other product terms, and they must be included in the final minimal expression.

b. Locate the 1's that are adjacent to only one other 1, and circle each such adjacent pair (these pairs are circled in the Karnaugh map in Figure 5.22).

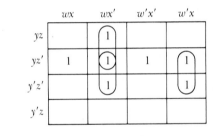

Figure 5.22

c. Circle a rectangular block of four 1's if the block is the unique rectangular block of four 1's that includes some 1 not yet circled during steps **a.** and **b.**

d. Circle a rectangular block of eight 1's if the block is the unique rectangular block of eight 1's that includes some 1 not yet circled during steps **a.**–**c.**

e. Circle the largest possible rectangular blocks of 1's (consisting of two, four, or eight 1's) needed to cover any remaining uncircled 1's.

We illustrate this procedure in the next four examples.

Example 5   We apply these steps to the Karnaugh map (Figure 5.23) of the Boolean expression

$$E = wxyz + wx'yz + w'xyz + wxyz' + wx'yz' + w'x'yz'$$
$$+ wx'y'z' + w'xy'z' + wx'y'z + w'x'y'z$$

From step **a.** we have Figure 5.24(a) and from step **b.** we obtain Figure 5.24(b).

	wx	wx'	w'x'	w'x
yz	1	1		1
yz'	1	1	1	
y'z'		1		1
y'z		1	1	

Figure 5.23

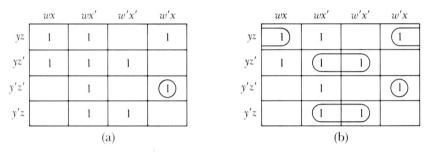

Figure 5.24

Finally, from step **c.** we obtain Figure 5.25.

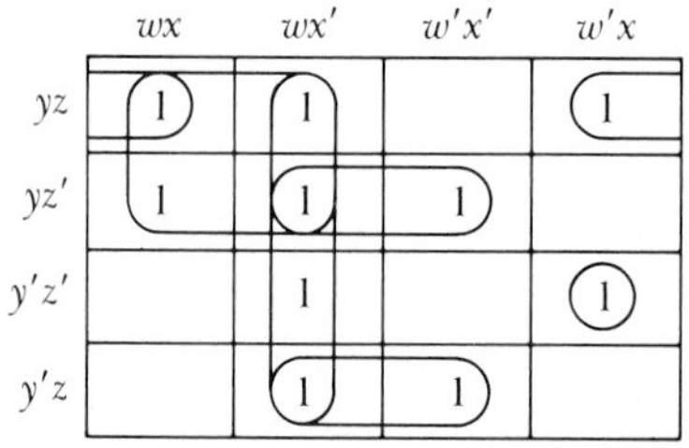

	wx	wx'	w'x'	w'x
yz	1	1		1
yz'	1	1	1	
y'z'		1		1
y'z		1	1	

Figure 5.25

All 1's have now been circled, and the resulting minimal representation of the original Boolean expression E is

$$w'xy'z' + xyz + x'yz' + x'y'z + wy + wx'$$ □

Example 6   We apply the above steps to the Karnaugh map (Figure 5.26) of the Boolean expression

$$E = wxyz + wx'yz' + w'x'yz' + w'xyz' + wx'y'z'$$
$$+ w'x'y'z' + w'xy'z' + wx'y'z$$

From step **a.** we have Figure 5.27(a) and from step **b.** we obtain Figure 5.27(b). Finally, as a result of step **c.** we have Figure 5.28.

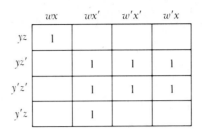

Figure 5.26

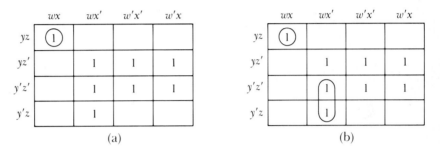

(a)   (b)

Figure 5.27

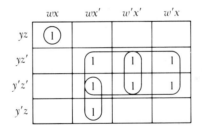

Figure 5.28

We find that a minimal representation of the Boolean expression E is

$$wxyz + wx'y' + x'z' + w'z' \qquad \square$$

Example 7   We apply the above steps to the Karnaugh map (Figure 5.29) of the Boolean expression

$$E = wx'yz + w'xyz + wx'yz' + w'x'yz + w'xyz' + wx'y'z'$$
$$+ w'x'y'z' + w'xy'z' + wx'y'z + w'xy'z$$

There are no 1's circled as a result of steps **a.** and **b.** From step **c.** we have Figure 5.30.

	wx	wx'	$w'x'$	$w'x$
yz		1		1
yz'		1	1	1
$y'z'$		1	1	1
$y'z$		1		1

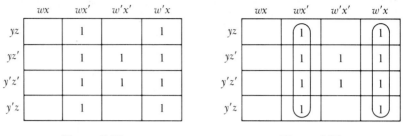

Figure 5.29

Figure 5.30

Then from step **e.** we obtain Figure 5.31.

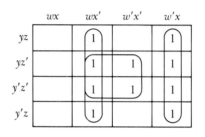

Figure 5.31

Thus a minimal representation of the Boolean expression E is

$$wx' + w'x + x'z'$$ □

Example 8 We apply the above steps to the Karnaugh map (Figure 5.32) of the Boolean expression

$$E = wxyz + wx'yz + w'xyz + wxyz' + wx'yz' + wxy'z' \\ + wx'y'z' + w'x'y'z' + wxy'z + wx'y'z$$

	wx	wx'	$w'x'$	$w'x$
yz	1	1		1
yz'	1	1		
$y'z'$	1	1	1	
$y'z$	1	1		

Figure 5.32

There are no 1's circled as a result of step **a.** From step **b.** we have Figure 5.33.

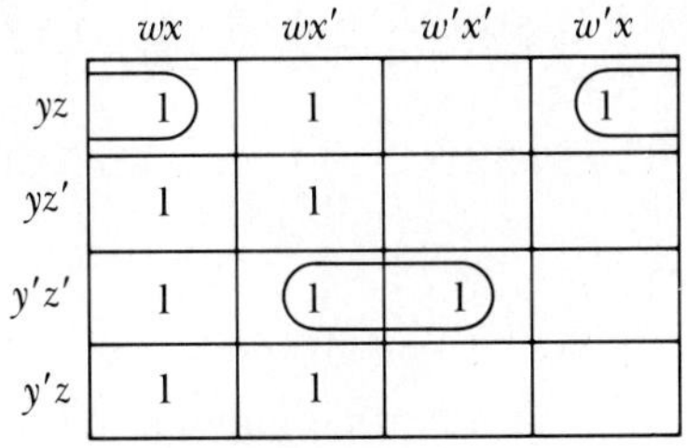

	wx	wx'	$w'x'$	$w'x$
yz	1	1		1
yz'	1	1		
$y'z'$	1	1	1	
$y'z$	1	1		

Figure 5.33

Finally, from step **d.** we obtain Figure 5.34.

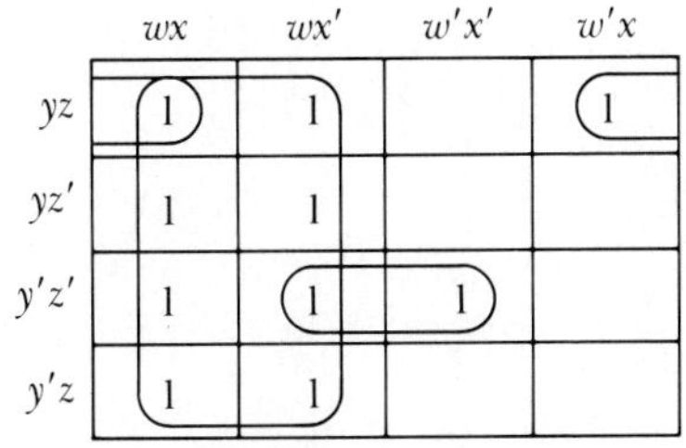

	wx	wx'	$w'x'$	$w'x$
yz	1	1		1
yz'	1	1		
$y'z'$	1	1	1	
$y'z$	1	1		

Figure 5.34

Thus, we see that a minimal representation of the Boolean expression E is

$$xyz + x'y'z' + w \qquad \square$$

5.4 Problems

In Problems 1–3 determine which product terms in the given Boolean expression are adjacent.

1. $wxy + w'xy'z + w'x'y + wx'y$
2. $w'xy'z + wxy'z + wxy'z' + w'xy'z'$
3. $wxy + wxz + wx'z' + xyz + wxy'$

In Problems 4–6 use identity (1) on p. 207 to simplify the given Boolean expression.

4. $wx'yz + wxyz$
5. $wx'yz + w'x'yz + wx'yz' + w'x'yz'$
6. $wxy + w'xz + xy'z' + wxz + xyz'$

In Problems 7–9 describe the Karnaugh map of the given Boolean expression and circle all pairs of adjacent 1's.

7. $w'x'yz + wxyz' + w'xyz' + wx'y'z' + w'x'y'z' + w'x'y'z$

8. $wxyz + w'xyz + wx'yz' + w'x'yz' + w'xyz' + w'x'y'z' + wxy'z$

9. $wxyz + w'xyz + wxy'z + w'xy'z$

10. Use a Karnaugh map to obtain a minimal representation of the Boolean expression given in Problem 7.

11. Use a Karnaugh map to obtain a minimal representation of the Boolean expression given in Problem 8.

12. Use a Karnaugh map to obtain a minimal representation of the Boolean expression given in Problem 9.

13. Use a Karnaugh map to obtain a minimal representation of the Boolean expression:

$$wxyz + wxyz' + wx'yz' + w'x'yz' + w'xyz' + wx'y'z' + w'x'y'z'$$
$$+ wxy'z + wx'y'z + w'x'y'z + w'xy'z$$

14. Use a Karnaugh map to obtain a minimal representation of the Boolean expression:

$$wx'yz + w'x'yz + wx'yz' + w'x'yz' + wxy'z' + wx'y'z' + w'x'y'z'$$
$$+ w'xy'z' + wx'y'z + w'x'y'z + w'xy'z$$

A Karnaugh map for a Boolean expression in three variables can be made with a 2×4 table:

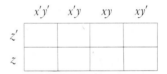

15. Determine which boxes are adjacent (in the sense of Definition 5.9) in the above 2×4 table.

16. Use Karnaugh maps to find minimal representations of the following Boolean expressions.
    (a) $xy'z + xy'z' + x'yz' + xyz$
    (b) $xyz + x'y'z' + xyz' + x'yz'$
    (c) $xyz + x'yz + xy'z' + x'y'z' + xy'z + x'yz'$

A Karnaugh map for a Boolean expression in five variables can be made with the following table:

	vwx	vwx'	$vw'x'$	$vw'x$	$v'w'x$	$v'w'x'$	$v'wx'$	$v'wx$
yz								
yz'		X					X	
$y'z'$								
$y'z$								

In addition to being adjacent to the customary boxes of a 4×4 table, each box in the left half of this table is also adjacent to a box symmetrically located (with respect to the double line) in the right half of the table. Thus, for instance, the boxes we have marked with an X are adjacent.

17. Use the above information and the techniques you have learned so far to find minimal expressions for the following Boolean expressions:

   (a) $vwxyz + vwxyz' + vw'x'yz' + v'w'x'yz' + v'wxyz' + vwx'y'z'$
   $+ vw'x'y'z' + v'w'x'y'z' + v'wx'y'z' + vwx'y'z + vw'x'y'z$
   $+ v'w'x'y'z + v'wx'y'z + v'wxy'z$

   (b) $vwxyz + v'wxyz + vwx'yz' + vw'x'yz' + vw'x'y'z' + vw'xy'z'$
   $+ v'w'x'y'z' + vw'x'y'z + vw'xy'z + v'w'xy'z + v'w'x'y'z$

   (c) $vwx'yz + vw'x'yz + v'w'x'yz + v'wx'yz + vwxyz' + vw'x'yz'$
   $+ v'wxyz' + vwxy'z' + vw'x'y'z' + v'w'x'y'z' + v'wxy'z' + vw'x'y'z$

5.5   PRIME IMPLICANTS

In Section 5.4 we saw that Karnaugh maps provide one method for finding minimal representations of Boolean expressions. In the next few sections we develop another approach to this problem. This approach is computer implementable and more versatile than the Karnaugh maps because we can apply it to Boolean expressions in many variables. The following definitions play a basic role in this approach.

Definition 5.10   Let f and g be Boolean functions in the variables $x_1, x_2, \ldots, x_n$ on the Boolean algebra S. We say that f is an *implicant* of g if whenever $f(x_1, x_2, \ldots, x_n) = 1$, then $g(x_1, x_2, \ldots, x_n) = 1$. If f is an implicant of g, we say that f *implies* g and we write $f \leq g$.

A similar definition holds for Boolean expressions.

Definition 5.11   Let F and G be Boolean expressions in variables $x_1, x_2, \ldots, x_n$ on the Boolean algebra S. We say that F is an *implicant* of G if whenever values are assigned to the variables $x_1, x_2, \ldots, x_n$ so that $F = 1$, then this same assignment of values results in $G = 1$. If F is an implicant of G, we say that F *implies* G, and we write $F \leq G$.

Example 1   Let f and g be Boolean functions in the variables x and y defined by

$$f(x, y) = xy + y$$

and

$$g(x, y) = x'y + xy + xy'$$

From Table 5.13 we see that f is an implicant of g ($f \leq g$), since whenever $f(x, y)$ equals 1, $g(x, y)$ also equals 1.

x	y	$f(x, y) = xy + y$	$g(x, y) = x'y + xy + xy'$
0	0	0	0
0	1	1	1
1	0	0	1
1	1	1	1

Table 5.13

It also follows from Table 5.13 that the Boolean expression $F = xy + y$ implies (is an implicant of) the Boolean expression $G = x'y + xy + xy'$.   □

Example 2   Suppose that

$$E = T_1 + T_2 + \cdots + T_k$$

is a Boolean expression in S and that each T_i is a product term. Then each of the terms T_i implies E, since if we assign values to the literals of T_i so that $T_i = 1$, then for these same assigned values, $E = 1$. For instance, if

$$E = xyz' + x'z + yz$$

then yz is an implicant of E (If $yz = 1$ only, then $E = 1$).   □

If f and g are Boolean functions in the variables $x_1, x_2, \ldots, x_n$, and if $f \leq g$ and $g \leq f$, then $f(x_1, x_2, \ldots, x_n) = 1$ if and only if $g(x_1, x_2, \ldots, x_n) = 1$. Consequently, the functions f and g are equal. Moreover, if f, g, and h are Boolean functions such that $f \leq g$, and $g \leq h$, then $f \leq h$. These observations are sufficiently important to be stated as a theorem.

Theorem 5.12   **a.** If f and g are Boolean functions on the Boolean algebra S, and if $f \leq g$ and $g \leq f$, then $f = g$.

b. If f, g, and h are Boolean functions on the Boolean algebra S, and if $f \leq g$ and $g \leq h$, then $f \leq h$.

Corollary 5.13   **a.** If F and G are Boolean expressions on the Boolean algebra S, and if $F \leq G$ and $G \leq F$, then $F = G$ (F and G are equivalent).

b. If F, G, and H are Boolean expressions on the Boolean algebra S, and if $F \leq G$ and $G \leq H$, then $F \leq H$.

Theorem 5.12 shows that in some respects the "order relation" $\leq$ is analogous to the usual "less than or equal to" relation for the real numbers. However, the analogy between the "implies" relation and the "less than or equal to" relation is not perfect. For example, it is possible that for some Boolean functions f and g, $f \not\leq g$ and $g \not\leq f$. For example, if the Boolean functions f and g defined by $f(x, y) = xy$ and $g(x, y) = x'y$, then (as you are asked to verify in Problem 5) $f \not\leq g$ and $g \not\leq f$.

Definition 5.14   Let P and Q be two Boolean expressions that are product terms. Then Q is said to *cover* P if every literal found in Q is also found in P.

Example 3   The product term $wx'z$ covers the product term $wx'yz$, but $wx'z$ does not cover the product term $wxy'z'$.   □

The term "cover" may seem a bit strange—if anything, its definition may appear to be the opposite of what it should be—but justification for this usage will soon be apparent. The term "is included in" frequently replaces "covers."

The next theorem establishes a relationship between "covers" and "implies."

Theorem 5.15   Suppose that P and Q are product terms on the Boolean algebra S. Then Q covers P if and only if P implies Q.

Proof: If Q covers P, then every literal occurring in Q also occurs in P. From the commutativity of multiplication (see Theorem 5.1) we can write

$$P = QR$$

where R is the product of literals in P that do not occur in Q. Now note that if we choose values for the variables so that $P = 1$, then this same choice of values will also result in $Q = 1$; consequently, it follows from the definition of "implies" that P implies Q.

To establish the converse (that if P implies Q, then Q covers P), we employ a proof by contradiction: We assume that P implies Q, but that Q does not cover P. Under this assumption there is a literal, say x, that appears in Q but does not appear in P. Assign this literal the value 0, and assign to all of the literals in P the value 1. Then we have $Q = 0$ and $P = 1$, which means that P does not imply Q. Since this contradicts our assumption, it must be the case that Q covers P.   ■

To minimize a Boolean expression, we also want to minimize the number of literals. This leads us to the following notion.

Definition 5.16   Suppose that E is a Boolean expression and that P is a product term. Then P is a *prime implicant* of E in case:

    a. P implies E $(P \leq E)$, and

    b. If Q is a product term that covers (is included in) P and $Q \neq P$, then Q does not imply E $(Q \nleq E)$.

Example 4   Suppose that

$$E = x'y'z' + xy'z' + xy'w$$

We show that $P = y'z'$ is a prime implicant of E. To see that P implies E suppose that $P = 1$. This means that $y' = 1$ and $z' = 1$, and therefore, either $x'y'z' = 1$ or $xy'z' = 1$ depending on whether $x = 0$ or $x = 1$. Note that in either of these cases it follows that $E = 1$, and hence, P implies E.

To show that P satisfies **b.** of Definition 5.16, proceed as follows. If Q covers $P = y'z'$ and $P \neq Q$, then either $Q = y'$ or $Q = z'$. Thus we must show that $y' \nleq E$ and $z' \nleq E$. To this end, let $y' = 1$ and let $x = 0$, $z = 1$, and $w = 0$. Then we have $y' = 1$, but $E = 0$, so y' does not imply E. Now let $z' = 1$, and let $y = 1$, $x = 0$, and $w = 0$. Then we have $z' = 1$, but $E = 0$, so z' also fails to imply E. It follows that $P = y'z'$ is a prime implicant of E.   □

The next result illustrates the significance of prime implicants in minimizing Boolean expressions.

Theorem 5.17   Suppose that

$$E = T_1 + T_2 + \cdots + T_k$$

is a Boolean expression on the Boolean algebra S and that each T_i is a product term. Then there are k prime implicants of E—$P_1, P_2, \ldots, P_k$—such that

$$E = P_1 + P_2 + \cdots + P_k$$

and for each i

$$T_i \leq P_i$$

Proof: Since $E = T_1 + T_2 + \cdots + T_k$ we know from Example 2 that for each i, $1 \leq i \leq k$, $T_i \leq E$. If T_i itself is a prime implicant, set $P_i = T_i$; otherwise, eliminate literals from T_i until the removal of a literal results in an expression that does not imply E. The resulting expression P_i is a prime implicant of E. Moreover, since for each i, P_i is included in (covers) T_i, we have $T_i \leq P_i$ for $1 \leq i \leq k$.

We now must show that

$$E = P_1 + P_2 + \cdots + P_k$$

First note that if we assign values to the literals of E so that $E = 1$, then

$$T_1 + T_2 + \cdots + T_k = 1$$

and, consequently, $T_i = 1$ for some i. Since $T_i \leq P_i$, we have $P_i = 1$, and it follows that

$$P_1 + P_2 + \cdots + P_k = 1$$

Therefore, we have shown

$$E \leq P_1 + P_2 + \cdots + P_k \qquad (1)$$

Conversely, if we assign values to literals so that $P_1 + P_2 + \cdots + P_k = 1$, then it must be the case that $P_i = 1$ for at least one i. Since $P_i \leq E$, we have $E = 1$. Therefore,

$$P_1 + P_2 + \cdots + P_k \leq E \qquad (2)$$

and it follows from (1), (2), and Corollary 5.13 that

$$E = P_1 + P_2 + \cdots + P_k$$

as we were to show. ∎

Note that it follows from Theorem 5.17 that we can replace any expression for E involving k summands by a sum of k prime implicants. Thus, only prime implicants are necessary in a minimal representation of E.

5.5 Problems

In Problems 1–4 determine whether or not the Boolean function f implies the Boolean function g.

1. $f(x, y) = x'y$; $g(x, y) = xy' + y$
2. $f(x, y) = xy' + x'y$; $g(x, y) = xy$
3. $f(x, y, z) = x'yz + xz$; $g(x, y, z) = xz' + yz + y'z$
4. $f(w, x, y, z) = wxy + wyz + wx'z + w'xy' + xy'z$;
   $g(w, x, y, z) = wxy + wz + w'xy'$
5. Show that if $f(x, y) = xy$ and $g(x, y) = x'y$, then $f \not\leq g$ and $g \not\leq f$.
6. Find an example, other than that given in Problem 5, of two Boolean functions f and g such that $f \not\leq g$ and $g \not\leq f$.

In Problems 7–9 determine which of the Boolean expressions F are prime implicants of the Boolean expression E.

7. $E = x'y' + xz' + yz$
   (a) $F = x'y'$ (b) $F = xy'$ (c) $F = y'z'$

8. $E = xy + x'y'z' + x'yz'$
   (a) $F = x'z'$ (b) $F = xyz$ (c) $F = xyz'$

9. $E = xyz' + x'yz' + xz + xy'z'$
   (a) $F = x$ (b) $F = y$ (c) $F = yz'$
   (d) $F = xyz'$

5.6 THE QUINE–MCCLUSKEY ALGORITHM: HOW IT WORKS

As we have previously indicated, determining prime implicants of a given Boolean expression is important in finding a minimal representation of the expression. In this section we will describe the Quine–McCluskey algorithm for finding prime implicants. First, however, we introduce the concept of a "canonical sum."

Recall from Theorem 5.8 that $xx = x$ for any literal x, so we do not need to repeat a literal in any product term. Moreover, since $xx' = 0$ for any literal x, it also follows that a literal and its complement will never both appear in a nonzero product term.

Definition 5.18   A *standard product term* in a Boolean expression E in n variables is a product term in E that consists of n literals, in which no literal is repeated and in which a literal and its complement do not both appear.

A Boolean expression that is written entirely as a sum of standard product terms is said to be a *canonical sum*, or a *sum in canonical form.*

Example 1   Suppose E is a Boolean expression in the variables x_1, x_2, x_3, and x_4 defined by

$$E = x_1'x_2x_4 + x_1x_2'x_3x_4' + x_2 + x_1x_2x_3x_4 + x_1x_3x_1'x_4$$

Then $x_1x_2'x_3x_4'$ and $x_1x_2x_3x_4$ are standard product terms while the product terms $x_1'x_2x_4$, x_2, and $x_1x_3x_1'x_4$ are not.   □

Example 2   The Boolean expression

$$wxy'z + wx'yz + w'x'y'z'$$

is in canonical form since all of the product terms consist of four literals, no literal is repeated in any product term, and no product term contains both a literal and its complement.   □

We can write any Boolean expression in canonical form. For instance, if we have

$$E = xyz' + xy' + x \tag{1}$$

then from the identity

$$Aq + Aq' = A$$

we can obtain an equivalent representation of E that consists of standard product terms as follows:

$$
\begin{aligned}
xyz' + xy' + x &= xyz' + xy'(z + z') + x(y + y') \\
&= xyz' + xy'z + xy'z' + xy + xy' \\
&= xyz' + xy'z + xy'z' + xy(z + z') + xy'(z + z') \\
&= xyz' + xy'z + xy'z' + xyz + xyz' + xy'z + xy'z'
\end{aligned}
$$

In addition, observe that any Boolean expression derived from the tabular definition of a Boolean function, as in Section 5.2, will be a canonical sum.

We are now ready to describe the Quine–McCluskey Algorithm. Although this algorithm is quite involved, it can be implemented on a digital computer and is reasonably practical if the number of variables is not too large (less than, say, 15).

Algorithm 5.19   *Quine–McCluskey Algorithm.* To generate the prime implicants of a Boolean expression $E = T(1) + T(2) + \cdots + T(K)$, written in canonical form.

Input: $T(1), T(2), \ldots, T(K)$
Output: $\{U\}$, the set of prime implicants

1. $I \leftarrow 1$
2. $L(I) \leftarrow \{T(1), T(2), \ldots, T(K)\}$

   In the course of this algorithm, several sets are formed. The first is the set $L(1)$, the standard product terms in E.

3. FOR $U \in L(1)$
   a. $F(U) \leftarrow 0$

   Initialize. Eventually $F(U) = 0$ will mark U as a prime implicant.

4. WHILE $|L(I)| \geq 2$ DO

   Perform the following steps as long as $L(I)$ has at least two elements.

   a. $L(I + 1) \leftarrow \varnothing$

   Initialize to prepare for the formation of the next set.

   b. FOR $U \in L(I)$ AND $V \in L(I)$
      1. IF $U = QS$ AND $V = QS'$ THEN
         a. $L(I + 1) \leftarrow L(I + 1) \cup \{Q\}$

   In step b. we check all pairs of elements U and V in $L(I)$. If U and V differ in only one literal (if $U = QS$ and $V = QS'$), then add Q to the next set and flag U and V to indicate that they are not prime implicants.

b. $F(U) \leftarrow 1$
c. $F(V) \leftarrow 1$
c. $I \leftarrow I + 1$ Steps c. and d. initialize to prepare
d. FOR $U \in L(I)$ for formation of the next set.
   1. $F(U) \leftarrow 0$
5. FOR $F(U) = 0$ Each element U in some $L(I)$ with
  a. OUTPUT U $F(U) = 0$ is a prime implicant of E.

In the next section we will examine why this algorithm works. For the moment, however, let us illustrate it with the following example

Example 3   To apply the Quine–McCluskey Algorithm, let

$$F = xyz' + x'yz' + xz + xy'z'$$

be a Boolean expression in the variables x, y, and z. We first convert the product term xz to a standard product term by using the identity

$$xz = xyz + xy'z$$

Then

$$E = xyz' + x'yz' + xyz + xy'z + xy'z'$$

is an equivalent Boolean expression in canonical form. Table 5.14 lists the results of applying the Quine–McCluskey Algorithm to E.

$L(1)$		$L(2)$		$L(3)$	
U	$F(U)$	U	$F(U)$	U	$F(U)$
xyz'	1	yz'	0	x	0
$x'yz'$	1	xy	1		
xyz	1	xz'	1		
$xy'z$	1	xz	1		
$xy'z'$	1	xy'	1		

Table 5.14

The set $L(1)$ consists of the product terms of E. When we consider the first pair of product terms, xyz' and $x'yz'$, we find that

$$xyz' = Qs$$

and

$$x'yz' = Qs'$$

where $Q = yz'$ and $s = x$. Thus, we place $Q = yz'$ in the set $L(2)$ and we set $F(xyz') = 1$ and $F(x'yz') = 1$.

Consideration of the pair of product terms xyz' and xyz shows that

$$xyz' = Qs$$

and

$$xyz = Qs'$$

where $Q = xy$ and $s = z'$. Thus, we add xy to $L(2)$ and we set $F(xyz') = 1$ and $F(xyz) = 1$.

Since the pair of product terms xyz' and $xy'z$ differ in more than one literal, we do nothing with this pair. Comparing all of the remaining pairs of product terms in $L(1)$ yields the values of $F(U)$ for $U \in L(1)$ and determines column $L(2)$. We then apply the same procedure to $L(2)$, and we find that $L(3)$ contains only one element, the literal x.

Since just one element occurs in $L(3)$, the procedure stops, and we see that the only product terms U with $F(U) = 0$ in the various sets $L(1), L(2), L(3)$ are yz' and x. These, then, form the complete set of prime implicants of the Boolean expression E.

5.6 Problems

1. In the following Boolean expression in variables $x_1, x_2,$ and x_3, which terms are standard product terms?

$$x_1'x_2 + x_1x_2'x_3 + x_3 + x_1x_2x_3 + x_1x_1x_3$$

2. In the following Boolean expression in variables $x_1, x_2,$ and x_3, which terms are standard product terms?

$$x_1x_2x_3 + x_1x_3 + x_2x_3' + x_2x_2'x_3$$

In Problems 3–8 find a Boolean expression in canonical form that is equivalent to the given Boolean expression.

3. $x + y$ 4. $x + y + z$ 5. $xy + y$

6. xy 7. $xz + xy' + y'$ 8. $xy + xz$

In Problems 9–14 use the Quine–McCluskey Algorithm to find the prime implicants of the given Boolean expression. (If necessary, first find an equivalent Boolean expression in canonical form.)

9. $xyz' + x'yz + xy'z' + x'yz'$

10. $xyz + x'yz' + x'y'z + xy'z + x'y'z'$

11. $wxy'z + wxyz' + xy'z + w'xyz'$

12. $x + yz + x'yz$

13. $w'x'y'z + w'x'yz + w'xyz' + wxyz' + w'xyz + wxyz' + wxyz$

14. $wxy'z + wx'yz' + wxy'z + w'x'yz + w'xyz' + w'xy'z + wxyz' + wx'y'z$

THE QUINE–MCCLUSKEY
*5.7 ALGORITHM: WHY IT WORKS

To see why the Quine–McCluskey Algorithm works, we use the following idea.

Definition 5.20   Let P be a product term in some of the variables $x_1, x_2, \ldots, x_n$. A product term Q in *all* of these variables is a *completion* of P if P is included in (covers) Q.

> **Example 1**   Let x_1, x_2, x_3, x_4, and x_5 be variables in the Boolean algebra S, and let $P = x_1 x_3' x_4$. Then $Q = x_1 x_2 x_3' x_4 x_5'$ is a completion of P, as is $Q = x_1 x_2' x_3' x_4 x_5'$. The product terms $x_1 x_2 x_3 x_4 x_5'$ and $x_1' x_2' x_3' x_4 x_5$ are not completions of P.   □

The following is the principal theorem used to establish the validity of the Quine–McCluskey Algorithm.

Theorem 5.21   Suppose that E is a sum of standard product terms in S in the variables $x_1, x_2, \ldots, x_n$. Let P be a product term in some of these variables. Then P implies E if and only if *each* completion of P (in the variables $x_1, x_2, \ldots, x_n$) is a summand of E.

Proof: Suppose that P implies E, and let Q be a completion of P. We will suppose that Q is not a product term found as a summand in E and will arrive at a contradiction. To each literal in Q, assign the value 1. Since P is included in Q, P clearly takes on the value 1, and since P implies E, it follows that E must also take on the value 1. But this gives us the desired contradiction: Since Q is not a product term in E, and since E is in canonical form, we have that each summand of E must contain at least one literal that is the complement of the corresponding literal in Q. Thus, each product term in E has value 0, and hence E itself is 0. But we also have that E is equal to 1. This contradiction leads us to conclude that Q is a standard product term in E.

Conversely, let P be a product term and suppose that *each* completion of P is found as a summand of E. We show that P implies E. Suppose, then, that $P = 1$; in other words, suppose that each literal in P has been assigned the value 1. Then it must be the case that *some* completion of P also has the value 1 (Why?). Since *each* completion of P is a summand of E, it follows that $E = 1$, and, therefore, P implies E.   ∎

We are now in a position to establish the validity of the Quine–McCluskey Algorithm.

Theorem 5.22   Suppose that we apply the Quine–McCluskey Algorithm to a Boolean expression E in canonical form.

(a) If P is a product term occurring in any set $L(I)$ generated by the Quine–McCluskey Algorithm, then P implies E.

(b) A product term P is a prime implicant of E if and only if $P \in L(I)$ for some set $L(I)$ generated by the Quine–McCluskey Algorithm and $F(P) = 0$.

Proof: (a) The proof is by induction on the number of the set in which P appears. If P appears in $L(1)$, then P is a standard product term in E and so must imply E. Assume now that every member in $L(I)$ implies E, and suppose that P appears in $L(I + 1)$. The product term P appears in $L(I + 1)$ only if for some literal s, the product terms sP and $s'P$ appear in $L(I)$. Observe that if P takes on the value 1, then either sP or $s'P$ must also take on this value. Since both sP and $s'P$ are in $L(I)$, it follows from the induction assumption that they both imply E, and hence, the value of E is also 1. Thus we have that whenever P is 1, E is also 1, and, consequently P implies E.

(b) Suppose that the product term P is a prime implicant of E. Since by Theorem 5.21 each completion of P is a summand of E, it follows from the Quine–McCluskey procedure that P must eventually appear in some set generated by this algorithm. To see this, suppose, for example, that P does not contain the two variables a and b. Then all of the following completions of P must appear in E and so in the set $L(1)$:

$$abP \qquad ab'P$$
$$a'bP \qquad a'b'P$$

By the Quine–McCluskey procedure, $L(2)$ must contain bP, $b'P$, aP, and $a'P$, and it follows that P itself will appear in $L(3)$. We can use an inductive argument on the number of variables missing from P to establish the general case that each prime implicant of E must appear in some set $L(I)$ (Problem 1).

We must now show that $F(P) = 0$. If $F(P) = 1$, then it would be the case there is a literal s (or s') in P and a product term $\hat{P}$ included in P such that

a. $P = s\hat{P}$ or $P = s'\hat{P}$, and

b. $s\hat{P}$ and $s'\hat{P}$ are both in the same list, say $L(K)$.

Thus, $\hat{P}$ will be in $L(K + 1)$ and from (a) of this theorem it follows that $\hat{P}$ implies E. However, this contradicts the assumption that P is a prime implicant. Consequently, we conclude that $F(P) = 0$.

Now suppose that P appears in one of the sets generated by the Quine–McCluskey Algorithm and that $F(P) = 0$. Then by part (a) of this theorem, P implies E. If P is not a prime implicant, then there is a product term Q included in P that also implies E. But this means that each completion of Q is a product term appearing as a summand in E. As before, it follows from the Quine–McCluskey procedure itself that Q appears in some set generated by this algorithm. Moreover, the appearance of Q in some set will result (perhaps after several steps) from product terms $s\hat{P}$ and $s'\hat{P}$, where $P = s\hat{P}$ or $P = s'\hat{P}$. But $\hat{P}$ will occur in the set succeeding the set containing $s\hat{P}$ and $s'\hat{P}$. Since both $F(s\hat{P}) = 1$ and $F(s'\hat{P}) = 1$, it must be the case that $F(P) = 1$, contradicting our assumption that $F(P) = 0$. Thus we conclude that P is a prime implicant of E.   ∎

5.7   Problem

1. Use an inductive argument to show that if P is a prime implicant of a Boolean expression E written in canonical form, then P will appear in some set $L(I)$ generated by the Quine–McCluskey Algorithm.

5.8   THE COVERING PROBLEM

Now that we have a method for finding prime implicants, we can turn to the fundamental problem of this chapter: to obtain a minimal equivalent of a Boolean expression where "minimal" is defined in Definition 5.7.

In Theorem 5.7 we showed that only prime implicants are necessary in a minimal representation of a Boolean expression. The following result will help us choose these prime implicants.

Theorem 5.23   Suppose that

$$E = T_1 + T_2 + \cdots + T_m$$

is a Boolean expression in canonical form, and let $P_1, P_2, \ldots, P_k$ be prime implicants of E. Then

$$E = P_1 + P_2 + \cdots + P_k$$

if and only if for each i, $1 \le i \le m$, there is a j, $1 \le j \le k$, such that P_j covers T_i. (In other words, each T_i is covered by some P_j.)

Proof: Suppose first that

$$E = P_1 + P_2 + \cdots + P_k$$

and let T_i be given. We must show that for some j, $1 \le j \le k$, P_j covers T_i. Assign each literal in T_i the value 1. Since T_i is a standard product term, this assignment determines the value of each literal in E. Moreover, as a consequence of this assignment we have $T_i = 1$ and it follows that

$$1 = T_1 + T_2 + \cdots + T_m = E = P_1 + P_2 + \cdots + P_k \qquad (1)$$

From (1) we see that $P_j = 1$ for some j, $1 \le j \le k$. Since we determine the values of the literals found in P_j by the values assigned to the literals of T_i, it follows that if $T_i = 1$, then $P_j = 1$. Thus, T_i implies P_j, and by Theorem 5.15 we have that P_j covers T_i, which concludes the first half of the proof of this theorem.

Now suppose that given any T_i, $1 \leq i \leq m$, there is a prime implicant P_j, $1 \leq j \leq k$, such that P_j covers T_i. We show under this assumption that

$$E = P_1 + P_2 + \cdots + P_k$$

We first establish that

$$P_1 + P_2 + \cdots + P_k \leq E \tag{2}$$

Note that if $P_1 + P_2 + \cdots + P_k = 1$, then $P_r = 1$ for some r, $1 \leq r \leq k$. Since P_r implies E, it follows that $E = 1$. Thus, $P_1 + P_2 + \cdots + P_k$ implies E, and we have established (2).

To conclude, we show that

$$E \leq P_1 + P_2 + \cdots + P_k \tag{3}$$

Suppose that we assign values to the literals of E so that $E = 1$. Then since

$$E = T_1 + T_2 + \cdots + T_m$$

it follows that $T_i = 1$ for some i, $1 \leq i \leq m$. By our hypothesis there is a prime implicant P_j, $1 \leq j \leq k$, which covers T_i. Therefore, by Theorem 5.15, T_i implies P_j, and hence $P_j = 1$. Consequently,

$$P_1 + P_2 + \cdots + P_k = 1$$

and this establishes (3). Finally, from (2), (3), and Corollary 5.13 we have

$$E = P_1 + P_2 + \cdots + P_k$$

the desired result.   ■

Example 1   We can use the Quine–McCluskey Algorithm to show that $P_1 = x'yz'$, $P_2 = wxy'$, and $P_3 = wy'z'$ are among the prime implicants of the Boolean expression

$$E = w'x'yz' + wx'y'z' + wxy'z' + wxy'z$$

Note that each product term in E is covered by at least one of the prime implicants P_1, P_2, and P_3. Therefore, by Theorem 5.23

$$E = x'yz' + wxy' + wy'z'   \qquad \square$$

Because of Theorem 5.23, we see that to obtain a minimal representation of a Boolean expression E, we must find the minimal number of prime implicants that will cover all of the product terms in the canonical representation of E. This problem is a special case of what is commonly referred to as the "covering problem."

It may be the case that certain prime implicants of a Boolean expression E *must* appear in any sum of prime implicant terms that is equivalent to E. These prime implicants are called essential prime implicants.

Definition 5.24   Suppose that E is a Boolean expression in canonical form. A prime implicant P of E is said to be an *essential prime implicant* of E if P is the only prime implicant that covers (is included in) some standard product term in E.

You can verify that the following result is an immediate consequence of this definition and Theorem 5.23.

Theorem 5.25   Suppose that

$$E = P_1 + P_2 + \cdots + P_n$$

is a Boolean expression such that each P_i is a prime implicant of E. If P is an essential prime implicant of E, then $P = P_i$ for some i, $1 \le i \le n$.

Theorem 5.25 is significant because it tells us that *every* essential prime implicant must appear in *any* sum of prime implicants that is equivalent to the given Boolean expression E. However, as we see in the next example, all of the essential prime implicants may not be *sufficient* to cover all of the product terms occurring in E.

Example 2   Applying the Quine–McCluskey Algorithm to the Boolean expression

$$E = w'x'yz' + wx'y'z' + wxy'z' + wx'yz' + wxy'z$$

yields the sets

	$L(1)$			$L(2)$	
U		$F(U)$	U		$F(U)$
$w'x'yz'$		1	$x'yz'$		0
$wx'y'z'$		1	$wy'z'$		0
$wxy'z'$		1	$wx'z'$		0
$wx'yz'$		1	wxy'		0
$wxy'z$		1			

and hence the prime implicants $x'yz'$, $wy'z'$, $wx'z'$, and wxy'. Note that wxy' is the only prime implicant that covers $wxy'z$, so wxy' is an essential prime implicant. Similarly $x'yz'$ is an essential prime implicant since it is the only prime implicant covering the term $w'x'yz'$. The remaining prime implicants are not essential.

Observe that neither of the essential prime implicants (wxy' and $x'yz'$) covers $wx'y'z'$. Thus, by Theorem 5.23 the sum of the essential prime implicants is not equivalent to E. Consequently, to represent E we must use at least one nonessential prime implicant to cover the product term $wx'y'z'$. Since it is clear that we could use either $wx'z'$ or $wy'z'$ for this purpose, we find that either

$$wxy' + x'yz' + wx'z'$$

or

$$wxy' + x'yz' + wy'z'$$

gives us the desired minimal representation of E.   □

The following steps summarize a procedure for obtaining a minimal representation of a Boolean expression E in canonical form.

a. Apply the Quine–McCluskey Algorithm to E to obtain a list of the prime implicants of E.

b. Determine the essential prime implicants. (These must appear in any sum of prime implicants that is equivalent to E.)

c. In addition to the essential prime implicants, include sufficient (but a minimal number of) additional prime implicants so that at least one prime implicant covers each product term in E.

To develop a more systematic way to determine essential prime implicants and covers for each of the standard product terms of a Boolean expression E, it is convenient to use a prime implicant table. We construct such a table by comparing the prime implicants of E with the standard product terms in E. From the Quine–McCluskey Algorithm we determine that the prime implicants of

$$E = w'x'yz' + wx'y'z' + wxy'z' + wx'yz' + wxy'z \qquad (4)$$

are $x'yz'$, $wy'z'$, $wx'z'$, and wxy'. Table 5.15 is the prime implicant table associated with E.

In this table, an X lies at the intersection of a row and a column whenever the prime implicant to the left covers the product term at the top. If only one X occurs in a given column, then the prime implicant this X represents is

	$w'x'yz'$	$wx'y'z'$	$wxy'z'$	$wx'yz'$	$wxy'z$
$x'yz'$	⊗			X	
$wy'z'$		X	X		
$wx'z'$		X		X	
wxy'			X		⊗

Table 5.15

essential: It is the only prime implicant that covers the product term at the top of the column. In Table 5.15 the circled X's indicate the two essential prime implicants, $x'yz'$ and wxy'.

Rows with circled X's are called *essential* rows. Note that any other X's in the essential rows represent standard product terms that are also covered by the essential prime implicant of that row. For example, in Table 5.15 the first row is essential, and there is another X in that row in the fourth column. The prime implicant $x'yz'$ also covers $wx'yz'$. Similarly, the fourth row is essential, and its uncircled X indicates that the prime implicant wxy' covers the standard product term $wxy'z'$. By drawing lines through all essential rows and through those columns that contain X's found in these rows, we obtain Table 5.16. These lines show that we have accounted for all of the essential prime implicants and have identified which standard product terms they cover.

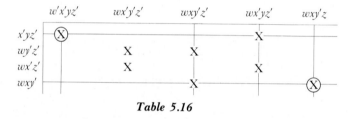

Table 5.16

It is apparent, however, that there is no essential prime implicant to cover the standard product term $wx'y'z'$. We must therefore look to the other prime implicants, $wy'z'$ and $wx'z'$, that cover this product term. Since both of these prime implicants contain the same number of literals, we can use either of them to obtain a minimal representation of E. Finally, combining all of the information from Table 5.16, we obtain both

$$x'yz' + wy'z' + wxy'$$

and

$$x'yz' + wx'z' + wxy'$$

as minimal representations of the Boolean expression (4).

Example 3 We will use a prime implicant table to obtain a minimal representation of the Boolean expression

$$E = wx'y'z' + w'xy'z' + wx'y'z + w'xyz' + wxy'z' + wx'yz$$
$$+ wxy'z + w'xyz + wxyz$$

Applying the Quine–McCluskey Algorithm yields the prime implicants $w'xz'$, $xy'z'$, $w'xy$, xyz, wy', and wz. Table 5.17 is the prime implicant table associated with E.

As before, the circled X's represent essential prime implicants, and the crossed out rows indicate essential rows. Columns not crossed out denote those

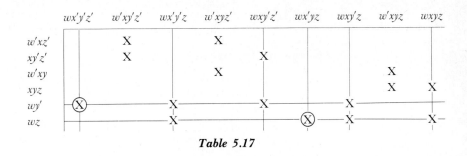

	wx'y'z'	w'xy'z'	wx'y'z	w'xyz'	wxy'z'	wx'yz	wxy'z	w'xyz	wxyz
w'xz'		X		X					
xy'z'		X			X				
w'xy				X				X	
xyz								X	X
wy'	Ⓧ		X		X	X			
wz			X			X	Ⓧ		X

Table 5.17

standard product terms that the essential prime implicants wy' and wz do not cover. These standard product terms are $w'xy'z'$, $w'xyz'$, and $w'xyz$. Although we have not developed any systematic procedure for selecting additional prime implicants to complete the desired minimal representation of E, it should be clear from Table 5.17 that $w'xz'$ and $w'xy$ cover the remaining three standard product terms. Thus,

$$wy' + wz + w'xz' + w'xy$$

is a minimal representation of the Boolean expression E.   □

Although we can use the procedure illustrated in these examples to simplify Boolean expressions in many variables, choosing nonessential prime implicants to complete a minimal representation can become a formidable task when there are many variables. A variety of algorithms deal with this problem, but they tend to be quite intricate and are beyond the scope of this text.

5.8   Problems

In the following problems use the Quine–McCluskey Algorithm and prime implicant tables to find minimal representations of the given Boolean expressions.

1. $xyz' + x'yz + xy'z' + x'yz'$
2. $y + xz + xy'z$
3. $xyz' + xy'z + xy'z' + x'yz' + x'y'z'$
4. $x'yz' + x'yz + xyz' + xyz$
5. $w'x'y'z' + w'x'yz' + w'xy'z' + w'xy'z + w'xyz' + wx'y'z' + wx'y'z$
6. $w'x'y'z' + w'x'y'z + w'x'yz' + wx'y'z' + wx'yz + wx'yz + wxyz' + wxyz$
7. $w'x'y'z + w'xy'z' + w'xyz' + w'xyz + wx'y'z' + wx'y'z + wxy'z' + wx'yz$
   $+ wxyz$
8. $w'x'y'z' + w'xyz + w'xy'z + w'xyz' + w'xyz + wx'y'z + wx'yz + wxy'z$
   $+ wxyz'$
9. $w'x'y'z' + w'x'yz' + w'xy'z' + w'xy'z + w'xyz' + w'xyz + wx'y'z'$
   $+ wx'y'z + wx'yz' + wx'yz$

Chapter 5   REVIEW

Concepts for Review

switch (p. 187)
switching system (p. 187)
parallel connection (p. 188)
series connection (p. 188)
complementary switch (p. 189)
Boolean algebra (p. 191)
Boolean function (p. 194)
Boolean expression (p. 194)
equivalent Boolean expressions (p. 203)
literal (p. 204)
minimal representation of a Boolean expression (p. 204)
Karnaugh maps (p. 206)
implies (p. 219)
covers (p. 221)
prime implicant (p. 222)
standard product term (p. 224)
sum in canonical form (p. 224)
Quine–McCluskey Algorithm (p. 225)
completion (p. 228)
the covering problem (p. 230)
essential prime implicant (p. 232)

Review Problems

1. Find a Boolean expression that corresponds to the Boolean function f:

x	y	z	$f(x, y, z)$
0	0	0	1
0	0	1	1
0	1	0	0
0	1	1	0
1	0	0	1
1	0	1	1
1	1	0	0
1	1	1	0

2. Diagram a switching system that corresponds to the Boolean function f given in Problem 1.

3. Let $E = x(x + y + z) + xy(z + y')$.
    (a) Diagram a switching system that corresponds to E as it is written.
    (b) Write E as a sum of products; then simplify and diagram the corresponding switching system.

4. List the variables and literals found in $xyz' + z' + y'$.

5. Construct a table to show that the Boolean expressions $(y + z)(x + y)$ and $y + zx$ are equivalent.

6. Use a Karnaugh map to obtain a minimal representation of the Boolean expression $wxyz' + wx'yz' + w'x'yz' + w'x'y'z'$.

7. Use a Karnaugh map to obtain a minimal representation of the Boolean expression $wxyz + wx'yz + w'xyz + wx'yz' + wx'y'z' + wx'y'z + wxy'z + w'xy'z$.

8. Let $g(x, y, z) = xyz + z$ and $f(x, y, z) = xy + xz + z$.
    (a) Does f imply g? (Prove your assertion.)
    (b) Does g imply f? (Prove your assertion.)

9. Write the expression $x + xyz'$ in its canonical form.

10. Use the definition of prime implicant to show that $xy'w$ is a prime implicant of $x'y'z' + xy'z' + xy'w$.

11. Use the Quine–McCluskey Algorithm to find the prime implicants of

    $$wx'yz + wxyz' + wx'yz' + w'x'yz' + w'xy'z' + w'x'y'z' + wx'y'z'$$

12. Identify the essential prime implicants of the expression given in Problem 11.

13. Write a minimal representation of the Boolean expression given in Problem 11.

14. Use the Quine–McCluskey Algorithm to find a minimal sum of products representation of the Boolean function f where

x	y	z	$f(x, y, z)$
0	0	0	1
0	0	1	1
0	1	0	0
0	1	1	1
1	0	0	0
1	0	1	0
1	1	0	0
1	1	1	1

6 Symbolic Logic and Logic Circuits

In the preceding chapter we saw how to apply Boolean algebra to design certain switching systems. In this chapter we will consider how to apply Boolean algebra to logic and logic circuit design.

After briefly introducing symbolic logic and logic circuits, we will explain how to use logic gates to represent various Boolean expressions. In particular, we will explore how to use these gates to design circuits that can efficiently perform binary addition.

6.1 COMPOUND STATEMENTS AND LOGICAL EQUIVALENCE

In the study of logic, a *statement* is an assertion to which we can assign a truth value (*T* for true, *F* for false). By this definition, "2 + 3" is not a statement because we cannot say it is true or false. However, "2 + 3 = 6" is a statement because we can label it as true or false: We can assign the truth value *F* to it.

We can use connectives ("and," "or," and "not") to build compound statements from given statements. For instance, from the statements

"2 + 3 = 6"
"The sun rises in the east."

we can build the statements

"2 + 3 = 6 *and* the sun rises in the east."
"2 + 3 = 6 *or* the sun rises in the east."

as well as the statements

"2 + 3 ≠ 6"
"The sun does *not* rise in the east."

To determine the truth value of a compound statement, we must examine the truth values assigned to its components. Tables 6.1, 6.2, and 6.3 identify the truth values of various compound statements. The symbols p, q represent given statements.

p	q	p or q
T	T	T
T	F	T
F	T	T
F	F	F

Table 6.1

p	q	p and q
T	T	T
T	F	F
F	T	F
F	F	F

Table 6.2

p	not p
T	F
F	T

Table 6.3

To write compound statements such as

$$p \text{ and } (q \text{ or } (\text{not } r)) \tag{1}$$

more compactly, we can substitute symbols for the words "or," "and," and "not":

$p \vee q$ means "p or q" ($p \vee q$ is called the *disjunction* of p, q)
$p \wedge q$ means "p and q" ($p \wedge q$ is called the *conjunction* of p, q)
$\sim p$ means "not p" ($\sim p$ is called the *negation* of p)

With these notational changes, truth Tables 6.1, 6.2, and 6.3 become the truth Tables 6.4, 6.5, and 6.6, respectively.

p	q	$p \vee q$
T	T	T
T	F	T
F	T	T
F	F	F

Table 6.4

p	q	$p \wedge q$
T	T	T
T	F	F
F	T	F
F	F	F

Table 6.5

p	$\sim p$
T	F
F	T

Table 6.6

We can use the statements $p \lor q$, $p \land q$, $\sim p$, and $\sim q$ to form other compound statements. For instance, (1) becomes $p \land (q \lor (\sim r))$. Truth tables can help us calculate the truth values of such complex statements, as the following example shows.

Example 1   In Table 6.7 we compute the truth values of $(\sim p) \lor (\sim q)$. We introduce the columns headed by $\sim p$ and $\sim q$ to aid in making these calculations.

p	q	$\sim p$	$\sim q$	$(\sim p) \lor (\sim q)$
T	T	F	F	F
T	F	F	T	T
F	T	T	F	T
F	F	T	T	T

Table 6.7   □

In Chapter 5 we saw that different Boolean expressions can represent the same Boolean function. In a similar fashion, different compound statements can yield the same truth values. The next definition formalizes this idea.

Definition 6.1   Two statements that have the same truth values are *logically equivalent* (or, more simply, they are *equivalent*).

To show that two statements are equivalent, we only need to construct a truth table for each statement and then compare the results.

Example 2   To show that the statements $(\sim p) \lor (\sim q)$ and $\sim (p \land q)$ are equivalent, we compare their truth values. Table 6.8 lists the truth values of $\sim (p \land q)$.

p	q	$p \land q$	$\sim (p \land q)$
T	T	T	F
T	F	F	T
F	T	F	T
F	F	F	T

Table 6.8

Comparing the truth values in the fourth column of Table 6.8 with those in the fifth column of Table 6.7, we see that the statements $(\sim p) \lor (\sim q)$ and $\sim (p \land q)$ are equivalent.   □

Two types of statements, tautologies and contradictions, are particularly important. A *tautology* is a statement that is true under all circumstances, and a *contradiction* is a statement that is false under all circumstances.

Example 3

(a) Table 6.9 shows a computation of the truth values of the statement $p \vee (\sim p)$. Because this statement is true in all cases, it is a tautology.

p	$\sim p$	$p \vee (\sim p)$
T	F	T
F	T	T

Table 6.9

(b) Table 6.10 shows that the statement $(p \wedge q) \wedge (\sim p)$ is false in all cases. It is therefore a contradiction. □

p	q	$\sim p$	$p \wedge q$	$(p \wedge q) \wedge (\sim p)$
T	T	F	T	F
T	F	F	F	F
F	T	T	F	F
F	F	T	F	F

Table 6.10

6.1 Problems

In Problems 1–8 show that the given statements are equivalent.

1. $\sim (p \vee q); (\sim p) \wedge (\sim q)$
2. $\sim (\sim p); p$
3. $(p \vee q) \vee r; p \vee (q \vee r)$
4. $p \vee p; p$
5. $(p \wedge q) \wedge r; p \wedge (q \wedge r)$
6. $p \wedge (q \vee r); (p \wedge q) \vee (p \wedge r)$
7. $p \vee (q \wedge r); (p \vee q) \wedge (p \vee r)$
8. $p \vee q; \sim (\sim p \wedge (\sim q))$
9. Show that $[\sim (p \vee q)] \vee [(\sim p) \wedge q] \vee p$ is a tautology.
10. Show that $(p \vee q) \wedge [(\sim p) \wedge (\sim q)]$ is a contradiction.
11. Construct your own tautology.
12. Construct your own contradiction.

6.2 IMPLICATION

In addition to a disjunction and a conjunction, a third type of compound statement, called an *implication*, is important. Suppose that p and q are statements. The implication $p \Rightarrow q$ (read as either "p implies q" or "if p, then q") has the truth values shown in Table 6.11.

p	q	$p \Rightarrow q$
T	T	T
T	F	F
F	T	T
F	F	T

Table 6.11

We use the following example to show that the entries in Table 6.11 are reasonable from a "common sense" standpoint. Suppose that before an election a candidate makes the statement,

$$\text{"If I am elected, then taxes will not be raised."} \tag{1}$$

This statement is of the form $p \Rightarrow q$, where p is the statement

"I am elected."

and q is the statement

"Taxes will not be raised."

We check each row of Table 6.11 for this example.

Row 1. If the candidate is elected (p is true), and if taxes are not raised (q is true), then clearly the candidate's campaign statement ($p \Rightarrow q$) is true.

Row 2. If the candidate is elected (p is true), and if taxes are raised (q is false), then the candidate's campaign promise ($p \Rightarrow q$) is obviously false.

Rows 3 and 4. If the candidate is not elected (p is false), then the candidate's statement is not false, regardless whether or not taxes are raised; thus, in both of these cases, the statement $p \Rightarrow q$ must be true, which is what Table 6.11 indicates.

In the following examples we investigate the truth values of various implications.

Example 1

(a) The statement

$$\text{"If } 1 + 1 = 2, \text{ then } 2 + 3 = 3.\text{"}$$

is of the form $p \Rightarrow q$, where

$$p \text{ is } \text{``}1 + 1 = 2\text{''} \quad \text{and} \quad q \text{ is } \text{``}2 + 3 = 3\text{''}$$

Since statement p is true and statement q is false, it follows from Table 6.11 that the statement $p \Rightarrow q$ is false.

(b) The statement

$$\text{``If } 2 + 3 = 3, \text{ then } 1 + 1 = 2.\text{''}$$

is of the form $p \Rightarrow q$, where

$$p \text{ is } \text{``}2 + 3 = 3\text{''} \quad \text{and} \quad q \text{ is } \text{``}1 + 1 = 2\text{''}$$

Since p is false and q is true, it follows from Table 6.11 that the statement $p \Rightarrow q$ is true.   □

In some cases we are able to determine an implication's truth value without knowing the truth values of its component parts, as the next example illustrates.

Example 2   The statement

$$\text{``If } x = 5, \text{ then } x^2 = 25.\text{''}$$

is of the form $p \Rightarrow q$, where

$$p \text{ is } \text{``}x = 5\text{''} \quad \text{and} \quad q \text{ is } \text{``}x^2 = 25\text{''}$$

Note that it is not possible to determine the truth value of "$x = 5$"; such a determination, however, is not necessary. It is sufficient to observe that if "$x = 5$" is true, then "$x^2 = 25$" is also true. Thus in this case it is impossible to have a situation in which p is true and q is false. According to the truth table, to have p true and q false is the only way an implication can be false, so the original statement must be true.   □

In the statement $p \Rightarrow q$, p is called the *hypothesis* and q the *conclusion*. Several statements related to $p \Rightarrow q$ result when we interchange or negate the hypothesis and conclusion. Table 6.12 lists the names of these new statements.

Statement: $p \Rightarrow q$

Converse: $q \Rightarrow p$
Contrapositive: $(\sim q) \Rightarrow (\sim p)$
Inverse: $(\sim p) \Rightarrow (\sim q)$

Table 6.12

Example 3   The converse, contrapositive, and inverse of the statement

$$\text{``If } x = 5, \text{ then } x \text{ is an odd integer.''}$$

are:

Converse: "If x is an odd integer, then $x = 5$."

Contrapositive: "If x is not an odd integer, then $x \neq 5$."

Inverse: "If $x \neq 5$, then x is not an odd integer."   □

You may have observed in Example 3 that both the statement and its contrapositive are true. This reflects the fact that a statement and its contrapositive always have the same truth value, as we see in the next theorem.

Theorem 6.2   The statements $p \Rightarrow q$ and $(\sim q) \Rightarrow (\sim p)$ are equivalent.

Proof: We construct the truth table for $(\sim q) \Rightarrow (\sim p)$ to compare the truth values of $(\sim q) \Rightarrow (\sim p)$ with those of $p \Rightarrow q$ (Table 6.13).

p	q	$p \Rightarrow q$	$\sim q$	$\sim p$	$(\sim q) \Rightarrow (\sim p)$
T	T	T	F	F	T
T	F	F	T	F	F
F	T	T	F	T	T
F	F	T	T	T	T

Table 6.13

Since the third and last columns of the table are identical, it follows that the statements $p \Rightarrow q$ and $(\sim q) \Rightarrow (\sim p)$ are equivalent.   ■

The preceding theorem is the logical basis for all proofs by contradiction. The next example illustrates how it is used.

Example 4   Suppose that we wish to prove:

$$\text{``If } A \subseteq B, \text{ then } A \setminus B = \emptyset.\text{''} \tag{2}$$

Rather than prove (2) directly, we find it is easier to prove its contrapositive,

$$\text{``If } A \setminus B \neq \emptyset, \text{ then } A \nsubseteq B.\text{''} \tag{3}$$

If $A \setminus B \neq \emptyset$, then there is some element x such that $x \in A \setminus B$. Thus, $x \in A$ and $x \notin B$. Since there is an element of A that is not an element of B, we have $A \not\subseteq B$. Therefore, we have established that (3) is true, and because (3) is equivalent to (2), we have also shown that (2) is true. $\qquad\square$

We can also state an implication in terms of the more fundamental connectives we studied in Section 6.1. For example, the next theorem (which you are asked to prove in Problem 28) states implication in terms of disjunction and negation.

Theorem 6.3 The statements $p \Rightarrow q$ and $(\sim p) \vee q$ are equivalent.

In the first two sections of this chapter we have seen how to compute the truth value of a compound statement. The design of circuits to mechanize such computations is a fundamental computer design problem. We will take up this problem in the next section.

6.2 Problems

In Problems 1–8 decide whether the given statement is true or false.
1. If $1 + 1 = 2$, then $2 + 2 = 5$.
2. If $2 + 2 = 5$, then $1 + 1 = 2$.
3. If $1 + 1 = 2$, then $2 + 3 = 5$.
4. If $2 + 2 = 5$, then $1 + 1 = 3$.
5. If $x = 2$, then $3x = 6$.
6. If $3x = 6$, then $x = 2$.
7. If $x = 0$, then $x^2 = x$.
8. If $x^2 = x$, then $x = 0$.

In Problems 9–12 show that the given statements are equivalent.
9. $(p \vee q) \Rightarrow r; \ [(\sim p) \wedge (\sim q)] \vee r$
10. $p \Rightarrow (q \wedge r); \ (p \Rightarrow q) \wedge (p \Rightarrow r)$
11. $p \Rightarrow (q \vee r); \ (p \Rightarrow q) \vee (p \Rightarrow r)$
12. $p \Rightarrow (\sim q); \ q \Rightarrow (\sim p)$

In Problems 13–20 show that the given statement is a tautology.
13. $p \Rightarrow (p \vee q)$
14. $[(p \Rightarrow q) \wedge (q \Rightarrow r)] \Rightarrow (p \Rightarrow r)$ (the law of syllogism)
15. $(p \wedge q) \Rightarrow p$
16. $[p \wedge (\sim p)] \Rightarrow q$

17. $(p \wedge q) \Rightarrow (p \vee q)$

18. $[(p \vee q) \wedge (\sim p)] \Rightarrow q$

19. $[(\sim q) \wedge (p \Rightarrow q)] \Rightarrow \sim p$

20. $[p \wedge (p \Rightarrow q)] \Rightarrow q$

21. Show that if r and $r \Rightarrow s$ are tautologies, then so is s.

22. Under what conditions is $p \Rightarrow (\sim p)$ a true statement?

In Problems 23 and 24 give

    (a) the converse

    (b) the contrapositive

    (c) the inverse

of the given statement and determine the truth value of each.

23. If $x = 6$, then x is an odd integer.

24. If $x^2 = x$, then x is zero.

25. The statement $p \Leftrightarrow q$, read as "p if and only if q," is defined to be $(p \Rightarrow q) \wedge (q \Rightarrow p)$. Construct a truth table for $p \Leftrightarrow q$.

26. Construct the truth table of $((\sim p) \wedge (q \vee r)) \Leftrightarrow q$.

27. Show that the following statement is a contradiction:

$$[(p \wedge r) \vee (q \wedge (\sim r))] \Leftrightarrow [((\sim p) \wedge r) \vee ((\sim q) \wedge (\sim r))]$$

28. Prove Theorem 6.3.

6.3 LOGIC CIRCUITS

In Chapter 5 we saw how to design circuits, built from switches, to mechanize the computation of the values of Boolean functions. Now we will consider designing circuits to mechanize the computation of truth values. The basic building blocks of these circuits are called *logic gates*. Although present technology allows the integration of many thousands of logic gates in a single silicon chip, the design of small-scale circuits remains an important problem.

    To simplify the logical notation we employed in the preceding two sections, we shall substitute $+$ for $\vee$, $'$ for $\sim$, and let $p \cdot q$ or pq denote $p \wedge q$. With these changes, the statement $(\sim (p \wedge q)) \vee s$ becomes $(pq)' + s$. Because we are interested in designing a circuit to implement logic computations, it is natural to let 0 and 1 designate "false" and "true," respectively. With these changes in notation, the truth Tables 6.4, 6.5, and 6.6 from Section 6.1 become the truth Tables 6.14, 6.15, and 6.16.

p	q	$p + q$
1	1	1
1	0	1
0	1	1
0	0	0

Table 6.14

p	q	$p \cdot q$
1	1	1
1	0	0
0	1	0
0	0	0

Table 6.15

p	p'
1	0
0	1

Table 6.16

Notice that if we compare Tables 6.14, 6.15, and 6.16 above with Tables 5.4, 5.5, and 5.6 of Chapter 5 (p. 189), we see that we actually have a Boolean algebra, in which each Boolean expression corresponds to a logic statement.

Logic gates carry out the operations of conjunction, disjunction, and negation. Figure 6.1 shows these three types of gates, along with the outputs that correspond to all the possible inputs into these gates.

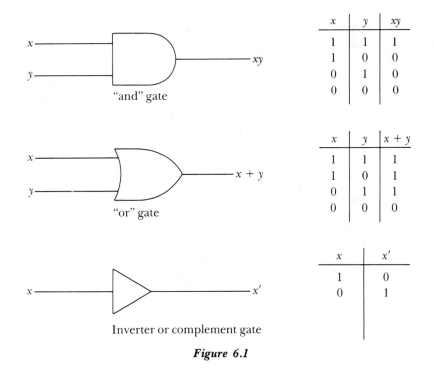

Figure 6.1

We can combine these gates to build logic circuits (gate implementations) of Boolean expressions. To do this, we pass the literals through various logic gates and eventually obtain the desired Boolean expression (compound logic statement). Examples 1 and 2 illustrate the procedure.

Example 1   To build a gate implementation of the Boolean expression

$$x + x'y$$

we must obtain an x', form an $x'y$ term, and then add x and $x'y$ together. As Figure 6.2 shows, we begin by passing the x through an inverter gate to obtain an x'. This gives us our three literals x, x', and y. Next we take the x' and y through an "and" gate to obtain the $x'y$ term. Finally, we can send the x and $x'y$ through an "or" gate to obtain the desired expression $x + x'y$. Figure 6.2 shows the complete gate implementation (logic circuit).   □

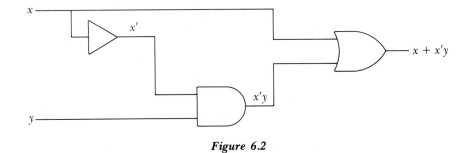

Figure 6.2

Example 2   Figure 6.3 shows the gate implementation that represents the Boolean expression

$$xy + y'(x + z)$$ □

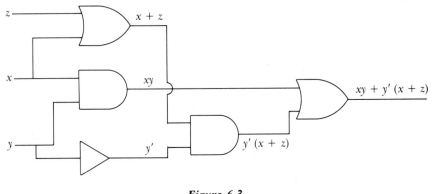

Figure 6.3

In the next section we will use logic gates to design a circuit that can perform binary addition. As preparation for this application we see in the following example how to use logic gates to design circuits with a specified output.

Example 3   We design a logic circuit that has the outputs given in Table 6.17.

x	y	Output
0	0	0
0	1	1
1	0	1
1	1	0

Table 6.17

By Section 5.2, we know that Table 6.17 represents a Boolean expression

$$x'y + xy'$$

Drawing its gate implementation, we obtain the desired circuit as shown in Figure 6.4.   □

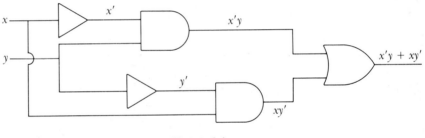

Figure 6.4

Because Boolean expressions represent logic circuits, we can simplify these circuits by simplifying (minimizing) the Boolean expressions, as we did in Chapter 5.

6.3   **Problems**

In Problems 1–6 give the gate implementations of the given Boolean expression.

1. $xy + x'$
2. $x(y' + x)$
3. $(x' + y)(y' + x)$
4. $x'y + yz' + x$
5. $xyz' + z$
6. $x'y' + xy'z$

In Problems 7–10 give the Boolean expression corresponding to the given logic circuit.

7.

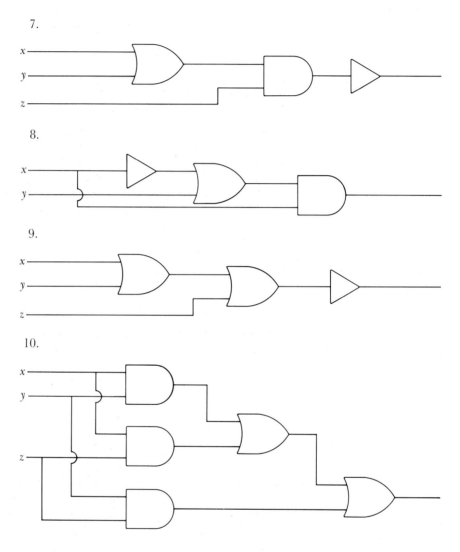

8.

9.

10.

In Problems 11–14 draw the gate implementations of the given function.

11. x	y	Output
0	0	1
0	1	0
1	0	0
1	1	1

12. x	y	Output
0	0	1
0	1	0
1	0	1
1	1	0

13. x	y	z	Output
0	0	0	1
0	0	1	0
0	1	0	1
1	0	0	0
1	1	0	0
1	0	1	0
0	1	1	1
1	1	1	0

14. x	y	z	Output
0	0	0	1
0	0	1	1
0	1	0	1
1	0	0	0
1	1	0	0
1	0	1	0
0	1	1	0
1	1	1	0

In this section we have used "and," "or," and "inverter" gates to construct gate implementations of Boolean expressions. In the following problems we introduce two new logic gates. Using either one of these gates as the *sole* building block, we can create a logic circuit of *any* Boolean expression.

15. The "not and" ("nand") gate is designated by

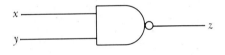

and is defined by $z = (xy)'$. Give the gate implementation of a "nand" gate using only the "and," "or," and "inverter" gates.

16. The "not or" ("nor") gate is designated by

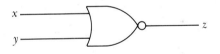

and is defined by $z = (x + y)'$. Give the gate implementation of a "nor" gate using only the "and," "or," and "inverter" gates.

17. Show that any Boolean expression can be implemented using only "nand" gates. (*Hint:* Show that the following basic expressions can be implemented using only "nand" gates: (a) $x + y$, (b) xy, (c) x'. Note for example that $(xy)' = x' + y'$.)

18. Show that any Boolean expression can be implemented using only "nor" gates. (*Hint:* Show that the following basic expressions can be implemented using only "nor" gates: (a) $x + y$; (b) xy; (c) x'.)

In Problems 19–22 implement the given Boolean expression using only "nand" gates. Then do the same using only "nor" gates.

19. $xy' + x'y$

20. $x(x' + y)'$

21. $y + (xz)'$

22. $(x + y)z'$

6.4   GATE IMPLEMENTATION OF BINARY ADDITION

In Table 6.18 we review the addition of two single-digit binary numbers, x and y.

x	y	$x + y$
0	0	0
0	1	1
1	0	1
1	1	10

Table 6.18

The binary sum $1 + 1 = 10$ introduces a sum that has more digits than either of its summands. Thus when designing a system that will perform binary addition, we must provide for the possibility of a "carry" bit. We do this by generating two outputs for each two inputs x and y: One output (denoted by s) corresponds to the right-hand digit of the sum of x and y, and the other output (denoted by c) corresponds to the carry bit. You can easily verify the values of s and c given in Tables 6.19 and 6.20.

x	y	s	x	y	c
0	0	0	0	0	0
0	1	1	0	1	0
1	0	1	1	0	0
1	1	0	1	1	1

Table 6.19   **Table 6.20**

Applying Section 5.2 methods to Tables 6.19 and 6.20, we see that these tables represent Boolean expressions

$$s = x'y + xy' \qquad (1)$$

and

$$c = xy \qquad (2)$$

respectively. Figure 6.5 shows a gate implementation of the Boolean expressions (1) and (2). This design allows us to add two single-digit binary numbers by providing a way to obtain a two-digit answer (s is the right-hand digit; c is the carry digit).

We can simplify the system described in Figure 6.5 by observing that the Boolean expressions $x'y + xy'$ and $(x + y)(xy)'$ are equivalent (see Problem 1).

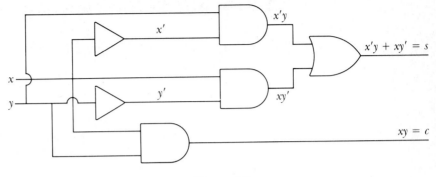

Figure 6.5

Replacing $s = x'y + xy'$ with $s = (x + y)(xy)'$, we can use fewer gates for the logic circuit. Figure 6.6 shows the simpler version.

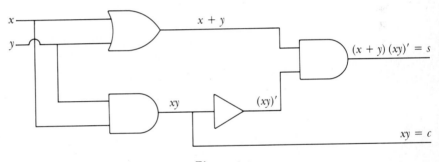

Figure 6.6

You should observe that though both designs yield the sum of two single-digit binary numbers, Figure 6.6 requires only four gates and thus less circuitry than the design in Figure 6.5. In Problems 2–5 you are asked to design the gate implementations of other Boolean expressions that can be used to add two single-digit binary numbers.

Gate implementations used to add two single-digit binary numbers are often referred to as *half adders* (the reason for the term "half" will become apparent shortly). The simplest half adder is the one illustrated in Figure 6.6. Figure 6.7 shows a convenient way to abbreviate a half adder.

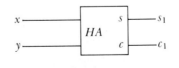

Figure 6.7

Suppose now that we wish to add two multiple-digit binary numbers, such as

$$\begin{array}{r} x_3 x_2 x_1 \\ + y_3 y_2 y_1 \\ \hline s_4 s_3 s_2 s_1 \end{array} \tag{3}$$

Note that as we add each column of digits, we will also need to account for the carry bit from the preceding column. Consequently, we need to be able to add three bits together, rather than just two. A *full adder* performs precisely this function.

The full adder can process the carry bit together with the two bits present in each column of a sum of two binary numbers. By using a suitable combination of full adders, we can add any two binary numbers:

$$\begin{array}{r} x_n x_{n-1} \cdots x_2 x_1 \\ + y_n y_{n-1} \cdots y_2 y_1 \end{array} \tag{4}$$

To describe the design of a full adder, we let x and y be the bits appearing in a column of (4), and we let z be the carry bit produced from the preceding column. As before, s denotes the rightmost digit of the sum $x + y + z$, and c denotes the carry digit.

As you might have guessed, a full adder consists of an appropriate combination of two half adders. To see how to combine these half adders, we first use them to compute the sum $x + y + z$. (In working this through we will introduce three half adders but will eventually use only two.) Suppose the outputs of a half adder that computes the binary sum $x + y$ are s_1 and c_1 (as in Figure 6.7). In base 10 we can write

$$x + y = 2c_1 + s_1$$

so we can substitute as follows:

$$x + y + z = (x + y) + z = (2c_1 + s_1) + z = 2c_1 + (s_1 + z)$$

Now, as Figure 6.8 indicates, we can use a second half adder to compute the

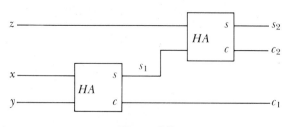

Figure 6.8

sum $s_1 + z$. (The terms s_2 and c_2 denote the outputs of this second half adder in the figure.) Consequently, we have (in base 10)

$$x + y + z = 2c_1 + (2c_2 + s_2) = 2(c_1 + c_2) + s_2$$

To write the sum $x + y + z$ in binary form we must write $2(c_1 + c_2)$ in powers of two. To accomplish this we can use a third half adder to calculate the sum $c_1 + c_2$ (Figure 6.9). The outputs of this third half adder are s_3 and c_3, and we have

$$x + y + z = 2(2c_3 + s_3) + s_2 = 2^2 c_3 + 2s_3 + s_2 \qquad (5)$$

Observe that since the sum of $x + y + z$ is at most a two-digit number, it must be the case that $c_3 = 0$. Because of this, we can simplify the design in Figure 6.9 by replacing the half adder farthest to the right by an "or" gate (see Problem 8). With this simplification, we obtain the full adder that will add three single-digit binary numbers (Figure 6.10).

To see how this system works, examine Figure 6.11, which traces the progress of the sum of $x = 0$, $y = 1$, and $z = 1$ through a full adder.

Now that we have created a system that will add three single-digit binary numbers, we can use it to add two n-digit binary numbers, as in (4). First we

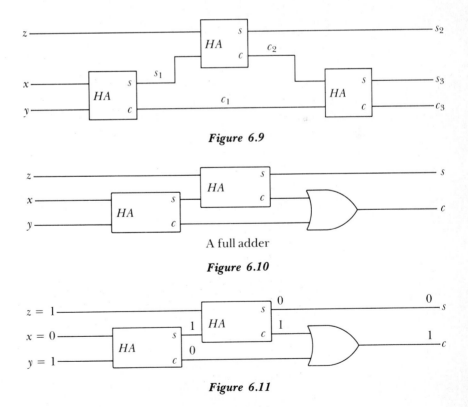

Figure 6.9

A full adder

Figure 6.10

Figure 6.11

assign a full adder to each of the n columns in (4). The full adder handles the bits in its column as well as the carry bit produced from the preceding column. Thus, for the ith column in (4) the input values x, y, and z become x_i, y_i, and c_{i-1}, respectively. If we denote a full adder as indicated in Figure 6.12, then the design in Figure 6.13 will enable us to add $x_n x_{n-1} \cdots x_2 x_1$ and $y_n y_{n-1} \cdots y_2 y_1$.

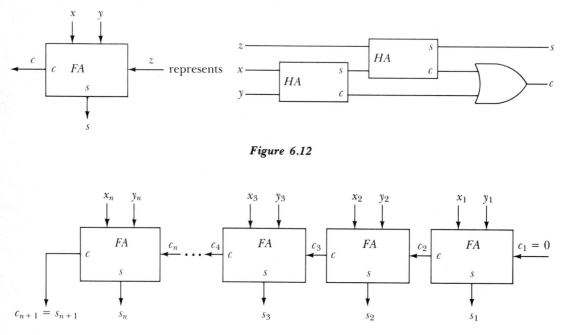

Figure 6.12

Figure 6.13

In the Problem set we introduce other gate implementations and different approaches to adding binary numbers.

6.4   Problems

1. Show that the two Boolean expressions $x'y + xy'$ and $(x + y)(xy)'$, used to give two different realizations of a half adder, are equivalent.

In Problems 2–5 show that the given Boolean expressions can be used to design a half adder, and give the corresponding gate implementation.

2. $s = (xy + x'y')'$; $c = (x' + y')'$
3. $s = (x + y')' + (x' + y)'$; $c = xy$
4. $s = (x + y)(x' + y')$; $c = (x' + y')'$
5. $s = (xy + x'y')'$; $c = xy$

6. Use Figure 6.6 to design a half adder that uses only "nor" gates. (See Problem 16 of Section 6.3 for the definition of a "nor" gate.)

7. Use Figure 6.6 to design a half adder that uses only "nand" gates. (See Problem 15 of Section 6.3 for the definition of a "nand" gate.)

8. Justify the simplification made in Figure 6.9 to obtain Figure 6.10.

9. Using only full adders and inverter gates, implement a circuit that will compute the difference of two four-digit binary numbers. (*Hint:* Use a "two's complement" to convert the subtraction problem to an addition problem.)

10. Figure 6.13 is a design that will perform the addition of $x_n x_{n-1} \cdots x_2 x_1$ and $y_n y_{n-1} \cdots y_2 y_1$. Show that

$$s_i = c_i(x_i y_i' + x_i' y_i)' + c_i'(x_i y_i' + x_i' y_i)$$

and

$$c_{i+1} = c_i(x_i y_i' + x_i' y_i) + x_i y_i$$

11. The following gates can be used to denote $x_1 x_2 \cdot \ldots \cdot x_k$ and $x_1 + x_2 + \cdots + x_k$, respectively.

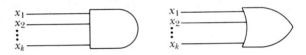

(a) Find the Boolean expressions corresponding to s and c in the following diagram.

(b) Show that the following diagram can be used as a full adder.

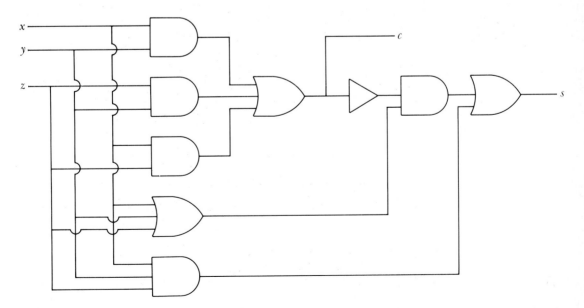

Chapter 6   REVIEW

Concepts for Review

statement (p. 238)
disjunction (p. 239)
conjunction (p. 239)
negation (p. 239)
truth table (p. 239)
logical equivalence (p. 240)
tautology (p. 241)
contradiction (p. 241)
implication (p. 242)
hypothesis (p. 243)
conclusion (p. 243)
converse (p. 243)

contrapositive (p. 243)
inverse (p. 243)
proof by contradiction (p. 244)
"and" gate (p. 247)
"or" gate (p. 247)
inverter gate (p. 248)
"nor" gate (p. 251)
"nand" gate (p. 251)
gate implementation
   of binary addition (p. 252)
half adder (p. 253)
full adder (p. 254)

Review Problems

In Problems 1 and 2 show that the given statements are equivalent.

1. $\sim(p \vee (q \wedge s))$; $((\sim p) \wedge (\sim q)) \vee ((\sim p) \wedge (\sim s))$
2. $\sim(p \wedge (q \wedge s))$; $((\sim p) \vee (\sim q)) \vee (\sim s)$
3. Prove that $[\sim(p \vee q)] \Rightarrow \sim p$ is a tautology.
4. Prove that $p \Leftrightarrow (q \wedge (\sim p))$ is a contradiction.
5. Show that the statement $p \Rightarrow q$ and its contrapositive are equivalent.

In Problems 6 and 7 give (a) the converse, (b) the contrapositive, and (c) the inverse of each statement. Decide the truth value of each statement.

6. If $x^4 = 1$, then $x = 1$.
7. If $x = 5$, then $2x - 10 = 0$.

In Problems 8 and 9 give the gate implementation of the given Boolean expression.

8. $xz + z'$
9. $x(y + xz' + zy)$
10. Draw the gate implementation of the function f defined by:

x	y	z	$f(x, y, z)$
0	0	0	1
0	0	1	1
0	1	0	0
0	1	1	1
1	0	0	1
1	0	1	1
1	1	0	0
1	1	1	1

11. Explain why any Boolean expression can be drawn using only "nand" gates.

12. a. Give the Boolean expressions for s and c where $x + y = (cs)_2$ and x, $y \in \{0, 1\}$.

13 b. Draw a gate implementation that will yield the two outputs s and c.

7 Difference Equations

Difference equations are used to develop mathematical models in many fields, including biology, economics, and computer science. These equations are the discrete counterpart of differential equations, and many results in difference equations parallel those obtained in the study of differential equations. In this chapter we first examine a number of situations that can be modeled by difference equations, and then we develop sufficient theory to solve these equations. Applications of difference equations are given throughout the chapter.

7.1 SOME EXAMPLES OF DIFFERENCE EQUATIONS

The following equations are difference equations:

$$f(n) = f(n - 1) + n$$
$$f(n + 2) - 6f(n + 1) + 3f(n) = 0$$
$$f(n) = 2f(n - 1) + 3f(n - 2)$$
$$f(n + 3) - n^2 f(n + 1) + 4f(n - 1) = 2^n$$
$$f(n)f(1) + f(n - 1)f(2) + \cdots + f(1)f(n) = 4$$

In each of these equations f is an unknown function; n is generally an integer. Our principal goal in this chapter is to solve such equations by determining functions f which make the equations true.

In the same way that we can substitute a number for a variable to verify that the number is a solution to a given algebraic equation, we can "plug" a function into a difference equation to check whether or not the function is a solution to the equation. If the resulting equation is true for suitable values of n, then the given function is a solution.

Example 1 We show that $f(n) = 3^n$ is a solution to the difference equation

$$f(n) = 2f(n - 1) + 3f(n - 2) \tag{1}$$

for all integers n. Observe that if $f(n) = 3^n$, then

$$f(n - 1) = 3^{n-1} \tag{2}$$

and

$$f(n - 2) = 3^{n-2} \tag{3}$$

Substituting (2) and (3) into (1) yields

$$\begin{aligned}
2f(n - 1) + 3f(n - 2) &= 2 \cdot 3^{n-1} + 3 \cdot 3^{n-2} \\
&= 2 \cdot 3^{n-1} + 3^{n-1} \\
&= 3 \cdot 3^{n-1} \\
&= 3^n \\
&= f(n)
\end{aligned}$$

Thus we have shown that if $f(n) = 3^n$, then

$$f(n) = 2f(n - 1) + 3f(n - 2)$$

for all integers n, so $f(n) = 3^n$ is a solution to (1). $\qquad\qquad\square$

Difference equations have varied applications. In the rest of this section we will examine four situations that are modeled by difference equations.

Compound Interest At the end of each year, an investor deposits k dollars into an account that earns interest rate r, compounded annually. Our problem is to find an expression that describes the amount of money in the account immediately after the nth deposit.

Let $A(n)$ denote the amount of money in the account after the nth deposit. Then $A(n)$ is the sum of $A(n - 1)$ (the amount in the account after the $(n - 1)$st deposit), $rA(n - 1)$ (the interest earned by the $A(n - 1)$ dollars during the year), and k, the last k-dollar deposit. Thus we obtain the difference equation

$$\begin{aligned}
A(n) &= A(n - 1) + rA(n - 1) + k, \quad \text{or equivalently,} \\
A(n) &= (1 + r)A(n - 1) + k
\end{aligned}$$

Solving this difference equation would allow us to determine the amount of money in the account at any future date.

Seed Production[1]   Certain plants flower and produce seeds either one year or two years after they germinate. After flowering, these plants die. Suppose that 0.2 of the seeds result in plants that flower after one year, while 0.5 of the seeds produce plants that flower after two years. (Note that 0.3 of the seeds fail to produce plants that survive to produce more seeds.) In addition, suppose that on the average each plant flowering after one year produces 375 seeds, while on the average each plant flowering after two years produces 800 seeds. Figure 7.1 presents this information schematically. We assume that each year all seeds in the seed bank are planted.

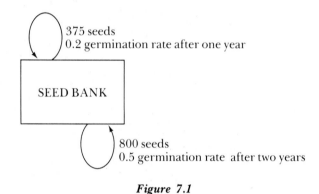

375 seeds
0.2 germination rate after one year

SEED BANK

800 seeds
0.5 germination rate after two years

Figure 7.1

We want to determine $S(n)$, the number of seeds produced by flowers during year n. Note that $S(n + 2)$ (the number of seeds produced during year $n + 2$) depends on $S(n + 1)$ (the number of seeds produced during year $n + 1$), and $S(n)$ (the number of seeds produced during year n). More explicitly, 0.2 of the seeds present at year $n + 1$ (that is, $0.2\,S(n + 1)$ seeds) will produce plants that flower, and each such seed will result in 375 seeds in year $n + 2$. Thus these "one-year seeds" contribute $(375)(0.2)S(n + 1)$ seeds to the seed bank in year $n + 2$. In a similar fashion, 0.5 of the seeds present at year n produce plants that flower in two years, and each such plant adds an additional 800 seeds to the seed bank in year $n + 2$. That is, the "two-year seeds" contribute $(800)(0.5)S(n)$ seeds to the seed bank in year $n + 2$. Because no other source contributes to the seed bank, we have

$$S(n + 2) = (375)(.2)S(n + 1) + (800)(.5)S(n)$$
$$= 75S(n + 1) + 400S(n)$$

A solution to this difference equation would enable us to predict the number of seeds available at any year n.

[1] Adapted from Stephen P. Hubbell and Patricia A. Warner, "On Measuring the Intrinsic Rate of Increase of Populations with Heterogeneous Life Cycles," *The American Naturalist*, 113, pp. 277–293.

Sums of the form $\Sigma_{j=1}^{n} j^{k}$ These sums arise regularly in many mathematical applications, and we can use difference equations to evaluate them. Observe that if

$$S(n) = \sum_{j=1}^{n} j^{k}$$

then

$$S(n + 1) = \sum_{j=1}^{n+1} j^{k}$$

Since

$$\sum_{j=1}^{n+1} j^{k} = \sum_{j=1}^{n} j^{k} + (n + 1)^{k}$$

it follows that

$$S(n + 1) = S(n) + (n + 1)^{k}$$

A solution to this difference equation would provide a simple expression for evaluating the sum $\Sigma_{j=1}^{n} j^{k}$ for each positive integer n.

Divide and Conquer Divide-and-conquer algorithms are based on the idea that many problems would be simpler if they were smaller. BINARY SEARCH (Algorithm 1.2) is a good example of such an algorithm. Recall that this algorithm searches a given list by first reducing the list to half its size, then to one quarter its size, and so on. An alternative to this algorithm would be TRINARY SEARCH, an algorithm which divides a list into thirds, ninths, and so on.

Searches are heavily used in data management. Thus it is important to know the relative complexities of search algorithms. In Chapter 1 we used induction to determine the complexity of the BINARY SEARCH. Here we set up a difference equation that can be used to determine its complexity.

We assume that the computer requires one unit of time to determine if a particular entry in a list of n entries falls in the first half or in the second half of the list. Let $B(n)$ denote the number of time units that the computer needs to execute BINARY SEARCH on a list with n entries. Then $B(n)$ is equal to the time the computer uses to determine whether the desired entry falls before or after the middlemost entry (1 unit of time) plus the time the computer requires to apply the algorithm to a list with $n/2$ entries ($B(n/2)$ units of time). This is expressed by the difference equation

$$B(n) = B(n/2) + 1 \tag{4}$$

if n is an even integer. If n is an odd integer, then the computer divides the original list into two almost equal portions, one with $(n - 1)/2$ entries and the other with $(n + 1)/2$ entries. Although (4) is not applicable in this case, this equation, nevertheless, does provide a good analysis of **BINARY SEARCH** and will yield a relatively accurate approximation of the time the computer requires to execute this algorithm.

A similar analysis of **TRINARY SEARCH** shows that the computer must make at least one, and perhaps two, comparisons to determine which third of the list contains the desired entry. Assume that the entry sought has an equal probability of appearing anywhere in the list; thus, the probability of finding the entry in the first third of the list is $\frac{1}{3}$, and the probability that it is in the last two thirds of the list is $\frac{2}{3}$. Since an entry in the first third of the list requires one comparison while an entry in the last two thirds requires two comparisons, we would expect to make (on the average) $\frac{1}{3} \cdot 1 + \frac{2}{3} \cdot 2 = \frac{5}{3}$ comparisons. If $T(n)$ represents the time necessary to execute the trisection algorithm for a list of n elements we have the difference equation

$$T(n) = T\left(\frac{n}{3}\right) + \frac{5}{3} \tag{5}$$

Solving the difference equations (4) and (5) will enable us to find the algorithms' complexity. Once we find the functions B (for **BINARY SEARCH**) and T (for **TRINARY SEARCH**), we can determine which of the two algorithms is more efficient.

In this section we have used difference equations to model several practical problems. As you progress through this chapter you will learn to solve these difference equations and to apply difference equations to a number of other problems.

7.1 Problems

1. Show that $6 \cdot 2^n + 4$ is a solution of the difference equation
   $f(n) = 3f(n - 1) - 2f(n - 2)$.
2. Show that $(3^{n+1} - 1)/2$ is a solution of the difference equation
   $f(n + 1) - f(n) = 3^{n+1}$.
3. Show that $2^n + n2^n$ is a solution of the difference equation
   $f(n + 2) - 4f(n + 1) + 4f(n) = 0$.
4. Show that $1 + n$ is a solution of the difference equation
   $f(n) - 2f(n - 1) + f(n - 2) = 0$.
5. Show that $n(-1)^n$ is a solution of the difference equation
   $f(n + 2) + 2f(n + 1) + f(n) = 0$.
6. Show that $n!$ is a solution of the difference equation
   $f(n) - n(n - 1)f(n - 2) = 0$.
7. Show that $n!$ is a solution of the difference equation
   $f(n + 2) - f(n + 1) - (n + 1)^2 f(n) = 0$.

8. Show that for any constants α and β, $\alpha 3^n + \beta n 3^n$ is a solution of the difference equation $f(n + 2) - 6f(n + 1) + 9f(n) = 0$.

9. Show that for any constants α and β, $\alpha 2^n + \beta n 2^n + 2^{n-1}$ is a solution of the difference equation $f(n) - 4f(n - 1) + 4f(n - 2) = 0$.

10. A computer may search a list by dividing it into four quarters and then determining which quarter-list contains the desired entry. The expected number of comparisons needed to determine which quarter-list contains the desired entry is $\frac{9}{4}$. (Why?) Write a difference equation for $Q(n)$, the time required to search a list of n entries. Assume that each comparison requires one time unit.

11. Figure 7.2 represents the seed production of plants that flower after one, two, or three years. Use this diagram to write a difference equation for $S(n)$, the number of seeds present in the seed bank at year n.

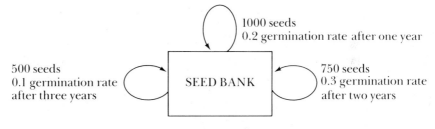

Figure 7.2

In Problems 12–17 write a difference equation for the given sum.

12. $S(n) = \displaystyle\sum_{j=1}^{n} j^3$

13. $S(n) = \displaystyle\sum_{j=1}^{n} \sin(j\pi/2)$

14. $S(n) = \displaystyle\sum_{j=1}^{n} \sqrt{j}$

15. $S(n) = \displaystyle\sum_{j=1}^{n} j^2$

16. $S(n) = \displaystyle\sum_{j=1}^{n} \cos(j\pi/4)$

17. $S(n) = \displaystyle\sum_{j=1}^{n} 3j^{5/2}$

18. A ball is dropped from a height of ten feet and always rebounds 0.7 of the height it has fallen. If $H(n)$ is the height it rebounds on the nth bounce, write a difference equation for $H(n)$.

19. The storage a computer requires to execute a certain algorithm on n data points is equal to the storage it requires to execute the algorithm on $n - 1$ points plus the storage it requires to execute the algorithm on $n - 2$ points. Write a difference equation for this algorithm's storage requirements.

20. Write a difference equation to describe the Fibonacci numbers (see Problem 14, Section 1.4).

21. Write a difference equation for the number of sequences of length n that can be formed using the symbols 00 and 1 (the symbol 00 has length two).

(*Hint:* Let $f(n)$ denote the number of sequences of length n, and consider two cases: (a) the last entry in the sequence is 0, and (b) the last entry in the sequence is 1.)

22. Write a difference equation for the number of sequences of length n that can be formed using the symbols 1, 2, 00 (the symbol 00 has length two).

23. Write a difference equation for the number of code words of length n that can be formed from a's and b's where no two b's are adjacent.

24. At each level of a chain letter a person is to send a chain letter to five additional people who then become involved in the next level. Write a difference equation for the number of people involved at level n.

25. At the end of year n of an investment plan, $50n$ dollars are deposited in an account that earns 10% interest compounded annually. Write a difference equation for the amount of money in the plan after the nth deposit.

LINEAR DIFFERENCE
7.2 EQUATIONS WITH CONSTANT COEFFICIENTS

Although many difference equations are extremely difficult to solve, there are certain broad classes of these equations that are relatively easy to handle. One such class consists of difference equations of the following kind.

Definition 7.1   A *linear difference equation with constant coefficients* is a difference equation that can be written in the form

$$f(n + k) + a_{n+k-1}f(n + k - 1) + \cdots + a_n f(n) = \phi(n) \quad a_n \neq 0$$

where the coefficients a_{n+k-1}, a_{n+k-2}, ..., a_n are constants. The positive integer k is called the *order* of the equation. If the function $\phi(n)$ is zero for all values of n, the equation is said to be *homogeneous*.

Example 1
(a) The equation

$$f(n + 4) - 3f(n + 2) + \tfrac{1}{3}f(n) = n^2 + 2n$$

is a fourth-order linear difference equation with constant coefficients. (Note that the coefficients a_{n+3} and a_{n+1} of $f(n + 3)$ and $f(n + 1)$, respectively, are zero.)

(b) The equation

$$f(n + 5) + 2f(n + 4) - 4f(n) = 0$$

is a fifth-order homogeneous linear difference equation with constant coefficients.   □

Example 2   In Section 7.1 we introduced a proposed model of seed production that had the difference equation

$$S(n + 2) = 75S(n + 1) + 400S(n)$$

If we write this equation as

$$S(n + 2) - 75S(n + 1) - 400S(n) = 0$$

we see that it is a second-order homogeneous linear difference equation with constant coefficients.   □

To find a solution to the general homogeneous equation

$$f(n + k) + a_{n+k-1}f(n + k - 1) + \cdots + a_n f(n) = 0 \qquad (1)$$

we make an "educated guess" that a function of the form

$$f(n) = A^n$$

will be a solution of (1) for some number $A = r$. To determine r, we substitute appropriate powers of A into (1) to obtain

$$A^{n+k} + a_{n+k-1}A^{n+k-1} + \cdots + a_n A^n = 0$$

or, equivalently,

$$A^n(A^k + a_{n+k-1}A^{k-1} + \cdots + a_n) = 0$$

which means $f(n) = r^n$ is a solution of (1) if and only if $A = 0$ or $A = r$ is a solution of

$$A^k + a_{n+k-1}A^{k-1} + \cdots + a_n = 0 \qquad (2)$$

Equation (2) is called the *characteristic equation* of (1). The root $A = 0$ results in the trivial solution $f(n) = 0^n = 0$. A root $A = r$ of the characteristic equation (2) results in the nontrivial solution $f(n) = r^n$ of (1). Thus, we have the following theorem.

Theorem 7.2   The function $f(n) = r^n$ is a nontrivial solution of

$$f(n + k) + a_{n+k-1}f(n + k - 1) + \cdots + a_n f(n) = 0$$

if and only if r is a solution of the characteristic equation

$$A^k + a_{n+k-1}A^{k-1} + \cdots + a_n = 0$$

The following examples illustrate the use of Theorem 7.2.

Example 3   To find a solution to the difference equation

$$f(n + 2) - f(n + 1) - 2f(n) = 0 \tag{3}$$

we substitute $f(n) = A^n$ into (3) to obtain

$$A^{n+2} - A^{n+1} - 2A^n = 0$$

or, equivalently,

$$A^n(A^2 - A - 2) = 0$$

The equation

$$A^2 - A - 2 = 0$$

is the characteristic equation of (3). Since we can factor this equation as $(A + 1)(A - 2) = 0$, it has roots $A = 2$ and $A = -1$. It therefore follows that $f(n) = 2^n$ and $f(n) = (-1)^n$ are nontrivial solutions of (3).   □

Example 4   You can easily verify that the characteristic equation of the difference equation

$$f(n + 3) - 3f(n + 2) - f(n + 1) + 3f(n) = 0 \tag{4}$$

is

$$A^3 - 3A^2 - A + 3 = 0 \tag{5}$$

By inspection we see that $A = 1$ is a root of (5), and therefore $(A - 1)$ is a factor of $A^3 - 3A^2 - A + 3$. Dividing by $(A - 1)$ yields

$$A^3 - 3A^2 - A + 3 = (A - 1)(A^2 - 2A - 3)$$

which, factored further, gives us

$$A^3 - 3A^2 - A + 3 = (A - 1)(A - 3)(A + 1)$$

The roots of this characteristic equation are 1, 3, and -1. Thus we find that $f(n) = 1^n = 1$, $f(n) = 3^n$ and $f(n) = (-1)^n$ are nontrivial solutions of (4).   □

Examples 3 and 4 showed us that a difference equation can have more than one solution. The next result indicates that homogeneous difference equations have infinitely many solutions.

Theorem 7.3   If $g(n)$ and $h(n)$ are solutions to the difference equation

$$f(n + k) + a_{n+k-1}f(n + k - 1) + \cdots + a_n f(n) = 0 \tag{6}$$

then for any constants α_1 and α_2, $\alpha_1 g(n) + \alpha_2 h(n)$ is also a solution.

Proof: The proof is quite easy; we simply substitute $f(n) = \alpha_1 g(n) + \alpha_2 h(n)$ into equation (6). This substitution yields

$$f(n + k) + a_{n+k-1} f(n + k - 1) + \cdots + a_n f(n)$$
$$= \alpha_1 g(n + k) + \alpha_2 h(n + k) + a_{n+k-1}(\alpha_1 g(n + k - 1)$$
$$+ \alpha_2 h(n + k - 1)) + \cdots + a_n(\alpha_1 g(n) + \alpha_2 h(n))$$
$$= \alpha_1 [g(n + k) + a_{n+k-1} g(n + k - 1) + \cdots + a_n g(n)]$$
$$+ \alpha_2 [h(n + k) + a_{n+k-1} h(n + k - 1) + \cdots + a_n h(n)]$$

Since g and h are solutions of (6), it follows that

$$\alpha_1 [g(n + k) + a_{n+k-1} g(n + k - 1) + \cdots + a_n g(n)]$$
$$+ \alpha_2 [h(n + k) + a_{n+k-1} h(n + k - 1) + \cdots + a_n$$
$$= \alpha_1 \cdot 0 + \alpha_2 \cdot 0 = 0$$

and this concludes the proof.   ∎

Example 5   To illustrate the use of Theorem 7.3, we recall that in Example 3 we saw that both $g(n) = 2^n$ and $h(n) = (-1)^n$ are solutions of

$$f(n + 2) - f(n + 1) - 2f(n) = 0 \tag{7}$$

It follows from Theorem 7.3 that $f(n) = \alpha_1(2)^n + \alpha_2(-1)^n$ is also a solution of (7) for any constants α_1 and α_2. Thus we have found an infinite number of solutions of (7).   □

Although in Example 5 we found an infinite number of solutions to (7), the question remains whether we have found *all* of the solutions. We consider this question in a more general context in the next few sections.

In many applications, kth-order linear difference equations may arise in a slightly different form than that given by

$$f(n + k) + a_{n+k-1} f(n + k - 1) + \cdots + a_n f(n) = 0 \tag{8}$$

For instance, although the equation

$$f(n + 1) - 6f(n) + 2f(n - 1) + 4f(n - 2) = 0 \tag{9}$$

is not written in the form given in (8), we can transform it into this form by substituting m for $n - 2$. With this substitution, equation (9) becomes the third-order linear difference equation

$$f(m + 3) - 6f(m + 2) + 2f(m + 1) + 4f(m) = 0$$

In a similar fashion we can easily see that

$$f(n) - 3f(n - 1) + 2f(n - 2) = 0$$

is a second-order linear difference equation.

To solve equations such as (9), however, it is not necessary to transform the given equation into standard form. We can simply apply the techniques we have already developed in this section.

Example 6   To find solutions to

$$f(n) - 3f(n-1) + 2f(n-2) = 0 \tag{10}$$

we substitute $f(n) = A^n$ into equation (10) to obtain

$$A^n - 3A^{n-1} + 2A^{n-2} = 0$$

Dividing both sides of this equation by A^{n-2} we have the characteristic equation

$$A^2 - 3A + 2 = 0$$

which factors to $(A - 1)(A - 2) = 0$. Thus $A = 1$ or $A = 2$, and therefore 1^n and 2^n are both solutions of (10). By Theorem 7.3

$$\alpha_1(1^n) + \alpha_2(2^n) = \alpha_1 + \alpha_2(2^n)$$

provides an infinite number of solutions to (10) since α_1 and α_2 are arbitrary constants.   □

7.2 Problems

In Problems 1–16 find an expression that will yield an infinite number of solutions to the given difference equation.

1. $f(n+2) + f(n+1) - 2f(n) = 0$
2. $f(n+2) - f(n+1) - 6f(n) = 0$
3. $4f(n+1) - 9f(n-1) = 0$
4. $2f(n) - 3f(n-1) - 5f(n-2) = 0$
5. $f(n+2) - 9f(n) = 0$
6. $f(n) + 4f(n-1) - 2f(n-2) = 0$
7. $f(n+2) - 3f(n+1) - 18f(n) = 0$
8. $f(n+1) + 2f(n) - 2f(n-1) = 0$
9. $f(n+3) + 7f(n+2) + 7f(n+1) - 15f(n) = 0$
10. $f(n+3) + 2f(n+2) - 5f(n+1) - 6f(n) = 0$
11. $9f(n+1) + 18f(n) - 7f(n-1) = 0$
12. $f(n+2) + 2f(n+1) - 3f(n) = 0$
13. $f(n) - 4f(n-1) + 4f(n-2) = 0$
14. $f(n) + f(n-1) - f(n-2) = 0$
15. $f(n+1) - f(n) = 0$

16. $f(n + 2) + f(n + 1) = 0$

17. Explain why Theorem 7.3 tells us that if g is a solution to $f(n + k) + a_{n+k-1}f(n + k - 1) + \cdots + a_n f(n) = 0$, then so is αg for any constant α.

7.3   THE GENERAL SOLUTION OF A HOMOGENEOUS LINEAR DIFFERENCE EQUATION

Most applications of difference equations require a specific solution. The easiest way to find such a solution is to find first *all* solutions of the given difference equation and then use additional information to determine the desired particular solution. Our objective now is to find the general form of all solutions (the *general solution*) of a given homogeneous linear difference equation with constant coefficients. In this section we will use determinants (discussed in Section 3 of the Appendix).

The following theorem (the proof is omitted) provides critical information regarding the general solution of the second-order difference equation

$$f(n + 2) + af(n + 1) + bf(n) = 0$$

Theorem 7.4   If $g(n)$ and $h(n)$ are solutions of the second-order homogeneous linear difference equation

$$f(n + 2) + af(n + 1) + bf(n) = 0 \tag{1}$$

and if the determinant

$$\begin{vmatrix} g(n) & h(n) \\ g(n - 1) & h(n - 1) \end{vmatrix} \tag{2}$$

is not equal to 0 for some integer n, then *every* solution of (1) can be written in the form

$$\alpha_1 g(n) + \alpha_2 h(n)$$

That is, $\alpha_1 g(n) + \alpha_2 h(n)$ is the general solution of (1).

The determinant (2) is called *Casorati's determinant* of g and h.

From Theorem 7.4 we see that to obtain the general solution of a second-order linear homogeneous difference equation, we only need to find two particular solutions of the equation that have a nonzero Casorati determinant.

Example 1   We have seen in Example 3 of Section 7.2 that $g(n) = 2^n$ and $h(n) = (-1)^n$ are solutions of

$$f(n + 2) - f(n + 1) - 2f(n) = 0 \tag{3}$$

To determine if $\alpha_1 2^n + \alpha_2(-1)^n$ is the general solution of (3) we calculate Casorati's determinant of g and h. Since

$$\begin{vmatrix} g(n) & h(n) \\ g(n-1) & h(n-1) \end{vmatrix} = \begin{vmatrix} 2^n & (-1)^n \\ 2^{n-1} & (-1)^{n-1} \end{vmatrix} = (-1)^{n-1}2^n - (-1)^n 2^{n-1}$$

$$= (-1)^{n-1}2^{n-1}(2-(-1))$$
$$= 3(-1)^{n-1}2^{n-1}$$
$$\neq 0$$

it follows by Theorem 7.4 that $\alpha_1 2^n + \alpha_2(-1)^n$ is the general solution of (3).

□

Consider now the general second-order linear homogeneous difference equation with constant coefficients

$$f(n+2) + af(n+1) + bf(n) = 0 \tag{4}$$

The characteristic equation associated with (4) is the quadratic equation

$$A^2 + aA + b = 0$$

If this equation has two distinct real roots, r_1 and r_2, then both $g(n) = r_1^n$ and $h(n) = r_2^n$ are solutions of (4).

Calculating the Casorati determinant of g and h yields

$$\begin{vmatrix} g(n) & h(n) \\ g(n-1) & h(n-1) \end{vmatrix} = \begin{vmatrix} r_1^n & r_2^n \\ r_1^{n-1} & r_2^{n-1} \end{vmatrix}$$

$$= r_1^n r_2^{n-1} - r_1^{n-1} r_2^n$$
$$= r_1^{n-1} r_2^{n-1}(r_1 - r_2)$$

Since $r_1 - r_2 \neq 0$, we have

$$\begin{vmatrix} g(n) & h(n) \\ g(n-1) & h(n-1) \end{vmatrix} \neq 0$$

Consequently, it follows from Theorem 7.4 that $\alpha_1 r_1^n + \alpha_2 r_2^n$ is the general solution of any second-order homogeneous difference equation whose characteristic equation has distinct roots.

Example 2   We find the general solution of

$$f(n+2) + 3f(n+1) + 2f(n) = 0 \tag{5}$$

The characteristic equation of (5) is

$$A^2 + 3A + 2 = (A+2)(A+1) = 0$$

The roots $A = -2$ and $A = -1$ of this equation are distinct, and, therefore

$$\alpha_1(-1)^n + \alpha_2(-2)^n$$

is the general solution of (5).

□

We can generalize Theorem 7.4 to higher-order homogeneous linear difference equations as follows.

Theorem 7.5   If $g_1(n)$, $g_2(n)$, ... , and $g_k(n)$ are solutions to the kth-order homogeneous difference equation

$$f(n + k) + a_{n+k-1}f(n + k - 1) + \cdots + a_n f(n) = 0 \qquad (6)$$

and if the $k \times k$ determinant

$$\begin{vmatrix} g_1(n) & g_2(n) & \cdots & g_k(n) \\ g_1(n - 1) & g_2(n - 1) & \cdots & g_k(n - 1) \\ \vdots & \vdots & \vdots & \vdots \\ g_1(n - k + 1) & g_2(n - k + 1) & \cdots & g_k(n - k + 1) \end{vmatrix}$$

is not equal to zero for some integer n, then every solution of (6) can be written in the form

$$\alpha_1 g_1(n) + \alpha_2 g_2(n) + \cdots + \alpha_k g_k(n)$$

where $\alpha_1, \alpha_2, \ldots, \alpha_k$ are constants.

The problem of finding the general solution to (6) is now reduced to that of finding k solutions $g_1(n), g_2(n), \ldots, g_k(n)$ that satisfy Theorem 7.5.

If the characteristic equation of (6) has k *distinct* roots, $r_1, r_2, \ldots, r_k$, it can be shown that the functions $g_1(n) = r_1^n, g_2(n) = r_2^n, \ldots, g_k(n) = r_k^n$ satisfy the hypothesis of Theorem 7.5 and hence, the general solution of (6) is

$$\alpha_1 r_1^n + \alpha_2 r_2^n + \cdots + \alpha_k r_k^n$$

Example 3   We find the general solution of the difference equation

$$f(n + 3) - 3f(n + 2) - f(n + 1) + 3f(n) = 0 \qquad (7)$$

The characteristic equation of this difference equation is

$$A^3 - 3A^2 - A + 3 = 0 \qquad (8)$$

Note that $A = 1$ is a root of (8), and, therefore, $(A - 1)$ is a factor of the left-hand side of the characteristic equation. Dividing by $(A - 1)$ yields

$$A^3 - 3A^2 - A + 3 = (A - 1)(A^2 - 2A - 3)$$

Thus,

$$A^3 - 3A^2 - A + 3 = (A - 1)(A + 1)(A - 3)$$

and the roots of (8) are $A = 1$, $A = -1$, and $A = 3$. Since these roots are distinct, it follows from the comments above that the general solution of (7) is

$$\alpha_1 1^n + \alpha_2(-1)^n + \alpha_3(3)^n = \alpha_1 + \alpha_2(-1)^n + \alpha_3(3)^n \qquad \square$$

In many applications we must select a particular solution of a difference equation that satisfies what are called *initial conditions*. In the case of a second-order difference equation we generally specify the initial conditions by two values of the desired solution.

Example 4   We find the particular solution to

$$f(n + 2) + 3f(n + 1) + 2f(n) = 0 \qquad (9)$$

that satisfies the initial conditions $f(0) = 0$ and $f(1) = 2$.

From Example 2 we know that *all* solutions to (9) have the form

$$f(n) = \alpha_1(-1)^n + \alpha_2(-2)^n$$

If we let $n = 0$ and use the initial condition $f(0) = 0$, we obtain

$$0 = \alpha_1(-1)^0 + \alpha_2(-2)^0 = \alpha_1 + \alpha_2 \qquad (10)$$

Similarly, letting $n = 1$, and using the initial condition $f(1) = 2$, we obtain

$$2 = -\alpha_1 - 2\alpha_2 \qquad (11)$$

From (10) and (11), we find that $\alpha_1 = 2$ and $\alpha_2 = -2$. Thus the desired particular solution is

$$\begin{aligned} f(n) &= 2(-1)^n - 2(-2)^n \\ &= 2(-1)^n + (-2)^{n+1} \end{aligned}$$ □

Example 5   Recall that the proposed model of seed production introduced in Section 7.1 led to the difference equation

$$S(n + 2) - 75S(n + 1) - 400S(n) = 0 \qquad (12)$$

We solve this difference equation under the assumption that we begin with ten seeds, all of which are planted. Flowers resulting after one year from these seeds will produce $10(.2)(375) = 750$ seeds. Consequently, $S(0) = 10$ and $S(1) = 750$. We can use this information to solve for $S(n)$.

We need to find the general solution of (12) before we can solve for the particular solution. We note that the characteristic equation of (12) is

$$A^2 - 75A - 400 = 0$$

Using the quadratic formula gives

$$A = 80 \quad \text{or} \quad A = -5$$

Thus

$$S(n) = \alpha_1(80)^n + \alpha_2(-5)^n$$

is the general solution of (12). Because we seek the particular solution that satisfies the conditions $S(0) = 10$ and $S(1) = 750$, we must select α_1 and α_2 so that

$$10 = \alpha_1 + \alpha_2 \quad \text{and} \quad 750 = 80\alpha_1 - 5\alpha_2$$

Solving for α_1 and α_2 gives

$$\alpha_1 = \frac{160}{17} \quad \text{and} \quad \alpha_2 = \frac{10}{17}$$

Consequently, we obtain the desired particular solution

$$S(n) = \left(\frac{160}{17}\right)(80)^n + \left(\frac{10}{17}\right)(-5)^n \qquad \square$$

This solution to Example 5 allows us to calculate the number of seeds available after n years. For example, after five years we will have

$$S(5) = \left(\frac{160}{17}\right)(80)^5 + \left(\frac{10}{17}\right)(-5)^5 \approx 3 \times 10^{10}$$

seeds.

In this section we have seen how to solve a homogeneous difference equation whose characteristic equation has distinct real roots. In the next section we consider the problem of repeated roots.

7.3   Problems

In Problems 1–8 find the general solution of the given difference equation. If initial conditions are given, find the particular solution that satisfies those conditions.

1. $f(n + 2) - f(n + 1) - 2f(n) = 0$; $f(0) = 0$, $f(1) = 1$
2. $f(n + 1) - 5f(n) = 0$
3. $f(n) = 7f(n - 1)$; $f(0) = 2$
4. $9f(n + 2) + 3f(n + 1) - 2f(n) = 0$; $f(0) = 1$, $f(1) = -1$
5. $2f(n + 2) + 3f(n + 1) - 2f(n) = 0$; $f(1) = 0$, $f(3) = -1$
6. $f(n + 3) - 6f(n + 2) + 11f(n + 1) - 6f(n) = 0$; $f(0) = 1$, $f(1) = 0$, $f(2) = -1$
7. $2f(n + 2) + 5f(n + 1) + f(n) - 2f(n - 1) = 0$
8. $f(n + 1) = \frac{1}{2}f(n)$; $f(1) = 1$
9. Show that $\alpha_1 3^n + \alpha_2 n 3^n$ is the general solution to the difference equation $f(n + 2) - 6f(n + 1) + 9f(n) = 0$.
10. Show that $\alpha_1 2^n + \alpha_2 n 2^n$ is the general solution to the difference equation $f(n + 2) - 4f(n + 1) + 4f(n) = 0$.

11. Show that $\alpha_1 + \alpha_2 n$ is the general solution to the difference equation $f(n) - 2f(n-1) + f(n-2) = 0$.

12. Show that $\alpha_1(-1)^n + \alpha_2 n(-1)^n$ is the general solution to the difference equation $f(n+2) + 2f(n+1) + f(n) = 0$.

13. Show that $\alpha_1 \cos(n\pi/4) + \alpha_2 \sin(n\pi/4)$ is the general solution of the difference equation $f(n+2) - \sqrt{2}f(n+1) + f(n) = 0$.

14. Show that $\alpha_1 \cos(n\pi/2) + \alpha_2 \sin(n\pi/2)$ is the general solution of the difference equation $f(n+2) + f(n) = 0$.

15. Compute $f(n)$, the number of sequences of length n that can be formed using the symbols 00 and 1 (the symbol 00 has length two). (*Hint:* See Problem 21, Section 7.1. Note that $f(1) = 1$ and $f(2) = 2$.)

16. Compute $f(n)$, the number of code words of length n that can be formed from a's and b's where no two b's are adjacent. (See Problem 23, Section 7.1.)

17. Compute $f(n)$, the number of people involved in the nth level of a chain letter begun at the first level by two friends. At each level a person must send the letter to five additional people. How many people are involved at the fifteenth level? (See Problem 24, Section 7.1.)

18. How many sequences of length 17 can be formed using the symbols 1, 2, 00? (The symbol 00 has length two.)

19. Find an expression for the number of sequences of length n that can be formed using the symbols 00, 1, 2, 3, 4, 55. (Both of the symbols 00 and 55 have length two.) How many such sequences of length twelve are there?

7.4 CHARACTERISTIC EQUATIONS WITH REPEATED ROOTS

In Section 7.3 we found that if the characteristic equation of

$$f(n+2) + af(n+1) + bf(n) = 0$$

has two *distinct* roots r_1 and r_2, then the general solution is $f(n) = \alpha_1 r_1^n + \alpha_2 r_2^n$ where α_1 and α_2 are arbitrary constants. If, however, the roots are not distinct—that is, if $r_1 = r_2$—then the general solution takes on a somewhat different form, which we examine in this section.

If $r_1 \neq 0$ is the only root of the characteristic equation

$$A^2 + aA + b = 0$$

then

$$A^2 + aA + b = (A - r_1)^2$$

Thus,

$$A^2 + aA + b = A^2 - 2r_1 A + r_1^2$$

and it follows that the difference equation above can be written in the form

$$f(n + 2) - 2r_1 f(n + 1) + r_1^2 f(n) = 0 \tag{1}$$

As in Section 7.2, $g(n) = r_1^n$ is one particular solution to (1). To find the general solution to (1), we can apply Theorem 7.3. First, however, we need a second solution. Note that if we define a function h by $h(n) = nr_1^n$, then

$$h(n + 1) = (n + 1)r_1^{n+1} \quad \text{and} \quad h(n + 2) = (n + 2)r_1^{n+2}$$

Substituting these expressions into (1) gives

$$\begin{aligned}
h(n + 2) - 2r_1 h(n + 1) + r_1^2 h(n) &= (n + 2)r_1^{n+2} - 2r_1(n + 1)r_1^{n+1} + r_1^2 nr_1^n \\
&= r_1^{n+2}[(n + 2) - 2(n + 1) + n] \\
&= r_1^{n+2} \cdot 0 = 0
\end{aligned}$$

so $h(n) = nr_1^n$ is also a solution of (1).

Note that if $r_1 \neq 0$, we have

$$\begin{aligned}
\begin{vmatrix} g(n) & h(n) \\ g(n - 1) & h(n - 1) \end{vmatrix} &= \begin{vmatrix} r_1^n & nr_1^n \\ r_1^{n-1} & (n - 1)r_1^{n-1} \end{vmatrix} \\
&= (n - 1)r_1^{2n-1} - nr_1^{2n-1} \\
&= -r_1^{2n-1} \\
&\neq 0
\end{aligned}$$

Therefore, by Theorem 7.3 the general solution of (1) is given by

$$\alpha_1 r_1^n + \alpha_2 nr_1^n \tag{2}$$

Example 1   We find the general solution of the difference equation

$$f(n) - 10f(n - 1) + 25f(n - 2) = 0 \tag{3}$$

The characteristic equation of (3) is

$$A^2 - 10A + 25 = 0$$

which has $r = 5$ as its only root. It follows from (2) that the general solution of (3) is

$$f(n) = \alpha_1 5^n + \alpha_2 n5^n \qquad \square$$

The following classical problem (known as the "gambler's ruin" problem) leads to a characteristic equation with repeated roots.

Example 2   Suppose a gambler bets one dollar on each play of a game in which the probability of winning is the same as the probability of losing. The gambler's goal is to quit the game with a total of k dollars. Let $P(n)$ denote the probability that whenever the gambler has n dollars, he will go broke before attaining his goal.

We show in Chapter 9 that $P(n)$ satisfies the second-order difference equation

$$\tfrac{1}{2}P(n + 1) - P(n) + \tfrac{1}{2}P(n - 1) = 0 \tag{4}$$

The characteristic equation of this difference equation is

$$\tfrac{1}{2}A^2 - A + \tfrac{1}{2} = 0$$

which has $r = 1$ as its only root. Consequently, by (2)

$$\begin{aligned} P(n) &= \alpha_1 1^n + \alpha_2 n 1^n \\ &= \alpha_1 + \alpha_2 n \end{aligned} \tag{5}$$

is the general solution to (4).

Note that

$P(0) = 1$ (If the gambler has 0 dollars, the probability of his going broke is one: He already *is* broke!)

$P(k) = 0$ (If the gambler obtains k dollars, the probability of his going broke is zero: He will quit and not go broke.)

Substituting $n = 0$ and $n = k$ back into (5) yields

$$1 = \alpha_1 + \alpha_2 \cdot 0 \quad \text{and} \quad 0 = \alpha_1 + \alpha_2 k$$

Thus, $\alpha_1 = 1$ and $\alpha_2 = -1/k$, and consequently

$$P(n) = 1 - n/k$$

In other words, the probability that the gambler will go broke before attaining the goal of quitting with k dollars is $1 - n/k$.   □

In higher-order homogeneous difference equations, if a root r appears t times (that is, it is a *root of multiplicity t*) in a characteristic equation, then the root r contributes t solutions

$$r^n, nr^n, n^2 r^n, \ldots, n^{t-1} r^n$$

to the general solution. Example 3 illustrates this.

Example 3   The characteristic equation of the difference equation

$$f(n + 3) - 3f(n + 2) + 3f(n + 2) - f(n) = 0$$

is

$$A^3 - 3A^2 + 3A - 1 = 0$$

Factoring this, we obtain

$$(A - 1)^3 = 0$$

Its root $A = 1$ is a root of multiplicity three, so the general solution of the difference equation is

$$\alpha_1 1^n + \alpha_2 n 1^n + \alpha_3 n^2 1^n = \alpha_1 + \alpha_2 n + \alpha_3 n^2 \qquad \square$$

Thus far we have seen how to obtain solutions to a difference equation whose characteristic equation has real roots. In the next section we obtain solutions to difference equations whose characteristic equations have complex roots.

7.4   Problems

In Problems 1–17 find the general solution of the given difference equation. If initial conditions are given, find the particular solution that satisfies those conditions.

1. $f(n + 2) + 4f(n + 1) + 4f(n) = 0$
2. $f(n + 2) + 6f(n + 1) + 9f(n) = 0; f(0) = 1, f(1) = 0$
3. $f(n + 1) - 4f(n) + 4f(n - 1) = 0$
4. $f(n + 2) - 8f(n + 1) + 16f(n) = 0$
5. $f(n) - 2f(n - 1) + f(n - 2) = 0; f(1) = 0, f(-1) = 4$
6. $4f(n + 2) - 4f(n + 1) + f(n) = 0$
7. $9f(n + 2) - 6f(n + 1) + f(n) = 0$
8. $f(n + 3) - 3f(n + 1) - 4f(n - 1) = 0$
9. $f(n + 2) - 3f(n + 1) - 4f(n - 1) = 0$
10. $f(n + 2) - 10f(n + 1) + 25f(n) = 0; f(0) = 0, f(1) = 2$
11. $f(n) + 14f(n - 1) + 49f(n - 2) = 0; f(0) = 0, f(1) = 0$
12. $f(n) - 10f(n - 1) + 25f(n - 2) = 0; f(0) = 0, f(1) = 0$
13. $f(n + 2) + 14f(n + 1) + 49f(n) = 0; f(0) = 1, f(1) = 2$
14. $f(n + 3) - 3f(n + 2) + 4f(n) = 0$
15. $f(n + 1) - 3f(n) + 3f(n - 1) - f(n - 2) = 0; f(0) = 0, f(1) = 2,$
    $f(2) = -1$
16. $f(n + 2) - f(n + 1) - f(n) + f(n - 1) = 0; \quad f(0) = 0, \quad f(1) = 1,$
    $f(2) = 0$
17. $f(n + 2) - 2f(n + 1) - 3f(n) + 4f(n - 1) + 4f(n - 2) = 0$
18. Find the general solution of the difference equation

$$f(n + 2) + af(n + 1) + bf(n) = 0$$

when the only root of the corresponding characteristic equation is $r_1 = 0$.

19. A sequence begins 2, 7, ... After the first two terms in the sequence, each successive term is found by doubling the immediately preceding term and then subtracting the term that appears two places previously. Thus, the

first three terms of this sequence are 2, 7, 12. Write a difference equation for the nth term of this sequence, and use the difference equation to find an expression for the nth term.

20. (a) Repeat Problem 19 given that the initial two terms of the sequence are a_1 and a_2.

    (b) What conditions on a_1 and a_2 will result in a constant sequence?

7.5 CHARACTERISTIC EQUATIONS WITH COMPLEX ROOTS

We now take up the problem of difference equations whose characteristic equations have complex roots.

The characteristic equation of a homogeneous linear difference equation with constant coefficients is a polynomial equation. Some or all of the roots of such an equation may be complex numbers. The complex roots always occur in pairs of a complex number and its complex conjugate. In other words, if $u + iv$ is a complex root of a polynomial equation with real coefficients, then its complex conjugate, $u - iv$, is also a root. In this section we solve difference equations that have characteristic equations with complex roots.

Assume that

$$f(n + 2) + af(n + 1) + bf(n) = 0 \qquad (1)$$

is a difference equation whose characteristic equation

$$A^2 + aA + b = 0 \qquad (2)$$

has complex roots $r_1 = u + iv$ and $r_2 = u - iv$, where $v \neq 0$. Since $r_1 \neq r_2$, it follows from Section 7.3 that the general solution is

$$\alpha_1 r_1^n + \alpha_2 r_2^n = \alpha_1(u + iv)^n + \alpha_2(u - iv)^n$$

Because the original equation has only real coefficients, it is reasonable to try to express the general solution in terms of real numbers. To this end we express $u + iv$ and $u - iv$ in polar coordinates. As Figure 7.3 shows, $u + iv$ and $u - iv$ are symmetrically located in the complex plane, and we see that if we let $\rho = \sqrt{u^2 + v^2}$ we can write $u = \rho \cos \theta$ and $v = \rho \sin \theta$.

Thus,

$$u + iv = \rho(\cos \theta + i \sin \theta)$$

and

$$u - iv = \rho(\cos \theta - i \sin \theta)$$

DeMoivre's Theorem states that

$$[\rho(\cos \theta \pm i \sin \theta)]^n = \rho^n(\cos n\theta \pm i \sin n\theta)$$

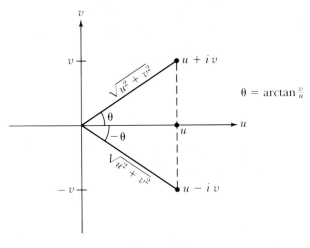

Figure 7.3

and it follows that we can write the general solution of equation (1) as

$$\begin{aligned}
\alpha_1(u + iv)^n + \alpha_2(u - iv)^n &= \alpha_1[\rho(\cos\theta + i\sin\theta)]^n \\
&\quad + \alpha_2[\rho(\cos\theta - i\sin\theta)]^n \\
&= \alpha_1\rho^n(\cos n\theta + i\sin n\theta) \\
&\quad + \alpha_2\rho^n(\cos n\theta - i\sin n\theta) \\
&= (\alpha_1 + \alpha_2)\rho^n\cos n\theta \\
&\quad + i(\alpha_1 - \alpha_2)\rho^n\sin n\theta \qquad (3)
\end{aligned}$$

If we substitute $\alpha_1 = \alpha_2 = \frac{1}{2}$ into (3) we see that $\rho^n\cos n\theta$ is a particular solution to the difference equation. Similarly, if we choose $\alpha_1 = -i/2$ and $\alpha_2 = i/2$ we see that

$$\begin{aligned}
i(-i/2 - i/2)\rho^n\sin n\theta &= -i^2\rho^n\sin n\theta \\
&= \rho^n\sin n\theta
\end{aligned}$$

is also a particular solution to equation (1).

Next we compute Casorati's determinant of $g(n) = \rho^n\sin n\theta$ and $h(n) = \rho^n\cos n\theta$.

$$\begin{aligned}
\begin{vmatrix} g(n) & h(n) \\ g(n-1) & h(n-1) \end{vmatrix} &= \begin{vmatrix} \rho^n\sin n\theta & \rho^n\cos n\theta \\ \rho^{n-1}\sin(n-1)\theta & \rho^{n-1}\cos(n-1)\theta \end{vmatrix} \\
&= \rho^{2n-1}[\sin n\theta\cos(n-1)\theta - \cos n\theta\sin(n-1)\theta]
\end{aligned}$$

$$(4)$$

Since $\sin\phi\cos\psi - \cos\phi\sin\psi = \sin(\phi - \psi)$, we can simplify (4) to obtain

$$\rho^{2n-1}\sin(n\theta - (n-1)\theta) = \rho^{2n-1}\sin\theta$$

Since we have assumed that $v \neq 0$, we see from Figure 7.3 that $0 < \theta < \pi$.

Consequently,

$$\begin{vmatrix} g(n) & h(n) \\ g(n-1) & h(n-1) \end{vmatrix} = \rho^{2n-1} \sin \theta \neq 0$$

By Theorem 7.4, therefore,

$$\alpha_1 \rho^n \sin n\theta + \alpha_2 \rho^n \cos n\theta \, \text{)} \tag{5}$$

is the general solution of (1).

In summary, we have shown that if $u \pm iv$ are roots of the characteristic equation (2), then the general solution of (1) in terms of real numbers is

$$\alpha_1 \rho^n \sin n\theta + \alpha_2 \rho^n \cos n\theta$$

where $\rho = \sqrt{u^2 + v^2}$ and $\theta = \arctan v/u$.

Example 1   To find the general solution of the difference equation

$$f(n) + 2f(n-1) + 2f(n-2) = 0 \tag{6}$$

we first note that its characteristic equation is

$$A^2 + 2A + 2 = 0$$

Using the quadratic formula we obtain the roots:

$$A = \frac{-2 \pm \sqrt{4-8}}{2} = -1 \pm i$$

Next, we plot the roots $r_1 = -1 + i$ and $r_2 = -1 - i$ as shown in Figure 7.4 and see that $\theta = 3\pi/4$ and $\rho = \sqrt{2}$.

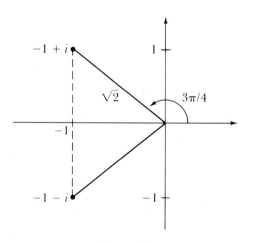

Figure 7.4

Finally, substituting these values into (5), we obtain

$$f(n) = \alpha_1(\sqrt{2})^n \sin \frac{3n\pi}{4} + \alpha_2(\sqrt{2})^n \cos \frac{3n\pi}{4}$$

as the general solution of the difference equation (6). ☐

Example 2 In a predator–prey relationship, an increase during one year in the number of prey often causes a corresponding increase during the next year in the number of predators. This increased number of predators will then cause a decrease in the number of prey in the subsequent year. We use difference equations to determine the number of predators and prey present during any year.

One of the simplest models describing the predator–prey relationship is based on the following two assumptions:

a. The change in the number of predators is proportional to the difference between the current number of prey and some base level B_1 of prey.

b. The change in the number of prey is proportional to the difference between the current number of predators and some base level B_2 of predators.

Let $f(n)$ be the difference between the number of prey present in year n and the base level B_2. Let $g(n)$ be the difference between the number of predators present in year n and the base level B_1. Since $g(n) - g(n - 1)$ is the change in the number of predators between year $n - 1$ and year n, statement **a.** above becomes

$$g(n) - g(n - 1) = k_1 f(n - 1) \qquad k_1 > 0 \qquad (7)$$

where k_1 is the constant of proportionality.

In a similar fashion, since $f(n - 1) - f(n)$ is the change in the number of prey, statement **b.** becomes

$$f(n) - f(n - 1) = -k_2 g(n - 1) \qquad k_2 > 0 \qquad (8)$$

Solving (8) for $g(n - 1)$, we have

$$g(n - 1) = \frac{1}{k_2} [f(n - 1) - f(n)]$$

Thus,

$$g(n) = \frac{1}{k_2} [f(n) - f(n + 1)]$$

Substituting these last two expressions into (7) gives

$$\frac{1}{k_2} [f(n) - f(n + 1)] - \frac{1}{k_2} [f(n - 1) - f(n)] = k_1 f(n - 1) \qquad (9)$$

Algebraic manipulation of (9) yields

$$2f(n) - f(n + 1) - f(n - 1) = k_1 k_2 f(n - 1)$$

and it follows that

$$f(n + 1) - 2f(n) + (k_1 k_2 + 1)f(n - 1) = 0 \qquad (10)$$

The characteristic equation of (10) is

$$A^2 - 2A + (k_1 k_2 + 1) = 0$$

and from the quadratic formula we find the roots

$$A = \frac{2 \pm \sqrt{-4k_1 k_2}}{2} = 1 \pm \sqrt{k_1 k_2}\, i$$

In Figure 7.5 we see that $\rho = \sqrt{1 + k_1 k_2}$ and $\theta = \arctan \sqrt{k_1 k_2}$.

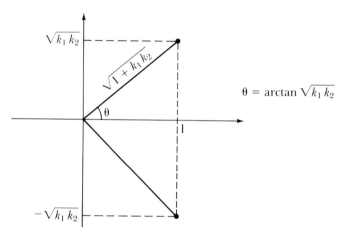

$$\theta = \arctan \sqrt{k_1 k_2}$$

Figure 7.5

Consequently, we write the general solution of (9) as

$$f(n) = (1 + k_1 k_2)^{n/2}(\alpha_1 \sin n\theta + \alpha_2 \cos n\theta)$$

where $\theta = \arctan \sqrt{k_1 k_2}$. Note that this solution is consistent with the expected oscillatory nature of the number of prey. □

7.5 Problems

In Problems 1–11 find the general solution of the given difference equation. If initial conditions are given, also find the particular solution that satisfies those conditions.

1. $f(n + 2) + 2f(n) = 0$
2. $f(n) - 2f(n - 1) + 2f(n - 2) = 0$
3. $f(n) + 4f(n - 1) + 8f(n - 2) = 0; \; f(0) = 2, \; f(1) = -1$
4. $f(n) - 4f(n - 1) + 8f(n - 2) = 0; \; f(2) = 1, \; f(3) = -2$
5. $f(n + 2) + 2f(n + 1) + 4f(n) = 0$
6. $f(n + 3) - f(n) = 0$
7. $f(n + 3) - f(n + 2) + f(n + 1) - f(n) = 0$
8. $f(n + 2) + f(n + 1) + f(n) + f(n - 1) = 0; \quad f(0) = 0, \quad f(1) = 2,$
   $f(2) = 1$
9. $f(n) + 4f(n - 1) + f(n - 2) + 4f(n - 3) = 0$
10. $f(n + 4) - 2f(n + 3) - f(n + 2) - 4f(n + 1) - 6f(n) = 0$
11. $f(n + 3) + 2f(n + 2) + 8f(n + 1) + 8f(n) + 16f(n - 1) = 0$
12. $f(n + 2) + 5f(n) + 6f(n - 2) = 0$
13. Assume the following about the predator–prey relationship:

   a. For every 5 prey above 80 there is an increase of one in the number of predators $(k_1 = 5)$.

   b. For every predator above 120 there is a decrease of 5 in the number of prey $(k_2 = \frac{1}{5})$.

   Find an expression for the number of prey present at year n. Suppose in addition that there are 100 prey present when $n = 0$, and 125 present when $n = 1$. Find the particular solution that describes this situation. How many prey are present in year 5?

7.6   NONHOMOGENEOUS DIFFERENCE EQUATIONS

Thus far we have considered only difference equations of the form

$$f(n + k) + a_{n+k-1}f(n + k - 1) + \cdots + a_n f(n) = \phi(n) \qquad (1)$$

where $\phi(n) = 0$ for all n. We now turn our attention to the nonhomogeneous case, where $\phi(n)$ is a nonzero function. In the next theorem we see that we can obtain the general solution of (1) by adding any particular solution of (1) to the general solution of the corresponding homogeneous equation.

Theorem 7.6   Let $p(n)$ be a particular solution to

$$f(n + k) + a_{n+k-1}f(n + k - 1) + \cdots + a_n f(n) = \phi(n) \qquad (2)$$

and let

$$\alpha_1 g_1(n) + \alpha_2 g_2(n) + \cdots + \alpha_k g_k(n)$$

be the general solution to

$$f(n + k) + a_{n+k-1}f(n + k - 1) + \cdots + a_n f(n) = 0$$

Then

$$\alpha_1 g_1(n) + \alpha_2 g_2(n) + \cdots + \alpha_k g_k(n) + p(n) \tag{3}$$

is the general solution to (2). That is, any solution of (2) can be expressed in the form (3).

Proof: Let $h(n)$ be any solution to (2). We must show that there are coefficients $\alpha_1, \alpha_2, \ldots, \alpha_k$ such that $h(n)$ can be written in the form (3).

Since both $p(n)$ and $h(n)$ are solutions of (2), we have

$$h(n + k) + a_{n+k-1}h(n + k - 1) + \cdots + a_n h(n) = \phi(n)$$

and

$$p(n + k) + a_{n+k-1}p(n + k - 1) + \cdots + a_n p(n) = \phi(n)$$

Subtracting one equation from the other we obtain

$$h(n + k) - p(n + k) + a_{n+k-1}(h(n + k - 1) - p(n + k - 1))$$
$$+ \cdots + a_n(h(n) - p(n)) = 0$$

Thus we see that $h(n) - p(n)$ is a solution to this homogeneous equation, so

$$h(n) - p(n) = \alpha_1 g_1(n) + \alpha_2 g_2(n) + \cdots + \alpha_k g_k(n)$$

for some choice of $\alpha_1, \alpha_2, \ldots, \alpha_k$. Consequently,

$$h(n) = \alpha_1 g_1(n) + \alpha_2 g_2(n) + \cdots + \alpha_k g_k(n) + p(n)$$

as we were to show.   ∎

Example 1   We use Theorem 7.6 to find the general solution to

$$f(n + 2) - 5f(n + 1) + 6f(n) = 2^{n+2} \tag{4}$$

First observe that $-n2^{n+1}$ is a particular solution of (4), since if

$$p(n) = -n2^{n+1}$$

then

$$p(n + 1) = -(n + 1)2^{n+2} \quad \text{and} \quad p(n + 2) = -(n + 2)2^{n+3}$$

and consequently,

$$p(n + 2) - 5p(n + 1) + 6p(n) = -(n + 2)2^{n+3} + 5(n + 1)2^{n+2} - 6n2^{n+1}$$
$$= 2^{n+1}[-2^2(n + 2) + 10(n + 1) - 6n]$$
$$= 2^{n+1} \cdot 2 = 2^{n+2}$$

(Note that we have merely verified that $-n2^{n+1}$ is a solution to (4); in the following section we will see how to find such solutions.)

Next we find the general solution of the associated homogeneous equation

$$f(n + 2) - 5f(n + 1) + 6f(n) = 0 \qquad (5)$$

The characteristic equation of (5) is

$$A^2 - 5A + 6 = 0$$

On factoring we find that $A = 2$ or $A = 3$. Thus, the general solution to the homogeneous equation (5) is $\alpha_1 2^n + \alpha_2 3^n$, and by Theorem 7.6, the general solution of the nonhomogeneous equation (4) is $\alpha_1 2^n + \alpha_2 3^n - n2^{n+1}$.   □

We can also apply Theorem 7.6 to higher order difference equations, as we illustrate in the next example.

Example 2   We use Theorem 7.6 to find the general solution to

$$f(n + 2) + 3f(n + 1) - 4f(n - 1) = 18n - 6 \qquad (6)$$

First observe that $p(n) = n^2 - n$ is a particular solution of (6) since

$$p(n + 2) = (n + 2)^2 - (n + 2), \qquad p(n + 1) = (n + 1)^2 - (n + 1),$$
$$\text{and} \quad p(n - 1) = (n - 1)^2 - (n - 1)$$

and consequently,

$$
\begin{aligned}
p(n + 2) &+ 3p(n + 1) - 4p(n - 1) \\
&= (n + 2)^2 - (n + 2) + 3(n + 1)^2 - 3(n + 1) - 4(n - 1)^2 \\
&\qquad\qquad\qquad\qquad\qquad\qquad\qquad\qquad\qquad + 4(n - 1) \\
&= n^2 + 4n + 4 - n - 2 + 3n^2 + 6n + 3 - 3n - 3 - 4n^2 \\
&\qquad\qquad\qquad\qquad\qquad\qquad\qquad\qquad + 8n - 4 + 4n - 4 \\
&= 18n - 6
\end{aligned}
$$

(Here we have merely verified that $n^2 - n$ is a solution to (6); in the next section we will see how to find such a solution.)

Next we find the general solution of the associated homogeneous equation

$$f(n + 2) + 3f(n + 1) - 4f(n - 1) = 0 \qquad (7)$$

The characteristic equation of (7) is

$$A^3 + 3A^2 - 4 = 0 \qquad (8)$$

Observe that 1 is a root of (8) and thus $(A - 1)$ is a factor of the left-hand side of (8). Dividing by $(A - 1)$ yields

$$A^3 + 3A^2 - 4 = (A - 1)(A^2 + 4A + 4) = (A - 1)(A + 2)^2$$

Therefore, the roots of (8) are $A = 1$, $A = -2$, $A = -2$, and it follows that the general solution of the homogeneous equation is

$$\alpha_1 1^n + \alpha_2 (-2)^n + \alpha_3 n(-2)^n = \alpha_1 + \alpha_2 (-2)^n + \alpha_3 n(-2)^n$$

By Theorem 7.6 the general solution of the nonhomogeneous equation (6) is

$$\alpha_1 + \alpha_2(-2)^n + \alpha_3 n(-2)^n + n^2 - n \qquad \square$$

We emphasize that in view of Theorem 7.6 we need to find only *one* particular solution to the nonhomogeneous equation (as well as the general solution to the homogeneous equation) in order to solve the nonhomogeneous equation.

The following theorem will prove useful when $\phi(n)$ in equation (1) is the sum of various terms.

Theorem 7.7   If

    a. $F_1(n)$ is a solution of

$$f(n + k) + a_{n+k-1}f(n + k - 1) + \cdots + a_n f(n) = \phi_1(n)\,_1(n) \quad (9)$$

and

    b. $F_2(n)$ is a solution of

$$f(n + k) + a_{n+k-1}f(n + k - 1) + \cdots + a_n f(n) = \phi_2(n) \qquad (10)$$

then $F_1(n) + F_2(n)$ is a solution of

$$f(n + k) + a_{n+k-1}f(n + k - 1) + \cdots + a_n f(n) = \phi_1(n) + \phi_2(n)$$

Proof: Since $F_1(n)$ is a solution of (9), we have

$$F_1(n + k) + a_{n+k-1}F_1(n + k - 1) + \cdots + a_n F_1(n) = \phi_1(n) \qquad (11)$$

Similarly, since $F_2(n)$ is a solution of (10), we have

$$F_2(n + k) + a_{n+k-1}F_2(n + k - 1) + \cdots + a_n F_2(n) = \phi_2(n) \qquad (12)$$

Adding (11) and (12) we obtain

$$F_1(n + k) + F_2(n + k) + a_{n+k-1}(F_1(n + k - 1) + F_2(n + k - 1))$$
$$+ \cdots + a_n(F_1(n) + F_2(n)) = \phi_1(n) + \phi_2(n)$$

as was to be shown.   ∎

Example 3   We find the general solution of

$$f(n + 2) - 3f(n + 1) + 2f(n) = 4(2^n) + n + 2 \qquad (13)$$

The characteristic equation of the corresponding homogeneous equation is $A^2 - 3A + 2 = (A - 2)(A - 1) = 0$. Since the roots of this equation are $A = 1$ and $A = 2$, the general solution of the homogeneous equation is $\alpha_1 + \alpha_2 2^n$.

Direct substitution shows that $F_1(n) = n2^{n+1}$ is a solution of the non-homogeneous difference equation

$$f(n + 2) - 3f(n + 1) + 2f(n) = 4(2^n) \tag{14}$$

and that $F_2(n) = -\frac{1}{2}n^2 - \frac{5}{2}n$ is a solution of the nonhomogeneous difference equation

$$f(n + 2) - 3f(n + 1) + 2f(n) = n + 2 \tag{15}$$

Thus, from Theorem 7.7 we see that

$$F_1(n) + F_2(n) = n2^{n+1} - \frac{1}{2}n^2 - \frac{5}{2}n$$

is a particular solution to (13). Adding this particular solution to the general solution of the homogeneous equation gives

$$\alpha_1 + \alpha_2 2^n + n2^{n+1} - \frac{1}{2}n^2 - \frac{5}{2}n$$

as the general solution of the nonhomogeneous difference equation (13).   □

7.6   Problems

In Problems 1–16 show that $p(n)$ is a particular solution of the given difference equation, and find the general solution to the difference equation. If initial conditions are given, find also the particular solution that satisfies those conditions.

1. $f(n + 2) - 3f(n + 1) + 2f(n) = 1; p(n) = -n$
2. $f(n + 2) - 2f(n + 1) + f(n) = 5 + 3n; p(n) = n^2 + \frac{1}{2}n^3$
3. $f(n + 2) - 3f(n + 1) + 2f(n) = 3^n; f(0) = 1, f(1) = 0; p(n) = (\frac{1}{2})3^n$
4. $f(n + 2) - 4f(n + 1) + 4f(n) = 2^n; p(n) = n(n - 1)2^{n-3}$
5. $f(n + 2) - 4f(n) = 2^n; f(1) = 3, f(2) = 1; p(n) = \frac{1}{8}n2^n$
6. $f(n + 2) - 4f(n) = n^2 - 1; p(n) = -n^2/3 - \frac{4}{9}n - \frac{11}{27}$
7. $f(n) - 10f(n - 1) + 25f(n - 2) = 2^n; p(n) = (\frac{1}{9})2^{n+2}$
8. $f(n) - 4f(n - 1) + 4f(n - 2) = 2^n; p(n) = n^2 2^{n-3}$
9. $f(n) - 3f(n - 1) + 2f(n - 2) = 2^n; p(n) = n2^{n+1}$
10. $f(n + 2) + 5f(n + 1) - 6f(n) = n; p(n) = \frac{1}{14}(n - \frac{9}{14})^2$
11. $f(n + 2) + f(n) = n + 1; f(0) = 1, f(1) = 0; p(n) = n/2$
12. $f(n + 2) + f(n + 1) - 3f(n) = -2n + 1; p(n) = 2n + 3$
13. $f(n + 1) = f(n) + 1; f(1) = 1; p(n) = n$
14. $f(n + 1) = 3f(n) + 2; p(n) = 3^n - 1$
15. $f(n + 2) + 5f(n + 1) + 6f(n) = 4^{n+1}; p(n) = (\frac{2}{21})4^n$
16. $f(n) + 2f(n - 1) + f(n - 2) = 3^{n-2}; p(n) = (3^{n-2})/16$

7.7 THE METHOD OF UNDETERMINED COEFFICIENTS

There are several methods for obtaining particular solutions to nonhomogeneous difference equations. Perhaps the simplest of these is the method of *undetermined coefficients*. Although this method is not the most widely applicable one, it is relatively easy to use and can be employed in a variety of problems.

To introduce the method of undetermined coefficients, we consider the difference equation

$$f(n + 2) + 2f(n + 1) - f(n) = 6n^2 + 24n + 25 \qquad (1)$$

The sum $f(n + 2) + 2f(n + 1) - f(n)$ will be equal to the polynomial $6n^2 + 24n + 25$ if and only if the function f is a second-degree polynomial. Thus we look for a solution of (1) of the form

$$f(n) = A_2 n^2 + A_1 n + A_0$$

where A_2, A_1, and A_0 are constants. We can determine these constants by substituting $f(n) = A_2 n^2 + A_1 n + A_0$ into (1). Observe that if $f(n) = A_2 n^2 + A_1 n + A_0$, then

$$\begin{aligned} f(n + 1) &= A_2(n + 1)^2 + A_1(n + 1) + A_0 \\ &= A_2 n^2 + (2A_2 + A_1)n + (A_2 + A_1 + A_0) \end{aligned}$$

and

$$\begin{aligned} f(n + 2) &= A_2(n + 2)^2 + A_1(n + 2) + A_0 \\ &= A_2 n^2 + (4A_2 + A_1)n + (4A_2 + 2A_1 + A_0) \end{aligned}$$

Consequently,

$$\begin{aligned} f(n + 2) + 2f(n + 1) - f(n) &= A_2 n^2 + (4A_2 + A_1)n + (4A_2 + 2A_1 + A_0) \\ &\quad + 2A_2 n^2 + 2(2A_2 + A_1)n \\ &\quad + 2(A_2 + A_1 + A_0) \\ &\quad - A_2 n^2 - A_1 n - A_0 \\ &= 2A_2 n^2 + (8A_2 + 2A_1)n \\ &\quad + (6A_2 + 4A_1 + 2A_0) \end{aligned}$$

Thus for $f(n) = A_2 n^2 + A_1 n + A_0$ to be a solution to (1) it must be the case that

$$2A_2 n^2 + (8A_2 + 2A_1)n + (6A_2 + 4A_1 + 2A_0) = 6n^2 + 24n + 25$$

Such an equation is valid for all n if and only if the corresponding coefficients are equal. That is, we must have

$$\begin{aligned} 2A_2 &= 6 \\ 8A_2 + 2A_1 &= 24 \\ 6A_2 + 4A_1 + 2A_0 &= 25 \end{aligned}$$

From these equations we find that $A_2 = 3$, $A_1 = 0$, and $A_0 = \frac{7}{2}$, and substituting these values into $f(n) = A_2 n^2 + A_1 n + A_0$, we find that $p(n) = 3n^2 + \frac{7}{2}$ is a particular solution to the nonhomogeneous equation (1).

To apply the method of undetermined coefficients it is necessary to select the correct form for a trial solution. Table 7.1 gives the form of a trial solution to

$$f(n + k) + a_{n+k-1} f(n + k - 1) + \cdots + a_n f(n) = \phi(n)$$

for various functions $\phi(n)$.

$\phi(n)$	Form of trial solution
$a_m n^m + a_{m-1} n^{m-1} + \cdots + a_1 n + a_0$	$A_m n^m + A_{m-1} n^{m-1} + \cdots + A_1 n + A_0$
ck^n	Ck^n
$k^n(a_m n^m + a_{m-1} n^{m-1} + \cdots + a_1 n + a_0)$	$k^n(A_m n^m + A_{m-1} n^{m-1} + \cdots + A_1 n + A_0)$

Table 7.1

We illustrate the use of Table 7.1 in the following examples.

Example 1   We find the general solution of

$$f(n) - 3f(n - 1) - 10f(n - 2) = 3(2^n) \tag{2}$$

The characteristic equation of the homogeneous equation

$$f(n) - 3f(n - 1) - 10f(n - 2) = 0$$

is

$$A^2 - 3A - 10 = (A - 5)(A + 2) = 0$$

Consequently, $A = 5$ or $A = -2$, and the general solution of the homogeneous equation is

$$\alpha_1 5^n + \alpha_2 (-2)^n$$

Since $\phi(n) = 3(2^n)$ is of the form ck^n from Table 7.1 (with $c = 3$ and $k = 2$), we select a trial solution of (2) of the form $f(n) = C2^n$. Substituting into (2) yields

$$C2^n - 3C2^{n-1} - 10C2^{n-2} = 3(2^n) \tag{3}$$

and dividing through by 2^{n-2}, we obtain

$$C2^2 - 3C2 - 10C = 3(2^2)$$

which, simplified, gives us

$$-12C = 12$$

from which it follows that $C = -1$. Therefore, substituting $C = -1$ into $f(n) = C2^n$, we find that $p(n) = -2^n$ is a particular solution of the nonhomogeneous equation, so by Theorem 7.6

$$\alpha_1 5^n + \alpha_2 (-2)^n - 2^n$$

is the general solution of the nonhomogeneous equation (2).   ☐

Example 2   We find the general solution of

$$f(n + 2) - f(n + 1) - 6f(n) = 6n^2 + 22n + 23 \tag{4}$$

The characteristic equation of the corresponding homogeneous difference equation is

$$A^2 - A - 6 = (A - 3)(A + 2) = 0$$

Therefore, $A = -2$ or $A = 3$, and the general solution of the homogeneous equation is

$$\alpha_1 (-2)^n + \alpha_2 3^n$$

Since $\phi(n) = 6n^2 + 22n + 23$ is of the form $a_2 n^2 + a_1 n + a_0$ in Table 7.1 (with $a_2 = 6$, $a_1 = 22$, and $a_0 = 23$), we select $A_2 n^2 + A_1 n + A_0$ as a trial solution. Substituting $f(n) = A_2 n^2 + A_1 n + A_0$ into (4) yields

$$[A_2(n + 2)^2 + A_1(n + 2) + A_0] - [A_2(n + 1)^2 + A_1(n + 1) + A_0]$$
$$- [6A_2 n^2 + 6A_1 n + 6A_0] = 6n^2 + 22n + 23$$

This eventually reduces to

$$-6A_2 n^2 + (2A_2 - 6A_1)n + 3A_2 + A_1 - 6A_0 = 6n^2 + 22n + 23$$

In order for this last equation to be valid for all n, the corresponding coefficients must be equal; hence,

$$-6A_2 = 6, \qquad 2A_2 - 6A_1 = 22, \quad \text{and} \quad 3A_2 + A_1 - 6A_0 = 23$$

From these equations we find that $A_2 = -1$, $A_1 = -4$, and $A_0 = -5$, and we determine that $p(n) = -n^2 - 4n - 5$ is a particular solution to the nonhomogeneous equation (4). The general solution, therefore, is

$$\alpha_1 (-2)^n + \alpha_2 3^n - n^2 - 4n - 5 \qquad\qquad ☐$$

A slight complication arises in using Table 7.1 whenever a term of the selected trial solution is also a solution to the homogeneous equation. If this occurs, then we must multiply every term of the trial solution by n. If the new trial solution still has a term that is a solution to the homogeneous equation, we again multiply it by n. This process continues until no term of the trial solution is a solution of the homogeneous equation.

For example, if $\phi(n) = n^2 + 1$, then the trial solution would ordinarily be of the form $A_2 n^2 + A_1 n + A_0$. Say, however, that the general solution of the homogeneous equation is $\alpha_1 n + \alpha_2$. Then both $A_1 n$ and A_0 are solutions of the homogeneous equation. Multiplying the original trial solution by n gives us $A_2 n^3 + A_1 n^2 + A_0 n$ as a trial solution. However, since $A_0 n$ is still a solution to the homogeneous equation, we must again multiply by n to obtain $A_2 n^4 + A_1 n^3 + A_0 n^2$. Since none of the terms $A_2 n^4$, $A_1 n^3$, or $A_0 n^2$ of this proposed solution is a solution to the homogeneous equation, $A_2 n^4 + A_1 n^3 + A_0 n^2$ is the desired trial solution.

Example 3   We find the general solution of

$$f(n) - 4f(n-1) + 4f(n-2) = 2^n \tag{5}$$

The characteristic equation of the corresponding homogeneous difference equation is

$$A^2 - 4A + 4 = 0$$

Thus $A = 2$ is a root of multiplicity two, and the general solution of the homogeneous equation is $\alpha_1 2^n + \alpha_2 n 2^n$.

Since $\phi(n) = 2^n$ we would ordinarily select $C2^n$ as the form of a trial solution. However, $C2^n$ and $Cn2^n$ are both solutions to the homogeneous equation and, therefore, we must multiply 2^n by n^2 and use $Cn^2 2^n$ as the form of the trial solution. Substituting $Cn^2 2^n$ into

$$f(n) - 4f(n-1) + 4f(n-2) = 2^n$$

yields

$$Cn^2 2^n - 4C(n-1)^2 2^{n-1} + 4C(n-2)^2 2^{n-2} = 2^n$$

and thus,

$$Cn^2 2^n - 4Cn^2 2^{n-1} + 8Cn 2^{n-1} - 4C2^{n-1}$$
$$+ 4Cn^2 2^{n-2} - 16Cn 2^{n-2} + 16C2^{n-2} = 2^n$$

Dividing by 2^{n-2} we obtain

$$4Cn^2 - 8Cn^2 + 16Cn - 8C + 4Cn^2 - 16Cn + 16C = 4$$

from which it follows that $8C = 4$, so $C = \frac{1}{2}$. Consequently, $p(n) = \frac{1}{2}n^2 2^n = n^2 2^{n-1}$ is a particular solution to the nonhomogeneous equation (5), and by Theorem 7.5 the general solution is

$$\alpha_1 2^n + \alpha_2 n 2^n + n^2 2^{n-1} \qquad\qquad \square$$

Example 4   As we indicated in Section 7.1, we can use difference equations to evaluate certain sums. For example, if

$$S(n) = \sum_{k=1}^{n} k^2$$

then

$$S(n) - S(n - 1) = n^2 \tag{6}$$

The homogeneous difference equation,

$$S(n) - S(n - 1) = 0$$

has as its general solution the constant α.

Since $\phi(n) = n^2$, we would ordinarily select $A_2 n^2 + A_1 n + A_0$ as a trial solution to the nonhomogeneous equation. However, since the constant A_0 is a solution to the homogeneous equation, we must multiply the trial solution by n. The resulting trial solution is

$$A_2 n^3 + A_1 n^2 + A_0 n$$

Substituting this trial solution into (6) gives us

$$A_2 n^3 + A_1 n^2 + A_0 n - A_2 (n - 1)^3 - A_1 (n - 1)^2 - A_0 (n - 1) = n^2$$

On simplifying this, we see that we must select A_2, A_1, and A_0 so that

$$3A_2 n^2 + (2A_1 - 3A_2)n + A_0 - A_1 + A_2 = n^2$$

for all integers n, from which it follows that

$$3A_2 = 1, \qquad 2A_1 - 3A_2 = 0, \quad \text{and} \quad A_0 - A_1 + A_2 = 0$$

Therefore, we find that $A_2 = \frac{1}{3}$, $A_1 = \frac{1}{2}$, and $A_0 = \frac{1}{6}$, so

$$\frac{1}{3}n^3 + \frac{1}{2}n^2 + \frac{1}{6}n$$

is a particular solution of (6). Adding this to the general solution of the homogeneous solution, we find that the general solution of (6) is

$$S(n) = \frac{1}{3}n^3 + \frac{1}{2}n^2 + \frac{1}{6}n + \alpha$$

Since $S(1) = 1$ we must have

$$1 = \frac{1}{3} + \frac{1}{2} + \frac{1}{6} + \alpha$$

which gives us $\alpha = 0$, so

$$S(n) = \sum_{k=1}^{n} k^2 = \frac{1}{3}n^3 + \frac{1}{2}n^2 + \frac{1}{6}n \quad \square \tag{7}$$

We can also establish this result by induction (see Problem 19). Note, however, that while an induction argument requires much less theory, it does not indicate how we formulated the original equation (7).

7.7   Problems

In Problems 1–12 find the general solution to the difference equation. If initial conditions are given, find the particular solution that satisfies those conditions.

1. $f(n + 2) - f(n + 1) - 2f(n) = 3^n$

2. $f(n) + 2f(n - 1) - 8f(n - 2) = 5^{n-2}$

3. $f(n) - 4f(n - 1) + 3f(n - 2) = -4$

4. $f(n + 2) - f(n + 1) - 2f(n) = 2^n; \ f(0) = 2, \ f(1) = 1$

5. $f(n + 2) - 2f(n + 1) + f(n) = -4; \ f(1) = 2, \ f(2) = -1$

6. $f(n + 2) - 3f(n + 1) - 10f(n) = 36n - 21$

7. $f(n) - 9f(n - 2) = n^2 - 4n - 1$

8. $f(n + 2) - 9f(n) = n^2 - 5n + 3^n$

9. $f(n + 2) + 2f(n + 1) - 8f(n) = -5n + 14; \ f(0) = 0, \ f(1) = -1$

10. $2f(n + 1) - 3f(n) - 5f(n - 1) = 5^{n-1} - 4$

11. $f(n) - 4f(n - 2) = -6n^2 + 32n - 23 - 2^{n-1}$

12. $f(n) + 6f(n - 1) + 12f(n - 2) + 8f(n - 3) = 3^n$

In Problems 13–18 use the techniques of Example 4 of this section to find a simple expression for the given sum.

13. $\displaystyle\sum_{k=1}^{n} k$

14. $\displaystyle\sum_{k=1}^{n} r^k$

15. $\displaystyle\sum_{k=1}^{n} (-1)^k k$

16. $\displaystyle\sum_{k=1}^{n} 2^k$

17. $\displaystyle\sum_{k=1}^{n} 2^k k$

18. $\displaystyle\sum_{k=1}^{n} (-1)^k k^2$

19. Use induction to show that for each positive integer n

$$\sum_{k=1}^{n} k^2 = \frac{1}{3}n^3 + \frac{1}{2}n^2 + \frac{1}{6}n$$

20. Explain (perhaps with an example) why there is a problem when a term of the selected trial solution is also a solution to the homogeneous equation.

21. Use a difference equation to determine the expression for $e(n)$, the number of sequences of length n where each element of the sequence is 0, 1, or 2, and an even number of 0's appears. (*Hint:* If the last term in the sequence is 1 or 2, then there are $e(n - 1)$ sequences. If the last term is 0, then there are $3^{n-1} - e(n - 1)$ sequences.)

22. An account initially contains $1000. At the end of year n, $50n$ dollars are deposited into the account. The account earns interest at the rate of 10% per year. Use a difference equation to determine the amount of money in the account after the nth deposit.

7.8 SOME ADDITIONAL APPLICATIONS OF DIFFERENCE EQUATIONS

In this section we once again take up the algorithms BINARY SEARCH and TRINARY SEARCH described in Section 7.1. We also introduce another "divide-and-conquer" algorithm and consider a few other applications of difference equations.

BINARY and TRINARY Search Algorithms   We will use difference equations to compare the efficiency of BINARY SEARCH and TRINARY SEARCH. In Section 7.1 of this chapter we found that $B(n)$, the number of comparisons required by BINARY SEARCH to locate a given entry in a list of n entries, satisfies the equation

$$B(n) = B(n/2) + 1 \tag{1}$$

and that $T(n)$, the number of comparisons required by TRINARY SEARCH, satisfies the equation

$$T(n) = T(n/3) + \tfrac{5}{3} \tag{2}$$

If we substitute $n = 2^m$ into (1) we obtain

$$B(2^m) = B(2^m/2) + 1 = B(2^{m-1}) + 1 \tag{3}$$

Let $F(m) = B(2^m)$. Then equation (3) becomes

$$F(m) = F(m - 1) + 1 \tag{4}$$

In Problem 1 you are asked to verify that the general solution to (4) is given by

$$F(m) = m + c \tag{5}$$

Note that since $F(m) = B(2^m)$, it follows that $F(0) = B(1)$, which is the number of comparisons required to find an entry in a list that has only one item. Since this list has only one element, and since we know that the desired element is in the list, no comparison is required; that is, $B(1) = 0$. Thus, on substituting $m = 0$ into (5), we have $c = 0$. Consequently,

$$B(2^m) = F(m) = m \tag{6}$$

Finally, since $n = 2^m$, we have $m = \log_2 n$, and therefore, the solution of equation (1) is

$$B(n) = B(2^m) = m = \log_2 n \tag{7}$$

In Problem 2 you are asked to use a similar argument to show that

$$T(n) = \tfrac{5}{3}\log_3 n \tag{8}$$

is the solution of (2).

In Problem 3 you are asked to show that $T(n) > B(n)$ for all $n > 1$. Thus, BINARY SEARCH is more efficient than TRINARY SEARCH.

Divide-and-Conquer Sort Algorithm   MERGESORT is a recursive sorting algorithm based on the divide-and-conquer concept. It splits a given list of numbers into two "halves," calls on itself to sort each half, and then merges the two ordered half-lists into a single ordered list. To measure its complexity, we will compute $f(n)$, the number of comparisons it makes to order a list of n numbers.

From the description of MERGESORT it follows that $f(n)$ satisfies the difference equation

$$f(n) = 2f\left(\frac{n}{2}\right) + (?) \tag{9}$$

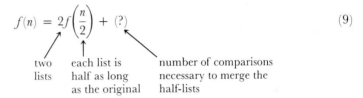

two   each list is   number of comparisons
lists   half as long   necessary to merge the
              as the original   half-lists

To solve for the unknown term, we must identify how many comparisons we need to make to combine the sorted lists L_1 and L_2 into a single list L. To perform this merge operation, we need only compare the first (largest) elements of L_1 and L_2, and then place the larger of the two in the first position in list L. Thus, each entry in list L (except the last, which is smallest by default) results from a single comparison. Consequently, we make $n - 1$ comparisons to merge the two lists. Substituting into (9) we obtain

$$f(n) = 2f(n/2) + (n - 1) \tag{10}$$

The solution of (10) that satisfies the initial condition $f(1) = 0$ gives us the complexity of MERGESORT. In Problem 4 you are asked to show that this solution is

$$f(n) = (n \log_2 n) - n + 1$$

Compound Interest   Suppose we invest a sum of k dollars at the end of each year into an account that earns interest at the rate r, compounded annually, and the account initially contains d dollars. We want to compute $A(n)$, the amount in the account immediately after the nth deposit.

As we saw in Section 7.1, $A(n)$ is the sum of $A(n - 1)$ (the amount in the account after the $(n - 1)$st deposit), $rA(n - 1)$ (the interest earned by the $A(n - 1)$ dollars during the year), and k (the last deposit of k dollars). This gives us the difference equation

$$A(n) = A(n - 1) + rA(n - 1) + k \tag{11}$$

or, equivalently,

$$A(n) - (1 + r)A(n - 1) = k$$

To solve this equation, we can use the techniques from Sections 7.6 and 7.7. We first find that the general solution of the homogeneous difference equation $A(n) - (1 + r)A(n - 1) = 0$ is $A(n) = \alpha(1 + r)^n$. To solve (11) we now need to find a particular solution to the nonhomogeneous equation

$$A(n) - (1 + r)A(n - 1) = k \tag{12}$$

Substituting the trial solution $A(n) = C$ in (12) yields

$$C - (1 + r)C = k$$

and, therefore,

$$C = -\frac{k}{r}$$

Consequently,

$$A(n) = \alpha(1 + r)^n - \left(\frac{k}{r}\right)$$

for an appropriate choice of α. Since $A(0) = d$ (the number of dollars in the account before any deposit is made) we find that

$$d = \alpha - \frac{k}{r}$$

and, hence,

$$\alpha = d + \frac{k}{r}$$

Thus, the amount of money in the account immediately after the nth deposit is given by

$$A(n) = \left(d + \frac{k}{r}\right)(1 + r)^n - \left(\frac{k}{r}\right) \tag{13}$$

Example 1   Upon the birth of their first child, the parents deposit \$500 into a bank account that earns 7% interest, compounded annually. Each year thereafter, on the child's birthday, they place an additional \$200 into the account. How much is in the account on the child's eighteenth birthday?

In this case, $n = 18$, $k = 200$, $d = 500$, and $r = 0.07$; substituting into equation (13), we have

$$A(18) = \left(500 + \frac{200}{0.07}\right)(1.07)^{18} - \left(\frac{200}{0.07}\right) = 8489.77$$

In other words, there will be \$8489.77 in the account when the child turns eighteen.   □

In this chapter we have looked at a number of problems that can be modeled by difference equations, and we have considered some basic techniques for solving these equations. In Chapter 10 we will take another approach to finding solutions to difference equations.

7.8 Problems

1. Show that $m + \alpha$ is the general solution to $F(m) = F(m - 1) + 1$.
2. Show that $T(n) = \frac{5}{3}\log_3 n$ is the solution to $T(n) = T(n/3) + \frac{5}{3}$, $T(1) = 0$.
3. Show that $T(n) > B(n)$ for each $n > 1$, where $B(n)$ and $T(n)$ are defined by $B(n) = \log_2 n$ and $T(n) = \frac{5}{3}\log_3 n$.
4. Show that the solution of $f(n) = 2f(n/2) + (n - 1)$ where $f(1) = 0$ is $f(n) = (n \log_2 n) - n + 1$.
5. (a) Show that the number of comparisons required to execute BUBBLESORT (described in Chapter 1) satisfies the difference equation

$$f(n) = f(n - 1) + (n - 1)$$

   (b) Show that the solution to this difference equation is

$$f(n) = \frac{1}{2}n^2 - \frac{1}{2}n$$

6. The nth power of a number a is to be calculated by multiplying $a^{n/2}$ by itself.
   (a) Show that the number of multiplications necessary to compute the nth power, $f(n)$, satisfies the difference equation

$$f(n) = f\left(\frac{n}{2}\right) + 1$$

   (b) Solve the difference equation of part (a).
7. Let $f(n)$ be the maximum number of regions into which n lines divide a plane.
   (a) Show that $f(n)$ satisfies the difference equation $f(n + 1) = f(n) + n + 1$.
   (b) Compute $f(n)$.
8. Suppose that a bank account earns 8% interest, compounded semi-annually. A customer deposits $1000 and adds $400 to the account every six months. How much money will be in the account after ten years?
9. The execution of an algorithm on n points requires the execution of the same algorithm once on $n - 1$ points and twice on $n - 2$ points. If the algorithm requires one second to execute on one point and four seconds to execute on two points, find an expression for the time required for n points.

*10. Let $B(n)$ denote the number of multiplications needed to compute the determinant of an $n \times n$ matrix by the cofactor expansion method. Show that $B(1) = 0$ and $B(2) = 2$, and that B satisfies the difference equation:

$$B(n) = nB(n - 1) + n$$

Solve this difference equation to find $B(n)$. (*Hint:* Note that

$$\begin{aligned}
B(n) = nB(n - 1) + n &= n[(n - 1)B(n - 2) + (n - 1)] + n \\
&= n[(n - 1)[(n - 2)B(n - 3) + (n - 2)] \\
&\quad + (n - 1)] + n \\
&= \ldots)
\end{aligned}$$

Chapter 7   REVIEW

Concepts for Review

solution of a difference equation (p. 261)
divide-and-conquer algorithms (p. 263)
linear difference equations with constant coefficients (p. 266)
order of a difference equation (p. 266)
homogeneous difference equation (p. 266)
characteristic equation of a homogeneous difference equation (p. 267)
general solution of a difference equation (p. 271)
Casorati's determinant (p. 271)
particular solution (p. 274)
general solution of a nonhomogeneous difference equation (p. 286)
method of undetermined coefficients (p. 290)
MERGESORT (p. 297)

Review Problems

1. Show that $(n + 1)3^n$ is a solution to the difference equation

$$f(n + 2) - 6f(n + 1) + 9f(n) = 0$$

2. Show that $(n!)^2$ is a solution to the difference equation

$$f(n + 2) - (n^2 + 3n + 2)^2 f(n) = 0$$

In Problems 3–14 find the general solution to the given difference equation. Where initial conditions are given, find also the particular solution that satisfies those conditions.

3. $f(n) - 3f(n - 1) - 10f(n - 2) = 0$; $f(0) = f(1) = 1$
4. $f(n + 2) - f(n + 1) - 6f(n) = 0$; $f(1) = 1, f(2) = 2$
5. $f(n + 2) + 4f(n + 1) - 5f(n) = 0$

6. $f(n) + 4f(n - 1) + 8f(n - 2) = 0$

7. $4f(n + 2) + 12f(n + 1) + 9f(n) = 0; f(0) = f(1) = 0$

8. $4f(n + 1) - 4f(n) + f(n - 1) = 0$

9. $f(n + 2) + 2f(n + 1) + 4f(n) = 0$

10. $f(n + 2) - f(n + 1) - 6f(n) = 5^n$

11. $f(n + 2) - f(n + 1) - 6f(n) = 2^n$

12. $f(n) - 2f(n - 1) + f(n - 2) = 2 + 3^n$

13. $f(n) + 2f(n - 1) + f(n - 2) = 2$

14. $f(n + 2) - 4f(n + 1) + 4f(n) = n2^n; f(1) = 0, f(2) = 1$

15. Define and explain the use of Casorati's determinant.

16. Compute Casorati's determinant for 3^n, $n3^n$, and $n^2 3^n$.

17. Show that Casorati's determinant for n^2 and n^3 is nonzero. What conclusion can you draw?

18. Set up a difference equation for the number of sequences of n digits formed from the patterns 1, 22, 333, and 4444. What are the initial conditions?

19. Use a difference equation to determine how many ways n red, white, and blue poker chips can be stacked so that no two red chips are adjacent.

20. Let $S(n) = \Sigma_{k=2}^{n} (k^2 - 3k)$. Write a difference equation for $S(n)$. Solve your equation to determine a simple expression for $S(n)$.

8 An Introduction to Enumeration

We can all count. However, as we shall see, there is far more to counting (*enumeration*) than one might suppose. In this chapter we take up a number of counting problems ranging from the patently obvious to the remarkably subtle.

Enumeration is essential to the analysis of the complexity of algorithms. It is also basic to many problems in probability, as we will see in the next chapter. The ideas and counting techniques in this chapter should give you sufficient background to deal with a wide variety of counting problems.

Some set theory is used in this chapter and you may wish to review Section 1 of the appendix before proceeding.

8.1 TWO BASIC PRINCIPLES

Consider the following "simple" counting problem. Suppose an identifier is to consist of a string of any three distinct letters of the alphabet. We want to know how many of these identifiers will contain either an *E* or an *F*. To solve this problem, we might attempt to list all such possible identifiers: *AEB*, *EAB*, *FAB*, *EAC*, *FAC*, and so on. Obviously, such an unsystematic approach is not only very time-consuming but also prone to error.

To handle counting problems of this nature, we need to pursue a somewhat more theoretical tack. We therefore begin with two fundamental counting principles whose simplicity belies their remarkable utility. In stating these principles we use the notation $|S|$ to designate the number of elements in the set S. The first basic principle gives the number of elements in the Cartesian product of two sets.

The Product Principle   If A and B are sets, then

$$|A \times B| = |A||B|$$

To apply the Product Principle to a counting problem, we need only identify and count elements in the sets A and B.

Example 1   Suppose we perform a simple experiment that consists of throwing a die and then drawing a card from a deck of fifty-two. We want to know the total number of possible outcomes of this experiment. Letting A be the set of outcomes of the die-throwing and B be the set of outcomes of the card-drawing, we find that $|A| = 6$ and $|B| = 52$. Each outcome of the random experiment of throwing a die *and* drawing a card corresponds in a natural way to an ordered pair (a, b) where $a \in A$ and $b \in B$. Consequently, the number of possible outcomes of this experiment is

$$|A \times B| = |A||B| = (6)(52) = 312 \qquad \square$$

A tree diagram such as Figure 8.1 can illustrate the Product Principle. Here the first six branches of the tree designate the six possible outcomes of rolling a die. From each of these six branches sprout fifty-two twigs, one for each possible outcome of the card-drawing experiment.

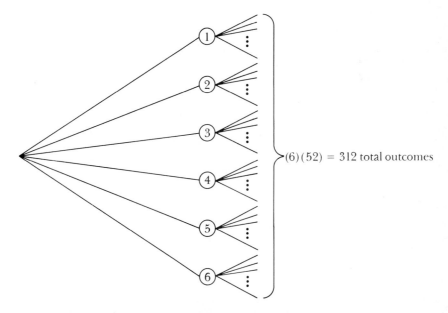

$(6)(52) = 312$ total outcomes

Figure 8.1

The second basic counting principle gives the number of elements in the union of two sets.

The Sum Principle   Suppose that A and B are sets. Then

$$|A \cup B| = |A| + |B| - |A \cap B|$$

Note that if we simply add $|A|$ and $|B|$ to obtain $|A \cup B|$, we would count all elements of $A \cap B$ twice. The Sum Principle therefore subtracts $|A \cap B|$ from $|A| + |B|$.

Example 2   In a Freshman class there are 212 computer science majors and 80 mathematics majors; 10 of these people are dual computer science–mathematics majors. The Registrar's Office wants to know how many students are majoring in either computer science or mathematics. If A denotes the set of computer science majors and B denotes the set of mathematics majors, then $|A \cup B|$ is the total number of students with either major. Since $|A| = 212$, $|B| = 80$, and $|A \cap B| = 10$, we have

$$|A \cup B| = |A| + |B| - |A \cap B| = 212 + 80 - 10 = 282$$

That is, there are 282 freshmen who have either math or computer science as a major.   □

Generalizing the Product Principle to accommodate any number of sets is an easy task. The following result can be proved by an induction argument (Problem 19).

Theorem 8.1   *The Generalized Product Principle.* If $A_1, A_2, \ldots, A_k$ are sets, then

$$|A \times A_2 \times \cdots \times A_k| = |A_1\|A_2|\cdots|A_k|$$

Example 3   Suppose that an identifier in a particular programming language consists of a sequence of characters in which the first character must be a letter of the English alphabet. The remaining characters can be either letters or digits. We want to know how many four-character identifiers are possible.

To count all identifiers of length four, we let A_1, A_2, A_3, and A_4 designate the sets of characters available for the first through fourth positions, respectively. Since each identifier $a_1 a_2 a_3 a_4$ corresponds in a natural way to an element of the set $A_1 \times A_2 \times A_3 \times A_4$, we count the number of members of this set.

By the Generalized Product Principle (Theorem 8.1), we have

$$|A_1 \times A_2 \times A_3 \times A_4| = |A_1\|A_2\|A_3\|A_4|$$

Since the set A_1 consists of the letters of the English alphabet, and since each of the sets A_2, A_3, and A_4 consists of twenty-six letters and ten digits, we have

$|A_1| = 26$ and $|A_2| = |A_3| = |A_4| = 36$. Consequently,

$$|A_1 \times A_2 \times A_3 \times A_4| = 26 \cdot 36 \cdot 36 \cdot 36 = 1{,}213{,}056$$

which means there are 1,213,056 identifiers of length four. □

The following theorem generalizes the Sum Principle to an arbitrary number of *disjoint sets*. You are asked to prove this result by induction in Problem 20.

Theorem 8.2 If $A_1, A_2, \ldots, A_n$ are sets such that $A_i \cap A_j = \emptyset$ for $1 \le i, j \le n$, and $i \ne j$, then

$$|A_1 \cup A_2 \cup \cdots \cup A_n| = |A_1| + |A_2| + \cdots + |A_n|$$

A generalization of the Sum Principle to nondisjoint sets is more difficult, and we defer it until Section 8.7.

Some counting problems are more easily solved indirectly. To illustrate this, we first note that if A and S are sets and if $A \subseteq S$, then

$$S = A \cup (S \setminus A)$$

We use the Sum Principle to write

$$|S| = |A| + |S \setminus A| - |A \cap (S \setminus A)|$$

Since $A \cap (S \setminus A) = \emptyset$, we have $|A \cap (S \setminus A)| = 0$. Thus

$$|S| = |A| + |S \setminus A|$$

so

$$|A| = |S| - |S \setminus A| \tag{1}$$

Equation (1) provides a way to determine $|A|$ when we know $|S|$ and $|S \setminus A|$. This indirect method of counting enables us in the next example to solve the problem proposed at the beginning of this section.

Example 4 An identifier is to consist of a string of any three distinct letters of the alphabet. We want to count those identifiers that contain either an E or an F. Let A denote the set of all identifiers that contain an E or an F, and let S denote the set of all three-letter identifiers. Then $S \setminus A$ is the set of all three-letter identifiers that do *not* contain either an E or an F. That is, $S \setminus A$ is the set of all three-letter identifiers formed only with the remaining twenty-four letters of the alphabet. By the Generalized Product Principle, we have

$$|S \setminus A| = 24 \cdot 24 \cdot 24 = (24)^3$$

Since S is the set of *all* three-letter identifiers, the Generalized Product Principle applied to S gives us

$$|S| = 26 \cdot 26 \cdot 26 = (26)^3$$

Thus substituting into (1), we obtain

$$|A| = |S| - |S \setminus A| = (26)^3 - (24)^3 = 3,752$$

which means that there are 3,752 identifiers (out of a total of $26^3 = 17,576$ possible identifiers) that will have either an E or an F.   □

In the remainder of this chapter we will encounter a number of ramifications of the Sum and Product Principles.

8.1 Problems

1. Compute $|A \cup B|$ if $|A| = 50$, $|B| = 20$, and $|A \cap B| = 7$.
2. Compute $|A \cap B|$ if $|A| = 10$, $|B| = 31$, and $|A \cup B| = 36$.
3. Suppose that out of 2,451 students entering a certain university, 500 failed mathematics, 300 failed English, and 150 failed both. How many failed either mathematics or English?
4. How many three-digit numbers can be formed using only the digits 5, 6, 7, 8, and 9?
5. How many four-letter sequences can be formed using the letters A, B, C, D, E, and F?
6. How many three-digit numbers can be formed without using the digits 0 or 9?
7. How many five-letter sequences do not use the letters E, F, G, or H?
8. How many five-letter sequences include at least one of the letters E, F, G, or H?
9. How many integers n, $1 \le n \le 1000$, include either a 5 or a 6?
10. License plates in a certain state consist of a sequence of two letters followed by four digits. How many such plates can be issued before the state must redesign the plates?
11. How many terms are there in the expansion of $(a + b)(c + d + e)(f + g + h + i)$?
12. How many terms containing a are there in the expansion of $(a + b)(c + d + e)(f + g + h + i)$?
13. How many terms containing either c or d are there in the expansion of $(a + b)(c + d + e)(f + g + h + i)$?
14. Use the Product Principle to prove that a set with n elements has 2^n subsets.
15. Each symbol in Morse code is represented by a sequence of at most four dots or dashes. For example, $\cdot\,-$ represents the letter a, and $-\cdot\cdot\cdot$ represents the letter b. How many symbols can be represented in this manner?

16. Compute the number of edges in K_n, the complete graph on n vertices. (See p. 123).

17. The *complete bipartite graph* $K_{m,n}$ is a graph G such that $V(G) = A \cup B$, $A \cap B = \varnothing$, $|A| = m$, $|B| = n$, and $E(G) = \{\{a, b\}|a \in A, b \in B\}$. Compute $|E(G)|$.

18. Generalize the Sum Principle to the union of three sets.

19. Use induction to prove Theorem 8.1.

20. Use induction to prove Theorem 8.2.

8.2   PERMUTATIONS

We can apply the Generalized Product Principle to count the number of possible ordered arrangements of a set of distinct objects. Although this problem presents no special difficulties, it is one of the most fundamental problems of enumeration. As we shall see, solving this problem is equivalent to counting the number of permutations of a set.

Definition 8.3   A *permutation* of a set S is a one-to-one mapping of the set S onto itself (see Appendix 2.8 and 2.9 for the definition of a one-to-one and an onto function).

Example 1   If $S = \{1, 2, 3\}$, then the following functions are all permutations of S.

(a) $f{:}S \to S$, defined by: $f(1) = 1$, $f(2) = 2$, $f(3) = 3$

(b) $f{:}S \to S$, defined by: $f(1) = 1$, $f(2) = 3$, $f(3) = 2$

(c) $f{:}S \to S$, defined by: $f(1) = 2$, $f(2) = 1$, $f(3) = 3$

(d) $f{:}S \to S$, defined by: $f(1) = 3$, $f(2) = 1$, $f(3) = 2$ □

Because the notation used in Example 1 is rather cumbersome, a permutation f of the set $\{a_1, a_2, \ldots, a_n\}$ is frequently designated by $(f(a_1), f(a_2), \ldots, f(a_n))$. Thus, for example, the permutation defined in (b) of Example 1 would be denoted by $(1, 3, 2)$ and the permutation defined in (c) would be represented by $(2, 1, 3)$. Note that when we write a permutation f in the form $(f(a_1), f(a_2), \ldots, f(a_n))$ it becomes apparent why a permutation can be considered as an ordered arrangement of the set $\{a_1, a_2, \ldots, a_n\}$ ($f(a_1)$ is the first element in this arrangement, $f(a_2)$ is the second element, and so on).

To count the number of permutations of a set, we can apply the Generalized Product Principle.

Theorem 8.4   If $|S| = n$, then S has

$$n(n - 1) \cdots (2)(1) = n!$$

different permutations.

Proof: Let $S = \{1, 2, \ldots, n\}$. In order to completely determine a permutation f of S we must specify $f(1), f(2), \ldots, f(n)$. There are n possible choices for $f(1)$; that is, $f(1)$ can be any one of the n numbers $1, 2, \ldots, n$. The choice of $f(2)$ is more restricted because we must choose $f(2)$ so that $f(2) \neq f(1)$ (f is a one-to-one function). Consequently, there are only $n - 1$ choices for $f(2)$. Similarly, we must choose $f(3)$ from $S \setminus \{f(1), f(2)\}$. Thus there are $n - 2$ choices for $f(3)$. Continuing in this fashion, we find that there are

$$n(n - 1)(n - 2) \cdots (2)(1)$$

ways of determining a permutation. That is, there are $n!$ permutations of a set consisting of n elements. ■

Example 2   An automated spot-welding machine makes its initial weld and then moves to ten different locations to make successive welds. The machine then returns to its initial position to begin the sequence again. We wish to determine the sequence of welds that the machine can traverse in the least amount of time. How many sequences will we have to examine?

Since the weld sites can be traversed in any order, each permutation of the sites represents a possible sequence of welds. Thus there are

$$10! = 3,628,800$$

such possibilities. Investigating each sequence individually to determine which sequence requires the least amount of time would obviously be a difficult task. □

We have seen that there are $n!$ possible permutations of n objects. Because there are so many possibilities, it would be helpful to be able to systematically determine these permutations. There are a number of ways to do this. One algorithm lists permutations in their lexicographic (dictionary) ordering. A dictionary would list the permutations formed from the letters a, b, c, and d as follows:

abcd	bacd	cabd	dabc
abdc	badc	cadb	dacb
acbd	bcad	cbad	dbac
acdb	bcda	cbda	dbca
adbc	bdac	cdab	dcab
adcb	bdca	cdba	dcba

More generally, lexicographic order is defined as follows.

Definition 8.5   In *lexicographic order* a word $s_1 s_2 \cdots s_n$ precedes a word $t_1 t_2 \cdots t_m$ if there is an integer k such that

    a. $s_i = t_i$ for $i < k$ and s_k precedes t_k, or

    b. $s_i = t_i$ for $i \leq n$ and $n < m$

Example 3   The word *bcad* precedes the word *bcda* since $b = b$, $c = c$, and a precedes d. The word *bcad* precedes the word *bcade* since $b = b$, $c = c$, $a = a$, $d = d$, and $4 < 5$.   □

To illustrate the idea of lexicographic ordering in the context of permutations, it is more convenient to deal with integers than with letters. We use the natural order of the positive integers 1, 2, 3, ... in place of the natural order of the alphabet a, b, c, ... The lexicographic ordering of "words" obtained from the first four positive integers is given below. Compare this list with the list of "words" based on the letters a, b, c, d on p. 308.

1234	2134	3124	4123
1243	2143	3142	4132
1324	2314	3214	4213
1342	2341	3241	4231
1423	2413	3412	4312
1432	2431	3421	4321

You should observe that the above list of "words" provides us with an ordered list of all permutations of the integers 1, 2, 3, 4. For a given permutation (or word) $a_1 a_2 \cdots a_n$, Algorithm 8.6 below enables us to find the next permutation in lexicographic order.

Before describing this algorithm we first observe that if $a_1 > a_2 > \cdots > a_n$ (for example, 54321), then we are at the end of the lexicographic list and there is no "next" permutation. Similarly, if $a_k > a_{k+1} > \cdots > a_n$, then $a_k a_{k+1} \cdots a_n$ cannot be rearranged to produce the permutation following $a_1 a_2 \cdots a_k a_{k+1} \cdots a_n$. For instance, the last three integers appearing in the permutation $p = 52431$ cannot be rearranged (without rearranging additional integers) to obtain the permutation that follows p in lexicographic order.

Suppose now that we are given the permutation $a_1 a_2 \cdots a_n$ of the first n positive integers. To obtain the next permutation in lexicographic order, observe that from the definition of this order, we would like to rearrange a minimum number of the rightmost integers of $a_1 a_2 \cdots a_n$. It follows from our previous observations that we must find the smallest integer k such that

$$a_k < a_{k+1} > \cdots > a_{n-1} > a_n \tag{1}$$

For this k it is impossible to rearrange $a_{k+1} a_{k+2} \cdots a_n$ to obtain the next permutation, but, as we shall see, we can obtain the next permutation by rearranging $a_k a_{k+1} \cdots a_n$. For instance, the permutation following 52431 is obtained by rearranging 2431.

We arrange $a_k a_{k+1} \cdots a_n$ by interchanging a_k with the smallest of the integers (call it a_m) $a_{k+1}, a_{k+2}, \ldots, a_n$ that exceeds a_k. We then put the remaining integers $(a_{k+1}, a_{k+2}, \ldots, a_n)$ into increasing order. For example, the permutation 52431 undergoes the following transformations:

$$52431 \qquad \text{results in} \qquad 53421$$

$$a_k = 2 \qquad\qquad a_m = 3$$

interchange

$$53421 \qquad \text{results in} \qquad 53124$$

place in increasing order

the next permutation in lexicographic order

In a similar way the permutation 635421 becomes 641235.

One final observation is in order. Since a_m is the smallest of the integers $a_{k+1}, a_{k+2}, \ldots, a_n$ that exceeds a_k, we have

$$a_{k+1} < a_{k+2} < \cdots < a_{m-1} < a_k < a_m < \cdots < a_n$$

Consequently, after interchanging a_k and a_m we can put the remaining integers into increasing order by simply reversing their order. We are now ready for the algorithm.

Algorithm 8.6   To generate in lexicographic order the permutation that follows a given permutation.

Input: $A(1)A(2) \cdots A(N)$, the given permutation

Output: $A(1)A(2) \cdots A(N)$, the next permutation in lexicographic order

1. $K \leftarrow N - 1$
2. WHILE $A(K) > A(K + 1)$ DO
   a. $K \leftarrow K - 1$

   We find $A(K)$, where K is the smallest index such that $A(K) < A(K + 1) > \cdots > A(N - 1) > A(N)$.

3. $M \leftarrow N$
4. WHILE $A(M) < A(K)$ DO
   a. $M \leftarrow M - 1$

   We find $A(M)$, the smallest of $A(K + 1), A(K + 2), \ldots, A(N)$ that exceeds $A(K)$.

5. TEMP $\leftarrow A(M)$

   In steps 5, 6, and 7, we interchange $A(M)$ and $A(K)$.

6. $A(M) \leftarrow A(K)$
7. $A(K) \leftarrow$ TEMP
8. FOR $H = K + 1$ TO N
   a. TEMP $(H) \leftarrow A(H)$

   In steps 8 and 9 we complete the next permutation by reversing the order of $A(K + 1), A(K + 2), \ldots, A(N)$.

9. FOR $H = K + 1$ TO N
  a. $A(H) =$
    $\text{TEMP}(N + K + 1 - H)$
10. FOR $I = 1$ TO N
  a. OUTPUT $A(I)$

The Generalized Product Principle also applies to counting the number of ordered arrangements of r distinct objects selected from a set of n objects.

Definition 8.7 An ordered arrangement of r distinct objects selected from a set of n objects is called an *r-permutation* of the n objects. The total number of such r-permutations of n objects is designated by $P(n, r)$.

Example 4 The following is a listing of all 2-permutations of the 5-element set $\{a, b, c, d, e\}$; that is, it is a listing of all ordered arrangements of two distinct elements selected from $\{a, b, c, d, e\}$.

ab	*ad*	*bc*	*be*	*ce*
ba	*da*	*cb*	*eb*	*ec*
ac	*ae*	*bd*	*cd*	*de*
ca	*ea*	*db*	*dc*	*ed*

$\square$

In Problem 13 you are asked to modify the proof of Theorem 8.4 to show that there are

$$n(n - 1)(n - 2) \cdots (n - r + 1)$$

r-permutations of n objects.

Note that if r and n are nonnegative integers and if $r < n$, then

$$n(n - 1)(n - 2) \cdots (n - r + 1)$$
$$= \frac{n(n - 1)(n - 2) \cdots (n - r + 1)(n - r) \cdots (2)(1)}{(n - r) \cdots (2)(1)} = \frac{n!}{(n - r)!}$$

and, therefore, $P(n, r) = n!/(n - r)!$, and thus we have the following theorem.

Theorem 8.8 The number of r-permutations of n objects is

$$P(n, r) = \frac{n!}{(n - r)!} \quad \text{for } r \leq n$$

Observe that when $r = n$, we have

$$\frac{n!}{(n - n)!} = \frac{n!}{0!} = n!$$

which by Theorem 8.4 is equal to $P(n, n)$.

Example 5   If $n = 10$ and $r = 6$, then

$$P(10, 6) = \frac{10!}{(10 - 6)!} = 10 \cdot 9 \cdot 8 \cdot 7 \cdot 6 \cdot 5 = 151{,}200 \qquad \square$$

Example 6   In how many ways can a president, a vice-president, and a secretary be elected from a group of nine persons?

In this problem, it is obvious that we want to do more than merely select three persons from the group of nine; we must also order the three selected people to determine who will be president, who will be vice-president, and who will be secretary. In other words, we must determine how many 3-permutations can be obtained from a 9-element set. By Theorem 8.8, we find that the number of possibilities is

$$P(9, 3) = \frac{9!}{(9 - 3)!} = 504 \qquad \square$$

Example 7   Suppose that we wish to develop a simple code, each word of which consists of five distinct letters. How many code words can we form from our alphabet of twenty-six letters?

For this problem, the order of the letters in any given code word is essential (for instance, the code word *cqrtv* is distinct from the code word *rtcvq*). Therefore, by Theorem 8.8 it follows that there are

$$P(26, 5) = \frac{26!}{(26 - 5)!} = 7{,}893{,}600$$

possible code words. $\qquad \square$

8.2   Problems

1. List all 3-permutations of $\{a, b, c, d\}$.
2. Evaluate using Theorem 8.8.
   - (a) $P(3, 2)$
   - (b) $P(7, 4)$
   - (c) $P(n, n - 1)$
   - (d) $P(n, 1)$
   - (e) $P(2n, n)$
3. In how many ways can six jobs be assigned to six workers?
4. In how many ways can five job applicants be ranked?
5. In how many ways can two applicants be selected for two distinct jobs from a pool of five applicants?
6. In how many ways can three applicants be selected for three distinct jobs from a pool of eight applicants?

7. How many one-to-one onto functions are there with domain $\{A, B, C, D\}$ and range $\{1, 2, 3, 4\}$?

8. How many one-to-one functions are there with domain $\{A, B, C, D\}$ and range $\{1, 2, 3, 4, 5\}$?

9. In how many ways can n couples stand in a single line that alternates woman–man–woman–man–$\cdots$?

10. In how many ways can n people be arranged in a circle? (Two arrangements are considered the same if one is a rotation of the other.)

11. In how many ways can r people be selected from a set of n people and then be arranged in a circle?

12. We would like to use our alphabet of 26 letters to form 30,000 words of equal length where no letter is repeated within a word. What is the minimum length word that will suffice?

13. Show that there are $n(n - 1)(n - 2) \cdots (n - r + 1)$ r-permutations of n objects.

8.3 COMBINATIONS

Many counting problems can be reduced to the problem of counting how many r-element subsets there are in a given n-element set. An r-element subset of an n-element set is often called an *r-combination of n objects*.

In contrast to an r-permutation, order is not considered in an r-combination. You should compare the following example with Example 4 of Section 8.2.

Example 1   The following is a listing of all 2-combinations of the 5-element set $\{a, b, c, d, e\}$; that is, it is a listing of all 2-element subsets of $\{a, b, c, d, e\}$.

$$\{a, b\} \qquad \{a, d\} \qquad \{b, c\} \qquad \{b, e\} \qquad \{c, e\}$$
$$\{a, c\} \qquad \{a, e\} \qquad \{b, d\} \qquad \{c, d\} \qquad \{d, e\} \qquad \square$$

The next theorem gives the number $C(n, r)$ of r-combinations of n objects.

Theorem 8.9   The number of r-element subsets (r-combinations) of an n-element set is

$$C(n, r) = \frac{n!}{r!(n - r)!}$$

Proof: Recall that an r-permutation of n objects is an ordered arrangement of r objects selected from the set of n objects. We can specify a particular r-permutation of n objects by using the following two steps:

1. Select an r-element subset from the n-element set. This step can be accomplished in $C(n, r)$ ways.

2. Arrange the *r* elements selected in step 1 into some particular order. By Theorem 8.4, this arrangement can be carried out in *r*! ways.

Since there are a total of $P(n, r)$ *r*-permutations of *n* objects, we may apply the Product Principle to obtain

$$P(n, r) = C(n, r)r!$$

Thus,

$$C(n, r) = \frac{P(n, r)}{r!}$$

Since by Theorem 8.8, $P(n, r) = n!/(n - r)!$, we substitute to obtain the desired result,

$$C(n, r) = \frac{n!}{r!(n - r)!}$$ ∎

Example 2   The number of 4-element subsets of a 9-element set is

$$C(9, 4) = \frac{9!}{4!(9 - 4)!} = \frac{9!}{4!5!} = 126$$ □

We often replace the notation $C(n, r)$ by the symbol $\binom{n}{r}$. For example, we can write $\binom{9}{4} = C(9, 4) = 126$.

It is critical, though at times difficult, to distinguish between permutations and combinations. You should keep in mind that *order* makes a difference in a permutation. Since a combination is a *set*, and since the order in which elements appear in a set is of no consequence, combinations do not rely on the elements' order. For example, the permutations *abcd* and *dabc* are distinct; the combinations $\{a, b, c, d\}$ and $\{d, a, b, c\}$, however, are the same.

Example 3   The number of ways that four objects can be selected from seven objects when the order of the four objects selected is ignored is

$$C(7, 4) = \frac{7!}{4!3!} = 35$$

That is, there are 35 possible 4-combinations of seven elements.   □

Example 4   The number of ways four objects can be selected from seven objects when the order of selected objects is accounted for is

$$P(7, 4) = \frac{7!}{(7 - 4)!} = 840$$

That is, there are 840 possible 4-permutations of seven elements.   □

Solutions to some problems depend on Theorem 8.9 and the Generalized Product Principle (Theorem 8.1).

Example 5 A basketball coach has thirteen players on his team—six forwards, three centers, and four guards. How many teams consisting of two forwards, one center, and two guards can he form from these players?

By Theorem 8.9 there are $\binom{6}{2}$ ways that the coach can select two forwards from his total of six forwards, there are $\binom{3}{1}$ ways he can select a center, and there are $\binom{4}{2}$ ways he can select two guards. Thus, by the Generalized Product Principle, we find that there are

$$\binom{6}{2}\binom{3}{1}\binom{4}{2} = 270$$

different teams that the coach can field. □

Example 6 In coding theory it is often necessary to determine the number of words that differ from a given code word in fewer than a specified number of positions. Suppose, for example, that the words are bit strings (strings of 0's and 1's) of length ten. We wish to find the number of bit strings that differ from the word $w = 0000000000$ in fewer than four positions.

For $0 \leq i \leq 3$, let A_i be the set of words that differ from w in precisely i positions. For instance, A_2 is the set of words that contain exactly two 1's. Observe that $A_0 \cup A_1 \cup A_2 \cup A_3$ is the set of words that differ from w in three or fewer positions; we must determine $|A_0 \cup A_1 \cup A_2 \cup A_3|$.

Since $A_i \cap A_j = \varnothing$ if $i \neq j$, it follows from Theorem 8.2 that

$$|A_0 \cup A_1 \cup A_2 \cup A_3| = |A_0| + |A_1| + |A_2| + |A_3|$$

Note that each element of A_i is uniquely determined once i positions have been selected for placing a 1. Since there are ten positions from which to select, we have for $i = 0, 1, 2, 3$

$$A_i = \binom{10}{i}$$

and consequently,

$$|A_0 \cup A_1 \cup A_2 \cup A_3| = \binom{10}{0} + \binom{10}{1} + \binom{10}{2} + \binom{10}{3}$$
$$= 1 + 10 + 45 + 105 = 161$$

which means there are 161 bit strings that differ from the word $w = 0000000000$ in fewer than four positions. □

The following examples involve somewhat more subtle applications of Theorem 8.9.

Example 7 Suppose that we have four boxes B_1, B_2, B_3, and B_4 and thirteen distinct objects. We wish to count the number of ways that the objects can be distributed into the four boxes so that there will be

3 objects in box B_1	6 objects in box B_3
2 objects in box B_2	2 objects in box B_4

Note that the number of ways to select the three objects for box B_1 is $C(13, 3)$. Once we have selected the contents for that box, only ten objects remain. From these ten, we must select two for box B_2, which can be accomplished in $C(10, 2)$ ways. Now eight objects remain, of which six are selected for box B_3; there are $C(8, 6)$ ways to select these items. Finally only two objects remain for box B_4, thus this box can be filled in just $C(2, 2) = 1$ way.

By the Product Principle, there are

$$C(13, 3) \cdot C(10, 2) \cdot C(8, 6) \cdot C(2, 2) = \frac{13!}{3!10!} \frac{10!}{2!8!} \frac{8!}{6!2!} = \frac{13!}{3!2!6!2!} = 360{,}360$$

ways to distribute the objects. □

Example 7 presents a counting result which we generalize in the next theorem. (You are asked to prove this theorem in Problem 22.) In this theorem we use the notation

$$\binom{n}{r_1, r_2, \cdots, r_k}$$

to denote

$$\frac{n!}{r_1!r_2! \cdots r_k!}$$

where $r_1 + r_2 + \cdots + r_k = n$.

Theorem 8.10 The number of ways that n distinct objects can be distributed into k boxes, B_1, B_2, ..., B_k such that there are r_1 distinct objects in box B_1, r_2 distinct objects in box B_2, ..., r_k distinct objects in box B_k is

$$\binom{n}{r_1, r_2, \cdots, r_k}$$

Example 8 To test the effectiveness of a certain drug, sixteen persons are chosen for an experiment. Seven of the persons are given the drug, five are not given the drug, and four are given a placebo instead of the drug. In how many ways can these persons be divided into groups of seven, five, and four to satisfy the requirements for the experiment?

The number of such divisions is the number of ways we can distribute sixteen distinct objects (people) into three boxes (categories), B_1, B_2, and B_3, so that

seven are in B_1 (given the drug)

five are in B_2 (given neither the drug nor the placebo)

four are in B_3 (given the placebo)

Thus, there are

$$\binom{16}{7,\,5,\,4} = \frac{16!}{7!5!4!} = 1{,}441{,}440$$

ways the group can be divided.   □

Example 9   Six people are playing a game of poker. Five cards from a deck of fifty-two are dealt to each player. We want to compute the total number of possible deals. We view each deal as a distribution of fifty-two labeled objects (cards) into seven boxes such that the first six boxes (hands) each contain five objects and the last box (the "stack") contains the remaining twenty-two objects. Thus, by Theorem 8.10 the number of possible deals is

$$\binom{52}{5,\,5,\,5,\,5,\,5,\,5,\,22} = \frac{52!}{5!5!5!5!5!5!22!} \approx 2.4 \times 10^{34}$$ □

Though our next example has little apparent relation to the distribution of objects into boxes, we can still apply Theorem 8.10.

Example 10   Suppose we wish to establish a signaling system with a string of lights that consists of five red lights, four green lights, and nine blue lights. How many different arrangements of these lights are possible? (We assume that all of the red lights are identical, all of the green lights are identical, and all of the blue lights are identical.)

To determine the number of different arrangements, we note that there are a total of eighteen positions, each of which is to contain either a red, a green, or a blue light. Thus we can think of an arrangement of the eighteen lights as a distribution of the eighteen positions (eighteen labeled objects) into three boxes (one for red, one for green, and one for blue) in such a way that the red box contains five positions, the green box contains four positions, and the blue box contains nine positions. Consequently, by Theorem 8.10 there are

$$\binom{18}{5,\,4,\,9} = \frac{18!}{5!4!9!} = 6{,}126{,}120$$

possible arrangements.   □

8.3 Problems

1. Compute
   (a) $C(7, 2)$ (b) $C(7, 4)$ (c) $C(n, 0)$
   (d) $C(n, n - 1)$ (e) $C(n, n)$ (f) $C(n, 1)$

2. List all 4-combinations of $\{a, b, c, d\}$.

3. How many ways can a hand of five cards be selected from a deck of 52 cards?

4. How many five-card hands contain only spades?

5. How many five-card hands contain only a single suit?

6. How many five-card hands contain two of one rank card and three of another rank?

7. How many five-card hands contain exactly one pair?

8. In bridge there are four hands of thirteen cards dealt from a deck of fifty-two cards. How many different bridge deals are possible?

9. In rummy each deal consists of four hands of nine cards dealt from a deck of fifty-two cards. How many different rummy deals are possible?

10. In how many ways can a committee of three be selected from a list of ten applicants?

11. In how many ways can a group of thirteen people be divided into two groups containing five people and a third group containing three people?

12. In how many ways can a group of seventeen people be divided into four teams consisting of five, two, seven, and three people respectively?

13. In how many ways can a group of seventeen people be divided into four disjoint teams that will form the executive, legal affairs, fundraising, and public relations committees of a local group? The executive, fundraising, and public relations committees are to consist of five people each, while the legal affairs committee is to have a membership of two.

14. Red, white, and blue lights are to be arranged in a straight line. How many arrangements are possible using three red lights, seven blue lights, and five white lights if all fifteen lights must be used?

15. How many different circular arrangements are there using the fifteen lights described in Problem 14? (Two arrangements are to be considered the same if one is a rotation of the other.)

16. How many different arrangements are there of the letters in the word MISSISSIPPI?

17. In how many ways can seven children arrange themselves in a circle?

18. In how many permutations of the letters of the word INSTRUCTOR do the vowels appear in alphabetical order?

19. How many n-digit sequences of zeros, ones, and twos contain exactly k zeros?

20. A man has a sufficient number of friends to invite a different set of four friends to a party each night for 365 nights. What is the minimum number of friends he has?

21. Prove that $\binom{2n}{n}$ is even.

22. Prove Theorem 8.10 by extending the argument used in Example 7.

23. Explain why

$$\binom{n}{r_1}\binom{n - r_1}{r_2} \cdots \binom{n - r_1 - r_2 - \cdots - r_{m-1}}{r_m} = \binom{n}{r_1, r_2, \cdots, r_m}$$

8.4   THE BINOMIAL THEOREM

The number $\binom{n}{r}$ is often referred to as a *binomial coefficient* because it appears as a coefficient in the binomial expansion of $(a + b)^n$, where n is a positive integer. For example, we have

$$(a + b)^1 = a + b = \binom{1}{0}a + \binom{1}{1}b$$

$$(a + b)^2 = a^2 + 2ab + b^2 = \binom{2}{0}a^2 + \binom{2}{1}ab + \binom{2}{2}b^2$$

and

$$(a + b)^3 = a^3 + 3a^2b + 3ab^2 + b^3 = \binom{3}{0}a^3 + \binom{3}{1}a^2b + \binom{3}{2}ab^2 + \binom{3}{3}b^3$$

In general, we have the following theorem.

Theorem 8.11   *Binomial Theorem.* For each positive integer n

$$(a + b)^n = \binom{n}{0}a^n + \binom{n}{1}a^{n-1}b^1 + \binom{n}{2}a^{n-2}b^2 + \cdots + \binom{n}{k}a^{n-k}b^k$$

$$+ \cdots + \binom{n}{n-1}a^1 b^{n-1} + \binom{n}{n}b^n$$

Proof: First we express $(a + b)^n$ as the product

$$(a + b)(a + b) \cdots (a + b) \tag{1}$$

where the factor $(a + b)$ appears n times. It follows from elementary rules of arithmetic that this product is equal to the sum of all possible products of a's and b's, where each such product arises from a selection of either an a or a b from each factor $(a + b)$ in (1). Thus, for instance, if $n = 5$, then a typical product term such as *abaab* will arise from the selection of an a from the first factor of (1), *a b* from the second factor, *a*'s from the third and fourth factors, and a *b* from the fifth factor.

Note that once we have selected the factors of (1) from which the *b*'s are chosen, an *a must* be selected from each of the remaining factors. In other words, the factors from which the *a*'s are chosen are automatically determined by the factors we choose for the *b*'s. Note, too, that each product term may contain anywhere from zero to *n b*'s. How many of these product terms will contain exactly *r b*'s? In other words, in how many ways can we choose *r* factors from which to select the *b*'s? The problem then is to determine the number of ways to select *r* objects from *n* objects, where the order of the *r* objects selected is ignored (the product term *abaab* is equal to the product term *ababa*). Since by Theorem 8.9 there are $\binom{n}{r}$ ways of making such a selection of *r* objects, it follows that there are $\binom{n}{0}$ product terms that contain no *b*'s; there are $\binom{n}{1}$ product terms that contain exactly one *b*; there are $\binom{n}{2}$ product terms that contain exactly two *b*'s; and so on. Therefore, we find that the product (1) is equal to

$$\binom{n}{0}a^n + \binom{n}{1}a^{n-1}b^1 + \binom{n}{2}a^{n-2}b^2 + \cdots + \binom{n}{n-1}a^1 b^{n-1} + \binom{n}{n}b^n$$

which is the desired result. ∎

In Problem 9 you are asked to give an alternative (and less transparent) induction proof of the Binomial Theorem.

Example 1 $(a + b)^5 = \binom{5}{0}a^5 + \binom{5}{1}a^4 b^1 + \binom{5}{2}a^3 b^2 + \binom{5}{3}a^2 b^3$

$$+ \binom{5}{4}a^1 b^4 + \binom{5}{5}b^5$$

$$= a^5 + 5a^4 b + 10a^3 b^2 + 10a^2 b^3 + 5ab^4 + b^5 \quad \square$$

We can use the Binomial Theorem for more than just determining the coefficients of a binomial expansion. The next theorems illustrate such possibilities.

Theorem 8.12

$$\binom{n}{0} + \binom{n}{1} + \binom{n}{2} + \cdots + \binom{n}{n} = 2^n$$

Proof: If we let $a = 1$ and $b = 1$ in the Binomial Theorem, we have

$$(1 + 1)^n = \binom{n}{0}1^n + \binom{n}{1}1^{n-1}1 + \binom{n}{2}1^{n-2}1^2 + \cdots + \binom{n}{n}1^n$$

from which it follows that

$$2^n = \binom{n}{0} + \binom{n}{1} + \binom{n}{2} + \cdots + \binom{n}{n} \qquad ∎$$

Theorem 8.13   If X is a set with n elements, then X has 2^n distinct subsets.

Proof: By Theorem 8.9 the number of subsets of X containing exactly r elements is $\binom{n}{r}$. Thus, by the Sum Principle the total number of subsets of X (including the empty set) is

$$\binom{n}{0} + \binom{n}{1} + \binom{n}{2} + \cdots + \binom{n}{n-1} + \binom{n}{n}$$

By Theorem 8.12, this sum is equal to 2^n. Consequently, an n-element set has 2^n subsets.   ∎

Theorem 8.14   For any positive integer n,

$$\binom{n}{0} - \binom{n}{1} + \binom{n}{2} - \binom{n}{3} + \cdots + (-1)^n\binom{n}{n} = 0$$

Proof: If we substitute $a = 1$ and $b = -1$ into

$$(a + b)^n = \binom{n}{0}a^n + \binom{n}{1}a^{n-1}b + \binom{n}{2}a^{n-2}b^2 + \cdots + \binom{n}{n}b^n$$

then we have

$$0 = (1 - 1)^n = \binom{n}{0} + \binom{n}{1}(-1) + \binom{n}{2}(-1)^2 + \cdots + \binom{n}{n}(-1)^n$$

which is the desired result.   ∎

8.4   Problems

1. What is the coefficient of a^6b^7 in the expansion of $(a + b)^{13}$?
2. What is the coefficient of x^5 in the expansion of $(2x + 3)^{10}$?
3. Prove that

(a) $4^n = \displaystyle\sum_{r=0}^{n} C(n, r)3^r$

(b) $k^n = \displaystyle\sum_{r=0}^{n} C(n, r)(k - 1)^r$, for $k = 2, 3, 4, \ldots$

(c) $(k - 1)^n = \displaystyle\sum_{r=0}^{n} (-1)^r C(n, r)k^{n-r}$, for $k = 2, 3, 4, \ldots$

4. Prove the "multinomial theorem":

$$(a_1 + a_2 + a_3 + \cdots + a_k)^n = \sum_{r_1+r_2+\cdots+r_k=n} \binom{n}{r_1, r_2, \ldots, r_k} a_1^{r_1} a_2^{r_2} \cdots a_k^{r_k}$$

5. What is the coefficient of $a^3 b^2 c^5$ in the expansion of $(a + b + c)^{10}$?

6. What is the coefficient of $a^3 b^4$ in the expansion of $(2a + 3b + 1)^{12}$?

7. Prove that

$$k^n = \sum_{r_1 + r_2 + \cdots + r_k = n} \binom{n}{r_1, r_2, \cdots, r_k}$$

8. Show that

$$\binom{n}{r_1, r_2} = \binom{n}{r_1} = \binom{n}{r_2}$$

9. Use an induction argument to prove the Binomial Theorem (Theorem 8.11).

10. Explain how the Sum Principle is used in the proof of Theorem 8.13.

11. Use induction to prove that an n-element set has 2^n subsets.

12. Give a proof of Theorem 8.14 by showing that the number of subsets with an even number of elements is the same as the number of subsets with an odd number of elements. (*Hint:* Let a be an element of the n-element set. For all subsets T consider the function:

$$f(T) = \begin{cases} T \setminus \{a\} \text{ if } a \in T \\ T \cup \{a\} \text{ if } a \notin T \end{cases}$$

13. Use induction to prove Theorem 8.14.

8.5 PASCAL'S TRIANGLE AND SOME BASIC IDENTITIES

The identities

$$\binom{n}{0} + \binom{n}{1} + \binom{n}{2} + \cdots + \binom{n}{n} = 2^n$$

and

$$\binom{n}{0} - \binom{n}{1} + \binom{n}{2} - \cdots + (-1)^n \binom{n}{n} = 0$$

derived in the previous section (Theorems 8.12 and 8.14) are but two of the many useful identities involving the binomial coefficients. Generally, such identities have two distinct types of proof: combinatorial (counting) and algebraic. Where possible, we shall present both types of proofs. Although an algebraic proof may be easier, a combinatorial proof is generally more enlightening.

In the following theorems we establish two fundamental identities involving binomial coefficients.

Theorem 8.15   If r and n are positive integers and if $r \leq n$, then

$$\binom{n}{r} = \binom{n}{n-r}$$

Proof (*Combinatorial*):   Combinatorial proofs usually result from counting the same set in two ways. For example, $\binom{n}{r}$ is the number of ways one can choose an r-element set from an n-element set.

Note that for each r-element set chosen, there corresponds a complementary set of $n - r$ elements that is not chosen. It follows that the number of ways of choosing an r-element set (which is $\binom{n}{r}$), is equal to the number of ways of choosing an $n - r$ element set (which is $\binom{n}{n-r}$); that is,

$$\binom{n}{r} = \binom{n}{n-r}$$

Proof (*Algebraic*):   By Theorem 8.9 we have

$$\binom{n}{r} = \frac{n!}{r!(n-r)!} \quad \text{and} \quad \binom{n}{n-r} = \frac{n!}{(n-r)!(n-(n-r))!} = \frac{n!}{(n-r)!r!}$$

Thus,

$$\binom{n}{r} = \binom{n}{n-r} \qquad\blacksquare$$

Theorem 8.16   If r and n are positive integers and if $r \leq n$, then

$$\binom{n}{r} = \binom{n-1}{r-1} + \binom{n-1}{r}$$

Proof (*Combinatorial*):   By Theorem 8.9, $\binom{n}{r}$ is the number of r-element subsets of an n-element set $\{a_1, a_2, \ldots, a_{n-1}, a_n\}$. Every r-element subset R of $\{a_1, \ldots, a_{n-1}, a_n\}$ is exactly one of the following two types:

$$\text{Type 1: } a_n \in R \qquad \text{Type 2: } a_n \notin R$$

If the subset R is of Type 1, then $a_n \in R$, and we need only select $r - 1$ additional elements from the set $a_1, a_2, \ldots a_{n-1}$ to complete R. Since we can carry out this selection process in $\binom{n-1}{r-1}$ ways, there are $\binom{n-1}{r-1}$ r-element subsets of Type 1.

If the subset R is of Type 2, then $a_n \notin R$, and we must select r elements from the set $a_1, a_2, \ldots, a_{n-1}$ to form R. Since this selection process can be carried out in $\binom{n-1}{r}$ ways, there are $\binom{n-1}{r}$ r-element subsets of Type 2.

Since every r-element subset is either of Type 1 or of Type 2 (but not of both), it follows from the Sum Principle that there are

$$\binom{n-1}{r-1} + \binom{n-1}{r}$$

r-element subsets of an n-element set. Thus,

$$\binom{n}{r} = \binom{n-1}{r-1} + \binom{n-1}{r}$$

Proof (*Algebraic*):   The algebraic proof is completely straightforward (though not particularly enlightening).

$$\binom{n-1}{r-1} + \binom{n-1}{r} = \frac{(n-1)!}{(r-1)!(n-1-(r-1))!} + \frac{(n-1)!}{r!(n-1-r)!}$$

$$= \frac{(n-1)!}{(r-1)!(n-r)!} + \frac{(n-1)!}{r!(n-r-1)!}$$

$$= \frac{(n-1)!r}{r!(n-r)!} + \frac{(n-1)!(n-r)}{r!(n-r)!}$$

$$= \frac{(n-1)!r + (n-1)!n - (n-1)!r}{r!(n-r)!}$$

$$= \frac{n!}{r!(n-r)!} = \binom{n}{r} \qquad \blacksquare$$

We next construct Pascal's triangle, which can be used to generate binomial coefficients and establish certain combinatorial identities. Figure 8.2 indicates the first few rows of this triangle.

We determine the entries in the triangle as follows

a. We place ones along the right and left sides of the triangle.

b. We obtain any entry q in the interior of any given row of the triangle by adding the two entries nearest to q that lie in the row just above the row containing q. For example, the circled 10 in the fifth row of Figure 8.2 is the sum of the entries 6 and 4 that lie closest to it in the fourth row.

Row 0					1				
Row 1				1		1			
Row 2			1		2		1		
Row 3		1		3		3		1	
Row 4	1		4		6		4		1
Row 5	1	5		10		⑩		5	1

Figure 8.2

To see how Pascal's triangle is related to the binomial coefficients, we assign a coordinate system to this triangle, so that the first coordinate refers to the row of the triangle and the second coordinate corresponds to the diagonal indicated in Figure 8.3. We will show that in this coordinate system the entry occupying the point (n, r) is $\binom{n}{r}$. This is clearly true for points on the boundary of the triangle where either $r = 0$ or $r = n$, since for these points we have either $\binom{n}{0} = 1$ or $\binom{n}{n} = 1$. It is also true for an interior point (n, r), since the nearest two numbers to (n, r) that lie in the $(n - 1)$st row are $\binom{n-1}{r}$ and $\binom{n-1}{r-1}$, and by Theorem 8.16

$$\binom{n}{r} = \binom{n-1}{r} + \binom{n-1}{r-1}$$

From these observations, we see that Pascal's triangle provides another way to calculate the binomial coefficients. More importantly, however, Pascal's triangle can be used as a theoretical tool to derive a variety of combinatorial identities. To obtain some of these identities, we first show that there are exactly $\binom{n}{r}$ paths in Pascal's triangle that start at $(0, 0)$, proceed downward, and terminate at the point (n, r).

To prove this, note that if we start at the point $(0, 0)$ then we have two ways to descend to a point in the first row: down to the left or down to the right. Once we have made this choice, we must proceed to the second row. Once again we are faced with the two possibilities: down to the left or down to the right. We continue in this manner for each succeeding row until we reach the point (n, r).

Note that whenever we go downward to the left, the n-coordinate increases by one, but the r-coordinate remains the same; if we proceed downward to the right, however, then both the n-coordinate and the r-coordinate increase by one. Thus, to descend from the point $(0, 0)$ to the point (n, r) we must take n

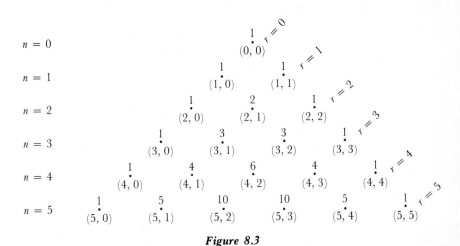

Figure 8.3

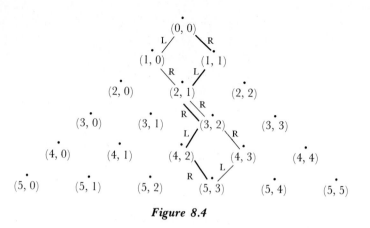

Figure 8.4

steps downward (increasing the first coordinate by n), of which r steps are downward to the right (increasing the second coordinate by r).

Two such paths from the point $(0, 0)$ to the point $(5, 3)$ are indicated in Figure 8.4; steps to the left have been labeled with an L, and steps to the right with an R.

It follows that for an arbitrary point (n, r) in the triangle, the number of downward paths from $(0, 0)$ to (n, r) is equal to the number of ways we can select r objects (right turns) from a set of n objects (opportunities to turn). This number is, of course, $\binom{n}{r}$. We use this observation to establish the following identity.

Theorem 8.17   For any two positive integers n and r with $r \leq n$

$$\binom{n}{0} + \binom{n+1}{1} + \binom{n+2}{2} + \cdots + \binom{n+r}{r} = \binom{n+r+1}{r}$$

Proof (*Combinatorial*):   From the above discussion there are $\binom{n+r+1}{r}$ descending paths from the point $(0, 0)$ to the point $(n + r + 1, r)$. As is customary in combinatorial proofs, we can count these paths in another way.

Let Q_k denote the set of those downward paths from $(0, 0)$ to $(n + r + 1, r)$ that have their *last* left turn at $(n + k, k)$. In Figure 8.5 a particular member of the set Q_k is illustrated.

Observe that $Q_i \cap Q_j = \varnothing$ if $i \neq j$ (a path cannot have two *last* left turns). Observe further that each descending path from $(0, 0)$ to $(n + r + 1, r)$ is in Q_k for some $k = 0, 1, \ldots, r$. Since the total number of such paths is $\binom{n+r+1}{r}$, we have from the Sum Principle that

$$|Q_0| + |Q_1| + \cdots + |Q_r| = \binom{n+r+1}{r} \tag{1}$$

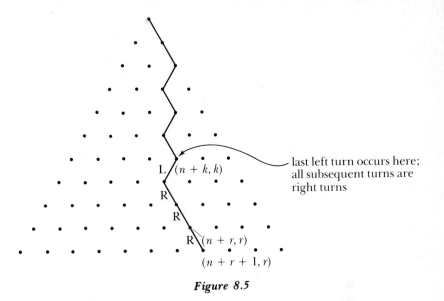

$L\ (n + k, k)$

last left turn occurs here; all subsequent turns are right turns

R

R

$R\ (n + r, r)$

$(n + r + 1, r)$

Figure 8.5

Note that, as shown in Figure 8.5, the route of any path in Q_k is uniquely determined once it leaves the point $(n + k, k)$: It must turn left at $(n + k, k)$ and then turn right at all subsequent points. Prior to its arrival at $(n + k, k)$ its downward path is arbitrary. Thus,

$$|Q_k| = \binom{n + k}{k}$$

Substituting this result into (1) gives

$$\binom{n}{0} + \binom{n + 1}{1} + \binom{n + 2}{2} + \cdots + \binom{n + r}{r} = \binom{n + r + 1}{r}$$

which is the desired result. ∎

Proof (*Algebraic*): Repeated applications of Theorem 8.16, beginning with

$$\binom{n + r + 1}{r} = \binom{n + r}{r - 1} + \binom{n + r}{r}$$

will yield a proof, the details of which are left as an exercise (Problem 7). ∎

8.5 Problems

1. Show that

$$\left[\binom{n}{0} + \binom{n}{1} + \cdots + \binom{n}{n} \right]^2 = \binom{2n}{0} + \binom{2n}{1} + \cdots + \binom{2n}{2n}$$

2. Show that if $n \geq 2$, then

$$\binom{n}{1} - 2\binom{n}{2} + 3\binom{n}{3} - \cdots + (-1)^{n-1}n\binom{n}{n} = 0$$

(*Hint:* Show that $r\binom{n}{r} = n\binom{n-1}{r-1}$)

3. Show that

$$C(n, 0) - 2C(n, 1) + 2^2 C(n, 2) + \cdots + (-1)^n 2^n C(n, n) = (-1)^n$$

4. The Sterling number of the second kind, $S(n, r)$, is the number of ways of partitioning an n-element set into r subsets, none of which can be empty.
   (a) Prove that

$$S(n, r) = S(n - 1, r - 1) + rS(n - 1, r)$$

   (*Hint:* Fix an element x in the n-element set, and consider the cases where $\{x\}$ is, and is not, an element of the partition.)
   (b) Use the fact that $S(n, 1) = S(n, n) = 1$ to compute $S(5, 2)$, $S(5, 3)$, and $S(5, 4)$.

5. Prove Vandermonde's Identity:

$$C(n + m, r) = C(n, 0)C(m, r) + C(n, 1)C(m, r - 1) + \cdots + C(n, r)C(m, 0)$$

   for integers $n \geq r \geq 0$ and $m \geq r \geq 0$. (*Hint:* Form a committee of r people from a group of n women and m men.)

6. In an n-cube (see Section 3.9, Problem 27) show that there are precisely $C(n, r)$ vertices that are a distance r from a given point. (Two vertices are a distance r apart if the shortest path connecting the vertices has length r.)

7. Finish the algebraic proof of Theorem 8.17.

8. (a) Use Theorem 8.17 to prove that

$$\binom{r}{r} + \binom{r + 1}{r} + \binom{r + 2}{r} + \cdots + \binom{n}{r} = \binom{n + 1}{r + 1}$$

   (b) Deduce that $1 + 2 + 3 + \cdots + n = [n(n + 1)]/2$. (*Hint:* Choose an appropriate value for r.)

9. (a) Show that $k^2 = 2\binom{k}{2} + \binom{k}{1}$
   (b) Deduce that $1 + 4 + 9 + \cdots + n^2 = 2\binom{n+1}{3} + \binom{n+1}{2}$.

8.6 COUNTING WITH REPETITIONS

Thus far we have been concerned with counting problems in which objects appeared at most once. In this section we take up the problem of counting when repetition of objects is permitted. As before, these problems can be divided into two categories: those for which order counts, and those for which order does not matter. We will first consider a problem of the former type.

Suppose we wish to build a code using the first ten letters of the alphabet. Our code is to consist entirely of four-letter "words," but any given letter can be used more than once within a word. Thus such words as *AEGF, AAAA, EEHB,* and *HEHB* would all be acceptable. How many code words can we form in this way?

This problem is easily resolved. First we suppose that each letter of each code word lies in one of four "boxes." Hence the word *EEHB* becomes

There are ten ways to fill the first box, and since repetitions are allowed, there are also ten ways to fill each of the remaining boxes. Thus, it follows from the Generalized Product Principle that there are $10 \cdot 10 \cdot 10 \cdot 10 = 10^4$ possible code words that can be formed in the manner just described. This problem is a special case of the following theorem, which you are asked to prove in Problem 17.

Theorem 8.18   *Redundant r-permutations.*   There are n^r arrangements of r objects selected from n objects if order counts and repetitions are allowed (that is, there are n^r redundant r-permutations of n objects).

You should compare the following example with Example 4 of Section 8.2 and Example 1 of Section 8.3.

Example 1   The $5^2 = 25$ redundant 2-permutations of the set $\{a, b, c, d, e\}$ are:

aa	*ba*	*ca*	*da*	*ea*
ab	*bb*	*cb*	*db*	*eb*
ac	*bc*	*cc*	*dc*	*ec*
ad	*bd*	*cd*	*dd*	*ed*
ae	*be*	*ce*	*de*	*ee*

□

Example 2   We would like to form at least 18,000 different code words using the twenty-six letters of the English alphabet. What is the shortest length word that we can use?

To answer this question, observe that each word is a redundant permutation (repetition allowed, order counts) of the twenty-six letters. Thus there are 26^n words of length n. Since

$$26^1 + 26^2 + 26^3 = 26 + 676 + 17{,}576 = 18{,}278$$

no word need contain more than three letters.   □

We now turn our attention to the problem of counting "redundant combinations," that is, selections in which order does not count and repetitions are allowed. As we shall see, this problem is most easily approached by considering an apparently unrelated question: In how many ways can we distribute n identical objects into m labeled boxes?

As a particular example, consider the problem of distributing four identical objects into ten boxes. To find the number of such distributions, it will be helpful to use diagrams consisting of nine vertical bars (|), four stars (*), and two blocks (■). The stars represent the four objects; the bars denote the sides of the boxes; and the blocks denote the left- and right-hand limits of the diagram. Thus, for instance, Figure 8.6 indicates that there is one object in the first box, two in the third box, and one in the tenth box.

Figure 8.7 shows the situation where three objects are in the fourth box and one is in the ninth box.

Viewed this way, we see that every arrangement of nine vertical bars and four stars (where the bars and stars are flanked by two blocks) corresponds to a unique distribution of four indistinguishable objects into ten boxes. Consequently, the number of such distributions is equal to the number of ways we can position the stars and bars.

In order to position the stars and bars, it is sufficient to select from the thirteen possible positions (four stars plus the nine bars) those positions that will contain the four stars. Figure 8.8(a) illustrates a selection of positions

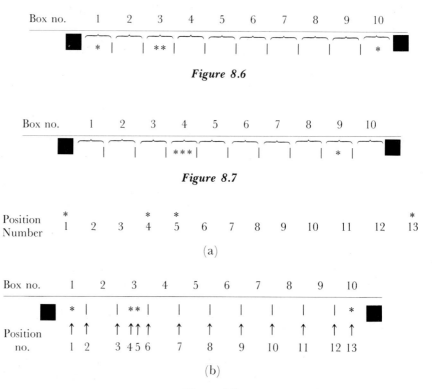

Figure 8.6

Figure 8.7

Figure 8.8

for stars that corresponds to the distribution shown in Figure 8.8(b) (and in Figure 8.6). Note that bars would be associated with the position numbers not having stars. Since there are $\binom{13}{4}$ ways to select positions for the stars, we conclude that there are $\binom{13}{4}$ ways to distribute four identical objects into ten labeled boxes. This result is a specific case of a very useful theorem, the proof of which parallels the argument just given (Problem 18).

Theorem 8.19 There are

$$\binom{n + r - 1}{r}$$

ways to distribute r identical objects into n distinct boxes.

Example 3 There are

$$\binom{5 + 7 - 1}{7} = \binom{11}{7} = 330$$

ways to distribute seven identical objects into five distinct boxes. □

We can use Theorem 8.19 to solve problems that have little apparent relation to the distribution of indistinguishable objects into distinct boxes.

Example 4 We count the number of solutions to the equation

$$r_1 + r_2 + r_3 + r_4 + r_5 = 10 \tag{1}$$

where each r_i is a nonnegative integer. For example, $r_1 = 2, r_2 = 0, r_3 = 0, r_4 = 5, r_5 = 3$ is one such solution.

To count the number of such solutions, think of distributing ten indistinguishable objects into five boxes labeled r_1, r_2, r_3, r_4, and r_5. The number of objects in box r_i represents the value of r_i. Note then that each distribution corresponds to a unique solution of (1) and conversely, that each solution to (1) corresponds to a unique distribution. From Theorem 8.19, then, there are

$$\binom{5 + 10 - 1}{10} = \binom{14}{10} = 1001$$

solutions to (1) where each r_i is a nonnegative integer. □

A slight shift in our point of view will enable us to count the total number of possible redundant r-combinations of n objects. (A redundant r-combination is a selection of r objects from a set of n objects for which repetition is allowed and order does not count.)

Consider n boxes, each of which is labeled to correspond to one of the n objects. We are to select r of these objects. Interpret the placement of one of

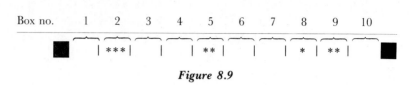

Figure 8.9

the r objects into the kth box as a selection of the kth object. If several objects are placed into the kth box, then the kth object is selected the appropriate number of times. Thus, for example, Figure 8.9 illustrates the selection of eight (r) objects from ten (n) objects where the second object is selected three times, the fifth twice, the eighth once, and the ninth twice.

From this point of view we see that the number of ways of selecting r objects from n objects where repetition is possible and order does not count is the same as the number of ways we can distribute r identical objects into n numbered boxes. Thus, we have the following result.

Theorem 8.20   *Redundant r-combinations.*   There are

$$\binom{n + r - 1}{r}$$

ways to select r objects from n objects if order does not count and repetitions are allowed (that is, there are $\binom{n+r-1}{r}$ redundant r-combinations of n objects).

Example 5   We want to count the number of possible outcomes if five identical dice are rolled. The number of possible outcomes is the number of ways we can select five objects from the set $\{1, 2, 3, 4, 5, 6\}$ where order does not count and repetitions are allowed. Thus there are

$$\binom{6 + 5 - 1}{5} = \binom{10}{5} = 252$$

possible outcomes.   □

	Repetitions allowed?	Order counts?	Number of selections of r objects from n objects
r-permutation	No	Yes	$\dfrac{n!}{(n - r)!}$
r-combination	No	No	$\dfrac{n!}{r!(n - r)!}$
Redundant r-permutation	Yes	Yes	n^r
Redundant r-combination	Yes	No	$\dbinom{n + r - 1}{r}$

Table 8.1

Table 8.1 summarizes the results from Theorems 8.8, 8.9, 8.18, and 8.20, which all relate to the selection of r objects from a set of n objects.

8.6   Problems

1. List all redundant 4-combinations of $\{1, 2\}$.

2. List all redundant 3-permutations of $\{a, b\}$.

3. We wish to form 1000 distinct words of length n from the set of symbols $\{a, b, c, d\}$. What is the minimum value of n?

4. We wish to form 5000 distinct words of length n or less from the set of symbols $\{a, b, c, d, e\}$. What is the minimum value of n?

5. Draw diagrams like those given in Figure 8.8(a) and 8.8(b) to illustrate the distribution of 7 objects into 8 boxes such that there are 2 objects in the first box, 3 objects in the fourth box, 1 object in the seventh box, and 1 object in the eighth box.

6. In how many ways can 10 identical objects be distributed into 15 boxes?

7. In how many ways can 9 identical objects be distributed into 8 boxes?

8. In how many ways can 10 distinct objects be distributed into 15 boxes?

9. In how many ways can 9 distinct objects be distributed into 8 boxes?

10. Show that if n identical dice are rolled, there are $\binom{n+5}{n}$ possible outcomes.

*11. There are 24 volumes of an encyclopedia on a shelf. In how many ways may we select 5 of these books if we do not select any two consecutive volumes? (*Hint:* Why is this equivalent to distributing 5 objects into 20 boxes so that no box contains more than one object?)

12. Show that m identical objects can be placed in n distinct boxes ($m \geq n$) in

$$\binom{m-1}{n-1}$$

ways so that no box is empty.

13. Show that the number of ways of arranging n a's and m b's, $m + 1 \geq n$, in a line so that no two a's are adjacent is $\binom{m+1}{n}$. (*Hint:* Place m objects in $n + 1$ boxes so that, except for possibly the two end boxes, no box is empty.)

14. Find the number of positive-integer solutions to $x_1 + x_2 + x_3 + x_4 = 13$.

15. Find the number positive-integer solutions to $x_1 + x_2 + x_3 + x_4 = 13$ if $x_1 \geq 2$ and $x_3 \geq 2$.

16. Find the number of integer solutions to $x_1 + x_2 + x_3 + x_4 + x_5 = 20$, where $x_1 \geq -2$ and $x_3 \geq -3$.

17. Prove Theorem 8.18.

18. Prove Theorem 8.19.

8.7 THE PRINCIPLE OF INCLUSION AND EXCLUSION

In Section 8.1 we introduced the Sum Principle

$$|A \cup B| = |A| + |B| - |A \cap B|$$

and used it to count the number of elements in the union of two sets. Then in Theorem 8.2 we generalized the Sum Principle to find the number of elements in the union of any number of disjoint sets. In this section we introduce the Principle of Inclusion and Exclusion, which generalizes the Sum Principle to the union of an arbitrary number of sets that are not necessarily disjoint.

The Principle of Inclusion and Exclusion for the union of any three sets A, B, C is given by:

$$|A \cup B \cup C| = |A| + |B| + |C| \\ - (|A \cap B| + |A \cap C| + |B \cap C|) + |A \cap B \cap C| \quad (1)$$

Examining Figure 8.10, we can see how equation (1) arises. In this figure each of the numbers 1 to 7 designates the smallest region it occupies; for example, the number 4 denotes only the indicated shaded region.

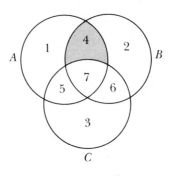

Figure 8.10

To find $|A \cup B \cup C|$, we must count each element of $A \cup B \cup C$ once and only once. Observe that any element in region 1, 2, or 3 is counted exactly once in the sum $|A| + |B| + |C|$. An element in region 4, however, is included twice in the sum $|A| + |B| + |C|$: once in $|A|$ and once in $|B|$. Subtracting (excluding) $|A \cap B|$ corrects this duplicate counting of the elements in region 4. Similarly each element in regions 5 and 6 has been included twice and the duplicates must be excluded by subtracting $|A \cap C|$ and $|B \cap C|$, respectively.

So far we have seen that any element in regions 1 through 6 is counted exactly once in the sum

$$|A| + |B| + |C| - (|A \cap B| + |A \cap C| + |B \cap C|) \quad (2)$$

However, consider the fate of an element in region 7. Such an element has been included three times (in $|A| + |B| + |C|$), but it has also been excluded three times (in $|A \cap B| + |A \cap C| + |B \cap C|$). Consequently, an element in region 7 has not been counted at all in (2). Thus to count these elements we must include the term $|A \cap B \cap C|$. We now have that every element of $A \cup B \cup C$ is counted exactly once on the right-hand side of (1).

This process can be generalized to establish the Principle of Inclusion and Exclusion.

Theorem 8.21   *Principle of Inclusion and Exclusion.*   For any sets $A_1, A_2, \ldots, A_n$

$$|A_1 \cup A_2 \cup \cdots \cup A_n| = \sum_{i=1}^{n} |A_i| - \sum_{\{i, j\}} |A_i \cap A_j|$$

$$+ \sum_{\{i, j, k\}} |A_i \cap A_j \cap A_k| - \cdots + (-1)^{n-1} |A_1 \cap A_2 \cap \cdots \cap A_n| \quad (3)$$

where the summation $\sum_{\{i, j\}}$ is to take place over all two-element subsets of $\{1, 2, \ldots, n\}$, the summation $\sum_{\{i, j, k\}}$ is to take place over all three-element subsets of $\{1, 2, \ldots, n\}$, and so on.

Proof: Suppose that $x \in A_1 \cup A_2 \cup \cdots \cup A_n$. Then x is counted once on the left-hand side of (3). We must show that x is also counted just once on the right-hand side of (3).

If x is in exactly m of the sets $A_1, A_2, \ldots, A_n$, then you can verify that

(1) x is in $\binom{m}{1} = m$ of the sets A_i, $i = 1, \ldots, n$

(2) x is in $\binom{m}{2}$ of the sets $A_i \cap A_j$, $\{i, j\} \subseteq \{1, 2, \ldots, n\}$

(3) x is in $\binom{m}{3}$ of the sets $A_i \cap A_j \cap A_k$, $\{i, j, k\} \subseteq \{1, 2, \ldots, n\}$

$\vdots$

(m) x is in $\binom{m}{m} = 1$ of the sets $\underbrace{A_i \cap A_j \cap A_k \cap \cdots \cap A_r}_{m \text{ sets}}$

$$\{i, j, k, \ldots r\} \subseteq \{1, 2, \ldots, n\}$$

Thus, we can conclude that:

(1) x is included $\binom{m}{1}$ times in the sum $\sum_{i=1}^{n} |A_i|$

(2) x is excluded $\binom{m}{2}$ times by $-\sum_{\{i,j\}} |A_i \cap A_j|$

(3) x is included $\binom{m}{3}$ times by $\sum_{\{i,j,k\}} |A_i \cap A_j \cap A_k|$

$\vdots$

(m) x is included (if m is odd) or excluded (if m is even) $\binom{m}{m} = 1$ time by

$$(-1)^{m-1} \underbrace{|A_i \cap A_j \cap A_k \cap \cdots \cap A_r|}_{m \text{ sets}}$$

It follows that x is counted a total of

$$\binom{m}{1} - \binom{m}{2} + \binom{m}{3} - \cdots + (-1)^{m-1} \binom{m}{m}$$

times on the right-hand side of (3). By Theorem 8.14

$$\binom{m}{1} - \binom{m}{2} + \binom{m}{3} - \cdots + (-1)^{m-1} \binom{m}{m} = \binom{m}{0} = 1$$

and therefore x is ultimately counted exactly once as we were to show. ∎

Example 1   At a certain university all new computer science students are to register in at least one of the following courses: Discrete Mathematics, Calculus, or Pascal Programming. Data from the Registrar's Office indicate that there are a total of 237 computer science students of which:

150 are registered in Discrete Mathematics.

120 are registered in Calculus.

100 are registered in Pascal Programming.

50 are registered in Discrete Mathematics and in Calculus.

20 are registered in Calculus and in Pascal Programming.

70 are registered in Discrete Mathematics and in Pascal Programming.

The Registrar wants to know how many students are registered in all three courses. We may proceed as follows.

Let D, C, and P designate the sets of students registered in the Discrete Mathematics, Calculus, and Pascal Programming classes, respectively. Since every new computer science student takes at least one of these courses, we know that $|D \cup C \cup P| = 237$. Moreover, the registration data indicates that: $|D| = 150$, $|C| = 120$, $|P| = 100$, $|D \cap C| = 50$, $|C \cap P| = 20$, and $|D \cap P| = 70$.

By the Principle of Inclusion and Exclusion (Theorem 8.21) we have

$$|D \cup C \cup P| = |D| + |C| + |P|$$
$$- |D \cap C| - |D \cap P| - |C \cap P| + |D \cap C \cap P|$$

and on substitution we obtain

$$237 = 150 + 120 + 100 - 50 - 70 - 20 + |D \cap C \cap P|$$

Consequently, $|D \cap C \cap P| = 7$, so exactly seven students are registered in all three courses.   □

Next we use the Principle of Inclusion and Exclusion to count derangements of the set $\{1, 2, \ldots, n\}$. A *derangement* of $\{1, 2, \ldots, n\}$ is a permutation in which no element appears in its natural location. For example, $(2, 3, 4, 1)$, $(3, 4, 1, 2)$, $(4, 1, 2, 3)$, $(4, 3, 2, 1)$, and $(3, 1, 2, 4)$ are all derangements of $\{1, 2, 3, 4\}$, while $(1, 4, 3, 2)$ is not because one element, 1, is in its natural location.

It is easiest to follow an indirect approach to the problem of counting the number of derangements. We will use the Principle of Inclusion and Exclusion to count the number of permutations that *fail* to be derangements. To this end, let A_i denote the set of all the permutations of $\{1, 2, \ldots, n\}$ in which at least the number i appears in its natural position. Note that the set $A_1 \cup A_2 \cup \cdots \cup A_n$ is the set of all permutations having at least one integer in its natural position. That is, $A_1 \cup A_2 \cup \cdots \cup A_n$ is the set of all permutations that fail to be derangements.

From the Principle of Inclusion and Exclusion we know that

$$|A_1 \cup A_2 \cup \cdots \cup A_n| = \sum_{i=1}^{n} |A_i| - \sum_{\{i, j\}} |A_i \cap A_j|$$
$$+ \sum_{\{i, j, k\}} |A_i \cap A_j \cap A_k| + \cdots + (-1)^{n-1}|A_1 \cap A_2 \cap \cdots \cap A_n|$$

Recall that the above sums take place over all possible subsets. Thus, there are as many terms in the sum $\sum_{\{i, j\}} |A_i \cap A_j|$ as there are two-element subsets of $\{1, 2, \ldots, n\}$; in other words, there are $\binom{n}{2}$ terms in $\sum_{\{i, j\}} |A_i \cap A_j|$. In a similar fashion we find that the third sum contains $\binom{n}{3}$ terms, and, in general, the mth sum contains $\binom{n}{m}$ terms.

We now turn to the problem of actually determining the value of the individual terms. Note that any permutation in A_i must have i in the ith position. Since the remaining $n - 1$ integers can be inserted in any order in the remaining $n - 1$ positions, we have

$$|A_i| = (n - 1)!$$

A permutation in $A_i \cap A_j$ must have i in the ith position and j in the jth position. The remaining $n - 2$ integers can be inserted in the remaining $n - 2$

positions in any order. Thus,

$$|A_i \cap A_j| = (n - 2)!$$

In a similar fashion we see that

$$\underbrace{|A_i \cap A_j \cap \cdots \cap A_k|}_{m \text{ sets}} = (n - m)!$$

Substituting these results into the Principle of Inclusion and Exclusion, we obtain:

$$|A_1 \cup A_2 \cup \cdots \cup A_n| = n(n - 1)! - \binom{n}{2}(n - 2)! + \binom{n}{3}(n - 3)!$$

$$+ \cdots + (-1)^{m-1}\binom{n}{m}(n - m)!$$

$$+ \cdots + (-1)^{n-1}\binom{n}{n}(n - n)!$$

$$= n! - \frac{n!}{2!} + \frac{n!}{3!} + \cdots + (-1)^{m-1}\frac{n!}{m!}$$

$$+ \cdots + (-1)^{n-1}\frac{n!}{n!}$$

Finally, since $A_1 \cup A_2 \cup \cdots \cup A_n$ is the set of all permutations that fail to be derangements, and since there is a total of $n!$ permutations, we can subtract the former from the latter to find that there are

$$D_n = n! - \left(n! - \frac{n!}{2!} + \frac{n!}{3!} + \cdots + (-1)^{n-1}\frac{n!}{n!}\right)$$

$$= \frac{n!}{2!} - \frac{n!}{3!} + \cdots + (-1)^n\frac{n!}{n!}$$

derangements of $\{1, 2, \ldots, n\}$.

Example 2 There are

$$D_4 = \frac{4!}{2!} - \frac{4!}{3!} + \frac{4!}{4!} = 9$$

derangements of $\{1, 2, 3, 4\}$, and there are

$$D_5 = \frac{5!}{2!} - \frac{5!}{3!} + \frac{5!}{4!} - \frac{5!}{5!} = 44$$

derangements of $\{1, 2, 3, 4, 5\}$.   □

Example 3 Ten different letters are placed randomly into ten addressed envelopes. We will count the number of ways in which

    a. no letter is placed into the proper envelope;

    b. exactly one letter is placed into the proper envelope;

    c. at least two letters are placed into the proper envelopes.

The number of ways in which no letter is placed into its proper envelope is D_{10}, the number of derangements of ten objects.

Placing exactly one of the ten letters into the proper envelope can be thought of as a two-step process. We first select the letter to be placed into the proper envelope (this can be done in 10 ways). Then we select a derangement of the remaining nine letters. It follows from the Product Principle that there are $10D_9$ ways to place exactly one letter into its proper envelope.

To calculate the number asked for in **c.**, we let A_i denote the set of arrangements in which exactly i letters are in their proper envelopes. Note that there are $\binom{10}{i}$ ways of selecting the i envelopes that are to be properly placed and D_{10-i} ways to select a derangement of the remaining $10 - i$ letters for $i \geq 2$. It follows that $|A_i| = \binom{10}{i}D_{10-i}, i \geq 2$. Since $A_i \cap A_j = \varnothing$ if $i \neq j$, we can use the Sum Principle to compute the number of ways in which at least two letters are properly placed into their envelopes:

$$|A_2 \cup A_3 \cup \cdots \cup A_{10}| = |A_2| + |A_3| + \cdots + |A_{10}|$$

$$= \binom{10}{2}D_8 + \binom{10}{3}D_7 + \cdots + \binom{10}{8}D_2 + 1 \quad \square$$

In this chapter we have considered a variety of fundamental counting techniques. In the next chapter we will apply these techniques to the solution of a number of probability problems.

8.7 Problems

In Problems 1–4 compute $|A \cup B \cup C|$ from the given information.

1. $|A| = 12$, $|B| = 11$, $|C| = 7$, $|A \cap B| = 5$, $|A \cap C| = 5$, $|B \cap C| = 3$, $|A \cap B \cap C| = 2$

2. $|A| = 8$, $|B| = 7$, $|C| = 7$, $|A \cap B| = 3$, $|A \cap C| = 3$, $|B \cap C| = 3$, $|A \cap B \cap C| = 2$

3. $|A| = 4$, $|B| = 4$, $|C| = 4$, $|A \cap B| = 2$, $|A \cap C| = 2$, $|B \cap C| = 2$, $|A \cap B \cap C| = 1$

4. $|A| = 3$, $|B| = 4$, $|C| = 3$, $|A \cap B| = 0$, $|A \cap C| = 1$, $|B \cap C| = 1$, $|A \cap B \cap C| = 0$

5. Compute $|A \cup B \cup C \cup D|$ if $|A| = 8$, $|B| = 11$, $|C| = 10$, $|D| = 11$, $|A \cap B| = 4$, $|A \cap C| = 2$, $|A \cap D| = 6$, $|B \cap C| = 5$, $|B \cap D| = 4$, $|C \cap D| = 6$, $|A \cap B \cap C| = 2$, $|A \cap B \cap D| = 3$, $|A \cap C \cap D| = 2$, $|B \cap C \cap D| = 3$, $|A \cap B \cap C \cap D| = 2$.

6. Compute the number of derangements of a set with six elements.

7. Draw a Venn diagram illustrating the Principle of Inclusion and Exclusion for four sets.

8. Find the number of permutations of the set $\{1, 2, \ldots, 10\}$ such that:
   (a) exactly 3 of these numbers are in their natural positions;
   (b) at least 6 of these numbers are in their natural positions;
   (c) at most 3 of these numbers are in their natural positions.

9. Find the number of permutations of the set $\{1, 2, 3, 4, 5, 6\}$ for which every even integer is in an even-numbered position and no odd integer is in its natural position.

10. Show that the number of derangements

$$D_n = \frac{n!}{2!} - \frac{n!}{3!} + \cdots + (-1)^n \frac{n!}{n!}, \quad n \geq 2$$

satisfies the difference equation $D_n = (n - 1)(D_{n-1} + D_{n-2})$.

*11. Explain why the number of derangements D_n satisfies the difference equation $D_n = (n - 1)(D_{n-1} + D_{n-2})$ for $n \geq 4$.

Chapter 8   REVIEW

Concepts for Review

Product Principle (p. 303)
Sum Principle (p. 304)
permutation (p. 307)
lexicographic order (p. 309)
r-permutation of n objects (p. 311)
$P(n, r)$ (p. 311)
r-combination of an n-element set (p. 313)
$C(n, r)$ (p. 313)
Binomial Theorem (p. 319)
binomial coefficients (p. 319)
Pascal's triangle (p. 324)
redundant permutations (p. 329)
redundant combinations (p. 332)
Principle of Inclusion and Exclusion (p. 335)
derangement (p. 337)

Review Problems

1. Compute $|A \cap B|$ if $|A| = 25$, $|B| = 32$, $|A \cup B| = 40$.

2. How many 6-digit numbers can be formed without using the digits 0 and 5?

3. How many 7-letter words include at least one of the letters C, A, T?

4. Compute

   (a) $P(6, 3)$ (b) $C(6, 3)$ (c) $\binom{7}{3, 2, 2}$

5. In how many ways can three people be selected for three distinct jobs from a pool of ten people?

6. In how many ways can twelve distinct symbols be arranged in a circle? (Two arrangements are the same if one is a rotation of the other.)

7. In how many ways can twelve people be divided into four groups containing two, three, four, and three people, respectively?

8. How many different arrangements are there of the word HAWAII?

9. How many five-card hands from a standard deck of cards contain two pairs (but do not contain a three-of-a-kind or four-of-a-kind)?

10. What is the coefficient of $x^4 y^8$ in the expansion of $(x + y)^{12}$?

11. What is the coefficient of x^4 in the expansion of $(2x + 3)^7$?

12. Use a combinatorial argument to explain why

$$\binom{n}{r_1, r_2} = \binom{n}{r_1}$$

13. State the identity which forms the basis of Pascal's triangle.

14. How many n-digit numbers contain exactly k ones?

15. How many n-digit numbers contain at least k ones?

16. Find the number of positive integer solutions to

$$x_1 + x_2 + x_3 + x_4 + x_5 = 12$$

17. Find the number of positive integer solutions to

$$x_1 + x_2 + x_3 + x_4 = 15 \quad \text{if } x_1 \geq 2 \text{ and } x_2 \geq 3$$

18. In how many ways can ten identical objects be placed into thirteen boxes?

19. How many outcomes are possible if seven identical dice are rolled?

20. Compute $|A \cup B \cup C|$ if $|A| = 13$, $|B| = 15$, $|C| = 21$, $|A \cap B| = 7$, $|A \cap C| = 8$, $|B \cap C| = 9$, $|A \cap B \cap C| = 3$.

21. Eleven gifts are to be exchanged in a Christmas gift exchange. In how many ways can this be done so that no person receives his or her own gift?

22. How many permutations of the integers $1, 2, 3, \ldots, 15$ have at least ten numbers in their natural positions?

9 Elementary Probability Theory

The notion of probability pervades much of our lives. Although we might be hard pressed to make this notion precise, we rarely hesitate to speak of the probability of rain tomorrow, the probability of nuclear war, the chances of winning a sweepstakes, and so on. The concept of probability is not new, and in fact, its origins appear to be tied to man's age-old penchant for gambling. The first treatise on probability has generally been attributed to Gerolamo Cardano (1501–1576), an inveterate gambler-physician, who, in order to enhance his gambling successes, investigated the probabilities of winning in a variety of games. The French mathematician Pascal (1623–1662) was also greatly interested in probability. Though he approached it from a somewhat broader perspective, Pascal did not neglect the "practical" applications of probability theory to gambling, and based much of his work on problems of this nature.

From these rather humble—if not slightly tainted—beginnings, probability theory has developed into one of the most useful and powerful of all the scientific disciplines. It has widespread application in areas ranging from physics to biology. In this text, we do not present a thorough study of probability theory. Rather, we shall content ourselves with a brief introduction to some of the more fundamental concepts of this theory and its applications.

9.1 BASIC CONCEPTS

The classical definition of probability is generally ascribed to the French mathematician Pierre Simon de Laplace (1749–1827). He defined the probability of an event to be the value obtained when the number of outcomes "favorable" to the event is divided by the total number of possible outcomes (where it is assumed that the occurrence of each outcome is equally likely). Thus, for example, the probability of drawing a heart from a deck of 52 cards is $13/52 = 1/4$, since there are 52 equally likely possible outcomes, 13 of which are favorable (drawing a heart).

Although modern probability theory is based to a significant extent on the classical definition of probability, this definition does present certain difficulties:

a. The classical definition allows us to deal only with problems involving a finite number of possible outcomes. We cannot use this definition to calculate, for example, the probability that a real number chosen from the interval $[0, 1]$ will lie within the interval $[\frac{1}{2}, \frac{3}{4}]$.

b. The classical definition depends on the idea of "equally likely," which is itself a probabilistic notion. Thus this definition is somewhat circular, a situation that we normally wish to avoid.

c. The classical definition does not allow us to deal with outcomes that are *not* equally likely.

Because of these difficulties, we see that there is a distinct need for a broader and more precise approach to probability theory. In this text, we do not go as far as we might because this would carry us too far afield and is not necessary for our rather limited purposes.

Much of our focus will be on experiments—procedures that have a (possibly infinite) number of distinguishable *outcomes*. In the context of probability theory, these experiments are often referred to as *random experiments*, and the set of all possible outcomes of a random experiment is called a *sample space*. A sample space may be either finite or infinite.

Example 1 *Finite sample space.* Suppose we roll two dice and record the numbers that appear. Then the sample space of this random experiment can be described as the set of ordered pairs $\{(x, y)|x \in \{1, 2, \ldots, 6\}$, $y \in \{1, 2, \ldots, 6\}\}$ where the first element in each ordered pair is the number showing on the first die, and the second element is the number showing on the second die. □

Example 2 *Infinite sample space.* We perform a random experiment by flipping a coin until we obtain two heads in a row, at which time we stop the experiment. The sample space of this experiment is the set of all sequences consisting of T's and H's that end immediately after the first appearance of two H's. A few of the possible outcomes of this random experiment are

$$HH, \; THH, \; HTHH, \; TTHH, \; TTTHH, \; THTHH, \; HTTHH \qquad \square$$

In this chapter, we restrict our attention to sample spaces that are finite.

Definition 9.1   Any subset of a sample space S is called an *event*.

Example 3   An event in the sample space given in Example 1 is the set

$$\{(1, 3), (2, 4), (3, 5), (4, 2)\} \qquad \square$$

We are now in a position to define rigorously the concept of probability.

Definition 9.2   Let S be a finite sample space. Then a *probability measure* on S is any function P that assigns a value $P(A)$, read as "the probability of A," to each event A in S and satisfies the following conditions:

a. $0 \le P(A) \le 1$

b. $P(S) = 1$

c. If $A_1, A_2, \ldots, A_n$ is a family of mutually disjoint events (subsets) of S, then

$$P(A_1 \cup A_2 \cup \cdots \cup A_n) = P(A_1) + P(A_2) + \cdots + P(A_n)$$

The pair (S, P) is called a *probability space*.

Example 4   Let $S = \{x_1, x_2, x_3\}$. The events of S are $\varnothing, \{x_1\}, \{x_2\}, \{x_3\}, \{x_1, x_2\}, \{x_1, x_3\}, \{x_2, x_3\}, \{x_1, x_2, x_3\}$. We define a probability measure P on the sample space S as follows:

$$P(\varnothing) = 0, \qquad P(\{x_1\}) = \frac{1}{2}, \qquad P(\{x_2\}) = \frac{1}{3}, \qquad P(\{x_3\}) = \frac{1}{6}$$

$$P(\{x_1, x_2\}) = \frac{5}{6}, \qquad P(\{x_1, x_3\}) = \frac{2}{3},$$

$$P(\{x_2, x_3\}) = \frac{1}{2}, \qquad P(\{x_1, x_2, x_3\}) = 1$$

Note that **b.** of Definition 9.2 is clearly satisfied since $P(S) = P(\{x_1, x_2, x_3\}) = 1$. Checking **c.** of Definition 9.2 requires considerably more work, as the following sample computations show:

$$P(\{x_1, x_2\} \cup \{x_3\}) = P(\{x_1, x_2\}) + P(\{x_3\})$$

since

$$1 = \frac{5}{6} + \frac{1}{6}$$

$$P(\{x_1, x_2\}) = P(\{x_1\}) + P(\{x_2\})$$

since

$$\frac{5}{6} = \frac{1}{2} + \frac{1}{3}$$

$$P(\{x_2, x_3\} \cup \varnothing) = P(\{x_2, x_3\}) \cup P(\varnothing)$$

since

$$\frac{1}{2} = \frac{1}{2} + 0$$

We leave it to you to check part **c.** of Definition 9.2 for the remaining families of mutually disjoint subsets of S. $\square$

It should be clear that for large sample spaces a direct verification of part **c.** of Definition 9.2 would be extremely tedious. Fortunately, there is another approach that avoids this difficulty.

Suppose that $S = \{x_1, x_2, \ldots, x_n\}$ is a sample space. For $i = 1, 2, \ldots, n$, define $P(\{x_i\})$ so that $P(\{x_1\}) + P(\{x_2\}) + \cdots + P(\{x_n\}) = 1$ and $P(\{x_i\}) \geq 0$. Then for each event A, define $P(A)$ by

$$P(A) = \sum_{x_i \in A} P(\{x_i\})$$

It is easy to show that a function defined in this fashion is a probability measure (see Problem 4). The next example illustrates this procedure.

Example 5 In a certain stream there are four species of fish: s_1, s_2, s_3, s_4. Studies have shown that 20% of the fish caught in this stream are of species s_1, 35% are of species s_2, 15% are of species s_3, and 30% are of species s_4. If we let $S = \{s_1, s_2, s_3, s_4\}$ and define a function P on the subsets of S by $P(\{s_1\}) = .20$, $P(\{s_2\}) = .35$, $P(\{s_3\}) = .15$, $P(\{s_4\}) = .30$ and for each $A \subseteq S$

$$P(A) = \sum_{s_i \in A} P(\{s_i\})$$

then (S, P) is a probability space. Note, for example, that $P(\{s_1, s_3\}) = P(\{s_1\}) + P(\{s_3\}) = .20 + .15 = .35$. $\square$

Now suppose that each outcome of a sample space $S = \{x_1, x_2, \ldots, x_n\}$ is equally likely. That is, assume that $P(\{x_i\}) = P(\{x_j\})$ for all $i, j = 1, \ldots, n$. Then since $P(S) = 1$ and

$$P(S) = P(\{x_1\}) + P(\{x_2\}) + \cdots + P(\{x_n\})$$

it follows that $P(\{x_i\}) = 1/n$ for $i = 1, \ldots, n$. An important consequence of this observation is that if $A \subseteq S$ and $|A| = k$, then

$$P(A) = \frac{|A|}{|S|} = \frac{k}{n} \tag{1}$$

Note that this essentially restates the classical definition of probability mentioned previously.

Example 6   We calculate the probability of obtaining a full house in a poker game played with fifty-two cards; a full house consists of three cards of one kind (for example three kings) and a pair of cards (for example two fives).

Here the sample space S consists of all 5-card hands; hence $|S| = n = \binom{52}{5}$. You should be able to use ideas you mastered in the previous chapter to verify that the number of ways of obtaining a full house is $k = 13\binom{4}{3} \cdot 12\binom{4}{2}$. If we assume that all 5-card hands are equally likely, then we see from (1) that the probability of a full house is

$$\frac{k}{n} = \frac{13\binom{4}{3} \cdot 12\binom{4}{2}}{\binom{52}{5}} \approx .0014 \qquad \square$$

Example 7   If it is assumed that births occur with equal likelihood on any of the 365 days of the year (ignoring leap years), what is the probability that in a room with k people no two people will have the same birthday?

In this case the experiment consists of determining the birthdays of the k people in the room. Thus an outcome of the experiment is a listing of the birthdays. If we number the days of the year consecutively, we see that we can take the sample space to be the set of all k-tuples, where each coordinate of each k-tuple is an integer from the set $\{1, 2, \ldots, 365\}$. Let $A \subseteq S$ be the event consisting of all k-tuples whose entries (coordinates) are distinct from one other. For instance, if $k = 5$, then $(243, 43, 17, 23, 56)$ and $(34, 243, 18, 6, 345)$ would belong to A while $(46, 67, 46, 89, 245)$ would not. Note that A is the event that no two people have the same birthday. We need to find $P(A)$. Since $|A| = (365)(364)(363) \cdots (365 - k + 1) = 365!/(365 - k)!$ and since $|S| = 365^k$, it follows that

$$P(A) = \frac{365!/(365 - k)!}{365^k}$$

We remark that if $k = 23$, then $P(A) = .493$, and it follows easily from a result in the next section that if there are more than twenty-two people in a room, then there is slightly better than a 50–50 chance that at least two of them will have the same birthday. Try it! $\square$

9.1   Problems

1. Five batches of a new drug for altering blood pressure are tested on rats. A batch is labeled with L, R, or S according to whether the blood pressure of a rat is lowered, raised, or stays the same, respectively, after the drug is ingested. Describe the sample space of this random experiment.

2. A bowl contains two red chips, two black chips, and one blue chip. Two chips are removed from the bowl (without replacement). What is the sample space of this random experiment? What would be the sample space if the first chip selected were replaced before the second chip is drawn?

3. A president, vice president, and secretary are to be selected from a list of five candidates. Define the sample space for this random experiment.

4. Let $S = \{x_1, x_2, \ldots, x_n\}$ be a finite sample space, and for $i = 1, 2, \ldots, n$ let $P(\{x_i\}) = p_i$. Suppose that $p_1 + p_2 + \cdots + p_n = 1$, and that for each $i, p_i \geq 0$. For each subset A of S, let $P(A) = \Sigma_{x_i \in A} P(\{x_i\})$. Show that P is a probability measure.

5. Show that if (S, P) is a probability space, then $P(\varnothing) = 0$.

6. For Problem 1, describe
   (a) the event that two of the batches are labeled L, two are labeled R, and one is labeled S.
   (b) the event that none of the batches is labeled L.

7. For Problem 2, describe the following events if the chips are drawn without replacement.
   (a) Exactly one of the two chips drawn is red.
   (b) At least one of the two chips drawn is red.
   (c) The chips drawn are of different colors.
   Also, describe these events if the first chip drawn is replaced before the second chip is drawn.

8. Find the probability of the occurrence of each of the events described in Problem 7.

9. A newly married couple plans to have five children. What is the probability that they will have three girls and two boys?

10. Find the probability of obtaining the following 5-card poker hands from a deck of fifty-two cards:
    (a) four of a kind (for example, four kings and one other card).
    (b) three of a kind (for example, three jacks and two other cards, neither of which is a jack and which do not form a pair).
    (c) two pairs (for example, two threes and two nines, and one card that is neither a three nor a nine).

11. If the numbers 1, 2, 3, 4, 5, and 6 are arranged randomly into the six boxes

what is the probability that the numbers 2 and 5 will be adjacent?

12. If two chips are drawn in succession with replacement (that is, the first is returned before the second is drawn) from a bowl containing four red chips and two white chips, what is the probability that the chips selected will be of the same color?

13. Two dice are rolled.
    (a) What is the probability that the sum of the two dice will be either seven or eleven?
    (b) What is the probability that the sum of the two dice is even?

14. Five workers currently hold five different jobs. The workers are reassigned to these jobs in a completely arbitrary fashion.
    (a) What is the probability that no worker is assigned the same job he previously held?
    (b) What is the probability that exactly one worker is reassigned the same job?
    (c) What is the probability that at least three workers are reassigned the same jobs?

9.2 SOME BASIC PROPERTIES OF THE PROBABILITY MEASURE

In this section we consider some of the more immediate consequences of Definition 9.2.

Theorem 9.3   If A is an event in a probability space (S, P) and if we let $A^c = S \setminus A$, then

$$P(A^c) = 1 - P(A)$$

Proof: Since $S = A \cup A^c$ and $A \cap A^c = \emptyset$, it follows from the definition of a probability space that

$$1 = P(S) = P(A \cup A^c) = P(A) + P(A^c)$$

and, hence

$$P(A^c) = 1 - P(A) \qquad \blacksquare$$

Example 1   To illustrate Theorem 9.3, we return to the birthday problem discussed at the end of Section 9.1. We found that the probability is .493 that in a group of twenty-three people no two people will have the same birthday. If we let A be the event that no two people in that group have the same birthday, then it is clear that A^c is the event that at least two people will have the same birthday, and from Theorem 9.3 we find that

$$P(A^c) = 1 - P(A) = 1 - .493 = .507 \qquad \square$$

The next result is the probabilistic equivalent of the Sum Principle (Section 8.1).

Theorem 9.4   If A and B are two events in a probability space (S, P), then

$$P(A \cup B) = P(A) + P(B) - P(A \cap B)$$

Proof: First note that, as Figure 9.1 shows, we can express the sets A and B as disjoint unions:

$$A = (A \cap B) \cup (A \cap B^c) \quad \text{and} \quad B = (A \cap B) \cup (A^c \cap B)$$

Hence, by **c.** of Definition 9.2,

$$P(A) = P(A \cap B) + P(A \cap B^c) \tag{1}$$

and

$$P(B) = P(A \cap B) + P(A^c \cap B) \tag{2}$$

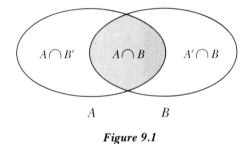

Figure 9.1

On adding equations (1) and (2) we obtain

$$P(A) + P(B) = P(A \cap B) + P(A \cap B) + P(A \cap B^c) + P(A^c \cap B) \tag{3}$$

Now observe that by Figure 9.1 we can express $A \cup B$ as the disjoint union:

$$A \cup B = (A \cap B) \cup (A \cap B^c) \cup (A^c \cap B)$$

and, therefore,

$$P(A \cup B) = P(A \cap B) + P(A \cap B^c) + P(A^c \cap B) \tag{4}$$

Substituting (4) into the right-hand side of (3) yields

$$P(A) + P(B) = P(A \cap B) + P(A \cup B),$$

and consequently we have

$$P(A \cup B) = P(A) + P(B) - P(A \cap B)$$

the desired result.   ■

Example 2   Let (S, P) be a probability space defined by $S = \{x_1, x_2, x_3, x_4, x_5\}$ and $P(\{x_1\}) = 1/10$, $P(\{x_2\}) = 1/5$, $P(\{x_3\}) = 2/5$,

$P(\{x_4\}) = 1/10$, $P(\{x_5\}) = 1/5$. Let $A = \{x_1, x_2, x_3\}$ and $B = \{x_2, x_4\}$. Then

$$P(A) = \frac{1}{10} + \frac{1}{5} + \frac{2}{5} = \frac{7}{10} \qquad P(B) = \frac{1}{5} + \frac{1}{10} = \frac{3}{10}$$

and

$$P(A \cap B) = P(\{x_2\}) = \frac{1}{5}$$

Thus, by Theorem 9.4 we have

$$P(A \cup B) = \frac{7}{10} + \frac{3}{10} - \frac{1}{5} = \frac{4}{5}$$

Note that a direct computation gives the same result:

$$P(A \cup B) = P(\{x_1, x_2, x_3, x_4\}) = \frac{1}{10} + \frac{1}{5} + \frac{2}{5} + \frac{1}{10} = \frac{4}{5} \qquad \square$$

Example 3 Suppose a survey of 200 people finds that 73 people smoke, 132 people drink, and 57 people smoke and drink. Based on this data what is the probability that a person either smokes or drinks?

Let A be the event that a person smokes and B be the event that a person drinks. We wish to find $P(A \cup B)$. By Theorem 9.4 we have

$$P(A \cup B) = P(A) + P(B) - P(A \cap B)$$

$$= \frac{73}{200} + \frac{132}{200} - \frac{57}{200} = \frac{37}{50}$$

which means that the probability is .74 that a person either smokes or drinks.

$\square$

The following theorem generalizes Theorem 9.4 to arbitrary (finite) unions of events and can be proved by induction (see Problem 15).

Theorem 9.5 Suppose that $A_1, A_2, \ldots, A_n$ are events in a probability space (S, P). Then

$$P(A_1 \cup A_2 \cup \cdots \cup A_n) = \sum_{i=1}^{n} P(A_i) - \sum_{1 \leq i < j \leq n} P(A_i \cap A_j)$$

$$+ \sum_{1 \leq i < j < k \leq n} P(A_i \cap A_j \cap A_k) + \cdots + (-1)^{n+1} P(A_1 \cap A_2 \cap \cdots \cap A_n)$$

Note the strong resemblance between Theorem 9.5 and the Principle of Inclusion and Exclusion from Chapter 8. In fact, if all outcomes are equally likely, then the Principle of Inclusion and Exclusion can be used to prove Theorem 9.5.

Example 4   A major construction firm is offering houses for sale. In addition to its basic offer, the firm will provide the following three options:

Option A_1: A two-car garage
Option A_2: Finished interior work
Option A_3: Outside landscaping

The firm has established that the public will respond to these options as follows:

Option	Probability of a buyer's choosing the option
A_1 only	.70
A_2 only	.60
A_3 only	.50
A_1 and A_2	.40
A_1 and A_3	.35
A_2 and A_3	.20
A_1, A_2, and A_3	.05

We wish to calculate the probability that a buyer will not be interested in any of the options; that is we want to find the probability of the event $(A_1 \cup A_2 \cup A_3)^c$. You should be able to verify that this probability is given by

$$
\begin{aligned}
P((A_1 \cup A_2 \cup A_3)^c) &= 1 - P(A_1 \cup A_2 \cup A_3) \\
&= 1 - [(P(A_1) + P(A_2) + P(A_3)) - (P(A_1 \cap A_2) \\
&\quad + P(A_1 \cap A_3) + P(A_2 \cap A_3)) + P(A_1 \cap A_2 \cap A_3)] \\
&= 1 - [(.7 + .6 + .5) - (.4 + .35 + .2) + .05] \\
&= .10 \qquad\qquad\qquad\qquad\qquad\qquad\qquad\qquad \square
\end{aligned}
$$

To conclude this section we once again consider the problem of the gambler's ruin, which we introduced in Section 7.4. A gambler begins betting with d dollars and has the objective of quitting when he has k dollars. To attain his objective, the gambler bets one dollar on the same game several times in succession. The probability that the gambler wins a dollar in one play of the game is p; the probability that he loses is $q = 1 - p$. What is the probability of the gambler's ruin, that is, what is the probability that he will lose the d dollars before he attains k dollars?

We will use difference equations to solve this problem. Let $P(n)$ denote the probability of eventual ruin when the gambler has n dollars. If the gambler has n dollars, he will, after the next play, have either $n + 1$ dollars or $n - 1$ dollars, so he has a probability of eventual ruin of $P(n + 1)$ or $P(n - 1)$, respectively.

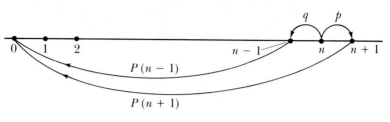

Figure 9.2

The probability, p, that he will win the next round, times the probability that he will go broke when he has $n + 1$ dollars, gives the probability that he will win one round and then go broke. Similarly, the product $qP(n - 1)$ is the probability of starting from n dollars, losing, and then going broke. Figure 9.2 schematically represents the situation.

Since these disjoint events are the only ways of going broke when he has n dollars, we have

$$P(n) = pP(n + 1) + qP(n - 1)$$

or, equivalently,

$$pP(n + 1) - P(n) + qP(n - 1) = 0$$

The characteristic equation of this difference equation is $pA^2 - A + q = 0$, and since $p + q = 1$, this quadratic can be factored as

$$(A - 1)(pA - q) = 0$$

Thus, $A = 1$ or $A = q/p$.

There are two cases to consider: $p = q$ and $p \neq q$. If $p = q$, then $q/p = 1$ and both roots of the characteristic equation are equal to 1. Thus, from Section 7.4 we have

$$P(n) = \alpha_1 + \alpha_2 n \tag{5}$$

for an appropriate choice of α_1 and α_2. Observe that $P(0) = 1$ (a gambler with 0 dollars is broke) and $P(k) = 0$ (the gambler will quit when he attains k dollars). Substituting these values into (5) shows that $\alpha_1 = 1$ and $\alpha_2 = -1/k$. Thus, if $p = q$

$$P(n) = 1 - \frac{n}{k}$$

is the probability that a gambler with n dollars will go broke.

If $p \neq q$ the roots 1 and q/p are distinct, so it follows from Section 7.3 that

$$P(n) = \alpha_1 + \alpha_2 \left(\frac{q}{p}\right)^n \tag{6}$$

for an appropriate choice of α_1 and α_2. To find α_1 and α_2 we again use the fact that $P(0) = 1$ and $P(k) = 0$ to obtain

$$\alpha_1 + \alpha_2 = 1 \quad \text{and} \quad \alpha_1 + \alpha_2 \left(\frac{q}{p}\right)^k = 0$$

from which it follows that

$$\alpha_1 = \frac{(q/p)^k}{(q/p)^k - 1} \quad \text{and} \quad \alpha_2 = \frac{1}{1 - (q/p)^k}$$

Therefore,

$$P(n) = \frac{(q/p)^k}{(q/p)^k - 1} + \frac{1}{1 - (q/p)^k}(q/p)^n \quad \text{if } q \neq p \tag{7}$$

Algebraic manipulation of (7) yields

$$P(n) = \frac{(q/p)^n - (q/p)^k}{1 - (q/p)^k} \quad \text{if } q \neq p \text{ and } n \leq k \tag{8}$$

which is the probability that a gambler with n dollars will go broke.

Example 5   A roulette wheel is divided into 37 sectors of which 18 are black, 18 are red, and 1 is green. A bet on red or black wins a sum equal to the bet whenever the outcome is a sector of the color chosen. Suppose a gambler has $100.00 and wants to double her money by making $1.00 bets. What is the probability of her losing the $100.00 before attaining her goal?

   In this case 18 out of the 37 outcomes are favorable and 19 out of 37 are unfavorable. Thus $p = 18/37$ and $q = 19/37$. Substituting into (8) gives

$$P(100) = \frac{(18/19)^{100} - (18/19)^{200}}{1 - (18/19)^{200}} \approx 0.0045$$

Thus, the gambler has about 45 chances in 10,000 of attaining her goal.   □

9.2   Problems

1. If a bowl contains three black chips, four red chips, and seven green chips, and if two chips are drawn from the bowl, what is the probability that at least one red chip is drawn? What is the probability that at least two green chips are selected?

2. Three fair coins are tossed. Find the probability that at least one of the coins lands heads-up.

3. Let A and B be events in a probability space (S, P).
   (a) Suppose $P(A) = 1/2$, $P(B) = 1/3$, and $P(A \cup B) = 2/3$. Find $P(A \cap B)$.

(b) Suppose $P(A) = 1/2$, $P(A \cup B) = 2/3$, and $P(A \cap B) = 1/4$. Find $P(B)$.

4. Show that if A and B are events in a probability space (S, P) and that if $A \subseteq B$, then $P(A) \leq P(B)$.

5. To be accepted into a certain college degree program, a student must successfully complete at least one of two language tests. Suppose that the probability of passing Test A is $2/3$ and the probability of passing the Test B is $1/2$. If the probability of passing both tests is $1/4$, find the probability of passing at least one of the tests.

6. Suppose that 40% of the population has high blood pressure, 60% of the population is overweight, and 30% of the population is both overweight and has high blood pressure. What percentage of the population is either overweight or has high blood pressure?

7. Use induction to show that if $A_1, A_2, \ldots, A_n$ are events in a probability space (S, P) then $P(A_1 \cup A_2 \cdots \cup A_n) \leq \Sigma_{i=1}^n P(A_i)$.

8. Three people play a game in which each person receives a numbered tag which is tossed into a hat. Each player then pulls out one of the tags; a player wins if she removes the same tag she placed into the hat.
   (a) Show that the probability that at least one player wins is $1 - 1/2! + 1/3!$.
   (b) Show that if n persons play this game, then the probability that at least one person will win is
   $$1 - \frac{1}{2!} + \frac{1}{3!} - \frac{1}{4!} + \cdots + (-1)^{n+1}\frac{1}{n!}$$

9. Find the probability that in a hand of five cards (taken from a 52-card deck), there will be
   (a) at least one ace.
   (b) exactly one ace.
   (c) at most one ace.

10. If two dice are rolled, what is the probability that either the sum of the dice will be a prime number or that at least one of the numbers showing will be odd?

11. What is the probability that in a room of 100 people, at least two people will have the same birthday?

12. Which is more likely:

    a. to obtain a 6 at least once if a die is tossed four times, or

    b. to obtain a sum of 12 at least once if two dice are tossed 24 times.

13. In a certain college class of 1000 students, 760 take math, 630 take English, 190 take a psychology course, 500 take both English and math, 120 take both psychology and math, 170 take both psychology and English, and 890 take at least one of these three subjects.

(a) What is the probability that a student will take all three courses?

(b) What is the probability that a student will take exactly two of these courses?

(c) What is the probability that a student will take exactly one the these courses?

(d) What is the probability that a student will not take any of these courses?

14. Suppose that you have a probability of .6 of winning a certain game. You start with 15 dollars and decide to play the game until you are either broke or have doubled your money. If each bet you make is one dollar, what is the probability that you double your money?

15. Use induction to prove Theorem 9.5.

16. The probability of winning a certain game is .45. You begin playing the game with 30 dollars and will play until you are broke or have doubled your money. If you place bets of one dollar, what is the probability of doubling your money? What will this probability be if you place bets of two dollars? Five dollars? Ten dollars? Thirty dollars?

9.3   CONDITIONAL PROBABILITY

We use the following rather simple situation to introduce the concept of conditional probability. If we flip a fair coin three times, the sample space of this random experiment is the set of outcomes

$$S = \{HHH, HHT, HTH, THH, HTT, THT, TTH, TTT\}$$

and each of these outcomes occurs with probability $1/8$. We compare the following probabilities:

a. the probability that exactly two heads occur in the three tosses,

b. the probability that exactly two heads occur if we know that at least one head occurs.

It is easy to obtain the probability asked for in part **a.** Let A be the event that exactly two heads occur. Then A is the event $\{HHT, HTH, THH\}$, and since each of the outcomes in A is equally likely, it follows from (1) of Section 9.1 that $P(A) = 3/8$.

To determine the probability asked for in part **b.** we make the following observation. If we know that at least one head has occurred, then we may take our sample space to be the event B consisting of all outcomes that contain at least one H. Thus

$$B = \{HHH, HHT, HTH, THH, HTT, THT, TTH\}$$

Observe that each of the seven outcomes in the sample space B is equally likely. The event that exactly two heads occur is

$$A \cap B = \{HHT, HTH, THH\}$$

and it follows that the probability that exactly two heads occur (given that at least one head occurs) is

$$\frac{|A \cap B|}{|B|} = \frac{3}{7}$$

This probability is generally referred to as a conditional probability.

To motivate the formal definition of this concept we rewrite $|A \cap B|/|B|$ as follows:

$$\frac{|A \cap B|}{|B|} = \frac{|A \cap B|/|S|}{|B|/|S|} = \frac{P(A \cap B)}{P(B)}$$

Definition 9.6 If A and B are events in a probability space (S, P), then the *conditional probability* of the event A given that the event B has occurred is defined to be

$$\frac{P(A \cap B)}{P(B)}$$

and is denoted by $P(A|B)$.

The following examples illustrate the computation of conditional probabilities.

Example 1 Suppose that (S, P) is the probability space where $S = \{x_1, x_2, x_3, x_4, x_5\}$ and P is defined by $P(\{x_1\}) = 1/10$, $P(\{x_2\}) = 1/5$, $P(\{x_3\}) = 2/5$, $P(\{x_4\}) = 1/10$, $P(\{x_5\}) = 1/5$. Let $A = \{x_1, x_2, x_3\}$ and $B = \{x_2, x_4\}$. Then

$$P(A \cap B) = P(\{x_2\}) = \frac{1}{5} \quad \text{and} \quad P(B) = \frac{1}{5} + \frac{1}{10} = \frac{3}{10}$$

and thus

$$P(A|B) = \frac{1/5}{3/10} = \frac{2}{3}. \qquad \square$$

Example 2 Suppose that in a sample population of 5000 people we find that 3250 people are overweight at age 40, and, of these 3250 people, 2600 were overweight at age 5. Moreover, of the 1750 people who are not overweight at age 40, 400 were overweight at age 5. Based on this data, what is the probability that if a person is overweight at age 5, then the person will be overweight at age 40?

Let A be the event that a person is overweight at age 40, and let B be the event that a person is overweight at age 5. We must find $P(A|B)$. Note that $A \cap B$ is the event that a person who was overweight at age 5 is also overweight at age 40. From the given data you can verify that

$$P(A \cap B) = \frac{2600}{5000}$$

You can also see from this data that $P(B) = (400 + 2600)/5000 = 3000/5000$, so we have

$$P(A|B) = \frac{2600/5000}{3000/5000} = \frac{13}{15}$$

That is, there is probability of .87 that a person who is overweight at age 5 will be overweight at age 40. □

Example 3 You might wish to challenge your friends with the following game, which reportedly dates back to the time of medieval fairs and gambling houses. You have three cards: one is red on both sides; one is red on one side and blue on the other; and one is blue on both sides. You and a friend are blindfolded and one of you selects one of the three cards and places it on a table. Both of you then remove the blindfolds and look at the card lying on the table. You offer your friend a dollar if the down side of the card is of a different color than the up side; otherwise your friend must pay you a dollar. If your friend has any doubts about this game (or about you), you point out that if, for instance, the color red is showing, then the card is obviously not the blue–blue card, and hence there is a 50–50 chance that the down side of the card is also red. Thus, you both have an equal chance of winning.

If your friend falls for this argument, don't hesitate to play, and be thankful that your friend is unaware of the pitfalls of conditional probabilities. To see what is happening here, we calculate the probability that if a particular color is showing (say, red) then the other side of the card is also red. If we let A denote the event "red is down" and B the event "red is up", then we want to find $P(A|B)$. It is helpful to label the cards as indicated in Figure 9.3.

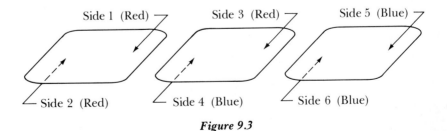

Figure 9.3

We can consider our sample space to be the set of ordered pairs

$$S = \{(1, 2), (2, 1), (3, 4), (4, 3), (5, 6), (6, 5)\}$$

where the entry in the first coordinate of each pair corresponds to the up side of the card lying on the table, and the second coordinate corresponds to the down side.

The set $A \cap B = \{(1, 2), (2, 1)\}$ is the event "red is down and red is up." Since all outcomes are equally likely, $P(A \cap B) = 2/6$. The set $B = \{(1, 2), (2, 1), (3, 4)\}$ is the event "red is up," and clearly $P(B) = 3/6$. Hence, the probability that you will win is

$$P(A|B) = \frac{P(A \cap B)}{P(B)} = \frac{2/6}{3/6} = \frac{2}{3}$$

The same argument holds for blue. Consequently, you may expect to win this game two times out of three.   □

To conclude this section, we examine situations where more than three events are involved. First note that from Definition 9.6 we have

$$P(A \cap B) = P(B)P(A|B) \qquad (1)$$

Suppose now we have events A_1, A_2, and A_3 in a probability space (S, P). Three applications of (1) yield

$$\begin{aligned}
P(A_1 \cap A_2 \cap A_3) &= P(A_3 \cap (A_2 \cap A_1)) \\
&= P(A_2 \cap A_1)P(A_3|A_2 \cap A_1) \\
&= P(A_1)P(A_2|A_1)P(A_3|A_2 \cap A_1) \qquad (2)
\end{aligned}$$

The next example illustrates an application of this computation.

Example 4   A box of 50 spark plugs is known to contain 8 defective plugs. If you select 3 plugs from the box, what is the probability that none of the 3 is defective? For $i = 1, 2, 3$, we let A_i be the event that the ith plug selected is not defective. We must find $P(A_1 \cap A_2 \cap A_3)$.

First note that $P(A_1) = 42/50$. Next observe that if we assume that the first spark plug selected was good, then there are 41 possibilities for choosing a nondefective plug from the remaining 49 plugs. Consequently, $P(A_2|A_1) = 41/49$. Finally, note that if we assume that the first two chosen plugs are good, then there are 40 possibilities for selecting a nondefective plug from the remaining 48 plugs, so $P(A_3|A_1 \cap A_2) = 40/48$. Thus, from (2)

$$P(A_1 \cap A_2 \cap A_3) = P(A_1)P(A_2|A_1)P(A_3|A_2 \cap A_1) = \frac{42}{50} \cdot \frac{41}{49} \cdot \frac{40}{48} = \frac{41}{70}$$

In other words, there is a probability of .59 that none of the three chosen plugs is defective.   □

In Problem 10 you are asked to use induction to obtain the following more general result.

Theorem 9.7 *General Law of Compound Probability.* If $A_1, A_2, \ldots, A_n$ are events in a probability space (S, P), then

$$P(A_1 \cap A_2 \cap \cdots \cap A_n) = P(A_1)P(A_2|A_1)P(A_3|A_1 \cap A_2)$$
$$\cdots P(A_n|A_1 \cap A_2 \cap \cdots \cap A_{n-1})$$

9.3 Problems

1. Suppose that A and B are events in a probability space (S, P) such that $P(A) = .7$, $P(B) = .4$, and $P(A \cap B) = .25$. Find $P(A|B)$ and $P(B|A)$.

2. Suppose that A and B are events in a probability space (S, P) such that $P(A) = .6$, $P(B) = .6$, and $P(A \cap B) = .6$. Find $P(A|B)$ and $P(B|A)$.

3. Under what conditions does $P(A|B) = P(B|A)$?

4. Two dice are rolled.
   (a) What is the probability that the sum of the dice is 6 if one of the dice shows a 4?
   (b) What is the probability that exactly one of the dice shows a 4 if the sum is 6?

5. A bowl contains 4 white chips and 2 red chips. Two chips are drawn from the bowl without replacement. What is the probability that the second chip drawn is red if the first chip drawn is white?

6. Two new products A and B are introduced into the market. It is estimated that the probability of success of product A is .75, the probability of success for product B is .6, and the probability that both will be successful is .45. If the product A is found to be successful, what is the probability of success for product B?

7. Suppose a fair coin is tossed 8 times and heads are obtained on the first 7 tosses. What is the probability of obtaining heads on the eighth toss?

8. It is known that 5% of the male population is color blind and that .25% of the female population is color blind. Assume that there are as many males as females in the population. If a given person is color blind, what is the probability that the person is female?

9. Suppose that $A_1, A_2, \ldots, A_n$ are events in a probability space (S, P) and that $\bigcup_{i=1}^{n} A_i = S$. Suppose further that for each $i \neq j$, $A_i \cap A_j = \emptyset$. Show that if B is any event in S such that $P(B) \neq 0$, then $P(A_1|B) + P(A_2|B) + \cdots + P(A_n|B) = 1$.

10. Prove Theorem 9.7 using an induction argument.

11. In a certain community 90% of the families have at least one car and 60% of the families have at least one bicycle. If 50% of the families have at least one car and one bicycle, what is the probability that a family with a car also owns a bicycle?

12. A hospital usually has a back-up generator system to deal with electrical failures. For a certain hospital, suppose that there is a probability of .002 that the main system will fail on any given day. Suppose further that if the main system fails then the probability that the back-up system will also fail is .03. What is the probability that on any given day both systems will fail?

13. Show that if A, B, and C are events in a probability space (S, P) then
    (a) $P(A \cap B|B) = P(A|B)$.
    (b) $P(A^c|B) = 1 - P(A|B)$.
    (c) $P(A \cup B|C) = P(A|C) + P(B|C) - P(A \cap B|C)$.

14. A shipment of twenty television sets is known to contain four defective sets. If five of these twenty sets are sold, what is the probability that at least one of these five is defective?

15. Suppose you take a test consisting of eight true–false questions.
    (a) If you simply guess at the answers, what is the probability that you will answer all of the questions correctly?
    (b) If you know the answers to exactly three questions and guess at the rest of the questions, what is the probability that you will answer all of the questions correctly?
    (c) Suppose you know that four of the answers are true and four are false. If you guess at each question, what is the probability that you will answer all of the questions correctly?

9.4   INDEPENDENT EVENTS

It is frequently the case that the probability of the occurrence of an event A is not affected by the occurrence or non-occurrence of an event B. This is equivalent to saying that the conditional probability of A given B is equal to the probability of A; that is,

$$P(A|B) = P(A) \tag{1}$$

If (1) holds, we say that the event A is *independent* of the event B.

If an event A is independent of an event B, then, as we show next, the event B is also independent of the event A. In other words, if $P(A|B) = P(A)$, then $P(B|A) = P(B)$. To see this, note that from Definition 9.6 we have

$$P(A \cap B) = P(A|B)P(B)$$

Thus,

$$P(B|A) = \frac{P(A \cap B)}{P(A)} = \frac{P(A|B)P(B)}{P(A)} \qquad (2)$$

Since $P(A|B) = P(A)$, it follows from (2) that

$$P(B|A) = \frac{P(A)P(B)}{P(A)} = P(B)$$

as was to be shown.

This result leads to the following important definition.

Definition 9.8   Events A and B in a probability space (S, P) are *independent* if $P(A|B) = P(A)$ (or, equivalently, if $P(B|A) = P(B)$).

Observe that since $P(A|B) = P(A \cap B)/P(B)$, we have $P(A|B) = P(A)$ if and only if $P(A \cap B) = P(A)P(B)$. This gives us the following fundamental characterization of independent events.

Theorem 9.9   Events A and B are independent if and only if $P(A \cap B) = P(A)P(B)$.

The next example shows how Theorem 9.9 can be used to test the independence of events.

Example 1   Let $S = \{x_1, x_2, x_3, x_4, x_5, x_6, x_7, x_8\}$ be a sample space in which all outcomes are equally likely; that is, $P(\{x_i\}) = 1/8$ for $i = 1, \ldots, 8$. Consider the following events:

$$A = \{x_1, x_2, x_3, x_4\}$$
$$B = \{x_1, x_2, x_5, x_6\}$$
$$C = \{x_5, x_6, x_7\}$$

Observe that

$$P(A \cap B) = P(\{x_1, x_2\}) = \frac{1}{4}$$

and

$$P(A)P(B) = \frac{1}{2} \cdot \frac{1}{2} = \frac{1}{4}$$

Consequently, by Theorem 9.9, A and B are independent events.

Note further that

$$P(A \cap C) = P(\varnothing) = 0$$

and

$$P(A)P(C) = \frac{1}{2} \cdot \frac{3}{8} = \frac{3}{16}$$

and, consequently, $P(A \cap C) \neq P(A)P(C)$, so the events A and C are not independent.

Moreover,

$$P(B \cap C) = \frac{1}{4}$$

and

$$P(B)P(C) = \frac{1}{2} \cdot \frac{3}{8} = \frac{3}{16}$$

and therefore, the events B and C also fail to be independent.   □

Often we assume that events separated by time or space are independent. As shown in the next example, this assumption simplifies the computation of certain probabilities.

Example 2   Suppose that an experiment consists of throwing a die and drawing a card from a deck of 52 cards. Let A be the event that an even number is rolled and let B be the event that a heart is drawn. We want to compute $P(A \cap B)$. It is a reasonable assumption that events A and B are independent. Under this assumption we have

$$P(A \cap B) = P(A)P(B) = \frac{1}{2} \cdot \frac{13}{52} = \frac{1}{8}   □$$

How might we extend Definition 9.8 to more than two events? The most evident way is to extend the characterization of independence obtained in Theorem 9.9 and say that events A_1, A_2, and A_3 are independent if

$$P(A_1 \cap A_2 \cap A_3) = P(A_1)P(A_2)P(A_3) \tag{3}$$

There is a problem here, however; if the events A_1, A_2, and A_3 are to be independent of each other, then it is certainly reasonable to expect that any pair of these events, say A_1 and A_2, also be independent of each other. That is

$$P(A_1 \cap A_2) = P(A_1)P(A_2)$$

Unfortunately, this does not necessarily follow from (3), as the next example illustrates.

Example 3   Two dice are rolled and the results are recorded as ordered pairs (i, j), where both i and j range from 1 to 6. Let A be the event that the first die is either 3, 4, or 5, let B be the event that the first die is either 1, 2, or 3, and let C be the event that the sum of the rolled dice is 9. Then you can easily verify the following probabilities:

$$P(A) = \frac{1}{2} \qquad P(A \cap B) = \frac{1}{6} \qquad P(A \cap B \cap C) = \frac{1}{36}$$

$$P(B) = \frac{1}{2} \qquad P(A \cap C) = \frac{1}{12}$$

$$P(C) = \frac{1}{9} \qquad P(B \cap C) = \frac{1}{36}$$

Note that $P(A)P(B)P(C) = (1/2)(1/2)(1/9) = 1/36 = P(A \cap B \cap C)$, but that $P(A)P(B) = (1/2)(1/2) = 1/4 \neq 1/6 = P(A \cap B)$; $P(A)P(C) = (1/2)(1/9) = 1/18 \neq 1/36 = P(A \cap C)$; and $P(B)P(C) = (1/2)(1/9) = 1/18 \neq 1/12 = P(B \cap C)$. We see then in this example that although (3) is satisfied, no pair of the events A, B, C is independent. $\qquad\square$

To rectify the problem arising in the preceding example, we might say that events A_1, A_2, and A_3 are independent if they are pairwise independent, that is, if $P(A_1 \cap A_2) = P(A_1)P(A_2)$, $P(A_1 \cap A_3) = P(A_1)P(A_3)$, and $P(A_2 \cap A_3) = P(A_2)P(A_3)$. However, as we see in the next example, these conditions are not sufficient to ensure that $P(A_1 \cap A_2 \cap A_3) = P(A_1)P(A_2)P(A_3)$.

Example 4 An experiment consists of tossing a fair coin twice. Let A_1 be the event that heads occurs on the first toss, let A_2 be the event that heads occurs on the second toss, and let A_3 be the event that exactly one head and one tail result from the two tosses. Then you can verify that $P(A_1) = 1/2$, $P(A_2) = 1/2$, $P(A_3) = 1/2$, and

$$P(A_1 \cap A_2) = \frac{1}{4} = P(A_1)P(A_2)$$

$$P(A_1 \cap A_3) = \frac{1}{4} = P(A_1)P(A_3)$$

$$P(A_2 \cap A_3) = \frac{1}{4} = P(A_2)P(A_3)$$

but that

$$P(A_1 \cap A_2 \cap A_3) = 0 \neq \frac{1}{8} = P(A_1)P(A_2)P(A_3) \qquad\square$$

The results of Examples 3 and 4 lead us to the following general definition of independence.

Definition 9.10   Let (S, P) be a probability space and let $\mathscr{A}$ be a set of events in this space. Then the events in $\mathscr{A}$ are *independent* if for each subset $\{A_1, A_2, \ldots, A_k\}$ of $\mathscr{A}$

$$P(A_1 \cap A_2 \cap \cdots \cap A_k) = P(A_1)P(A_2) \cdots P(A_k)$$

Example 5   Let $S = \{x_1, x_2, x_3, x_4, x_5, x_6, x_7\}$ be a sample space with the probability of each outcome as given in Table 9.1. Consider the events $A_1 = \{x_1, x_3, x_4, x_5\}$, $A_2 = \{x_1, x_2, x_4, x_6\}$, and $A_3 = \{x_1, x_2, x_3, x_7\}$. Direct computation gives $P(A_1) = 1/3$, $P(A_2) = 1/2$, $P(A_3) = 1/4$, $P(A_1 \cap A_2) = 1/6$, $P(A_1 \cap A_3) = 1/12$, $P(A_2 \cap A_3) = 1/8$, and $P(A_1 \cap A_2 \cap A_3) = 1/24$. Then since

$$P(A_1 \cap A_2) = P(A_1)P(A_2)$$
$$P(A_1 \cap A_3) = P(A_1)P(A_3)$$
$$P(A_2 \cap A_3) = P(A_2)P(A_3)$$
$$P(A_1 \cap A_2 \cap A_3) = P(A_1)P(A_2)P(A_3)$$

it follows from Definition 9.10 that the events A_1, A_2, and A_3 are independent. □

Outcome	Probability
x_1	1/24
x_2	1/12
x_3	1/24
x_4	1/8
x_5	1/8
x_6	1/4
x_7	1/12

Table 9.1

As is the case for two events, assuming the independence of several events simplifies the computation of certain probabilities.

Example 6   Three tests, T_1, T_2, and T_3, have been developed to detect the presence of a certain disease. It is known that test T_1 gives a positive reaction in 90% of the cases where a person has the disease, test T_2 gives a positive reaction in 85% of such cases, and test T_3 gives a positive reaction in 80% of these cases. We want to know the probability that all three tests will fail to detect the presence of the disease.

For $i = 1, 2, 3$, let A_i be the event that test T_i fails to give a positive reaction in the case that a person actually has the disease. Then $P(A_1) = .10$, $P(A_2) = .15$, and $P(A_3) = .20$. If we assume that these events are independent, we find the that the probability that all three tests fail to detect the presence of disease is

$$P(A_1 \cap A_2 \cap A_3) = P(A_1)P(A_2)P(A_3) = .003 \qquad \square$$

9.4 Problems

1. Suppose A and B are events such that $P(A) = .4$, $P(B) = .2$, and $P(A \cap B) = .09$. Are A and B independent events?

2. Show that if A and B are independent events in a probability space (S, P), then A^c and B^c are also independent events.

3. Suppose that a bear is tagged on each ear with a plastic tag and is then set free. Let q be the probability that the tag placed on the left ear is lost, and assume that q is also the probability that the tag on the right ear is lost. Find the probability that exactly one tag is lost given that at most one is lost (assume that losing a tag from the left ear is independent of losing a tag from the right ear).

4. Suppose that the probability of sighting a particular species of bird on a given day is $1/4$. If you spend six days bird watching, what is the probability that you will sight this species at least once?

5. A coin is flipped three times. For $i = 1, 2, 3$, let A_i be the event that tails is obtained on the ith toss. Use the definition of independence to show that the events A_1, A_2, and A_3 are independent.

6. Show that if A and B are independent events in a probability space (S, P), then so are the events A and B^c. (*Hint:* First show that $P(B^c|A) = 1 - P(B|A)$.)

7. If A and B are events in a probability space (S, P) such that $P(A) = .6$, $P(B) = .4$, and $P(A \cup B) = .7$, are the events A and B independent? Why or why not?

8. A bowl contains 19 red chips and 6 blue chips. A person removes 2 of the chips, notes what they are, and then puts them back into the bowl. A second person then repeats this process. What is the probability that at least one of the people will find at least one blue chip? What is the probability that at least one of them will find exactly one blue chip?

9. The probability that a pet dog will live at least 14 years is $1/3$. The probability that a pet cat will live at least 14 years is $1/6$.
   (a) What is the probability both pets will live at least 14 years?
   (b) What is the probability that neither pet will live at least 14 years?
   (c) What is the probability that exactly one of the pets will live at least 14 years?

9.5   BERNOULLI PROCESSES

The notion of independence plays an important role in Bernoulli processes.[1] These processes arise naturally in many mathematical models that involve experiments whose outcomes fall into two disjoint sets, which may be classified by "success" and "failure."

Definition 9.11   A random experiment E is called a *Bernoulli trial* if E has exactly two outcomes (usually called *success* or *failure*).

Definition 9.12   A *Bernoulli process* is a random experiment that consists of a finite sequence of Bernoulli trials such that

   a. the outcome of any one trial is independent of the outcomes of the other trials, and

   b. the probability of success (and, hence, failure) is the same for each trial.

Example 1   A fair coin is tossed 6 times. Each toss is considered to be a Bernoulli trial where a "success" is obtained if the result of the toss is heads; if the outcome is tails, the trial is considered to be a failure. Note that the trials are independent of each other and that the probability of success is the same for each trial (though many persons would incorrectly maintain that if heads have occurred five times in a row, then it is "more likely" that tails will occur on the sixth toss).   ☐

Example 2   A *binary channel* is a channel over which two symbols (commonly 0 and 1) are transmitted. A binary channel is called a *binary symmetric channel* if the probability of receiving a 0 when a 1 is transmitted is the same as the probability of receiving a 1 when a 0 is transmitted. That is, as shown in Figure 9.4, the probability of error (failure) is q in both cases, while the probability of correctly receiving the transmitted signal (success) is $p = 1 - q$.

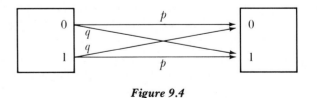

Figure 9.4

[1] Named for Jakob Bernoulli (1654–1705), member of one of mathematics' most prolific families.

In many applications, a sequence of zeros and ones is transmitted and it is assumed that the probability of error (failure) is the same for each trial (transmission). Moreover, in some cases we can assume that the symbols transmitted are independent events. Consequently, the transmission of a sequence of zeros and ones over a binary symmetric channel is a Bernoulli process. □

Example 3   It is known that a box of 100 fuses contains 7 defective fuses and 93 nondefective fuses. Four inspectors remove samples from the contents of the box as follows. Each inspector in turn removes 2 fuses from the box, examines them, and then returns them to the box. A "success" is recorded if both of the fuses selected are without defects. This occurs with a probability of $(93/100)(92/100) = .8556$ for each inspector. The selections made by the inspectors can be considered as a Bernoulli process consisting of 4 Bernoulli trials such that for each trial the probability of success is .8556 and the probability of failure is $1 - .8556 = .1444$. □

Perhaps the principal question arising in connection with Bernoulli processes is that of determining the probability of r successes in a sequence of n Bernoulli trials. For example, we might wish to compute the probability of 18 successes in the transmission of 20 symbols over a binary symmetric channel where the probability of success on each trial is $p = .95$. To compute the probability of 18 successes we let E be the Bernoulli process consisting of the 20 Bernoulli trials, for which each trial is the transmission of a single bit.

Since the trials are independent of each other, it follows that the probability of *each* outcome of E consisting of 18 successes and 2 failures is

$$(.95)^{18}(.05)^2$$

Moreover, since there are $\binom{20}{18}$ ways to select the positions for 18 successes to occur in 20 trials, we see that there are $\binom{20}{18}$ outcomes consisting of 18 successes and 2 failures. Consequently, the probability of exactly 18 successes in a sequence of 20 transmissions is

$$\binom{20}{18}(.95)^{18}(.05)^2 \approx .1887 \qquad \square$$

We can generalize this specific example to obtain the following theorem, which you will be asked to prove in Problem 11.

Theorem 9.13   The probability of exactly r successes in a sequence of n Bernoulli trials with probability p of success on each trial and with probability $q = 1 - p$ of failure on each trial is

$$\binom{n}{r}p^r q^{n-r} \tag{1}$$

Recall that $\binom{n}{r}p^r q^{n-r}$ is a typical term found in the binomial expansion of $(p + q)^n$; for this reason (1) is often referred as a *binomial probability* and a Bernoulli process is often called a *binomial process*.

Example 4 A weighted coin is tossed seven times. If, for this coin, the probability of heads is $2/3$ and the probability of tails is $1/3$, what is the probability that exactly three heads are obtained?

This experiment can be considered as a sequence of seven Bernoulli trials with $p = 2/3$ and $q = 1/3$. Hence, by Theorem 9.13 the probability of obtaining exactly three heads is

$$\binom{7}{3}\left(\frac{2}{3}\right)^3\left(\frac{1}{3}\right)^4 \approx .1280 \qquad \square$$

Example 5 A manufacturer estimates that 3% of the items produced at a particular factory are defective. Using this estimate, we compute the probability in a random selection of ten items that there are:

(a) exactly two defectives.

(b) no defectives.

(c) at least one defective.

The selection of the ten items can be considered as a Bernoulli process in which a "success" is the selection of a defective item. The probability of success is .03.

(a) We are to compute the probability of exactly two successes in a sequence of ten trials. By Theorem 9.13 this probability is

$$\binom{10}{2}(.03)^2(.97)^8 \approx .0317$$

(b) We compute the probability of exactly zero successes in a sequence of ten trials. By Theorem 9.13 this probability is

$$\binom{10}{0}(.03)^0(.97)^{10} = (.97)^{10} \approx .7374$$

Alternately we may view this as the selection of ten nondefective items in succession. Since these selections are independent events, each with probability .97, the probability of ten consecutive selections of nondefective items is $(.97)^{10}$.

(c) If A is the event that "at least one defective is selected," then A can be expressed as the union of ten mutually exclusive events $A_1, A_2, \ldots, A_{10}$, where for each i, A_i is the event that exactly i defectives are selected. Since

$P(A_i) = \binom{10}{i}(.03)^i(.97)^{10-i}$, we have

$$P(A) = \binom{10}{1}(.03)^1(.97)^9 + \binom{10}{2}(.03)^2(.97)^8 + \binom{10}{3}(.03)^3(.97)^7$$

$$+ \binom{10}{4}(.03)^4(.97)^6 + \binom{10}{5}(.03)^5(.97)^5 + \binom{10}{6}(.03)^6(.97)^4$$

$$+ \binom{10}{7}(.03)^7(.97)^3 + \binom{10}{8}(.03)^8(.97)^2 + \binom{10}{9}(.03)^9(.97)^1$$

$$+ \binom{10}{10}(.03)^{10}(.97)^0 \approx .2626$$

Because of the difficulty of this computation, it is preferable to compute the desired probability indirectly. Observe that the event "at least one defective is selected" is the complement of the event B, "no defectives are selected." Thus,

$$P(A) = 1 - P(B)$$

Since B is the event that all nondefectives are selected, we have

$$P(B) = (.97)^{10}$$

Thus, $P(A) = 1 - (.97)^{10} \approx .2626$. □

Example 6   A sequence of twenty symbols is transmitted over a binary symmetric channel for which the probability of error is $q = .001$. In this case the probability of zero errors is

$$P_0 = \binom{20}{0}(.001)^0(.999)^{20} \approx .9802$$

The probability of exactly one error is

$$P_1 = \binom{20}{1}(.001)(.999)^{19} \approx .0196$$

The probability of two or more errors is

$$1 - (P_0 + P_1) \approx .0002$$

Note that a single error is far more likely than two or more errors. For this reason, error-correcting procedures often concentrate on single errors. □

The most obvious way to generalize the notion of a Bernoulli process is to maintain the idea of a sequence of n independent trials, but rather than allowing just two outcomes (success or failure) for each trial, we permit k possible outcomes $s_1, s_2, \ldots, s_k$ with probabilities $p_1, p_2, \ldots, p_k$, respectively. Here, of course, $p_1 + p_2 + \cdots + p_k = 1$ and we assume the same probabilities apply to each trial.

Suppose, for instance, that a target consists of three concentric rings, C_1, C_2, and C_3. Suppose further that the probability of a dart thrower's hitting ring C_1 is $1/6$, ring C_2 is $1/2$, and ring C_3 is $1/3$. In this case we have three outcomes:

$$s_1 \text{—the dart hits ring } C_1$$
$$s_2 \text{—the dart hits ring } C_2$$
$$s_3 \text{—the dart hits ring } C_3$$

with probabilities $p_1 = 1/6$, $p_2 = 1/2$, and $p_3 = 1/3$, respectively.

The principal question that arises in such a setting may be formulated as follows:

Given a "generalized" Bernoulli process consisting of n trials, each of which has outcomes $s_1, s_2, \ldots, s_k$ (with probabilities $p_1, p_2, \ldots, p_k$, respectively), what is the probability that for each i, $1 \le i \le k$, the outcome s_i will occur n_i times? (Here $n_1 + n_2 + \cdots + n_k = n$.)

For instance, in the dart throwing example just discussed, we might wish to find the probability that ring C_1 will be hit three times, that ring C_2 will be hit twice, and that ring C_3 will be hit four times if a dart is thrown nine times.

To answer the general question posed above, for each i, $1 \le i \le k$, let A_i be the set of trials that result in outcome s_i. Then $|A_i| = n_i$. Note that the sets $A_1, A_2, \ldots, A_k$ are disjoint and that their union is the sample space associated with the generalized Bernoulli process. It follows from the independence of the trials in a Bernoulli process that the probability of each outcome that consists of n_1 trials with outcome s_1, n_2 trials with outcome s_2, $\ldots$, n_k trials with outcome s_k is

$$p_1^{n_1} p_2^{n_2} \cdots p_k^{n_k}$$

To compute the probability of the event consisting of *all* such outcomes, we must count the number of these outcomes. That is, we must calculate the number of ways n_1 trials can result in outcome s_1, n_2 trials can result in outcome s_2, $\ldots$, n_k trials can result in outcome s_k.

Recall from Theorem 8.10 that the "locations" of the n_1 occurrences comprising A_1, the n_2 occurrences comprising A_2, $\ldots$, the n_k occurrences comprising A_k can be selected from a total of n locations in

$$\binom{n}{n_1, n_2, \cdots, n_k} = \frac{n!}{n_1! n_2! \cdots n_k!}$$

ways. We therefore find that the probability that the Bernoulli process will result in an outcome with s_1 occurring n_1 times, s_2 occurring n_2 times, $\ldots$, s_k occurring n_k times is

$$\frac{n!}{n_1! n_2! \cdots n_k!} p_1^{n_1} p_2^{n_2} \cdots p_k^{n_k} \tag{2}$$

From this result we see that in our dart throwing example the probability of hitting ring C_1 three times, ring C_2 twice, and ring C_3 four times in nine throws is

$$\frac{9!}{3!2!4!}\left(\frac{1}{6}\right)^3\left(\frac{1}{2}\right)^2\left(\frac{1}{3}\right)^4 \approx .0180$$

Example 7   At a certain intersection the probability that a driver will go straight ahead is .6, will turn left is .2, will turn right is .15, and will make a U-turn is .05. It follows from (2) that the probability that if ten cars approach this intersection, five will go straight ahead, four will turn right, and one will make a U-turn is

$$\frac{10!}{5!0!4!1!}\,(.6)^5(.2)^0(.15)^4(.05)^1 \approx .0025 \qquad \square$$

In this section we have seen how Bernoulli processes can be used to handle situations where the outcomes of a random experiment are classified in terms of success or failure. Examples of such experiments are many and they have widespread applications.

9.5   Problems

1. Suppose six fair coins are tossed.
   (a) Find the probability of obtaining exactly four heads.
   (b) Find the probability of obtaining more than one head.
2. If in searching for oil a company drills four dry holes for every successful hole, what is the probability that the company will strike oil first on the fifth try?
3. Suppose there is a probability of .2 that a new car will have a defective headlight. What is the probability that at least five cars in a shipment of ten new cars will have defective headlights?
4. If in a dart game your probability of throwing a bull's-eye is .2, what is the probability that in ten throws you will get exactly three bull's-eyes? No bull's-eyes? At least one bull's-eye?
5. Four dice are thrown.
   (a) Find the probability that two 4's, one 6, and one 3 are obtained.
   (b) Find the probability that at least two 4's and one 6 are obtained.
   (c) Find the probability that no 6's are obtained.
6. Products passing through a production line are classified as: acceptable, blemished (but saleable), or unacceptable. Suppose that 85% of the units going through the line are acceptable, 10% are blemished, and 5% are unacceptable. A sample of 15 units is taken off the line.

(a) What is the probability that 10 of the units are acceptable, 3 are blemished, and 2 are unacceptable?

(b) What is the probability that all of the 15 units are acceptable?

7. A manufacturer of spark plugs employs the following quality control plan. For each lot of 1000 plugs, 15 are selected to be tested. If at least two of the 15 are found to be defective, then the entire lot is rejected. Find the probability that an entire lot will be rejected if:

(a) 2% of the plugs in the lot are defective.

(b) 5% of the plugs in the lot are defective.

(c) 10% of the plugs in the lot are defective.

8. Suppose that in a given election 40% of the voters will vote Democratic and 60% will vote Republican. In a sample of 10 voters what is the probability that 4 persons will vote Democratic and 6 will vote Republican? What is the probability that at least 4 of the 10 persons sampled will vote Republican?

9. The pointer on the following dial is spun.

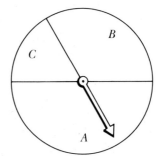

The probabilities of the pointer's stopping in the various regions of the dial are as follows:

$$\text{in region } A: \frac{1}{2} \qquad \text{in region } B: \frac{1}{3} \qquad \text{in region } C: \frac{1}{6}$$

If the pointer is spun ten times, find the probability of each of the following events:

(a) The pointer stops five times in region A, three times in region B and twice in region C.

(b) The pointer never stops in region C.

(c) The pointer stops at least five times in region A and at least twice in region B.

10. A die is thrown four times.

(a) Find the probability of obtaining two 3's, one 6, and one 4.

(b) Find the probability of obtaining at least two 2's and one 3.

(c) Find the probability of not obtaining either a 2 or a 5.

11. Prove Theorem 9.13.

*9.6   BAYES' THEOREM

In previous sections, we have used the probabilities of certain events to calculate conditional probabilities. We now examine how to use conditional probabilities to calculate other probabilities. As an illustration, consider the following example.

Example 1   A test has been developed for determining the presence of a certain disease. The test gives a positive reaction in 95% of the cases where a person actually has the disease. However, the test also gives a positive reaction in 15% of the cases where a person does not have the disease. It is known that 10% of the population has this disease. What is the probability that if a person reacts positively to the test, then the person actually has the disease?

We define the following events:

Event A—the event that a person reacts positively to the test
Event B_1—the event that a person has the disease
Event B_2—the event that a person does not have the disease

In these terms, we are interested in finding the conditional probability $P(B_1|A)$. We have the following information:

$$P(A|B_1) = .95 \qquad P(B_1) = .10$$
$$P(A|B_2) = .15 \qquad P(B_2) = 1 - P(B_1) = .9$$

Now we work towards expressing $P(B_1|A)$ in terms of the probabilities listed above. We have previously observed that

$$P(A|B_1)P(B_1) = P(A \cap B_1) = P(B_1|A)P(A) \tag{1}$$

and consequently,

$$P(B_1|A) = \frac{P(A|B_1)P(B_1)}{P(A)} \tag{2}$$

Note that $A = (A \cap B_1) \cup (A \cap B_2)$, and since $B_1 \cap B_2 = \varnothing$, it follows that $(A \cap B_1) \cap (A \cap B_2) = \varnothing$. Therefore, by Theorem 9.4 and (1)

$$P(A) = P(A \cap B_1) + P(A \cap B_2) = P(A|B_1)P(B_1) + P(A|B_2)P(B_2) \tag{3}$$

and by substituting (3) into (2) we obtain the somewhat surprising result

$$P(B_1|A) = \frac{P(A|B_1)P(B_1)}{P(A|B_1)P(B_1) + P(A|B_2)P(B_2)} = \frac{(.95)(.10)}{(.95)(.10) + (.15)(.90)}$$
$$= .4130$$

as the probability that a person who reacts positively to the test actually has the disease.   □

The arguments used in the previous example may be generalized to obtain the following result (see Problem 8).

Theorem 9.14   *Bayes' Theorem.* Suppose that $B_1, B_2, \ldots, B_m$ are mutually disjoint events in a probability space (S, P) and that $B_1 \cup \cdots \cup B_m = S$. Let A be any event in S. Then for each i, $1 \le i \le m$

$$P(B_i|A) = \frac{P(A|B_i)P(B_i)}{P(A|B_1)P(B_1) + P(A|B_2)P(B_2) + \cdots + P(A|B_m)P(B_m)}$$

(4)

The following two examples illustrate how Bayes' Theorem can be applied.

Example 2   Seeds of three varieties of tomatoes are scattered throughout a plot of land. Exactly 65% of the seeds are of variety S_1, 25% are of variety S_2, and 10% are of variety S_3. Research has shown that 40% of the seeds of variety S_1 will germinate, 50% of variety S_2 will germinate, and 60% of variety S_3 will germinate. If a particular seed germinates, what is the probability that it is of variety S_2?

For $i = 1, 2, 3$, let B_i be the event that a seed is of variety S_i; let A be the event that a seed germinates. Then by (4) we have

$$P(B_2|A) = \frac{P(A|B_2)P(B_2)}{P(A|B_1)P(B_1) + P(A|B_2)P(B_2) + P(A|B_3)P(B_3)}$$

$$= \frac{(.5)(.25)}{(.4)(.65) + (.5)(.25) + (.6)(.1)} \approx .2809 \qquad \square$$

Example 3   Three automobile plants each produce a certain make of car. Plant 1 produces 40% of these cars, Plant 2 produces 35%, and Plant 3 manufactures 25%. It is known that 5% of the cars manufactured in Plant 1 have a certain defect, 3% of the cars produced in Plant 2 have this same defect, and 1% of the cars originating in Plant 3 have the defect. If a car is found to have this defect, what is the probability that the car was manufactured in Plant 3?

To determine this probability let E be the event that the car has a defect, and for $i = 1, 2, 3$, let C_i be the probability that the car was manufactured in Plant i. We want to find $P(C_3|E)$.

By Bayes' Theorem we have

$$P(C_3|E) = \frac{P(E|C_3)P(C_3)}{P(E|C_1)P(C_1) + P(E|C_2)P(C_2) + P(E|C_3)P(C_3)}$$

$$= \frac{(.01)(.25)}{(.05)(.4) + (.03)(.35) + (.01)(.25)} \approx .0758 \qquad \square$$

9.6 Problems

Use Bayes' Theorem to solve the following problems.

1. Forty percent of all families in a certain city own their own homes. Of the families that own their own homes, 70% have cars; 40% of the families that do not own their own homes have cars. If a family owns a car, what is the probability that the family also owns its own home?

2. A car dealer sells four models, A, B, C, and D, of a certain car. The dealer has found that in any shipment he receives of these cars, 5% of the model A cars are defective, 8% of the model B cars are defective, 3% of the model C cars are defective, and 10% of the model D cars are defective. The dealer receives a shipment of 40 model A cars, 60 model B cars, 70 model C cars, and 30 model D cars. If one of these cars is sold and proves to be defective, what is the probability that the car is a model C car?

3. Three weighted coins A, B, and C have the following probability of heads — coin A: .6; coin B: .45; and coin C: .7. One of these coins is selected at random and tossed twice. If tails are obtained on both tosses, what is the probability that coin B was tossed?

4. In a certain stream it is known that 60% of the fish are rainbow trout, 30% are cutthroats, and 10% are golden trout. Furthermore, it is known that 40% of the rainbows weigh more than 1 lb., 70% of the cutthroat weigh more than 1 lb., and 30% of the goldens weigh more than 1 lb. If a fish is caught that weighs more than 1 lb., what is the probability that the fish is a cutthroat?

5. Three different methods, A, B, and C, of quality control are used in the testing of light bulbs. Method A is used to inspect 60% of the bulbs, method B is used to inspect 25% of the bulbs, and method C is used to inspect 15% of the bulbs. Method A is known to fail 7% of the time; method B, 4% of the time; and method C, 3% of the time. If you purchase a bulb that is found to be defective, what is the probability that method B was used to inspect it? If you purchase a bulb that proves to be good, what is the probability that method C was used to inspect it?

6. Three percent of the people living in a certain community have tuberculosis. Suppose that a doctor is correct ninety percent of the time in diagnosing a patient as having tuberculosis when the patient has the disease. Suppose further that four percent of the time the doctor diagnoses a patient as having tuberculosis even though the patient does not suffer from this malady. What is the probability that if a person is diagnosed as having tuberculosis, the patient actually has this disease?

7. A question on a multiple-choice exam has five possible answers, only one of which is correct. The probability that a student knows the correct answer is .7. If the student answers the question correctly, what is the probability that

the student knew the correct answer and did not just make a lucky guess? (Assume that if the student did not know the correct answer she is equally likely to pick any of the five answers.)

8. Prove Theorem 9.14.

9.7 RANDOM VARIABLES AND PROBABILITY DENSITY FUNCTIONS

One of the most fundamental concepts in probability theory is that of a random variable. Random variables will help us classify and systematically solve a number of seemingly unrelated problems.

Definition 9.15   Let (S, P) be a probability space. A function $X:S \rightarrow \mathbf{R}^1$ is called a *random variable* on S.

In other words, a random variable on S is any function that assigns a numerical value to each outcome in S.

Example 1   Consider the wheel illustrated in Figure 9.5. When the pointer is spun, there is a probability of $1/2$ that it will stop in region A, $1/3$ in region B, and $1/6$ in region C. If the pointer stops in region A, a player wins \$2; if the marker lands in region B, a player loses \$3; and if the marker lands in region C, a player wins \$1.

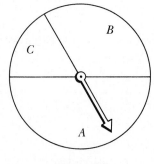

Figure 9.5

A player spins the pointer twice and the results are recorded as ordered pairs. We define our sample space S to be the set of all of these ordered pairs. Thus, $S = \{(A, A), (A, B), (B, A), (A, C), (C, A), (B, B), (B, C), (C, B), (C, C)\}$. We define a random variable X on S by letting X be the amount won (or lost) after two spins of the pointer; for instance, $X(A, B) = -\$1$, $X(B, C) = -\$2$, $X(C, C) = +\$2$, $X(C, A) = +\$3$, and so on.   □

Next we define a random variable that is commonly used in connection with Bernoulli processes. Suppose that a Bernoulli process consists of a sequence of n Bernoulli trials. Then we can view the outcomes of this random experiment as n-tuples $(x_1, x_2, \ldots, x_n)$ where each x_i is either 1 or 0; a 1 denotes a success and a 0 denotes a failure. We define a random variable on the sample space of this experiment by setting

$$X(x_1, x_2, \cdots, x_n) = x_1 + x_2 + \cdots + x_n \qquad (1)$$

Note that for each outcome (a sequence of n trials) X gives the number of successes.

Example 2   We roll a die five times. On each roll we record a success if the number showing is even, and a failure if the number is odd. Thus the outcome $(1, 0, 0, 1, 0)$ indicates that an even integer came up on the first and fourth rolls of the die, and odd integers came up on the second, third, and fifth rolls. The number of successes (even numbers showing) is

$$X(1, 0, 0, 1, 0) = 1 + 0 + 0 + 1 + 0 = 2 \qquad \square$$

Closely associated with the notion of a random variable is the idea of a probability density function.

Definition 9.16   Let (S, P) be a probability space and let $X: S \to \mathbf{R}^1$ be a random variable. For each $x \in \mathbf{R}^1$, let $A_x = \{y \mid y \in S \text{ and } X(y) = x\}$. The *probability density function* (*pdf*) associated with X is the function $f: \mathbf{R}^1 \to [0, 1]$ defined by

$$f(x) = P(A_x)$$

The value of f at x, $P(A_x)$, is the probability that the random variable will take on the value x. For fairly obvious reasons $P(A_x)$ is frequently denoted by $P(X = x)$ and read as "the probability that X is x."

Example 3   We find the probability density function f associated with the random variable X given in Example 1. This random variable is defined by

$$\begin{array}{llll}
X(A, A) = & 4 & X(A, B) = -1 & X(B, A) = -1 \\
X(A, C) = & 3 & X(C, A) = 3 & X(B, B) = -6 \\
X(B, C) = & -2 & X(C, B) = -2 & X(C, C) = 2
\end{array} \qquad (2)$$

The range of X is the set $\{-6, -2, -1, 2, 3, 4\}$. By the definition of a pdf, we have that $f(x) = 0$ if x does not belong to this set. To calculate $f(-6)$, $f(-2)$, $f(-1)$, $f(2)$, $f(3)$, and $f(4)$, we can proceed as follows. By Definition 9.16, $f(-6) = P(A_{-6})$ where $A_{-6} = \{y \mid y \in S \text{ and } X(y) = -6\}$. From the equations in (2) we see that $y = (B, B)$ is the only outcome for which $X(y) = -6$. Therefore, $A_{-6} = \{(B, B)\}$, and since the probability that the outcome (B, B) occurs is $(1/3)(1/3) = 1/9$, we obtain $f(-6) = P(A_{-6}) = 1/9$.

To find $f(-1)$ we note that there are two ways that the random variable X can take on the value -1: $X(A, B) = -1$ and $X(B, A) = -1$. Thus, $A_{-1} = \{(A, B), (B, A)\}$. Since the outcomes (A, B) and (B, A) each occur with a probability $1/6$, it follows that $f(-1) = P(A_{-1}) = 1/6 + 1/6 = 1/3$. You may use similar arguments to obtain the remaining values of $f(x)$ found below (see Problem 6).

$$f(x) = \begin{cases} 1/9 & \text{if } x = -6 \\ 1/9 & \text{if } x = -2 \\ 1/3 & \text{if } x = -1 \\ 1/36 & \text{if } x = 2 \\ 1/6 & \text{if } x = 3 \\ 1/4 & \text{if } x = 4 \\ 0 & \text{otherwise} \end{cases} \qquad (3) \qquad \square$$

Next we consider uniform distributions. Let (S, P) be a finite probability space and suppose $X : S \to \mathbf{R}^1$ is a random variable with range $\{x_1, x_2, \ldots, x_n\}$. If $P(X = x_i) = 1/n$ for each i, then the random variable X is said to have a *uniform distribution* with pdf, f, defined by

$$f(x) = \begin{cases} 1/n & \text{if } x = x_1, x_2, \ldots, x_n \\ 0 & \text{otherwise} \end{cases}$$

Example 4 Suppose that a random experiment consists of rolling a die once. The sample space for this random experiment is $S = 1, 2, 3, 4, 5, 6$. Let $X : S \to \mathbf{R}^1$ be random variable on S defined by $X(i) = 3i$. Then X has a uniform distribution with pdf

$$f(x) = \begin{cases} 1/6 & \text{if } x = 3, 6, 9, 12, 15, 18 \\ 0 & \text{otherwise} \end{cases} \qquad \square$$

To conclude this section we find the pdf associated with the random variable defined by (1) for Bernoulli processes. Suppose the probability of success on each Bernoulli trial is p, and the probability of failure is $q = 1 - p$. The random variable X defined by (1) gives the number of successes in a sequence of n Bernoulli trials, and hence, can take on values $0, 1, \ldots, n$. Since, as we have seen, the probability of exactly r successes in a sequence of n Bernoulli trials is $\binom{n}{r} p^r q^{n-r}$, it follows that the pdf associated with X is defined by

$$f(x) = \begin{cases} \binom{n}{x} p^x q^{n-x} & \text{if } x = 0, 1, \ldots, n \\ 0 & \text{otherwise} \end{cases} \qquad (4)$$

A random variable X with this pdf is said to have a *binomial distribution*.

Example 5 Suppose it is known that a certain make of airplane engine will fail with probability $1 - p$. Assume that a plane's engines function independently and that the plane can still fly if at least one half of its engines are operating. We use the pdf in (4) to find the probability that a four-engine plane will successfully complete a flight.

We can regard the performance of the plane's engines as a Bernoulli process with $n = 4$, and probability p. Thus, from (4) the pdf associated with the random variable giving the number of successes (the number of engines that function properly) is

$$f(x) = \begin{cases} \binom{4}{x} p^x (1 - p)^{4-x} & \text{if } x = 0, 1, 2, 3, 4 \\ 0 & \text{otherwise} \end{cases}$$

Since the plane can fly if either 2, 3, or 4 engines are working properly, we find that the probability of a safe flight is

$$f(2) + f(3) + f(4) = \binom{4}{2} p^2 (1 - p)^2 + \binom{4}{3} p^3 (1 - p) + \binom{4}{4} p^4 (1 - p)^0$$

$\square$

In Problem 14 you are asked to make an analysis similar to that of Example 5 for a two-engine plane, and to determine the values of p for which flights are safer in a four-engine plane than in a two-engine one.

9.7   Problems

1. A fair coin is tossed four times. For each head that appears you receive \$3 and for each tail you lose \$2.
   (a) Define and find the values of the corresponding random variable.
   (b) Find the associated pdf.

2. Repeat Problem 1 using a weighted coin for which the probability of getting heads is .6.

3. A fair coin is tossed three times. A random variable X is defined as the number of the toss on which heads first occurs. If no head appears, set $X = 0$.
   (a) Find the values for this random variable.
   (b) Find the pdf associated with this random variable.

4. A bowl contains four red chips and eight blue chips. Three chips are drawn from the bowl, without replacement, and a random variable is defined as the number of red chips that are drawn.
   (a) Find the values of this random variable.
   (b) What is the pdf associated with this random variable?

5. If you throw two dice and define a random variable as the sum of the dice, what is the associated pdf?

6. Verify the values of $f(x)$ given in (3) of this section.

7. Give a reason why the name "binomial distribution" is appropriate to label a random variable with the pdf in equation (4).

8. Show that if f is a pdf, then $\Sigma f(x) = 1$, where the sum is taken over all x such that $f(x) \neq 0$.

9. *Geometric distribution.* Suppose an experiment is performed that satisfies all of the conditions of a Bernoulli process except that *any* number of trials is allowed. Each trial has two possible outcomes, success or failure, with probabilities p and $1 - p$, respectively. A random variable is defined as the number of trials necessary to obtain the first success.
   (a) What values can this random variable take on?
   (b) Show that the pdf of this random variable is defined by

$$f(x) = \begin{cases} p(1 - p)^{x-1} & \text{if } x = 1, 2, 3, \ldots \\ 0 & \text{otherwise} \end{cases}$$

10. Let X be a random variable with a geometric distribution with $p = .7$. Find the following probabilities:
    (a) $P(X = 4)$
    (b) $P(X > 3)$
    (c) $P(2 \leq X < 4)$

11. Let X be a random variable with a geometric distribution with $p = .8$. Find the following probabilities.
    (a) $P(X = 3)$
    (b) $P(X \leq 5)$
    (c) $P(X > 4)$
    (d) $P(1 \leq X < 4)$

12. In a game of pool a player continues to shoot until she misses. Suppose the probability that the player misses any given shot is .2.
    (a) Find the probability that she will continue playing for four shots before she misses.
    (b) Find the probability that she will continue playing for at least three shots.
    (c) Find the probability that she will continue to shoot for at most five shots.

13. Five students are randomly asked a total of twenty questions. Let X be the random variable defined as the number of questions directed to any one student.
    (a) What values does X take on?
    (b) What is the pdf of X?

14. Use an argument similar to the one used in Example 5 to find the probability that a two-engine plane will make a safe flight. Assume that two-engine planes and four-engine planes use the same kind of engines and find the values of p for which it is preferable to use a four-engine plane rather than a two-engine plane.

9.8 EXPECTED VALUES OF RANDOM VARIABLES

Consider the following game. Two dice are rolled and the sum of the dice is noted. Depending on the sum obtained, you win (or lose) according to the payoffs in Table 9.2 (a negative payoff indicates a loss). Is it to your advantage to play this game?

Sum	Payoff	Probability that the sum occurs
2	-2	$1/36$
3	$+3$	$2/36$
4	$+2$	$3/36$
5	-2	$4/36$
6	-4	$5/36$
7	$+1$	$6/36$
8	$+5$	$5/36$
9	-4	$4/36$
10	0	$3/36$
11	0	$2/36$
12	$+1$	$1/36$

Table 9.2

To determine the amount you would expect to win or lose we proceed as follows. First note from Table 9.2 that the probability of losing \$2 is $(1/36 + 4/36) = 5/36$; this would account for an average loss of $2(5/36)$ dollars each time you play. In a like fashion we see that the probability of winning \$3 is $2/36$, thus, this would account for an average gain of $3(2/36)$ dollars each time you play. Similar reasoning determines the average losses or gains associated with the payoffs 2, -4, 1, 5, and 0 dollars (which are $2(3/36)$, $(-4)(9/36)$, $1(7/36)$, $5(5/36)$ and $0(5/36)$ dollars, respectively). Adding these results, we find that on the average you could expect to finish with

$$-2(5/36) + 3(2/36) + 2(3/36) - 4(9/36)$$
$$+ 1(7/36) + 5(5/36) + 0(5/36) = -2/36 \quad (1)$$

dollars for each game played; consequently, you're better off looking for another game to play.

Let us slightly abstract the situation just discussed. Let S be the sample space consisting of the pairs $\{(i,j) | i \in \{1, 2, 3, 4, 5, 6\}, j \in \{1, 2, 3, 4, 5, 6\}\}$, where i corresponds to the number rolled on the first die, and j corresponds to the number rolled on the second die. Define a random variable X on this sample space by setting $X(i, j)$ equal to the payoff associated with $i + j$ found in Table 9.2. Then the range of X is the set $\{-4, -2, 0, 1, 2, 3, 5\}$. It is easily verified that the pdf, f, associated with X is given by

$$f(x) = \begin{cases} 9/36 & \text{if } x = -4 \\ 5/36 & \text{if } x = -2 \\ 5/36 & \text{if } x = 0 \\ 7/36 & \text{if } x = 1 \\ 3/36 & \text{if } x = 2 \\ 2/36 & \text{if } x = 3 \\ 5/36 & \text{if } x = 5 \\ 0 & \text{otherwise} \end{cases}$$

Note now that the "expected" result, (1), of playing this game once can be expressed as

$$(-4)f(-4) + (-2)f(-2) + 0f(0) + 1f(1) + 2f(2) + 3f(3) + 5f(5)$$

These considerations lead us to the following definition.

Definition 9.17 Let (S, P) be a probability space, let X be a random variable on S with range $\{x_1, x_2, \ldots, x_n\}$, and let f be the pdf associated with X. Then the *expected value* of X, $E(X)$, is defined by

$$\begin{aligned} E(X) &= x_1 f(x_1) + x_2 f(x_2) + \cdots + x_n f(x_n) \\ &= x_1 P(X = x_1) + x_2 P(X = x_2) + \cdots + x_n P(X = x_n) \end{aligned}$$

Example 1 Suppose that X is a random variable with range $\{x_1, x_2, \ldots, x_n\}$ and with a uniform pdf, f. Then $f(x_i) = 1/n$ for each x_i in the range of X and is 0 otherwise. The expected value of X is

$$E(X) = x_1\left(\frac{1}{n}\right) + x_2\left(\frac{1}{n}\right) + \cdots + x_n\left(\frac{1}{n}\right) = \frac{x_1 + x_2 + \cdots + x_n}{n} \qquad \square$$

In Example 1, we see that the expected value of a random variable with uniform distribution is simply the average or mean of the numbers $x_1, x_2, \ldots, x_n$. For this reason, the expected value of a random variable X is occasionally referred to as the "mean" of X. In general, the expected value of a random variable can be interpreted as a "weighted" average of the values found in the range of X. The weights are determined by the "concentration" or "density" of probability at each of these values.

Example 2 Suppose you draw three coins from a sack containing 5 dimes, 3 nickels, and 2 quarters. If all of the coins are drawn with equal likelihood, how much money do you "expect" to take out? Define a sample space S by

$$S = \{\{N, N, N\}, \{N, N, D\}, \{N, D, D\}, \{D, D, D\}, \{N, N, Q\}, \{N, D, Q\},$$
$$\{Q, D, D\}, \{Q, Q, N\}, \{Q, Q, D\}\}$$

Let X be the random variable on this space defined by the total value of the coins drawn, for example, $X(N, N, D) = 20$. You may verify (see Problem 3) that the pdf associated with X is defined by

$$f(x) = \begin{cases} 1/120 & \text{if } x = 15 \\ 15/120 & \text{if } x = 20 \\ 30/120 & \text{if } x = 25 \\ 10/120 & \text{if } x = 30 \\ 6/120 & \text{if } x = 35 \\ 30/120 & \text{if } x = 40 \\ 20/120 & \text{if } x = 45 \\ 3/120 & \text{if } x = 55 \\ 5/120 & \text{if } x = 60 \\ 0 & \text{otherwise} \end{cases}$$

Therefore, by Definition 9.17 the expected value of each drawing is

$$E(X) = 15\left(\frac{1}{120}\right) + 20\left(\frac{15}{120}\right) + 25\left(\frac{30}{120}\right) + 30\left(\frac{10}{120}\right) + 35\left(\frac{6}{120}\right)$$
$$+ 40\left(\frac{30}{120}\right) + 45\left(\frac{20}{120}\right) + 55\left(\frac{3}{120}\right) + 60\left(\frac{5}{120}\right)$$
$$= 34.5¢ \qquad \qquad \square$$

Next we will compute the expected value of the random variable X that gives the number of successes in a Bernoulli process consisting of n Bernoulli trials. If you were to flip a coin ten times, you would probably expect to obtain five heads; that is, $E(X) = 5$ if flipping a coin ten times is viewed as a Bernoulli process with $n = 10$ and $p = 1/2$. Note in this case that $E(X) = np$. We now show that $E(X) = np$ for any Bernoulli process consisting of n trials and with probability p of success on each trial.

Theorem 9.18 Suppose a Bernoulli process consists of n Bernoulli trials, with probability p of success on each trial. If X is the random variable giving the number of successes of the Bernoulli process, then $E(X) = np$.

Proof: First recall that if the probability of success in each trial of a Bernoulli process is p, then the pdf of the associated random variable X is

$$f(x) = \begin{cases} \binom{n}{x} p^x (1-p)^{n-x} & \text{if } x = 0, 1, \ldots, n \\ 0 & \text{otherwise} \end{cases}$$

Thus, by the definition of the expected value of a random variable, we have

$$E(X) = \sum_{i=0}^{n} i\binom{n}{i} p^i (1 - p)^{n-i}$$

To determine this value, first observe that

$$E(X) = \sum_{i=0}^{n} i\binom{n}{i} p^i (1 - p)^{n-i} = \sum_{i=1}^{n} i\binom{n}{i} p^i (1 - p)^{n-i}$$

$$= \sum_{i=1}^{n} \frac{in(n-1)!}{i!(n-i)!} p p^{i-1} (1 - p)^{n-i} = np \sum_{i=1}^{n} \binom{n-1}{i-1} p^{i-1} (1 - p)^{n-i}$$

To evaluate

$$np \sum_{i=1}^{n} \binom{n-1}{i-1} p^{i-1} (1 - p)^{n-i}$$

it is convenient to make the substitutions

$$k = i - 1 \quad \text{and} \quad m = n - 1$$

Note that if $i = 1$, then $k = 0$, and if $i = n$, then $k = n - 1 = m$. With this in mind, we see that

$$np \sum_{i=1}^{n} \binom{n-1}{i-1} p^{i-1} (1 - p)^{n-i} = np \sum_{k=0}^{m} \binom{m}{k} p^k (1 - p)^{m-k}$$

But by the Binomial Theorem 8.11, $\sum_{k=0}^{m} \binom{m}{k} p^k (1 - p)^{m-k} = (p + 1 - p)^m = 1^m = 1$, and it follows that $E(X) = np$, as we wished to show.   ∎

Example 3 Suppose you roll a pair of dice five times. How many times would you expect to be successful in obtaining either a seven or an eleven? Since the probability of obtaining a seven or an eleven on any one roll of the dice is $8/36$, it follows from Theorem 9.18 that the expected number of successes is $5(8/36) = 40/36$.   □

Next, we see how expected values can be used in the analysis of an algorithm's complexity. Thus far, we have measured the complexity of an algorithm in terms of a worst-case analysis. Such an analysis can be very useful in establishing an upper bound on the running time of an algorithm. However, if the worst case occurs infrequently, then an *average*-case analysis may provide a more realistic gauge of an algorithm's efficiency.

The average-case complexity of an algorithm is the expected number of executions of some specified step of the algorithm. To illustrate this concept we compute the average-case complexity of SEQUENTIAL SEARCH, which was discussed in Chapter 1 (Algorithm 1.1). Recall that the worst-case complexity of this algorithm was equal to the total number of entries in a stored list.

To make an average-case analysis of SEQUENTIAL SEARCH, we let S denote the sample space consisting of all possible outputs of this algorithm when it is applied to a stored list of n entries. Thus, $S = \{0, 1, \ldots, n\}$, where 0 denotes NOT FOUND. Let p denote the probability that an input Y is in the list and assume that if Y is in the list, then it is equally likely to be in any one of the positions $1, 2, \ldots, n$. Therefore, we have

$$P(i) = \frac{p}{n} \qquad i = 1, 2, \ldots, n \qquad (2)$$

where $P(i)$ is the probability that Y is in the ith position in the list.

Note that the output 0 will occur only if the input is not found in the list. Thus,

$$P(0) = 1 - p \qquad (3)$$

To compute the expected number of comparisons SEQUENTIAL SEARCH makes, we define a random variable X on S, where $X(i)$ is the necessary number of comparisons this algorithm makes to produce the output i. Observe that only one comparison is needed to produce the output 1, while two comparisons are necessary to obtain the output 2. In general, $X(i) = i$ for $i = 1, 2, \ldots, n$. However, $X(0) = n$ because all n items in the list must be compared with the given input to produce an output 0.

By Definition 9.17 the expected value of X, the number of comparisons, is

$$E(X) = \sum_{i=0}^{n} iP(X = i) = \sum_{i=1}^{n-1} iP(X = i) + nP(X = n) \qquad (4)$$

From (2) it follows that

$$P(X = i) = \frac{p}{n} \qquad i = 1, 2, \ldots, n \qquad (5)$$

Note that

$$P(X = n) = P(\{0, n\}) = P(0) + P(n) = 1 - p + \frac{p}{n} \qquad (6)$$

Substituting (5) and (6) into (4), we obtain

$$E(X) = \sum_{i=1}^{n-1} i\left(\frac{p}{n}\right) + n\left(1 - p + \frac{p}{n}\right) = \left(\frac{p}{n}\right)\sum_{i=1}^{n-1} i + n\left(1 - p + \frac{p}{n}\right)$$

From Example 1 of Section 1.3 we have $\sum_{i=1}^{n-1} i = n(n-1)/2$, and therefore,

$$E(X) = \left(\frac{p}{n}\right)\frac{n(n-1)}{2} + n\left(1 - p + \frac{p}{n}\right) = n - (n-1)\left(\frac{p}{2}\right)$$

Thus, for example, if an input has a probability of $3/4$ of being found in the given list, then

$$E(X) = \left(\frac{5}{8}\right)n + \frac{3}{8}$$

is the expected number of comparisons needed to locate an entry in a list of n items.

9.8 Problems

1. Suppose that the range of a random variable X is

$$x_1 = 2, \quad x_2 = -3, \quad x_3 = \frac{1}{2}, \quad x_4 = 3$$

and suppose that the pdf f associated with X is defined by

$$f(x_1) = \frac{1}{6}, \quad f(x_2) = \frac{1}{12}, \quad f(x_3) = \frac{1}{2}, \quad f(x_4) = \frac{1}{4}$$

Find $E(X)$.

2. Suppose that if you fish for an hour on a certain river the probability of catching exactly one fish is .35, the probability of catching exactly two fish is .20, the probability of catching exactly three fish is .10, the probability of catching exactly four fish is .07, the probability of catching five fish or more is .03, and the probability of catching no fish is .25. How many fish can you expect to catch in an hour?

3. Verify the values of the pdf, f, given in Example 2.

4. A bowl contains two white chips and one red chip. Chips are successively drawn from the bowl (without replacement) until a red chip is obtained. What is the expected number of draws required?

5. A bowl contains four white chips and three red chips. Chips are successively drawn from the bowl (without replacement) until a red chip is obtained. What is the expected number of draws required?

6. In Problem 4, what is the expected number of draws to first draw a white chip?

7. What is the expected value of the random variable defined in Problem 1 of Section 9.7?

8. What is the expected value of the random variable defined in Problem 2 of Section 9.7?

9. Suppose you draw three chips from a bowl containing three red chips, two blue chips, and two white chips. Each red chip has a value of 5¢, each blue chip is valued at 10¢, and each white chip is worth 25¢. If the sum of the

values of the chips is a multiple of 10, you win a dollar; otherwise, you lose the sum of the values of the three chips you drew. Would you expect to win money playing this game?

10. A fair coin is tossed until either tails occurs or two consecutive heads occur. What is the expected number of tosses?

*11. The following is a classic probability problem. You throw a fair coin until heads occurs. If heads occurs on the first toss you receive x dollars. If the first head occurs on the second toss you receive x^2 dollars, and in general if the first head occurs on the nth toss you receive x^n dollars.
   (a) What is the expected value of this game (note that you will have to find the sum of an infinite series—see Chapter 10).
   (b) If $x = \$1.00$, what is the expected value of this game?
   (c) What happens if $x = \$2.00$?

12. In certain regions of the country smudge pots are used to protect farm plants from the cold. Suppose that if the temperature drops to 25°F, twenty percent of a crop will be lost if smudge pots are not used. The probability that such a temperature drop will occur is .3. The cost of setting out the smudge pots reduces profits by 5¢ on each dollar. Under these conditions, would you set out the smudge pots?

13. If seven fair coins are tossed, what is the expected number of heads that will result?

14. If a pair of dice is rolled twenty times, what is the expected number of times that the sum of the dice will be even?

15. SEQUENTIAL SEARCH is to be applied in a situation where an input has a probability of .98 of being in the list. What is the expected number of comparisons needed to execute this algorithm?

*16. SEQUENTIAL SEARCH is to be applied in a situation where an input Y has a probability of p of being in the list. The list is arranged so that the probability that Y is in the ith position in the list is p_i, where $p_1 + p_2 + \cdots + p_n = p$.
   (a) Write an expression for $E(X)$, where $X(i)$ is the number of comparisons needed to produce the output i.
   (b) Show that $E(X)$ is minimal if $p_1 \geq p_2 \geq \cdots \geq p_n$.

*9.9 POISSON DISTRIBUTIONS[1]

Suppose we would like to determine the probability that a certain number of calls will arrive at a switchboard during some unit interval of time (an hour, a minute, or whatever). We make the following assumptions about these calls:

[1] In this section some use is made of limits and infinite series (see Chapter 10).

a. The number of incoming calls is independent of the number of calls that have already been made.

b. There is a number λ (called a *Poisson parameter*) such that over any small interval of time of duration h, the probability of an incoming call during this interval is λh. The value λ may be interpreted as the "average" number of incoming calls during a unit interval of time.

c. The probability that two calls will arrive during a small interval of time is essentially equal to 0 (and for our purposes is assumed to be 0).

Let X denote the number of calls that arrive at the switchboard during a given unit interval of time I. We wish to determine $P(X = k)$, the probability that there are exactly k calls during the interval of time I.

To this end, we first divide I into n equal subintervals, $I_1, I_2, \ldots, I_n$, where we assume that the subintervals are sufficiently small so that assumptions **b.** and **c.** apply to them. Note that since the unit interval I is assumed to have length 1, it follows that each subinterval I_i has length $1/n$, and from **b.** the probability of a call occurring in any one of these subintervals is λ/n.

By **a.** we can view this problem as a Bernoulli process consisting of n Bernoulli trials, where the ith trial corresponds to the ith subinterval, I_i. A success is recorded if a call occurs during this interval of time; otherwise, we have a failure. From **b.** the probability of a success is λ/n, and the probability of a failure is $1 - \lambda/n$. Observe that $P(X = k)$ is the probability of having k successes, and from (4) of Section 9.7 we find that

$$P(X = k) = \binom{n}{k}\left(\frac{\lambda}{n}\right)^k\left(1 - \frac{\lambda}{n}\right)^{n-k}$$

Assumptions **b.** and **c.** above are reasonable for small units of time, or, equivalently, for large values of n. This leads us to consider what happens to $P(X = k)$ as n increases without bound. Standard algebraic manipulations yield

$$\lim_{n \to \infty} P(X = k) = \lim_{n \to \infty} \frac{n!}{k!(n-k)!}\left(\frac{\lambda}{n}\right)^k\left(1 - \frac{\lambda}{n}\right)^{n-k}$$

$$= \lim_{n \to \infty} \frac{n(n-1)(n-2)\cdots(n-k+1)}{k!n^k}$$

$$\times \lambda^k\left(1 - \frac{\lambda}{n}\right)^n\left(1 - \frac{\lambda}{n}\right)^{-k}$$

$$= \lim_{n \to \infty} \frac{1(1 - 1/n)(1 - 2/n)\cdots(1 - (k-1)/n)}{k!}$$

$$\times \lambda^k\left(1 - \frac{\lambda}{n}\right)^n\left(1 - \frac{\lambda}{n}\right)^{-k} \tag{1}$$

Now we use two obvious limits and one that is not so obvious. For fixed k,

(a) $\lim\limits_{n \to \infty} (1(1 - 1/n)(1 - 2/n) \cdots (1 - (k - 1)/n)) = 1$

(b) $\lim\limits_{n \to \infty} (1 - \lambda/n)^{-k} = 1$

(c) $\lim\limits_{n \to \infty} (1 - \lambda/n)^n = e^{-\lambda}$

where $e = 2.714 \ldots$ (this is an irrational number that frequently occurs in calculus). If we substitute these limits into (1) we find that

$$\lim_{n \to \infty} P(X = k) = \frac{\lambda^k e^{-\lambda}}{k!} \qquad \text{for } k = 0, 1, 2, \cdots \tag{2}$$

Definition 9.19   A random variable X has a *Poisson* pdf if there is a $\lambda > 0$ such that

$$f(x) = \begin{cases} \dfrac{\lambda^x e^{-\lambda}}{x!} & \text{if } x = 0, 1, 2, \cdots \\ 0 & \text{otherwise} \end{cases}$$

The Poisson distribution is named for the French mathematician and physicist Siméon Poisson (1781–1840), who studied, among other topics, how this distribution applies to lawsuits, trials, and so on. Poisson distributions occur naturally in such diverse situations as

1. the number of patrons entering a bank.

2. the number of atoms disintegrating each second in a quantity of radioactive material.

3. the number of chocolate chips found in a batch of cookies.

4. the number of misprints found on a page.

5. the number of defects occurring per unit length of wire.

Example 1   A switchboard operator estimates that 7 calls enter the switchboard every 10 minutes. If a Poisson distribution is assumed, what is the probability that from 8 to 10 calls will be received during a 20 minute period? The unit interval of time can be taken to be 20 minutes and, for this interval of time, $\lambda = 14$. We need to find $P(X = 8) + P(X = 9) + P(X = 10)$. By (2) we have

$$P(X = 8) + P(X = 9) + P(X = 10) = \frac{14^8 \cdot e^{-14}}{8!} + \frac{14^9 \cdot e^{-14}}{9!}$$

$$+ \frac{14^{10} \cdot e^{-14}}{10!} \approx .15$$

which means there is a .15 probability that there will be 8 to 10 incoming calls during any 20 minute interval.   □

Next we see how a Poisson distribution can be used to approximate a binomial distribution.

Observe that since

$$\lim_{n \to \infty} \frac{n!}{k!(n-k)!} \left(\frac{\lambda}{n}\right)^k \left(1 - \frac{\lambda}{n}\right)^{n-k} = \frac{\lambda^k e^{-\lambda}}{k!}$$

it follows that for large values of n

$$\binom{n}{k}\left(\frac{\lambda}{n}\right)^k \left(1 - \frac{\lambda}{n}\right)^{n-k} \approx \frac{\lambda^k e^{-\lambda}}{k!}$$

Thus a Poisson distribution may be used to approximate a binomial distribution for large values of n. The advantage of this approach lies in the fact that, for large n, it is computationally easier to work with

$$\frac{\lambda^k e^{-\lambda}}{k!} \tag{3}$$

than with

$$\binom{n}{k}\left(\frac{\lambda}{n}\right)^k \left(1 - \frac{\lambda}{n}\right)^{n-k} \tag{4}$$

We illustrate this idea in the next example.

Example 2   Suppose that the probability of a transmission error over a binary symmetric channel is .03. We find the probability that between 0 and 5 errors occur during the transmission of 1000 symbols. This problem can be viewed as a Bernoulli process consisting of 1000 Bernoulli trials. Consequently the probability of between 0 and 5 errors is

$$\sum_{i=0}^{5} \binom{1000}{i} (.03)^i (.97)^{1000-i}$$

However, it is computationally easier to use a Poisson distribution with a parameter $\lambda = (1000)(.03) = 30$ to obtain

$$\sum_{i=0}^{5} \frac{30^i e^{-30}}{i!} = e^{-30}\left(1 + 30 + \frac{30^2}{2!} + \frac{30^3}{3!} + \frac{30^4}{4!} + \frac{30^5}{5!}\right) \approx 2.3 \times 10^{-8}$$

as an approximate solution.   □

To conclude this section we find the expected value of a random variable X that has a Poisson distribution. By the definition of a Poisson parameter, we might anticipate that if the Poisson distribution has parameter λ, then

$E(X) = \lambda$. To see that this is the case, first recall that if X has a Poisson distribution with Poisson parameter λ, then the pdf of X is defined by

$$f(x) = \begin{cases} \dfrac{e^{-\lambda}\lambda^x}{x!} & \text{if } x = 0, 1, 2, \cdots \\ \\ 0 & \text{otherwise} \end{cases}$$

Therefore,

$$E(X) = \sum_{x=0}^{\infty} xf(x) = \sum_{x=0}^{\infty} \frac{xe^{-\lambda}\lambda^x}{x!}$$

$$= \lambda e^{-\lambda} \sum_{x=1}^{\infty} \frac{x\lambda^{x-1}}{x(x-1)!} = \lambda e^{-\lambda} \sum_{x=1}^{\infty} \frac{\lambda^{x-1}}{(x-1)!}$$

If we let $k = x - 1$, then we obtain

$$E(X) = \lambda e^{-\lambda} \sum_{x=1}^{\infty} \frac{\lambda^{x-1}}{(x-1)!} = \lambda e^{-\lambda} \sum_{k=0}^{\infty} \frac{\lambda^k}{k!}$$

From calculus we obtain

$$\sum_{k=0}^{\infty} \frac{\lambda^k}{k!} = e^{\lambda}$$

Thus,

$$E(X) = \lambda e^{-\lambda} e^{\lambda} = \lambda$$

which is the expected value of a random variable X that has a Poisson distribution.

As we have seen, Poisson distributions are useful in a number of diverse situations. They are also important in approximating binomial distributions. In the next section we will see how Poisson distributions can be applied to queueing theory, which deals with the problems of waiting for a given service.

9.9. **Problems**

1. Explain why assumption **c.** on page 388 is reasonable for small intervals of time.

2. Suppose that a random variable X has a Poisson distribution with Poisson parameter $\lambda = 3$. Find
   (a) $P(X = 3)$ (b) $P(2 \leq X < 4)$ (c) $P(X \geq 2)$

3. Suppose that a random variable X has a Poisson distribution with Poisson parameter $\lambda = 2$. Find
   (a) $P(X = 0)$ (b) $P(X \leq 3)$ (c) $P(X > 1)$

4. Suppose that the number of earthquakes occurring in a certain country has a Poisson distribution and that the average number of earthquakes per year is four.
   (a) Find the probability that there will be four earthquakes in any given year.
   (b) Find the probability that there will be either three or four earthquakes in any given year.
   (c) Find the probability that there will be more than two earthquakes in any given year.

5. Suppose that in a certain stream a fisherman can expect to catch 2.2 fish an hour. If the number of fish caught has a Poisson distribution, find
   (a) the probability of catching three fish in an hour.
   (b) the probability of not catching any fish in an hour.
   (c) the probability of catching five fish in two hours.

6. Suppose that the number of defects in a roll of wire occur on the average at the rate of one defect per 600 feet of wire. Assume that the number of defects has a Poisson distribution. Find
   (a) the probability that there are exactly three defects in 600 feet of wire.
   (b) the probability that there are more than two defects in 600 feet of wire.
   (c) the probability that there are exactly three defects in 1800 feet of wire.
   (d) the probability that there are fewer than four defects in 1200 feet of wire.

7. Suppose that α-particles are emitted from a radioactive substance at an average rate of two per second. Assume that the number of α-particles emitted has a Poisson distribution, and find
   (a) the probability that no α-particles are emitted in a given second.
   (b) the probability that three, four, or five α-particles are emitted in one second.
   (c) the probability that no α-particles are emitted during a half second time interval.
   (d) the probability that two α-particles are emitted during a two second time interval.

8. Compare the binomial and the Poisson distributions in each of the following cases.
   (a) $P(X = 3)$ if $n = 7, p = .2$
   (b) $P(X = 0)$ if $n = 5, p = .3$
   (c) $P(2 \leq X \leq 4)$ if $n = 20, p = .05$

9. Suppose that a certain manufacturing process produces 4% defective items. Use the Poisson distribution to approximate the probability that in a batch of 100 items, there are fewer than three defective items.

10. Suppose that the probability of a transmission error over a binary symmetric channel is .02. Use the Poisson distribution to approximate the probability that more than three errors occur during the transmission of 500 symbols.

*9.10 INTRODUCTION TO QUEUEING THEORY[1]

In this section we deal with problems of the following sort. Suppose you wish to install a "Quick-Photo" machine in a local supermarket. Your research has shown that on the average 12 persons per hour will wish to use the machine. If the time required to operate the machine is 4 minutes, what is the average amount of time that a person will have to wait in line to be photographed?

Problems of this kind belong to the domain of queueing theory—literally, the theory of waiting in lines (queues). Although these problems have many variations, they can usually be characterized in terms of three factors: **a.** the arrival process, **b.** the service process, and **c.** queue discipline. We expand briefly on these notions.

    a. *The arrival process.* Arrivals may occur in a regular and orderly fashion (for example, every 30 seconds); it is more likely, however, that the arrivals will be random, though it may be possible to establish the average number of arrivals per unit time. It is also important to consider whether or not there is a maximal queue length beyond which new arrivals will not join the queue. (For example, if the parking lot of a restaurant is full, potential customers [arrivals] may go elsewhere.)

    b. *The service process.* This factor depends in part on the average length of time it takes to service an arrival and on how many service stations are available.

    c. *Queue discipline.* Despite the terminology, this does not deal with the behavior of the persons in the line, but rather it involves how the arrivals are processed: on first come, first served basis, on a last come, first served basis, by random ordering of service, and so on.

We shall restrict ourselves to fairly simple situations involving queues. In particular, we shall consider only those cases where arrivals occur in a random fashion and where the average number of arrivals is known. We also assume that individual service time may vary, but that an average service rate can be ascertained. More precisely, we assume that the number of arrivals has a Poisson distribution with Poisson parameter λ, and that the number of people serviced also has a Poisson distribution with Poisson parameter μ. Thus, we can interpret λ as the average number of arrivals per unit interval of time (the average arrival rate), and we can let μ denote the average service rate.

In terms of queue discipline, we treat only the first come, first served case, and, finally, we suppose that the entire system has achieved an equilibrium which is independent of time. That is, the distribution of arrivals and the distribution of service times are the same at, say, 2 P.M. as they are at 3 P.M. The entire process is summarized by Figure 9.6.

---

[1] This section makes use of infinite series (see Chapter 10).

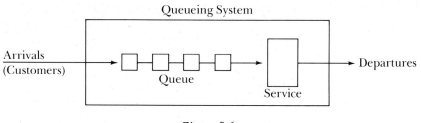

Figure 9.6

We say that a "customer" is in the system if the customer is either in the queue or is being serviced. (A "customer" could be a person in a barbershop, a car waiting to pass through a toll station, an airplane waiting to be serviced, a student wishing to use a computer terminal, a telephone call entering a switchboard, and so on.) The focus in the remainder of this section will be on finding answers to the following questions:

(1) What is P_n, the probability that there are n customers in the system?

(2) What is L_S, the average number of customers in the system?

(3) What is L_Q, the average number of customers in the queue?

(4) What is W_S, the average time that a customer spends in the system?

(5) What is W_Q, the average time that a customer must wait in the queue?

The answer to question (1) is the most difficult to obtain. We attack this problem by first comparing the probability that there are n people in the system at time t with the probability that there are n people in the system at a slightly later time, $t + \Delta t$, where Δt is a small number. Since we are dealing with Poisson processes, the following conditions are satisfied.

a. If λ is the Poisson parameter used to describe the arrival process (λ is the average number of arrivals per unit interval of time), then for suitably small values of Δt, $\lambda \Delta t$ is the probability that there is one arrival in the time interval extending from time t to time $t + \Delta t$. (In the example that began this section, an average of 12 persons per hour wanted to use the machine, and so $\lambda = 12$.)

b. If μ is the Poisson parameter used to denote the average number of customers served per unit time, then for suitably small values of Δt, $\mu \Delta t$ is the probability that one customer is serviced in the time interval extending from time t to time $t + \Delta t$. (In the example of the Quick-Photo machine, four minutes were required to operate the machine. Thus, $60/4 = 15$ customers can be served each hour, so $\mu = 15$.)

c. We can assume that the probability of two or more arrivals in the time interval $[t, t + \Delta t]$ is so small that it is negligible. Similarly, we can suppose that the probability that two or more customers are serviced in the interval $[t, t + \Delta t]$ can be neglected.

For each nonnegative integer n, let $p_n(t)$ be the probability that there are n customers in the system at time t. To calculate $p_n(t)$ we make use of $p_n(t + \Delta t)$, the probability that there are n persons in the system at time $t + \Delta t$. We first observe that there are four ways that n customers can be in the system at time $t + \Delta t$.

A. There are n customers in the system at time t and no one enters or leaves the system in the time interval $[t, t + \Delta t]$. The probability of this occurring is

$$p_n(t)(1 - \lambda \Delta t)(1 - \mu \Delta t)$$

where $1 - \lambda \Delta t$ is the probability that no one enters the system and $1 - \mu \Delta t$ is the probability that no one leaves the system (is serviced) during this interval.

B. There are $n - 1$ customers in the system at time t, but an additional customer arrives during the time interval $[t, t + \Delta t]$ and no one leaves (is serviced) during this interval. The probability of this occurring is

$$p_{n-1}(t)(\lambda \Delta t)(1 - \mu \Delta t)$$

Here, of course, we have assumed that $n \geq 1$.

C. There are $n + 1$ customers in the system at time t, and one customer leaves the system, but no one arrives, during the time interval $[t, t + \Delta t]$. This situation occurs with probability

$$p_{n+1}(t)(1 - \lambda \Delta t)(\mu \Delta t)$$

D. There are n customers in the system at time t, and one customer enters and one customer leaves the system during the time interval $[t, t + \Delta t]$. The probability of this happening is

$$p_n(t)(\lambda \Delta t)(\mu \Delta t) = p_n(t)(\lambda \mu)(\Delta t)^2$$

Because of assumption **c.** above, we do not consider the cases where there are two or more arrivals or two or more departures (customers serviced) during the time interval $[t, t + \Delta t]$.

The events A, B, C, D are mutually exclusive, and it follows that if $n \geq 1$, then the probability that there are n customers in the system at time $t + \Delta t$ is given by

$$
\begin{aligned}
p_n(t + \Delta t) &= P(A) + P(B) + P(C) + P(D) \\
&= p_n(t)(1 - \lambda \Delta t)(1 - \mu \Delta t) + p_{n-1}(t)(\lambda \Delta t)(1 - \mu \Delta t) \\
&\quad + p_{n+1}(t)(1 - \lambda \Delta t)(\mu \Delta t) + p_n(t)(\lambda \mu)(\Delta t)^2 \qquad (1)
\end{aligned}
$$

In the case that $n = 0$, similar considerations yield

$$p_0(t + \Delta t) = p_0(t)(1 - \lambda \Delta t)(1) + p_1(t)(1 - \lambda \Delta t)(\mu \Delta t) \qquad (2)$$

(see Problem 8). Standard algebraic manipulations applied to (1) result in

$$\begin{aligned}
p_n(t + \Delta t) = {} & p_{n-1}(t)(\lambda \Delta t - \lambda \mu (\Delta t)^2) \\
& + p_n(t)(1 - \lambda \Delta t - \mu \Delta t + 2\lambda \mu (\Delta t)^2) \\
& + p_{n+1}(t)(\mu \Delta t - \lambda \mu (\Delta t)^2)
\end{aligned} \qquad (3)$$

for $n \geq 1$. For $n = 0$, equation (2) can be written as

$$p_0(t + \Delta t) = p_0(t)(1 - \lambda \Delta t) + p_1(t)(\mu \Delta t - \lambda \mu (\Delta t)^2) \qquad (4)$$

Since $(\Delta t)^2$ is considerably smaller than Δt for small values of Δt, little harm is done if we ignore all terms containing $(\Delta t)^2$; of course, a small error is introduced by doing so, but acceptance of this error gives the following simplified versions of (3) and (4):

$$\begin{aligned}
p_n(t + \Delta t) = {} & p_{n-1}(t)(\lambda \Delta t) + p_n(t)(1 - \lambda \Delta t - \mu \Delta t) \\
& + p_{n+1}(t)(\mu \Delta t) \qquad \text{if } n \geq 1
\end{aligned} \qquad (5)$$

and

$$p_0(t + \Delta t) = p_0(t)(1 - \lambda \Delta t) + p_1(t)(\mu \Delta t) \qquad \text{if } n = 0 \qquad (6)$$

From (5) and (6) we obtain

$$\frac{p_n(t + \Delta t) - p_n(t)}{\Delta t} = p_{n-1}(t)\lambda - p_n(t)(\lambda + \mu) + p_{n+1}(t)\mu \qquad \text{if } n \geq 1$$

$$\tag{7}$$

and

$$\frac{p_0(t + \Delta t) - p_0(t)}{\Delta t} = -\lambda p_0(t) + \mu p_1(t) \qquad \text{if } n = 0 \qquad (8)$$

Now we use the assumption that we are in a state of equilibrium. That is, we assume that the probability that there are n customers in the system at time t is the same as the probability that there are n customers in the system at time $t + \Delta t$: $p_n(t) = p_n(t + \Delta t)$. It follows from this assumption that the left-hand side of each of the equations (7) and (8) is equal to 0. Moreover, since these probabilities are time independent, we can replace $p_n(t)$ (and $p_n(t + \Delta t)$) with a constant P_n. Thus, from (7) and (8) we obtain the difference equations

$$\mu P_{n+1} - (\lambda + \mu)P_n + \lambda P_{n-1} = 0 \qquad \text{if } n \geq 1 \qquad (9)$$

$$\mu P_1 - \lambda P_0 = 0 \qquad \text{if } n = 0 \qquad (10)$$

Note that from (10) we have

$$P_1 = \frac{\lambda}{\mu} P_0 \qquad (11)$$

Using methods developed in Chapter 7, we find that the solution of the difference equation (9) is

$$P_n = \frac{\lambda^n}{\mu^n} P_0 \quad \text{for } n \geq 1$$

Consequently, in view of (11),

$$P_n = \frac{\lambda^n}{\mu^n} P_0 \quad \text{for } n \geq 0 \tag{12}$$

Since

$$P_0 + P_1 + P_2 + \cdots = 1 \tag{13}$$

(Why?) we can use (12) and (13) to solve for P_0:

$$1 = \sum_{n=0}^{\infty} P_n = \sum_{n=0}^{\infty} \frac{\lambda^n}{\mu^n} P_0 = P_0 \sum_{n=0}^{\infty} \frac{\lambda^n}{\mu^n} \tag{14}$$

The series $\sum_{n=0}^{\infty} (\lambda^n/\mu^n) = \sum_{n=0}^{\infty} (\lambda/\mu)^n$ is a geometric series with common ratio λ/μ. Thus, if $0 < \lambda < \mu$, we have $|\lambda/\mu| < 1$, so the series converges with sum $1/(1 - \lambda/\mu)$. Consequently, from (14) we have

$$1 = \frac{P_0}{1 - \lambda/\mu} \quad \text{if } 0 < \lambda < \mu$$

Hence

$$P_0 = 1 - \frac{\lambda}{\mu}$$

Therefore, from (12) we have that the probability of having n customers in the system at any time is

$$P_n = \left(1 - \frac{\lambda}{\mu}\right)\left(\frac{\lambda}{\mu}\right)^n \quad \text{for } n \geq 0$$

The cases $\lambda = \mu$ and $\lambda > \mu$ are dealt with in Problem 10.

Example 1   In the Quick-Photo machine example that introduced this section, an average of twelve customers per hour wish to use a machine that requires four minutes to operate. As we have already noted, in this situation $\lambda = 12$ and $\mu = 15$. Thus, the probability that there are n customers in this system is

$$P_n = \left(1 - \frac{12}{15}\right)\left(\frac{12}{15}\right)^n$$

In particular, the probability that there are three customers in the system at any one time is

$$P_3 = \left(1 - \frac{12}{15}\right)\left(\frac{12}{15}\right)^3 = .1024 \qquad \square$$

The average or expected number of customers in the system is

$$L_S = \sum_{n=0}^{\infty} nP_n = \sum_{n=1}^{\infty} n\left(\frac{\lambda}{\mu}\right)^n\left(1 - \frac{\lambda}{\mu}\right)$$

$$= \left(1 - \frac{\lambda}{\mu}\right)\sum_{n=1}^{\infty} n\left(\frac{\lambda}{\mu}\right)^n$$

But by Example 5 of Section 10.3,

$$\sum_{n=1}^{\infty} n\left(\frac{\lambda}{\mu}\right)^n = \frac{\lambda/\mu}{(1 - \lambda/\mu)^2}$$

and therefore,

$$L_S = \sum_{n=0}^{\infty} nP_n = \left(1 - \frac{\lambda}{\mu}\right)\frac{\lambda/\mu}{(1 - \lambda/\mu)^2} = \frac{\lambda}{\mu - \lambda} \qquad (15)$$

Example 2   The average number of customers in the Quick-Photo system with $\lambda = 12$ and $\mu = 15$ is

$$L_S = \frac{12}{15 - 12} = 4 \qquad \square$$

Next we determine the expected number, L_Q, of customers waiting in the queue at any particular time. This number is equal to the expected number of customers in the system minus the expected number of customers being serviced. What then is the expected number of customers in service? Note that if there is any customer at all in the system, then some customer is being serviced (no customer would stand in line for nothing). Since P_0 is the probability that there is no one at all in the system, it follows that $1 - P_0$ is the probability that there is at least one customer in the system. If we assume that there is just one service station, then we have that

$$(1 - P_0) \cdot 1 = \left(1 - \left(1 - \frac{\lambda}{\mu}\right)\right) = \frac{\lambda}{\mu}$$

is the expected number of customers being serviced. Consequently, the expected number L_Q of customers in the queue is given by

$$L_Q = \frac{\lambda}{\mu - \lambda} - \frac{\lambda}{\mu} = \frac{\lambda^2}{\mu(\mu - \lambda)}$$

Example 3 The average number of customers waiting in line to use the Quick-Photo machine is

$$L_Q = \frac{12^2}{15(15 - 12)} = 3.2 \qquad \square$$

To calculate the average time, W_S, spent in the system we make the following observation. The average number of customers found in the system is equal to the product of the arrival rate (λ) and the average time spent in the system (W_S); thus,

$$\frac{\lambda}{\mu - \lambda} = \lambda \cdot W_S$$

and hence

$$W_S = \frac{1}{\mu - \lambda} \qquad (16)$$

Similarly, we can obtain the average time, W_Q, spent in the queue by noting that the average number of customers in the queue (L_Q) is equal to the arrival rate multiplied by the average amount of time spent waiting in the queue. That is,

$$L_Q = \frac{\lambda^2}{\mu(\mu - \lambda)} = \lambda W_Q$$

and hence

$$W_Q = \frac{\lambda}{\mu(\mu - \lambda)} \qquad (17)$$

Example 4 From (16) we have that in the Quick-Photo machine example, the average time that a customer must spend in the system is

$$W_S = \frac{1}{15 - 12} = \left(\frac{1}{3}\right) \text{ hour} = 20 \text{ minutes}$$

From (17) we see that the average time the customer must spend in line is

$$W_Q = \frac{12}{15(15 - 12)} = \left(\frac{4}{15}\right) \text{ hour} = 16 \text{ minutes} \qquad \square$$

In this section we have only introduced a few of the most elementary ideas underlying queueing theory. Further topics for study include situations where waiting lines can accommodate only a certain number of arrivals, where there are multiple serving stations, where arrivals do not follow a Poisson process, and where there are various combinations of these and other situations. Topics such as these are treated in many specialized texts on queueing theory.

9.10   Problems

1. Suppose that a toll booth on a bridge is capable of handling 150 cars an hour and that on the average 120 cars an hour use the bridge. Assume that both the number of arrivals and the number of cars passing through the toll booth (the number of cars served) have a Poisson distribution. Find
   (a) the average time a car must wait in line.
   (b) the average number of cars waiting in line to pass through the toll booth at any one time.
   (c) the probability that there are 6 cars in the system (in line or at the toll booth).

2. A small bank has a single teller. Suppose that on the average the number of persons entering the bank is 8 per hour and the teller can serve 10 customers per hour. Assume the appropriate Poisson distributions and find
   (a) the average waiting time before being served in this bank.
   (b) the average number of customers waiting in line to be served.
   (c) the average number of time a customer will spend in the bank.

3. The manager of the bank described in Problem 2 will hire a second teller when the average number of persons waiting in line is more than four. How much must the average number of arrivals increase for this to happen?

4. Suppose that the capacity of a communication line is 1600 bits/second. Eight-bit characters are transmitted over this line so the line can handle 200 characters/second. Suppose that the volume on this line is 150 characters/second. Find the average delay in using the line and the average number of characters waiting to be sent. (Assume the appropriate Poisson distributions.)

5. A mechanic who works five days a week in a garage can repair an average of 20 cars per week. If, on the average, 16 people a week would like to have their cars repaired, what is the average number of days a person must wait to have his car serviced? On the average, how many people are waiting to have their car enter the garage for repairs? (Assume the appropriate Poisson distributions.)

6. The owner of the garage described in Problem 5 will hire another mechanic if, on the average, more than seven persons are waiting for their cars to be serviced. How many people per week must want to use this garage before the owner will employ an additional mechanic?

7. Interpret the quotient λ/μ.

8. Verify equation (2) of this section.

9. Solve the difference equation (9) in this section.

10. For the results obtained in this section it was assumed that $\lambda < \mu$. Discuss the queueing situation if $\lambda = \mu$ and if $\lambda > \mu$.

It can be shown that if there are $m > 1$ service stations in a queueing situation and if the service rate μ is the same at each station, then

$$P_0 = \frac{1}{\left[\sum_{n=0}^{m-1}\left(\frac{1}{n!}\right)\left(\frac{\lambda}{\mu}\right)^n\right] + \frac{1}{m!(1 - \lambda/\mu m)}\left(\frac{\lambda}{\mu}\right)^m}$$

$$P_n = \left(\frac{1}{n!}\right)\left(\frac{\lambda}{\mu}\right)^n P_0 \qquad \text{if } n = 0, 1, \cdots, m - 1$$

$$P_n = \frac{1}{m!m^{n-m}}\left(\frac{\lambda}{\mu}\right)^n P_0 \qquad \text{if } n \geq m$$

$$L_Q = \frac{(\lambda/\mu)^{m+1} P_0}{m \cdot m!(1 - \lambda/\mu m)^2}$$

$$L_S = L_Q + \frac{\lambda}{\mu}$$

$$W_S = \frac{L_S}{\lambda}$$

$$W_Q = \frac{L_Q}{\lambda}$$

Use this information to solve the following problems.

11. A bank has three tellers, each capable of servicing 15 customers/hour. On the average 40 customers enter the bank each hour.
    (a) What is the average time spent waiting in line in this bank?
    (b) What is the average number of customers in the bank at any one time?
    (c) What is the probability that there are exactly four customers in the bank?
    (d) How much time would you expect to spend in the bank?

12. There are five customs officers at a certain border crossing. Each officer can check, on the average, three cars a minute. If, on the average, 800 cars an hour cross the border, find
    (a) the average amount of time needed to make the border crossing.
    (b) the average number of cars waiting in line to go through customs.
    (c) the average wait in line for each car.
    (d) the probability that there are exactly four cars waiting in line or being checked by an officer.

Chapter 9   REVIEW

Concepts for Review

random experiment (p. 343)
outcome (p. 343)
sample space (p. 343)

event (p. 344)
probability measure (p. 344)
probability space (p. 344)
conditional probability (p. 356)
independent events (p. 361)
Bernoulli (binomial) process (p. 366)
binomial probability (p. 368)
Bayes' Theorem (p. 374)
random variable (p. 376)
probability density function (pdf) (p. 377)
uniform distribution (p. 378)
binomial distribution (p. 378)
expected value of a random variable (p. 382)
average-case analysis (p. 384)
Poisson distribution (p. 389)
queueing theory (p. 393)

Review Problems

1. A pair of dice, one red and one green, are rolled. Describe the sample space of this random experiment.

2. For the random experiment given in Problem 1, describe the following events.
   (a) The red die is even.
   (b) The sum of the dice is seven.
   (c) The red die is even or the green die is 6.

3. Determine the probability of each of the events described in Problem 2.

4. Let A and B be events such that $P(A \cup B) = .75$, $P(A) = .3$, and $P(B) = .6$; find $P(A \cap B)$.

5. A card is drawn from a standard deck of fifty-two cards. Let A be the event that a spade is drawn and let B be the event that an ace is drawn. Compute $P(A)$, $P(B)$, and $P(A \cup B)$.

6. Suppose that A and B are events such that $P(A) = .3$, $P(B) = .2$, and $P(A \cap B) = .1$. Compute $P(A|B)$ and $P(B|A)$. Are A and B independent events?

7. A bowl contains five quarters, three dimes, and seven nickels. Three coins are drawn without replacement. What is the probability that the third coin drawn is a quarter if the first two coins drawn are a quarter and a dime?

8. Two dice are rolled. What is the probability that the sum is seven if one of the dice shows a 4?

9. A bowl contains four slips of paper numbered 1 to 4, and all slips are equally likely to be drawn. Two slips are drawn. Let A be the event that

the slips numbered 1 and 2 are drawn, that is, $A = \{1, 2\}$. Similarly, let $B = \{1, 3\}$ and $C = \{1, 4\}$. Are the events A and B independent? Are the events A, B, and C independent?

10. A bowl contains 8 red chips and 5 blue chips. Three chips are drawn without replacement from the bowl and the number of blue chips drawn is noted.
    (a) What is the random variable X associated with this random experiment?
    (b) What is the pdf of X?
    (c) What is the expected value of X?

11. Five thousand tickets are sold in a lottery at two dollars each. First prize in the lottery is $4000, second prize is $3000, and there are two other prizes of $1000 each. What is the expected value of a ticket for this lottery?

12. Suppose that the probability of being struck by lightning is .000002 and that in a certain community of 20,000 people there are usually twelve electrical storms a year. Assume the appropriate Poisson distribution and find the probability that lightning will strike at least one member of the community during any given year.

13. Suppose that in an office building the average number of lightbulbs that need to be replaced during any given year is 200. Assume the appropriate Poisson distribution and find the probability that 100 light bulbs will burn out in a given year.

14. A computer terminal is available 18 hours a day. On the average, twenty-five students would like to use the terminal during this period and the average time spent on the terminal is thirty minutes. Assume the appropriate Poisson distributions and find:
    (a) the average time a student must wait in line to gain access to the terminal.
    (b) the average number of students waiting in line.
    (c) the probability that there are three students either waiting in line or using the terminal.

15. On the average, eight people an hour would like to use a telephone booth. The average length of a call is three minutes. Find:
    (a) the average amount of time a person must wait to use the telephone.
    (b) the average number of persons waiting in line to use the telephone.
    (c) the probability that there are two persons waiting in line to use the telephone.
    The telephone company will install another phone if the average waiting time is more than three minutes. How many more people per hour must wish to use the telephone before the company will put in a new phone?

10 Generating Functions

Generating functions provide a remarkably versatile mathematical tool that can be used in an unusual variety of contexts, ranging from probability and statistics to solving difference equations. As we shall see, generating functions can also be useful in solving certain combinatorial problems.

Generating functions are defined in terms of infinite series, which are important in many areas of mathematics and are commonly introduced in elementary calculus courses. In this text, we shall limit ourselves to the results involving series necessary to develop the concept of a generating function.

Since infinite series are tied closely to the idea of a sequence, we begin with a look at some of the basic properties of sequences.

10.1 THE LIMIT OF A SEQUENCE

Definition 10.1 An *infinite sequence* is a function whose domain is the set of all integers greater than or equal to some fixed integer s. Commonly, s will be either 0 or 1.

Example 1
(a) The function $f(n) = n^2$, $n = 0, 1, 2, \ldots$ is a sequence with $s = 0$.

(b) The function $g(n) = n^3 - 4$, $n = 3, 4, 5, \ldots$ is a sequence with $s = 3$.

(c) The function $h(n) = 3^n$, $n = 1, 2, \ldots$ is a sequence with $s = 1$. □

A sequence f is often denoted by an ordered listing of the values taken on by the function. For instance, in Example 1 the first sequence can be described by the ordered listing 0, 1, 4, 9, 16, ... , and the second sequence can be described by the ordered listing 23, 60, 121, ...

Another common way to denote a sequence whose domain is the set of integers greater than or equal to s is to write $\{f(n)\}_s^\infty$. Thus, the sequence $f(n) = n^2, n = 0, 1, 2, \ldots$ can be denoted by $\{n^2\}_0^\infty$. If in context the value s is clear, we write $\{n^2\}$.

Definition 10.2   (a) A sequence f is said to be *bounded above* if there is a number A such that $f(n) \leq A$ for each integer n in the domain of f; the number A is called an *upper bound* of f.

(b) A sequence f is said to be *bounded below* if there is a number B such that $f(n) \geq B$ for each integer n in the domain of f; the number B is called a *lower bound* of f.

(c) A sequence is said to be *bounded* if it has both an upper bound and a lower bound.

Example 2   In each of the following examples, we assume that $n = 1, 2, \ldots$.

(a) Let $f(n) = 1/n$. Since $1/n > 0$ for each n, it follows that f is bounded below by 0, and since $1/n \leq 1$ for each n, it follows that f is bounded above by 1. Note that any number less than 0 is also a lower bound for f, and any number greater than 1 is an upper bound for f.

(b) If $f(n) = (-1)^n n$, then f takes on arbitrarily large positive and negative values. Consequently, this sequence has neither an upper bound nor a lower bound.

(c) If $f(n) = n$, then f has no upper bound. Since $f(n) = n \geq 1$ for each n, it follows that any number less than or equal to 1 is a lower bound of f.

(d) If $f(n) = -n$, then f has no lower bound. Since $f(n) = -n \leq -1$ for each n, it follows that any number greater than or equal to -1 is an upper bound for f.
□

If a sequence has an upper bound, then clearly it has infinitely many upper bounds. Of particular interest is one such upper bound—the least, or smallest, upper bound.

Definition 10.3   A number A is said to be a *least upper bound* of a sequence f if

a. A is an upper bound of f, and

b. if C is also an upper bound of f, then $A \leq C$.

In other words, A is a least upper bound of f if A is the smallest of all of the upper bounds of f.

Example 3 The least upper bound of the sequence $f(n) = 4 - 1/n$, $n = 1, 2, \ldots$, is 4, since 4 is an upper bound of f and no number smaller than 4 is an upper bound of f. $\square$

The greatest lower bound of a sequence is defined analogously.

Definition 10.4 A number B is said to be a *greatest lower bound* of a sequence f if

    a. B is a lower bound of f, and

    b. if C is also a lower bound of f, then $C \leq B$.

In other words, B is a greatest lower bound of f if B is the largest of all of the lower bounds of f.

Example 4 The greatest lower bound of the sequence $\{n^2\}$, $n = 2, 3, \ldots$ is 4, since 4 is a lower bound of f and no number larger than 4 is a lower bound of f. $\square$

The real numbers have the property that if a sequence $\{f(n)\}$ is bounded above, then f has a least upper bound. This property is known as the *Axiom of Completeness.*

Axiom 10.5 *Axiom of Completeness.* If a sequence of real numbers has an upper bound, then it has a least upper bound.

The Axiom of Completeness may be used to obtain the following analogous result (see Problem 22).

Theorem 10.6 If a sequence of real numbers has a lower bound, then it has a greatest lower bound.

Often we will be particularly interested in the values $f(n)$ of a sequence f that correspond to large values of n. It may occur, for example, that for sufficiently large n the values $f(n)$ become arbitrarily close to a fixed number L. If this is the case, we say that the limit of the sequence is L, and we write

$$\lim_{n \to \infty} f(n) = L$$

Thus, for instance, we have

$$\lim_{n \to \infty} \frac{1}{n} = 0$$

since we can make $1/n$ arbitrarily close to 0 by choosing n sufficiently large.

The expressions "arbitrarily close" and "sufficiently large" are admittedly somewhat vague. The following fundamental definition remedies this lack of mathematical precision.

Definition 10.7   A number L is the *limit* of a sequence f if, given $\varepsilon > 0$ (the measure of "closeness"), there is an integer N (the measure of "sufficiently large") such that $|f(n) - L| < \varepsilon$ whenever $n > N$. If such a number L exists we write

$$\lim_{n \to \infty} f(n) = L$$

If a sequence f has a limit we say that f is *convergent*. Otherwise, the sequence is said to be *divergent*.

Let us examine in some detail the content of this definition (which is one of the most important definitions in mathematics and is the result of several centuries of struggle with the notion of a limit). If we wish to verify that a number L is the limit of a sequence f, we must show that "after a while" the values of f will be arbitrarily close to L. The positive (but arbitrarily small) number ε is given first. Once this number is given, we must find an integer N such that the distance between L and $f(n)$ is less than ε for all integers $n > N$. You would naturally expect (and it is generally the case) that the smaller the number ε, the larger N will be.

Example 5   Let $f(n) = 1/n^2$, $n = 1, 2, \ldots$. It is intuitively clear that the limit of this sequence is $L = 0$. Note, for instance, that if $\varepsilon = 1/100$ then if $N = 10$, we have for each $n > N$

$$|f(n) - L| = \left| \frac{1}{n^2} - 0 \right| < \frac{1}{100}$$

If, however, we begin with a smaller value of ε, say $\varepsilon = 1/10,000$, then we need a larger value of N to satisfy Definition 10.7. For example, if $N = 100$, we have .

$$|f(n) - L| = \left| \frac{1}{n^2} - 0 \right| < \frac{1}{10,000}$$

whenever $n > N$. $\square$

Example 6   The limit of the sequence $f(n) = (-1)^n$, $n = 1, 2, \ldots$ is not $L = 1$, since if $\varepsilon = 1/2$, no matter how large we choose N there will be an integer $n > N$ (any odd integer $n > N$ will do) such that $|f(n) - L| = |(-1)^n - 1| = |-2| = 2 > 1/2$. In fact, it is easy to see that the sequence $f(n) = (-1)^n$ does not have any limit at all (see Problem 12).   □

In Figure 10.1 we illustrate what happens geometrically when a sequence f has limit L. Observe that $L - \varepsilon < f(n) < L + \varepsilon$, or equivalently, $|f(n) - L| < \varepsilon$, for $n > N$.

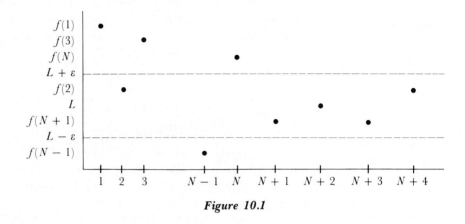

Figure 10.1

Example 7   To show formally that

$$\lim_{n \to \infty} \frac{1}{n} = 0$$

we substitute $1/n$ for $f(n)$, and 0 for L in Definition 10.7. We must show that for an arbitrary $\varepsilon > 0$ there is an integer N with the property that if $n > N$, then $|1/n - 0| < \varepsilon$. In other words, we must find an integer N such that if $n > N$, then $1/n < \varepsilon$.

To find N, observe that if $n > 1/\varepsilon$, then $1/n < \varepsilon$; therefore, we can let N be any integer larger than $1/\varepsilon$.   □

In the next theorem we see that for a sequence f to have a limit, f must be bounded (however, the converse of this result—if f is bounded, then f has a limit—does not necessarily hold; see Problem 18).

Theorem 10.8   If a sequence $\{f(n)\}$ converges, then it is bounded.

Proof: Let $\lim_{n \to \infty} f(n) = L$ and select $\varepsilon = 1$. Then by Definition 10.7 there is an integer N such that $|f(n) - L| < 1$ whenever $n > N$. Thus, if $n > N$, we have

$$L - 1 < f(n) < L + 1$$

Consequently, $L - 1$ and $L + 1$ serve as a lower and an upper bound, respectively, for the "tail," $f(N + 1)$, $f(N + 2)$, ..., of the sequence f. It follows that to obtain an upper bound for the entire sequence we need only choose a number larger than each of the numbers $f(1)$, $f(2)$, ..., $f(N)$, and $L + 1$. Similarly, any number less than each of the numbers $f(1), f(2), \ldots, f(N)$ and $L - 1$ will serve as a lower bound for this sequence. ∎

The contrapositive of Theorem 10.8 is frequently employed and we state it as a corollary.

Corollary 10.9   If a sequence f is not bounded, then f is divergent.

For example, since the sequence $\{-n^2\}$ is not bounded, it is not convergent.

We conclude this section with the following useful definition and result.

Definition 10.10   A sequence $\{f(n)\}$ is said to be an *increasing sequence* if for each n, $f(n) \le f(n + 1)$. A sequence $\{f(n)\}$ is said to be a *decreasing sequence* if for each n, $f(n) \ge f(n + 1)$.

For instance, the sequence $\{1 - 1/n^2\}$ is an increasing sequence and the sequence $\{1/n^2\}$ is a decreasing sequence.

Theorem 10.11   Suppose that $\{f(n)\}$ is an increasing sequence that is bounded above. Then the sequence $\{f(n)\}$ is convergent and converges to its least upper bound.

Proof: Since $\{f(n)\}$ is bounded above, it follows from the Axiom of Completeness that this sequence has a least upper bound, B. We show that $\{f(n)\}$ converges to B. Let $\varepsilon > 0$ be given. Since B is an upper bound for $\{f(n)\}$, we have that $f(n) \le B$ for each n. Since B is a *least* upper bound for $\{f(n)\}$ it follows that $B - \varepsilon$ is not an upper bound for f. Thus, there is an integer N such that $B - \varepsilon < f(N) \le B$. Now observe that since $\{f(n)\}$ is an increasing sequence, it follows that $f(N) \le f(n)$ for each $n > N$. Therefore, if $n > N$, we have

$$B - \varepsilon < f(N) \le f(n) \le B$$

and, hence, for $n > N$, $|f(n) - B| < \varepsilon$. Consequently, by the definition of convergence, the sequence $\{f(n)\}$ converges to its least upper bound B. ∎

Example 8   Since the sequence $\{1 - 1/n^2\}$ is increasing and bounded above, it follows from Theorem 10.11 that this sequence has a limit. □

10.1 Problems

For each of the sequences given in Problems 1–10:

   a. Find two upper bounds or state that none exists.

   b. Find the least upper bound or state that none exists.

   c. Find two lower bounds or state that none exists.

   d. Find the greatest lower bound or state that none exists.

1. $f(n) = (-2)^n; n = 0, 1, \ldots$
2. $g(n) = 1/n; n = 1, 2, 3, \ldots$
3. $k(n) = 1 - 1/n; n = 2, 3, 4, \ldots$
4. $f(n) = n; n = 1, 2, 3, \ldots$
5. $f(n) = 1 + (-1)^n; n = 1, 2, 3, \ldots$
6. $h(n) = 6; n = 3, 4, 5, \ldots$
7. $f(n) = \dfrac{n + 1}{n}; n = 2, 3, 4, \ldots$
8. $g(n) = 2^{-n}; n = 2, 3, \ldots$
9. $r(n) = 2^n; n = 1, 2, 3, \ldots$
10. $f(n) = \dfrac{3}{2 - n}; n = 3, 4, 5, \ldots$

11. A ball is dropped from a height of 10 feet and always rebounds .6 of the height from which it falls. Find a sequence that describes how high the ball bounces on each successive bounce.

12. Show that the sequence $\{(-1)^n\}$ does not have a limit.

13. Determine the limit (if it exists) of each of the sequences given in Problems 1–10.

14. Show that $\{1 - 1/n^2\}$ is an increasing sequence.

15. Show that $\{1/n^2\}$ is a decreasing sequence.

16. Prove or find a counterexample: If a sequence converges, then it has a lower bound.

17. Show that if a decreasing sequence has a lower bound, then it converges to its greatest lower bound.

18. Find a sequence that is bounded but which fails to have a limit.

19. It is true that $\lim_{n \to \infty} (1 - 1/n) = 1$. As in Definition 10.7 find the value of N that corresponds to:
    (a) $\varepsilon = 2$ (c) $\varepsilon = \frac{1}{4}$
    (b) $\varepsilon = \frac{1}{2}$ (d) arbitrary $\varepsilon > 0$

20. It is true that $\lim_{n \to \infty} ((3n^2 + 1)/n^2) = 3$. As in Definition 10.7 find the value of N that corresponds to:

(a) $\varepsilon = 2$

(c) $\varepsilon = \frac{1}{4}$

(b) $\varepsilon = \frac{1}{2}$

(d) arbitrary $\varepsilon > 0$

21. Show that a sequence has at most one limit.

*22. Use the Axiom of Completeness to show that if a sequence of real numbers has a lower bound, then the sequence has a greatest lower bound.

23. Studies have indicated that among all arithmetic operations, the operation of division occurs with a frequency of less than 10%. For this reason, some early computers did not have a divide instruction and instead divided by means of a subroutine. One commonly used subroutine to compute $1/a$ was based on the following:

$$a_0 = \frac{a + 1}{2} \qquad a_{n+1} = a_n(2 - a_n a), \quad n = 0, 1, 2, \ldots$$

Show that if $\lim_{n \to \infty} a_n$ exists, then $\lim_{n \to \infty} a_n = 1/a$.

10.2   COMPUTATION OF LIMITS

A little experimentation will show you that establishing the existence or value of a limit of a sequence directly from the definition of a limit can be an extremely difficult task. The following theorem will allow us to find many limits from just a few.

Theorem 10.12   *Limit Theorem.*   Suppose that $\{f(n)\}$ and $\{g(n)\}$ are sequences. If $\lim_{n \to \infty} f(n)$ and $\lim_{n \to \infty} g(n)$ both exist, then:

(a) $\displaystyle \lim_{n \to \infty} (f(n) + g(n)) = \lim_{n \to \infty} f(n) + \lim_{n \to \infty} g(n)$

(b) $\displaystyle \lim_{n \to \infty} (f(n) - g(n)) = \lim_{n \to \infty} f(n) - \lim_{n \to \infty} g(n)$

(c)   For any constant α, $\displaystyle \lim_{n \to \infty} \alpha f(n) = \alpha \lim_{n \to \infty} f(n)$

(d) $\displaystyle \lim_{n \to \infty} [f(n)g(n)] = \lim_{n \to \infty} f(n) \lim_{n \to \infty} g(n)$

(e)   If $\lim g(n) \neq 0$, then $\displaystyle \lim_{n \to \infty} \frac{f(n)}{g(n)} = \frac{\displaystyle \lim_{n \to \infty} f(n)}{\displaystyle \lim_{n \to \infty} g(n)}$

This theorem says that the limit of the sum (or difference) of two sequences is equal to the sum (or difference) of the limits; a similar statement holds for products and quotients. This result, though intuitively obvious, is surprisingly difficult to prove. We establish parts (a) and (d) of the theorem; proofs of the remaining parts are left to the more intrepid reader in Problems 11–13.

Proof: (a) Suppose that $\lim_{n \to \infty} f(n) = K$ and $\lim_{n \to \infty} g(n) = L$. Given $\varepsilon > 0$, we must find an integer N such that if $n > N$, then

$$|f(n) + g(n) - (K + L)| < \varepsilon \tag{1}$$

Since

$$|f(n) + g(n) - (K + L)| = |(f(n) - K) + (g(n) - L)|$$

and since, for any real numbers a and b,

$$|a + b| \le |a| + |b| \tag{2}$$

we have

$$|f(n) + g(n) - (K + L)| \le |f(n) - K| + |g(n) - L|$$

Thus, to establish (1) it suffices to show that there is an integer N such that if $n > N$, then

$$|f(n) - K| < \frac{\varepsilon}{2} \quad \text{and} \quad |g(n) - L| < \frac{\varepsilon}{2}$$

Note that since

$$\lim_{n \to \infty} f(n) = K$$

it follows that corresponding to the positive number $\varepsilon/2$, there is an integer N_1 such that if $n > N_1$, then

$$|f(n) - K| < \frac{\varepsilon}{2}$$

Similarly, since

$$\lim_{n \to \infty} g(n) = L$$

it follows that corresponding to the positive number $\varepsilon/2$, there is an integer N_2 such that if $n > N_2$, then

$$|g(n) - L| < \frac{\varepsilon}{2}$$

Let N be the larger of the two integers N_1 and N_2. Then if $n > N$, it follows that $n > N_1$ and $n > N_2$, and consequently,

$$|f(n) + g(n) - (K + L)| = |(f(n) - K) + (g(n) - L)| \le |f(n) - K| + |g(n) - L| < \varepsilon/2 + \varepsilon/2 = \varepsilon$$

which is what we wished to show.

(d) Suppose that $\lim_{n \to \infty} f(n) = K$ and $\lim_{n \to \infty} g(n) = L$. Given $\varepsilon > 0$, we must find an integer N such that if $n > N$, then

$$|f(n)g(n) - KL| < \varepsilon$$

Note that since

$$|f(n)g(n) - KL| = |f(n)g(n) - f(n)L + f(n)L - KL|$$

we have from (2) that

$$|f(n)g(n) - KL| \leq |f(n)g(n) - f(n)L| + |f(n)L - KL| \tag{3}$$

Moreover, since $|ab| = |a||b|$, it follows from (3) that

$$|f(n)g(n) - KL| \leq |f(n)||g(n) - L| + |L||f(n) - K| \tag{4}$$

Since $\{f(n)\}$ is a convergent sequence, it is bounded. Thus, there is a number B such that $|f(n)| < B$, for each integer n in the domain of f. From (4) we now have

$$|f(n)g(n) - KL| \leq B|g(n) - L| + |L||f(n) - K| \tag{5}$$

Since $\lim_{n \to \infty} g(n) = L$, it follows that corresponding to the positive number $\varepsilon/(2B)$, there is an integer N_1 such that if $n > N_1$, then

$$|g(n) - L| < \frac{\varepsilon}{2B} \tag{6}$$

Similarly, since $\lim_{n \to \infty} f(n) = K$, corresponding to the positive number $\varepsilon/(2|L|)$, there is an integer N_2 such that if $n > N_2$, then

$$|f(n) - K| < \varepsilon/(2|L|) \tag{7}$$

Observe now that if N is the larger of the integers N_1 and N_2, then (6) and (7) are true for each $n > N$. Consequently, on substituting (6) and (7) into (5) we have for each $n > N$

$$|f(n)g(n) - KL| < B \cdot \frac{\varepsilon}{2B} + |L| \cdot \frac{\varepsilon}{2|L|} = \frac{\varepsilon}{2} + \frac{\varepsilon}{2} = \varepsilon$$

the desired result. ∎

Example 1 To compute

$$\lim_{n \to \infty} \frac{2n^2 + 3n - 1}{5n^2 - 7n}$$

we divide the numerator and denominator by n^2 and use Theorem 10.12 to obtain

$$\lim_{n \to \infty} \frac{2n^2 + 3n - 1}{5n^2 - 7n} = \lim_{n \to \infty} \frac{2 + 3/n - 1/n^2}{5 - 7/n}$$

$$\overset{(e)}{=} \frac{\lim_{n \to \infty} (2 + 3/n - 1/n^2)}{\lim_{n \to \infty} (5 - 7/n)}$$

$$\overset{(a), (b), (c), (d)}{=} \frac{\lim_{n \to \infty} 2 + 3 \lim_{n \to \infty} 1/n - \lim_{n \to \infty} 1/n \lim_{n \to \infty} 1/n}{\lim_{n \to \infty} 5 - 7 \lim_{n \to \infty} 1/n}$$

$$= \frac{2 + 3(0) - 0}{5 - 7(0)} = \frac{2}{5} \qquad \square$$

The technique illustrated in Example 1 can be used to compute any limit of the form

$$\lim_{n \to \infty} \frac{P(n)}{Q(n)}$$

where $P(n)$ and $Q(n)$ are polynomials. Simply divide both the numerator and denominator by the highest power of n appearing in those polynomials, and then apply Theorem 10.12. (Actually, as you are asked to show in Problem 14, you need only consider the highest power terms in $P(n)$ and $Q(n)$ to compute $\lim_{n \to \infty} (P(n)/Q(n))$.)

10.2 Problems

In Problems 1–10 compute the given limit or state that the limit does not exist.

1. $\lim_{n \to \infty} \dfrac{3n^2 - 7n}{2n^2 - 1}$

2. $\lim_{n \to \infty} \dfrac{7n^3 - 3n^2 + 1}{2n^3 - 2}$

3. $\lim_{n \to \infty} \dfrac{n^2 - 2n + 1}{2n^3 - 3n^2}$

4. $\lim_{n \to \infty} \dfrac{n^3 + 10n^2 + 100}{2n^5 + n^3 - n^2}$

5. $\lim_{n \to \infty} \dfrac{2n^2 - 3n + 1}{n - 7}$

6. $\lim_{n \to \infty} \dfrac{2n^3 - 7n^2 + 4n - 1}{n^4 + 1}$

7. $\lim_{n \to \infty} \dfrac{n^3 + 7n}{n^2 + 2}$

8. $\lim_{n \to \infty} (-1)^n$

9. $\lim_{n \to \infty} (3 + (-1)^{n+1})$

10. $\lim_{n \to \infty} (-n)^n$

11. Prove part (c) of Theorem 10.12.

12. Use parts (a) and (c) of Theorem 10.12 to prove part (b) of this theorem.

13. Prove part (e) of Theorem 10.12. (*Hint:* First show that $\lim_{n \to \infty}$ $1/g(n) = 1/L$ where $\lim_{n \to \infty} g(n) = L$; note that $|1/g(n) - 1/L| = (1/(|L||g(n)|)) |g(n) - L|$.)

14. Show that

$$\lim_{n \to \infty} \frac{a_k n^k + a_{n-1} n^{k-1} + \cdots + a_1 n + a_0}{b_j n^j + b_{n-1} n^{j-1} + \cdots + b_1 n + b_0} = \begin{cases} 0 & \text{if } k < j \\ \dfrac{a_k}{b_k} & \text{if } k = j \end{cases}$$

What is the value of this limit if $k > j$?

10.3 INFINITE SUMS AND SERIES

Infinite sums play an important role in discrete mathematics. Intuitively, an infinite sum is the sum of all the terms appearing in a sequence of numbers. For example, the infinite sum of the terms of the sequence $\frac{1}{2}, \frac{1}{4}, \ldots, 1/2^n, \ldots$ is $\frac{1}{2} + \frac{1}{4} + \frac{1}{8} + \cdots + 1/2^n + \cdots$.

But how do we compute an infinite sum? The usual elementary school tactic of adding each term to the sum of the preceding terms fails since this "algorithm" would never terminate. The solution to this problem is to *define* an infinite sum as the limit of a sequence of *finite* sums.

With each sequence $a_1, a_2, \ldots$ we can associate a sequence of *partial sums*, $\{s_n\}$, defined by

$$\begin{aligned} s_1 &= a_1 \\ s_2 &= a_1 + a_2 \\ s_3 &= a_1 + a_2 + a_3 \\ &\vdots \\ s_n &= a_1 + a_2 + \cdots + a_n \\ &\vdots \end{aligned}$$

The *infinite sum* of the sequence of terms $a_1, a_2, \ldots$ is defined to be

$$\lim_{n \to \infty} s_n$$

In other words, the *sum* of the terms $a_1, a_2, \ldots$ is defined to be the limit of the *sequence* of partial sums. This limit is frequently denoted by $a_1 + a_2 + \cdots$ or by $\Sigma_{k=1}^{\infty} a_k$. As you know, the limit of a sequence may or may not exist, and therefore, an infinite sum, in contrast to a finite sum, may not exist.

Infinite sums are generally discussed in the context of infinite series.

Definition 10.13   If $\{a_n\}$ is a sequence with partial sums $\{s_n\}$, then the *infinite series* associated with the sequence $\{a_n\}$ is the pair of sequences $(\{a_n\}, \{s_n\})$. If

$\lim_{n \to \infty} s_n$ exists, we say that the infinite series $(\{a_n\}, \{s_n\})$ *converges* and has *sum* $\lim_{n \to \infty} s_n$. If the sequence $\{s_n\}$ fails to converge, we say that the infinite series $(\{a_n\}, \{s_n\})$ *diverges*.

It is a common (but perhaps unfortunate) practice to denote the infinite series $(\{a_n\}, \{s_n\})$ by the symbol $\sum_{k=1}^{\infty} a_k$, which, as we have seen, is the same notation as that used to denote the infinite sum of the terms of the sequence $\{a_n\}$. This may be a bit confusing to you at first, but it is usually clear from the context whether $\sum_{k=1}^{\infty} a_k$ is being used to denote an infinite series or an infinite sum.

Example 1   We compute $\sum_{k=1}^{\infty} 1/2^k$. The nth partial sum of this series is given by

$$s_n = \frac{1}{2} + \frac{1}{4} + \frac{1}{8} + \cdots + \frac{1}{2^{n-1}} + \frac{1}{2^n}$$

If we use $f(n)$ to designate s_n, we have

$$f(n) = \frac{1}{2} + \frac{1}{4} + \frac{1}{8} + \cdots + \frac{1}{2^{n-1}} + \frac{1}{2^n} \tag{1}$$

and

$$f(n-1) = \frac{1}{2} + \frac{1}{4} + \frac{1}{8} + \cdots + \frac{1}{2^{n-1}} \tag{2}$$

On subtracting (2) from (1) we obtain the first-order difference equation

$$f(n) - f(n-1) = \frac{1}{2^n} \tag{3}$$

The corresponding homogeneous difference equation

$$f(n) - f(n-1) = 0$$

has $A - 1 = 0$ as its characteristic equation. Consequently, $A = 1$, and the constant function α is the general solution to the homogeneous equation.

To find a particular solution of (3), we substitute the trial solution $F(n) = C(1/2^n)$ into (3) to obtain

$$C\frac{1}{2^n} - C\frac{1}{2^{n-1}} = \frac{1}{2^n}$$

Multiplying this equation by 2^n yields

$$C - 2C = 1$$

from which it follows that $C = -1$ and that $p(n) = -1/2^n$ is a particular solution of the nonhomogeneous equation (3). Thus, by Theorem 7.5

$$f(n) = \alpha - \frac{1}{2^n} \tag{4}$$

for some choice of α.

To find α, note that since $f(1) = \frac{1}{2}$, substituting $n = 1$ into (4) yields $\alpha = 1$. Thus, the nth partial sum s_n of the series $\sum_{k=1}^{\infty} 1/2^k$ is given by

$$s_n = f(n) = 1 - \frac{1}{2^n}$$

By Definition 10.13 the sum of the series is

$$S = \lim_{n \to \infty} s_n = \lim_{n \to \infty} \left(1 - \frac{1}{2^n} \right) = 1 \qquad \square$$

Next we give an example of an infinite series that fails to have a sum.

Example 2   The infinite series $\sum_{k=1}^{\infty} (-1)^k = -1 + 1 - 1 + 1 - \cdots$ is divergent since the sequence of partial sums

$$s_1 = 1$$
$$s_2 = 1 - 1 = 0$$
$$s_3 = 1 - 1 + 1 = 1$$
$$\vdots$$

continues to oscillate between 1 and 0 and, consequently, fails to converge.
$\square$

Example 3   To find the sum of the infinite series

$$\sum_{k=0}^{\infty} ax^k = a + ax + ax^2 + \cdots + ax^n + \cdots$$

we apply the technique used in Example 1. If we let $f(n)$ denote the nth partial sum, s_n, we have

$$f(n) = a + ax + ax^2 + \cdots + ax^{n-2} + ax^{n-1}$$

and

$$f(n-1) = a + ax + ax^2 + \cdots + ax^{n-2}$$

On subtraction we obtain the first-order difference equation

$$f(n) - f(n-1) = ax^{n-1} \tag{5}$$

As in Example 1, the solution of the corresponding homogeneous equation

$$f(n) - f(n-1) = 0$$

is the constant function α. To find a particular solution of (5) we use the trial solution $p(n) = Cx^{n-1}$ Substitution into (5) yields

$$Cx^{n-1} - Cx^{n-2} = ax^{n-1} \tag{6}$$

and on dividing through by x^{n-2} we find that

$$Cx - C = ax$$

Thus,

$$C(x - 1) = ax$$

and consequently, if $x \neq 1$, we have

$$C = \frac{ax}{x-1}$$

Therefore,

$$p(n) = \frac{ax}{x-1} x^{n-1} = \frac{ax^n}{x-1}$$

is a particular solution to (5), and thus

$$f(n) = \alpha + \frac{ax^n}{x-1} \tag{7}$$

for an appropriate choice of α.

To find α note that since $f(1) = a$, substituting $n = 1$ into (7) yields

$$a = \alpha + \frac{ax}{x-1}$$

from which it follows that

$$\alpha = a - \frac{ax}{x-1} = -\frac{a}{x-1}$$

On substituting this expression back into (7), we find that

$$f(n) = -\frac{a}{x-1} + \frac{ax^n}{x-1}$$

is the nth partial sum.

Since the sum S of the series $\Sigma_{k=0}^{\infty} ax^k$ is the limit of the sequence of partial sums, we have

$$S = \lim_{n \to \infty} \left(-\frac{a}{x-1} + \frac{ax^n}{x-1} \right) = -\frac{a}{x-1} + \frac{a}{x-1} \lim_{n \to \infty} x^n$$

Since $\lim_{n \to \infty} x^n = 0$, if $|x| < 1$ and does not exist for $|x| > 1$, we see that

$$S = -\frac{a}{x-1} = \frac{a}{1-x} \qquad \text{if } |x| < 1$$

We leave it to you to show that the series diverges if $|x| \geq 1$. $\square$

An infinite series of the form $\sum_{k=0}^{\infty} ax^k$ is called a *geometric series*. The result involving the geometric series obtained in the preceding example is sufficiently important to merit restating it as a theorem.

Theorem 10.14   If a is a constant and if $|x| < 1$, then the geometric series $\sum_{k=0}^{\infty} ax^k$ converges and has sum $a/(1 - x)$; if $|x| \geq 1$, this series diverges.

Theorem 10.14 enables us to find the sum of a geometric series by means of a simple calculation.

Example 4   Since $|-\frac{1}{3}| < 1$, the geometric series

$$\sum_{k=0}^{\infty} 2\left(-\frac{1}{3}\right)^k$$

converges and has sum

$$\frac{2}{1 - (-\frac{1}{3})} = \frac{3}{2} \qquad \square$$

Example 5   We again make use of the technique described in Example 1 to find the sum of $\sum_{k=1}^{\infty} kx^k = x + 2x^2 + 3x^3 + \cdots$ for those values of x for which this series converges.

If

$$f(n) = x + 2x^2 + 3x^3 + \cdots + (n - 1)x^{n-1} + nx^n$$

then

$$f(n - 1) = x + 2x^2 + 3x^3 + \cdots + (n - 1)x^{n-1}$$

and so

$$f(n) - f(n - 1) = nx^n \tag{8}$$

As in the preceding examples, the homogeneous equation

$$f(n) - f(n - 1) = 0$$

has the constant function α as its general solution.

Substituting the trial solution

$$p(n) = (A_1 n + A_0)x^n \tag{9}$$

into (8) gives

$$(A_1 n + A_0)x^n - (A_1(n - 1) + A_0)x^{n-1} = nx^n$$

Division by x^{n-1} yields

$$(A_1 n + A_0)x - A_1(n - 1) - A_0 = nx$$

Therefore, we have

$$(A_1 x - A_1)n + A_1 + A_0(x - 1) = nx$$

Equating coefficients of like powers of n gives

$$A_1 x - A_1 = x \quad \text{and} \quad A_1 + A_0(x - 1) = 0$$

Therefore,

$$A_1 = \frac{x}{x - 1} \quad \text{and} \quad A_0 = -\frac{x}{(x - 1)^2}$$

and from (9) we have

$$p(n) = \left(\frac{x}{x - 1} n - \frac{x}{(x - 1)^2} \right) x^n = \frac{x^{n+1}}{x - 1} \left(n - \frac{1}{x - 1} \right)$$

Consequently,

$$f(n) = \alpha + \frac{x^{n+1}}{x - 1} \left(n - \frac{1}{x - 1} \right) \tag{10}$$

for an appropriate choice of α. Since $f(1) = x$, substituting $n = 1$ into (10) gives

$$x = \alpha + \frac{x^2}{x - 1} \left(1 - \frac{1}{x - 1} \right) = \alpha + \frac{x^2(x - 2)}{(x - 1)^2}$$

and, hence,

$$\alpha = x - \frac{x^2(x - 2)}{(x - 1)^2} = \frac{x}{(x - 1)^2} \tag{11}$$

From (10) and (11) we see that the nth partial sum is

$$f(n) = \frac{x}{(x - 1)^2} + \frac{x^{n+1}}{x - 1} \left(n - \frac{1}{x - 1} \right)$$

Thus, we find that the sum S of the series $\sum_{k=1}^{\infty} kx^k$ is given by

$$S = \lim_{n \to \infty} f(n) = \frac{x}{(x - 1)^2} + \frac{1}{x - 1} \lim_{n \to \infty} (x^{n+1} n) - \frac{1}{(x - 1)^2} \lim_{n \to \infty} x^{n+1}$$

If $|x| < 1$, then $\lim_{n \to \infty} (x^{n+1}) = 0$, and it can be shown that $\lim_{n \to \infty} (x^{n+1} n) = 0$; therefore,

$$\sum_{k=1}^{\infty} kx^k = \frac{x}{(x - 1)^2} \qquad \text{if } |x| < 1$$

If $|x| \geq 1$, then $\lim_{n \to \infty} x^{n+1} n$ does not exist, and hence, the series $\sum_{k=1}^{\infty} kx^k$ diverges. $\qquad\Box$

Example 6   From Example 5 we have

$$\sum_{k=1}^{\infty} \frac{k}{3^k} = \sum_{k=1}^{\infty} k\left(\frac{1}{3}\right)^k = \frac{\frac{1}{3}}{(\frac{1}{3} - 1)^2} = \frac{3}{4} \qquad \square$$

The following theorem, whose proof will be left as an exercise (see Problems 20–22), is often quite useful in determining the sum of a series.

Theorem 10.15   If $\sum_{k=1}^{\infty} a_k = S_1$ and $\sum_{k=1}^{\infty} b_k = S_2$ are two convergent series, then the following series are convergent and have the indicated sums:

(a) $\displaystyle\sum_{k=1}^{\infty} (a_k + b_k) = S_1 + S_2$

(b) $\displaystyle\sum_{k=1}^{\infty} \alpha a_k = \alpha S_1$

(c) $\displaystyle\sum_{k=N}^{\infty} a_k = S_1 - (a_1 + a_2 + \cdots + a_{N-1})$

Example 7   We use Theorem 10.15 to compute the sum of the series $\sum_{k=3}^{\infty} 10k(\frac{1}{3})^k$. By part (b) of Theorem 10.15 we have

$$\sum_{k=3}^{\infty} 10k\left(\frac{1}{3}\right)^k = 10 \sum_{k=3}^{\infty} k\left(\frac{1}{3}\right)^k \tag{12}$$

It now follows from (12) and part (c) of Theorem 10.15 that

$$\sum_{k=3}^{\infty} 10k\left(\frac{1}{3}\right)^k = 10\left(\sum_{k=1}^{\infty} k\left(\frac{1}{3}\right)^k - \left(\frac{1}{3} + \frac{2}{9}\right)\right)$$

Since by Example 6, $\sum_{k=1}^{\infty} k(\frac{1}{3})^k = \frac{3}{4}$, we have

$$\sum_{k=3}^{\infty} 10k\left(\frac{1}{3}\right)^k = 10\left(\frac{3}{4} - \frac{5}{9}\right) = \frac{35}{18} \qquad \square$$

10.3 Problems

In Problems 1–17 find the sum of the given series or state that the series is divergent.

1. $\displaystyle\sum_{k=0}^{\infty} 5\left(\frac{1}{3}\right)^k$

2. $\displaystyle\sum_{k=0}^{\infty} k\left(-\frac{1}{2}\right)^k$

3. $\displaystyle\sum_{k=0}^{\infty} 3^k$

4. $\displaystyle\sum_{k=2}^{\infty} \frac{5}{2^k}$

5. $\displaystyle\sum_{k=1}^{\infty} 7\left(-\frac{1}{2}\right)^k$

6. $\displaystyle\sum_{k=1}^{\infty} 2^k$

7. $\displaystyle\sum_{k=1}^{\infty} \frac{k}{2^k}$

8. $\displaystyle\sum_{k=1}^{\infty} \frac{k}{5^k}$

9. $\displaystyle\sum_{k=2}^{\infty} k\left(\frac{1}{3}\right)^k$

10. $\displaystyle\sum_{k=1}^{\infty} (-1)^k k \left(\frac{1}{3}\right)^k$ 11. $\displaystyle\sum_{k=0}^{\infty} \frac{1+k}{3^k}$ 12. $\displaystyle\sum_{k=0}^{\infty} \left(-\frac{1}{6}\right)^k (1+k)$

13. $\displaystyle\sum_{k=3}^{\infty} (-1)^k \frac{k+2}{2^k}$ 14. $\displaystyle\sum_{k=2}^{\infty} \left(\frac{1}{7}\right)^k (3-k)$ 15. $\displaystyle\sum_{n=1}^{\infty} \left(\frac{1}{n} - \frac{1}{n+1}\right)$

16. $\displaystyle\sum_{n=1}^{\infty} \left(\frac{1}{n+2} - \frac{1}{n}\right)$ 17. $\displaystyle\sum_{n=1}^{\infty} \ln \frac{n}{n+1}$

18. Show that the series $\sum_{k=1}^{\infty} k$ is divergent.

19. Show that the series $\sum_{k=0}^{\infty} ax^k$, $a \neq 0$, is divergent for $x = 1$.

In Problems 20–22 assume that the series $\sum_{k=1}^{\infty} a_k$ and $\sum_{k=1}^{\infty} b_k$ are convergent.

20. Show that $\displaystyle\sum_{k=1}^{\infty} (a_k + b_k) = \sum_{k=1}^{\infty} a_k + \sum_{k=1}^{\infty} b_k$.

21. Show that $\displaystyle\sum_{k=1}^{\infty} \alpha a_k = \alpha \sum_{k=1}^{\infty} a_k$ where α is constant.

22. Show that $\displaystyle\sum_{k=N}^{\infty} a_k = \sum_{k=1}^{\infty} a_k - (a_1 + a_2 + \cdots + a_{N-1})$.

23. Let K be a positive integer. Show that $\sum_{k=1}^{\infty} a_{k+K}$ converges if and only if $\sum_{k=1}^{\infty} a_k$ converges.

24. Find the total distance travelled by the ball described in Problem 11 of Section 1.

25. Assume that 50% of every dollar expended in a particular community is again spent in that community and the remainder is spent outside the community.
    (a) Explain why $1 + \frac{1}{2} + \frac{1}{4} + \frac{1}{8} + \cdots$ measures the impact of one dollar spent in the community.
    (b) Find the sum of the infinite series given in (a).

26. Two people play a simple game that consists of flipping a fair coin. The first person to obtain "heads" wins. According to probability theory, the probability of winning this game is

$$\frac{1}{2} + \frac{1}{8} + \frac{1}{32} + \cdots + \frac{1}{2^{2n-1}} + \cdots$$

Compute this probability by calculating the sum of this infinite series.

10.4   GENERATING FUNCTIONS

Although, historically, generating functions arose from the study of certain questions in the theory of probability, these functions are perhaps most naturally introduced as a method of solving difference equations.

Generating functions are associated with sequences, and are defined in terms of infinite series.

Definition 10.16   Let $\{f(n)\}_0^\infty$ be a sequence. The *generating function* $F: \mathbf{R}^1 \to \mathbf{R}^1$ of f is defined by

$$F(x) = \sum_{k=0}^{\infty} f(k)x^k$$

At times we will want to emphasize the fact that the generating function F arises from a particular sequence f. In this case we use $X(f)$ in place of F; that is, we write

$$X(f)(x) = \sum_{k=0}^{\infty} f(k)x^k$$

Example 1   We compute the generating function F of the sequence f where

$$f(n) = \begin{cases} 1 & \text{if } n = 0 \text{ or } 1 \\ 0 & \text{otherwise} \end{cases}$$

By definition

$$F(x) = \sum_{k=0}^{\infty} f(k)x^k = f(0) + f(1)x + f(2)x^2 + \cdots$$
$$= 1 + x + 0x^2 + 0x^3 + \cdots$$
$$= 1 + x$$

Thus, $F(x) = X(f)(x) = 1 + x$.   □

Example 2   We compute $X(f)(x)$ if $f(n) = 1$ for all n. By definition

$$X(f)(x) = \sum_{k=0}^{\infty} f(k)x^k$$

Since $f(k) = 1$ for all k, we have

$$X(f)(x) = \sum_{k=0}^{\infty} x^k$$

This is a geometric series with sum $1/(1 - x)$ whenever $|x| < 1$; hence,

$$X(f)(x) = \frac{1}{1 - x} \quad \text{for } |x| < 1$$ □

Example 3   We compute the generating function of the sequence $f(n) = \alpha^n$. By definition

$$X(f)(x) = \sum_{k=0}^{\infty} \alpha^k x^k = \sum_{k=0}^{\infty} (\alpha x)^k$$

This is a geometric series with sum $1/(1 - \alpha x)$ whenever $|\alpha x| < 1$. Thus, if $f(n) = \alpha^n$, then

$$X(f)(x) = \frac{1}{1 - \alpha x} \quad \text{for } |x| < \frac{1}{|\alpha|} \qquad \square$$

Example 4 We compute the generating function associated with $f(n) = n\alpha^n$ where $\alpha > 0$. From the definition of a generating function we have

$$X(f)(x) = \sum_{k=0}^{\infty} k\alpha^k x^k = \sum_{k=0}^{\infty} k(\alpha x)^k$$

From Example 5 of Section 10.3 we know that this series converges with sum $\alpha x/(\alpha x - 1)^2$ if $|\alpha x| < 1$. It follows that if $f(n) = n\alpha^n$, then

$$X(f)(x) = \frac{\alpha x}{(\alpha x - 1)^2} \quad \text{for } |x| < \frac{1}{\alpha} \qquad \square$$

Since the sum of an infinite series does not always exist, there is some question about the existence of generating functions. In general, if a sequence f does not grow "too rapidly," then its generating function will exist in some interval centered at 0. This is the content of the following theorem, whose proof is omitted.

Theorem 10.17 If a sequence f has the property that for each n, $|f(n)| < AR^n$ for some positive numbers R and A, then $X(f)(x)$ exists for each x such that $|x| < 1/A$.

Before we can exploit generating functions to solve difference equations, we must develop some of the elementary properties of these functions. Our first result states that the generating function of a sequence whose value at n is $f(n) + g(n)$ is the sum of the generating functions of f and g.

Theorem 10.18 Let f and g be sequences. If $X(f)(x)$ and $X(g)(x)$ both exist, then $X(f + g)(x)$ exists and $X(f + g)(x) = X(f)(x) + X(g)(x)$.

Proof: The proof follows easily from the definition of a generating function and Theorem 10.15 since

$$
\begin{aligned}
X(f + g)(x) &= \sum_{k=0}^{\infty} (f(k) + g(k))x^k \\
&= \sum_{k=0}^{\infty} f(k)x^k + \sum_{k=0}^{\infty} g(k)x^k \\
&= X(f)(x) + X(g)(x)
\end{aligned}
$$

as was to be shown.   ■

An equally straightforward proof establishes that constants can be "factored across X," as indicated in the next theorem.

Theorem 10.19 If f is a sequence and if $X(f)(x)$ exists, then for each constant α, $X(\alpha f)(x) = \alpha X(f)(x)$.

In what follows it will often be convenient to denote the generating function of a sequence such as $f(n) = 7 \cdot 3^n$ by $X(7 \cdot 3^n)$.

Example 5 We compute the generating function of $f(n) = 5 \cdot 2^n + 3n$. From Theorems 10.18 and 10.19 we have

$$
\begin{aligned}
X(f)(x) &= X(5 \cdot 2^n + 3n)(x) \\
&= X(5 \cdot 2^n)(x) + X(3n)(x) \\
&= 5X(2^n)(x) + 3X(n)(x)
\end{aligned}
\tag{1}
$$

If in Example 3 we set $\alpha = 2$, then we find that

$$
X(2^n)(x) = \frac{1}{1 - 2x} \quad \text{for } |x| < \frac{1}{2}
$$

and if in Example 4 we set $\alpha = 1$, then we find that

$$
X(n)(x) = \frac{x}{(x - 1)^2} \quad \text{for } |x| < 1
$$

On substituting these expressions into (1) we see that

$$
X(5 \cdot 2^n + 3n)(x) = \frac{5}{1 - 2x} + \frac{3x}{(x - 1)^2} \quad \text{for } |x| < \frac{1}{2} \qquad \square
$$

We need one final result, which relates the generating function of a sequence f to the generating function of the sequence g defined by $g(n) = f(n + 1)$. Such "shifting" results provide the key to using generating functions to solve difference equations.

Theorem 10.20 *Shifting Theorem.* Let f be a sequence and let g be the sequence defined by $g(n) = f(n + 1)$. If $X(f)(x)$ exists, then

$$
X(g)(x) = \frac{1}{x} [X(f)(x) - f(0)]
$$

or, equivalently

$$
X(f(n + 1))(x) = \frac{1}{x} [X(f(n))(x) - f(0)]
$$

Proof: By definition

$$X(g)(x) \; = \; \sum_{k=0}^{\infty} g(k)x^k$$

Thus, since $g(n) = f(n + 1)$, we have

$$
\begin{aligned}
X(g)(x) \; = \; X(g(n))(x) \; = \; X(f(n + 1))(x) \; &= \; \sum_{k=0}^{\infty} f(k + 1)x^k \\
&= \; f(1) + f(2)x + f(3)x^2 + \cdots + f(k + 1)x^k + \cdots \\
&= \; \frac{1}{x} [f(1)x + f(2)x^2 + f(3)x^3 \cdots + f(k + 1)x^{k+1} \cdots] \\
&= \; \frac{1}{x} (f(0) + [f(1)x + f(2)x^2 \\
&\quad\;\; + \cdots + f(k + 1)x^{k+1} + \cdots] - f(0)) \\
&= \; \frac{1}{x} [X(f)(x) - f(0)]
\end{aligned}
$$

as was to be shown.   ∎

The next theorem, a generalization of the preceding result, can be proven either by induction or by generalizing the proof of Theorem 10.20. We leave this choice to you (see Problem 17).

Theorem 10.21   Let f be a sequence and let g be the sequence defined by $g(n) = f(n + q)$ where q is a positive integer. Then

$$X(g)(x) \; = \; \frac{1}{x^q} [X(f)(x) - f(0) - f(1)x - f(2)x^2 - \cdots - f(q - 1)x^{q-1}]$$

In Table 10.1 we list a number of sequences and their corresponding generating functions. For convenience, we have listed generating functions that we have already derived, some that will be established later, and some whose derivation requires calculus and so are beyond the scope of this text. Table 10.1, while far from complete, is adequate for solving the exercises in this text.

The next example illustrates the use of Table 10.1.

Example 6   If $f(n) = 6n^2 3^n - 4^n$, then it follows from 5, 8, 14, and 15 of Table 10.1 that

$$X(f)(x) \; = \; 6 \, \frac{9x^2 + 3x}{(1 - 3x)^3} - \frac{1}{1 - 4x} \qquad\qquad \square$$

Sequence, value at n	Generating Function, value at x
1. $f(n)$	$\displaystyle\sum_{n=0}^{\infty} f(n)x^n$
2. 0	0
3. 1	$\dfrac{1}{1-x}$
4. $(-1)^n$	$\dfrac{1}{1+x}$
5. α^n	$\dfrac{1}{1-\alpha x}$
6. $n\alpha^n$	$\dfrac{\alpha x}{(1-\alpha x)^2}$
7. $\alpha^n + n\alpha^n$	$\dfrac{1}{(1-\alpha x)^2}$
8. $n^2\alpha^n$	$\dfrac{\alpha^2 x^2 + \alpha x}{(1-\alpha x)^3}$
9. $\dfrac{\alpha^{n+1} - \beta^{n+1}}{\alpha - \beta}$	$\dfrac{1}{(1-\alpha x)(1-\beta x)}$
10. $\alpha^n \cos n\theta$	$\dfrac{1 - \alpha x \cos \theta}{1 + \alpha^2 x^2 - 2\alpha x \cos \theta}$
11. $\alpha^n \sin n\theta$	$\dfrac{\alpha x \sin \theta}{1 + \alpha^2 x^2 - 2\alpha x \cos \theta}$
12. $f(n) = \begin{cases} 1 & 0 \le n \le k \\ 0 & n \ge k+1 \end{cases}$	$\dfrac{1 - x^{k+1}}{1-x}$
13. $f(n) = \dbinom{n+k-1}{n}$	$(1-x)^{-k}$
14. $f(n) + g(n)$	$X(f)(x) + X(g)(x)$
15. $\alpha f(n)$	$\alpha X(f)(x)$
16. $f(n+q)$	$\dfrac{1}{x^q}\,[X(f)(x) - f(0) - f(1)x$ $- \cdots - f(q-1)x^{q-1}]$
17. $f(0)g(n) + f(1)g(n-1)$ $+ \cdots + f(n)g(0)$	$X(f)(x)X(g)(x)$

Table 10.1

In the next section we will often need to work backwards: We will start with a generating function and then find the sequence associated with it. The next example illustrates this process.

Example 7   Given the function

$$F(x) = \frac{1}{1 - 7x + 12x^2} = \frac{1}{(1 - 3x)(1 - 4x)}$$

we find from 9 of Table 10.1 that F is the generating function of the sequence

$$f(n) = \frac{3^{n+1} - 4^{n+1}}{3 - 4} = 4^{n+1} - 3^{n+1} \qquad \square$$

10.4   Problems

In Problems 1–4 use Definition 10.16 to compute $X(f)(x)$.

1. $f(n) = \begin{cases} 1 & 0 \le n \le 5 \\ 0 & n \ge 6 \end{cases}$

2. $f(n) = \begin{cases} 0 & 0 \le n < 2 \\ 1 & 3 \le n \le 5 \\ 0 & n > 5 \end{cases}$

3. $f(0) = 5, f(1) = 4, f(2) = 3, f(3) = 2, f(4) = 1, f(n) = 0$, for $n \ge 5$.

4. $f(n) = \begin{cases} 0 & n \ne k \\ 1 & n = k \end{cases}$

In Problems 5–16 use Table 10.1 to compute the generating function of the given sequence.

5. $f(n) = \begin{cases} 1 & 0 \le n \le 20 \\ 0 & n \ge 21 \end{cases}$

6. $g(n) = (-1)^n n$

7. $f(n) = 2^n + n3^n$

8. $k(n) = \cos \dfrac{n\pi}{2}$

9. $h(n) = \sin \dfrac{n\pi}{2}$

10. $f(n) = n$

11. $f(n) = n^2$

12. $g(n) = \dbinom{n + 5}{n}$

13. $g(n) = \dbinom{n + 3}{n}$

14. $f(n) = \dfrac{3^{n+1} - 1}{2}$

15. $k(n) = 3^{-n} \cos \dfrac{n\pi}{4}$

16. $r(n) = 5^{-n} 2^n 3n$

17. Either use induction or generalize the proof of Theorem 10.20 to prove Theorem 10.21.

18. Show that if α is a constant then $X(\alpha f)(x) = \alpha X(f)(x)$.

19. Use 5 and 6 of Table 10.1 to establish 7 of Table 10.1.

20. Establish 2 of Table 10.1.

21. Establish 4 of Table 10.1.

22. Establish 12 of Table 10.1.

23. Prove Theorem 10.19.

10.5   GENERATING FUNCTIONS AND DIFFERENCE EQUATIONS

In this section we use generating functions to find particular solutions of difference equations. We can break the basic technique for doing this into five steps.

Step 1. Compute the generating functions of both sides of the difference equation

Step 2. Use Table 10.1 to simplify the equation resulting from Step 1.

Step 3. Substitute the initial conditions.

Step 4. Solve for $X(f)$, the generating function of the solution f.

Step 5. Use Table 10.1 to determine the solution f from the generating function $X(f)$.

The next example illustrates this technique.

Example 1   We find the particular solution to the homogeneous difference equation

$$f(n + 2) - 3f(n + 1) + 2f(n) = 0 \qquad (1)$$

that satisfies $f(0) = 3$ and $f(1) = 6$.

By first taking the generating function of both sides of (1) we obtain

$$X(f(n + 2) - 3f(n + 1) + 2f(n))(x) = X(0)(x)$$

Next we apply 14, 15, and 2 of Table 10.1 from the previous section to obtain

$$X(f(n + 2))(x) - 3X(f(n + 1))(x) + 2X(f(n))(x) = 0$$

Then from 16 of Table 10.1, it follows that

$$\frac{1}{x^2}[X(f(n))(x) - f(0) - f(1)x]$$

$$-\frac{3}{x}[X(f(n))(x) - f(0)] + 2X(f(n))(x) = 0 \quad (2)$$

If we substitute $f(0) = 3$ and $f(1) = 6$ into (2), and if we denote $X(f(n))$ by F, then we have

$$\frac{1}{x^2}[F(x) - 3 - 6x] - \frac{3}{x}[F(x) - 3] + 2F(x) = 0$$

We next solve for $F(x)$.
Since

$$F(x)\left[\frac{1}{x^2} - \frac{3}{x} + 2\right] = \frac{3}{x^2} + \frac{6}{x} - \frac{9}{x}$$

we have

$$F(x) = \frac{\dfrac{3}{x^2} - \dfrac{3}{x}}{\dfrac{1}{x^2} - \dfrac{3}{x} + 2} = \frac{3 - 3x}{1 - 3x + 2x^2} = \frac{3(1 - x)}{(1 - 2x)(1 - x)} = \frac{3}{1 - 2x}$$

or, equivalently,

$$X(f)(x) = \frac{3}{1 - 2x}$$

Finally, we see from 5 and 15 of Table 10.1 that if $X(f)(x) = 3/(1 - 2x)$, then $f(n) = 3 \cdot 2^n$ and hence, $3 \cdot 2^n$ is the solution of (1) that satisfies the given initial conditions.   ◻

Generating functions are equally useful for solving nonhomogeneous difference equations.

Example 2   We find the solution to the difference equation

$$f(n) - 4f(n + 1) + f(n + 2) = -9n2^n \tag{3}$$

satisfying the initial conditions $f(0) = 0$ and $f(1) = 6$.
The notation will be simplified if we immediately let $F(x) = X(f)(x)$. On taking generating functions of both sides of (3) we obtain

$$F(x) - \frac{4}{x}[F(x) - f(0)] + \frac{1}{x^2}[F(x) - f(0) - f(1)x] = -9\frac{2x}{(1 - 2x)^2}$$

Substituting $f(0) = 0$ and $f(1) = 6$ into this equation yields

$$F(x)\left(1 - \frac{4}{x} + \frac{1}{x^2}\right) - \frac{6}{x} = \frac{-18x}{(1 - 2x)^2}$$

Therefore,

$$F(x)\left(1 - \frac{4}{x} + \frac{1}{x^2}\right) = \frac{6(1 - 2x)^2 - 18x^2}{x(1 - 2x)^2} = \frac{6 - 24x + 6x^2}{x(1 - 2x)^2}$$

and consequently,

$$F(x) = \frac{6(1 - 4x + x^2)}{x(1 - 2x)^2 \left(1 - \dfrac{4}{x} + \dfrac{1}{x^2}\right)} = \frac{6x(1 - 4x + x^2)}{(1 - 2x)^2 (x^2 - 4x + 1)}$$

$$= \frac{6x}{(1 - 2x)^2} = 3\frac{2x}{(1 - 2x)^2}$$

From 6 of Table 10.1 we find that $f(n) = 3n2^n$ is the desired solution.

$\square$

Example 3   We find the solution to the difference equation

$$f(n + 2) + f(n) = 0 \tag{4}$$

that satisfies $f(0) = f(1) = 1$.

Let $F(x) = X(f)(x)$. On taking the generating function of both sides of (4) we have

$$\frac{1}{x^2}[F(x) - f(0) - f(1)x] + F(x) = 0$$

Substitution of $f(0) = f(1) = 1$ gives

$$\frac{1}{x^2}[F(x) - 1 - x] + F(x) = 0$$

Thus,

$$F(x)\left(\frac{1}{x^2} + 1\right) = \frac{x + 1}{x^2}$$

and hence,

$$F(x) = \frac{x + 1}{x^2\left(\dfrac{1}{x^2} + 1\right)} = \frac{x + 1}{1 + x^2} = \frac{x}{1 + x^2} + \frac{1}{1 + x^2} \tag{5}$$

Now note that if we let $\alpha = 1$ and $\theta = \pi/2$ in 10 of Table 10.1, then we see that the generating function of $\cos(n\pi/2)$ is $1/(1 + x^2)$. Moreover, if $\alpha = 1$ and $\theta = \pi/2$, then from 11 of the same table we have that the generating function of $\sin(n\pi/2)$ is $x/(1 + x^2)$. Therefore, it follows from (5) that $f(n) = \sin(n\pi/2) + \cos(n\pi/2)$ is the solution of (4) that satisfies the given initial conditions.

$\square$

At this point it is only possible to solve those difference equations that have solutions resulting in easily-recognizable generating functions. In the next sections you will learn techniques that will greatly increase your ability to determine the function from which a given generating function is derived.

10.5   Problems

In Problems 1–19 use generating functions to find the indicated particular solution to the difference equation.

1. $f(n + 2) - 4f(n + 1) + 4f(n) = 0$; $f(0) = 0$, $f(1) = 2$
2. $f(n + 2) + 6f(n + 1) + 9f(n) = 0$; $f(0) = 0$, $f(1) = -3$
3. $g(n + 2) + 2g(n + 1) + g(n) = 0$; $g(0) = 0$, $g(1) = -1$
4. $f(n + 2) + 8f(n + 1) + 16f(n) = 0$; $f(0) = 0$, $f(1) = 4$
5. $f(n + 1) + f(n - 1) = 0$; $f(0) = f(1) = 0$
6. $f(n) - 3f(n - 1) + 2f(n - 2) = 0$; $f(0) = -2$, $f(1) = -2$
7. $k(n + 2) - k(n) = 0$; $k(0) = k(1) = 1$
8. $f(n) - 3f(n - 1) + 2f(n - 2) = 0$; $f(0) = -2$, $f(1) = -4$
9. $f(n + 2) - 3f(n + 1) - 10f(n) = 0$; $f(0) = 2$, $f(1) = 10$
10. $h(n + 2) - 8h(n + 1) + 16h(n) = 0$; $h(0) = 1$, $h(1) = 8$
11. $f(n + 2) - 5f(n + 1) - 6f(n) = 0$; $f(0) = 1$, $f(1) = 6$
12. $f(n + 2) - 5f(n + 1) - 6f(n) = 0$; $f(0) = 1$, $f(1) = -1$
13. $g(n + 1) - 13g(n) + 9g(n - 1) = -7(n3^n)$; $g(0) = 0$, $g(1) = 3$
14. $f(n + 2) + f(n + 1) + f(n) = 31(5^n)$; $f(0) = 1$, $f(1) = 5$
15. $f(n + 2) + f(n + 1) + f(n) = 7(-3)^n$; $f(0) = 1$, $f(1) = -3$
16. $h(n) + 3h(n - 1) - 2h(n - 2) = 2n + 5$; $h(0) = 0$, $h(1) = 1$
17. $f(n + 2) + 3f(n + 1) - 2f(n) = 4n + 10$; $f(0) = 0$, $f(1) = 2$
18. $f(n) - 3f(n - 1) + 5f(n - 2) = 3n + 2$; $f(0) = 1$, $f(1) = 2$
19. $f(n + 2) - 4f(n + 1) + 4f(n) = 0$; $f(0) = 1$, $f(1) = 4$

10.6   PARTIAL FRACTIONS AND THE PROBLEM OF INVERSION

Using generating functions to solve difference equations depends on our ability to find a sequence (the solution to the difference equation) from its generating function. Although many of the solutions of the "inversion problem" are beyond the scope of this text, the one we shall study, partial fractions, will significantly improve our ability to recognize generating functions.

A *rational function* is a function that can be expressed as the quotient of two polynomials. If the degree of the denominator is less than or equal to that of the numerator, the rational function is called an *improper fraction*; otherwise it is called a *proper fraction*.

For example,

$$f(x) = \frac{x^2 + 3x - 1}{x^3 + 2} \qquad g(x) = \frac{x^3 + 2x - 1}{5x} \qquad h(x) = \frac{x^4 - 7x^2}{x + 1}$$

are all rational functions; f is a proper fraction and g and h are improper fractions.

As the next example illustrates, an improper fraction can be expressed as the sum of a polynomial and a proper fraction.

Example 1 Using long division we obtain

(a) $\dfrac{x^5 + 3x^4 - 3x^3 - 6x^2 + 3x + 1}{x^2 + 3x - 1} = x^3 - 2x + \dfrac{x + 1}{x^2 + 3x - 1}$

(b) $\dfrac{5x^2 - 9x + 7}{x^2 - 2x + 1} = 5 + \dfrac{x + 2}{x^2 - 2x + 1}$ □

It can be shown that every polynomial with real coefficients may be factored as a product of linear factors (factors of the form $ax + b$), and quadratic factors (factors of the form $ax^2 + bx + c$) that cannot be further factored into linear factors with real coefficients.

Example 2 The polynomial $x^3 - 8$ can be written as $(x - 2)(x^2 + 2x + 4)$ where $(x - 2)$ is a linear factor and $(x^2 + 2x + 4)$ is a quadratic factor that cannot be further factored into real linear factors since it has no real zeros. □

It can be shown that every proper fraction can be expressed as the sum of fractions (called partial fractions). This sum, which is called a *partial fraction decomposition*, can be obtained as follows.

Let $f(x)$ be a proper fraction and assume that the denominator of $f(x)$ has been factored into linear and quadratic factors where the quadratic factors cannot be factored into linear factors with real coefficients. For each linear factor $L(x)$ that occurs r times in the denominator of f, we use a sum of the form

$$\frac{a_1}{L(x)} + \frac{a_2}{[L(x)]^2} + \cdots + \frac{a_r}{[L(x)]^r}$$

For each quadratic factor $q(x)$ that occurs s times in the denominator of f, we use a sum of the form

$$\frac{c_1 x + d_1}{q(x)} + \frac{c_2 x + d_2}{[q(x)]^2} + \cdots + \frac{c_s x + d_s}{[q(x)]^s}$$

The following examples illustrate this idea.

Example 3
(a) The form of the partial fraction decomposition of

$$\frac{3x + 4}{(x - 1)(x + 2)}$$

is

$$\frac{a}{x - 1} + \frac{b}{x + 2}$$

where the constants a and b are to be determined.

(b) The form of the partial fraction decomposition of

$$\frac{x^4 - 2x^2 + 3x - 5}{(x - 1)^2(x + 2)(x^2 + 4)^2}$$

is

$$\frac{a}{x - 1} + \frac{b}{(x - 1)^2} + \frac{c}{x + 2} + \frac{dx + e}{x^2 + 4} + \frac{gx + h}{(x^2 + 4)^2}$$

corresponds
to the factor
$(x - 1)^2$

corresponds
to the factor
$x + 2$

corresponds
to the factor
$(x^2 + 4)^2$

where the constants a, b, c, d, e, g, and h are to be determined.

(c) The form of the partial fraction decomposition of

$$\frac{x^3 + 3x + 1}{(x + 1)^3(x + 3)(x - 2)}$$

is

$$\frac{a}{x + 1} + \frac{b}{(x + 1)^2} + \frac{c}{(x + 1)^3} + \frac{d}{x + 3} + \frac{e}{x - 2}$$

where the constants a, b, c, d, and e are to be determined.   □

In each of the above examples there remains the problem of determining the constants. Examples 4 and 5 illustrate the two commonly used methods for obtaining these values.

Example 4   From part (a) of Example 3 we have

$$\frac{3x + 4}{(x - 1)(x + 2)} = \frac{a}{x - 1} + \frac{b}{x + 2} \tag{1}$$

where the constants a and b are to be determined. To find these values we first multiply both sides of equation (1) by the common denominator $(x - 1)(x + 2)$ to obtain

$$3x + 4 = a(x + 2) + b(x - 1) \tag{2}$$

We need to determine values of the constants a and b such that equation (2) is valid for *all* real numbers x. To find these values we choose two convenient values for x, $x = -2$ and $x = 1$, to obtain

$$3(-2) + 4 = a(-2 + 2) + b(-2 - 1)$$

and

$$3(1) + 4 = a(1 + 2) + b(1 - 1)$$

from which it follows that

$$b = \frac{2}{3} \quad \text{and} \quad a = \frac{7}{3}$$

Thus, we see that

$$\frac{3x + 4}{(x - 1)(x + 2)} = \frac{\frac{7}{3}}{x - 1} + \frac{\frac{2}{3}}{x + 2} \qquad \Box$$

The second method for determining the desired constants is to use the fact that if two polynomials take on the same value for each value of x, then their corresponding coefficients must be equal.

Example 5   As in the previous example, we can write

$$\frac{3x + 4}{(x - 1)(x + 2)} = \frac{a}{x - 1} + \frac{b}{x + 2}$$

and hence

$$3x + 4 = a(x + 2) + b(x - 1) \tag{3}$$

From (3) we have

$$3x + 4 = (a + b)x + (2a - b)$$

Setting corresponding coefficients equal, we obtain the system of equations

$$a + b = 3$$
$$2a - b = 4$$

On solving this system we find that $a = \frac{7}{3}$ and $b = \frac{2}{3}$, as before. $\Box$

In the next example we apply the technique of partial fractions to the inversion problem: the problem of finding the function whose generating function is given.

Example 6   Suppose that $X(f)(x) = (3x + 4)/(x - 1)(x + 2)$.   Then from Example 4 we have

$$X(f) = \frac{\frac{7}{3}}{x - 1} + \frac{\frac{2}{3}}{x + 2} = \frac{-7}{3}\left(\frac{1}{1 - x}\right) + \frac{1}{3}\left(\frac{1}{1 - (-\frac{1}{2}x)}\right)$$

It now follows from 5 and 15 of Table 10.1 that

$$f(n) = \frac{-7}{3}(1)^n + \frac{1}{3}\left(-\frac{1}{2}\right)^n = \frac{-7}{3} + \frac{1}{3(-2)^n} \qquad \square$$

These techniques can be used to solve certain difference equations, as is illustrated in the next example.

Example 7   We use generating functions to solve the difference equation

$$f(n + 2) - 10f(n + 1) + 25f(n) = 0 \tag{4}$$

with initial conditions $f(0) = 1$ and $f(1) = 15$. Taking the generating function of both sides of (4) we have

$$X(f(n + 2))(x) - 10X(f(n + 1))(x) + 25X(f(n))(x) = 0$$

If we let $F(x) = X(f(n))(x)$, then from Theorem 10.21 we have

$$\frac{1}{x^2}(F(x) - f(0) - f(1)x) - \frac{10}{x}(F(x) - f(0)) + 25F(x) = 0$$

Substituting $f(0) = 1$ and $f(1) = 15$ into the above equation gives

$$\frac{1}{x^2}(F(x) - 1 - 15x) - \frac{10}{x}(F(x) - 1) + 25F(x) = 0$$

and solving for $F(x)$ we obtain

$$F(x) = \frac{\dfrac{1}{x^2} + \dfrac{5}{x}}{\dfrac{1}{x^2} - \dfrac{10}{x} + 25} = \frac{5x + 1}{25x^2 - 10x + 1} = \frac{5x + 1}{(5x - 1)^2}$$

The form of the partial fraction decomposition of $F(x)$ is

$$\frac{5x + 1}{(5x - 1)^2} = \frac{a}{5x - 1} + \frac{b}{(5x - 1)^2}$$

Using either of the methods given in Examples 4 and 5, we find that $a = 1$ and $b = 2$. Consequently,

$$F(x) = \frac{1}{5x - 1} + \frac{2}{(5x - 1)^2} = \frac{-1}{1 - 5x} + \frac{2}{(1 - 5x)^2}$$

and from Table 10.1 we see that

$$f(n) = -(5^n) + 2(5^n + n5^n) = 5^n + 2n5^n$$

is the desired solution of (4). $\qquad \square$

10.6   Problems

In Problems 1–8 give the form of the partial fraction decomposition of the given fractions (without computing the constants).

1. $\dfrac{2x - 1}{(x + 7)(x - 3)(x - 2)}$

2. $\dfrac{3x}{(x + 1)(x - 1)(x + 2)}$

3. $\dfrac{7x + 10}{x(x - 3)(x - 4)^2}$

4. $\dfrac{2x - 11}{x(x + 4)^2(x - 1)}$

5. $\dfrac{3}{(x^2 - 2x + 2)(x - 1)^2}$

6. $\dfrac{2}{(x^2 - 7x + 20)^2(x + 1)}$

7. $\dfrac{5}{(x - 1)^3(x^2 + 2x + 2)^2}$

8. $\dfrac{10x - 11}{(x + 2)^2(x^2 - 3x + 4)^2}$

In Problems 9–16 compute the partial fraction decomposition of the given fraction (compute the required constants.)

9. $\dfrac{3x + 7}{(x - 3)(x + 5)}$

10. $\dfrac{2x + 11}{(x + 1)(x - 2)}$

11. $\dfrac{-2x + 2}{x(x + 2)}$

12. $\dfrac{6x - 2}{x(x - 1)}$

13. $\dfrac{-3x + 19}{x^2 - 9x + 20}$

14. $\dfrac{-7x + 40}{x^2 - 5x - 14}$

15. $\dfrac{x^2 - 7}{(x + 2)(x^2 + 3x + 5)}$

16. $\dfrac{x^2 - 8x - 8}{(x - 1)(x^2 + 2x + 2)}$

In Problems 17–23 use generating functions to solve the given difference equation.

17. $f(n + 2) - f(n + 1) - 6f(n) = 0; f(0) = 3, f(1) = -1$
18. $f(n + 2) - 3f(n + 1) + 2f(n) = 0; f(0) = 0, f(1) = 1$
19. $f(n + 2) - f(n + 1) - 6f(n) = 0; f(0) = 3, f(1) = 11$
20. $f(n + 2) - f(n + 1) - 6f(n) = 0; f(0) = -1, f(1) = -2$
21. $f(n + 2) - f(n - 1) - 6f(n) = 0; f(0) = 1, f(1) = 2$
22. $f(n + 2) - 3f(n + 1) - 10f(n) = 0; f(0) = -2, f(1) = 6$
23. $f(n + 2) - 4f(n) = 0; f(0) = 1, f(1) = 4$
24. Use generating functions to determine the sum $\sum_{k=3}^{n} k^2$.
25. Use generating functions to determine how many bit strings of length n have no two adjacent 1's.

10.7 CONVOLUTION

It is natural to ask whether the product of two generating functions $X(f)$ and $X(g)$ is the generating function of a function that can be easily expressed in terms of f and g. In other words, if

$$(X(f)(x))(X(g)(x)) = X(h)(x)$$

what is h?

The product of two generating functions is, by definition, the product of two infinite series. Therefore, in order to multiply generating functions, we must first find a reasonable way for multiplying series. The product of two infinite series $\sum_{k=0}^{\infty} a_k x^k$ and $\sum_{k=0}^{\infty} b_k x^k$ should be an infinite series $\sum_{k=0}^{\infty} c_k x^k$ with the property that if for any given x, $\sum_{k=0}^{\infty} a_k x^k = A$ and $\sum_{k=0}^{\infty} b_k x^k = B$, then $\sum_{k=0}^{\infty} c_k x^k = AB$. It can be shown that if we use the following definition for multiplying infinite series, then this property will hold. Note that in this definition, series are multiplied together as if they were finite polynomials, and then like terms are gathered.

Definition 10.22 Let $\sum_{k=0}^{\infty} a_k x^k$ and $\sum_{k=0}^{\infty} b_k x^k$ be infinite series. Then the *product* of these series is defined to be the infinite series $\sum_{k=0}^{\infty} c_k x^k$ where for each k,

$$c_k = a_k b_0 + a_{k-1} b_1 + a_{k-2} b_2 + \cdots + a_0 b_k \tag{1}$$

Observe in Definition 10.22 that the sum of the subscripts in each term in the right-hand side of (1) is equal to k.

Example 1 The product of the infinite series

$$x - \frac{x^3}{3!} + \frac{x^5}{5!} - \frac{x^7}{7!} + \cdots$$

and

$$1 + x + x^2 + x^3 + \cdots$$

is

$$x + x^2 + \left(1 - \frac{1}{3!}\right)x^3 + \left(1 - \frac{1}{3!}\right)x^4 + \left(1 - \frac{1}{3!} + \frac{1}{5!}\right)x^5$$

$$+ \left(1 - \frac{1}{3!} + \frac{1}{5!}\right)x^6 + \left(1 - \frac{1}{3!} + \frac{1}{5!} - \frac{1}{7!}\right)x^7$$

$$+ \left(1 - \frac{1}{3!} + \frac{1}{5!} - \frac{1}{7!}\right)x^8 + \cdots \qquad \square$$

We now return to the original question of this section: If

$$(X(f)(x))(X(g)(x)) = X(h)(x) \qquad (2)$$

what is the function h? From the definition of a generating function, equation (2) becomes

$$\left(\sum_{k=0}^{\infty} f(k)x^k\right)\left(\sum_{k=0}^{\infty} g(k)x^k\right) = \sum_{k=0}^{\infty} h(k)x^k$$

It follows from Definition 10.22 that

$$h(k) = f(0)g(k) + f(1)g(k-1) + \cdots + f(k)g(0) \qquad (3)$$

The sequence h defined by (3) is called the *convolution* of the sequences f and g. This leads us to the following result.

Theorem 10.23   *Convolution Theorem.*   The product

$$(X(f)(x))(X(g)(x))$$

is the generating function of the convolution of f and g.

In the next two examples we see how to use the Convolution Theorem to invert generating functions.

Example 2   We use the Convolution Theorem to find h if

$$X(h)(x) = \frac{1}{(1-x)^2}$$

Note that

$$X(h)(x) = \left(\frac{1}{1-x}\right)\left(\frac{1}{1-x}\right)$$

and if we let $X(f)(x) = 1/(1-x) = X(g(x))$, then we have

$$X(h)(x) = X(f)(x)X(g)(x) = \frac{1}{(1-x)^2}$$

From Table 10.1 of Section 10.4 we have that $f(n) = 1 = g(n)$, and hence by the Convolution Theorem

$$h(n) = f(0)g(n) + f(1)g(n-1) + \cdots + f(n)g(0)$$
$$= 1 \cdot 1 + 1 \cdot 1 + \cdots + 1 \cdot 1 = n + 1 \qquad \square$$

Example 3   We invert the generating function

$$H(x) = \frac{1}{(1-\alpha x)(1-\beta x)}$$

Let

$$X(f)(x) = \frac{1}{1 - \alpha x} \quad \text{and} \quad X(g)(x) = \frac{1}{1 - \beta x}$$

Then it follows from Table 10.1 that $f(n) = \alpha^n$ and $g(n) = \beta^n$. Thus, $H(x)$ is the product of the generating functions of α^n and β^n. Consequently, by Theorem 10.23, H is the generating function of the convolution, h, of these two functions. It follows that

$$\begin{aligned}
h(n) &= f(0)g(n) + f(1)g(n-1) + f(2)g(n-2) + \cdots + f(n)g(0) \\
&= \alpha^0 \beta^n + \alpha^1 \beta^{n-1} + \alpha^2 \beta^{n-2} + \cdots + \alpha^n \beta^0 \\
&= \beta^n \left[1 + \frac{\alpha}{\beta} + \left(\frac{\alpha}{\beta}\right)^2 + \cdots + \left(\frac{\alpha}{\beta}\right)^n \right] \\
&= \beta^n \frac{\left(\dfrac{\alpha}{\beta}\right)^{n+1} - 1}{\dfrac{\alpha}{\beta} - 1} = \frac{\dfrac{\alpha^{n+1}}{\beta} - \beta^n}{\dfrac{\alpha}{\beta} - 1} \\
&= \frac{\alpha^{n+1} - \beta^{n+1}}{\alpha - \beta}
\end{aligned}$$

$\square$

Inverting generating functions is not the only way to use the Convolution Theorem. It is also useful when solving certain difference equations, as we illustrate in the following classical combinatorial problem.

Example 4   Suppose $2n$ points are taken on a circle. In how many ways can the points be joined in pairs by n nonintersecting chords?

Figure 10.2 shows the five distinct ways that six points can be joined in pairs.

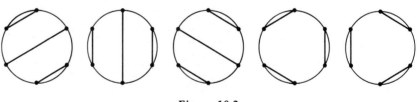

Figure 10.2

Let $f(n)$ denote the number of ways that $2n$ points can be joined by n nonintersecting chords. Now consider the problem for $2n + 2$ points. Starting at the top and proceeding in a clockwise direction around the circle, we designate the points by $P_1, P_2, \ldots, P_{2n+2}$ as shown in Figure 10.3.

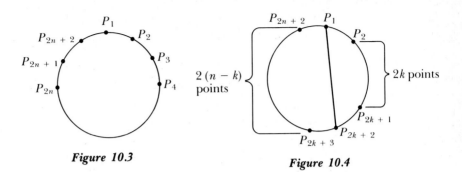

Figure 10.3   **Figure 10.4**

It is not difficult to see that any chord originating at P_1 must terminate at P_j for some j, where j is even (if a chord were to join, say, P_1 and P_3, then P_2 would be isolated and could not be joined by a nonintersecting chord to any other point). Assume then that a chord joins P_1 to P_{2k+2} as shown in Figure 10.4. This chord now separates the points on this circle into two sets, one containing $2k$ points and the other containing $2(n-k)$ points.

There are $f(k)$ ways of joining the set of $2k$ points in pairs using nonintersecting chords. Once these points are connected, there remain $f(n-k)$ ways of pairing the set of $2(n-k)$ points with nonintersecting chords. Consequently, if P_1 is joined to P_{2k+2}, then there are $f(k)f(n-k)$ ways to complete the pairing. Of course P_1 can be joined to *any* even-numbered point. That is, P_1 can be joined to P_2, P_4, P_6, $\ldots$, or P_{2n+2}. Table 10.2 summarizes the situation described thus far.

Point joined to P_1	k	Number of ways of joining remaining points with nonintersecting chords
P_2	0	$f(0)f(n)$
P_3	—	0
P_4	1	$f(1)f(n-1)$
P_5	—	0
P_6	2	$f(2)f(n-2)$
$\vdots$	$\vdots$	$\vdots$
P_{2n+2}	n	$f(n)f(0)$

Table 10.2

By the definition of the function f, there are $f(n+1)$ ways of joining the points $P_1, P_2, \ldots, P_{2n+2}$ with nonintersecting chords. On the other hand, if we add the entries in the third column of Table 10.2, then we also have the total number of ways that these points can be so joined. Thus, we obtain the difference equation

$$f(n + 1) = f(0)f(n) + f(1)f(n - 1) + \cdots + f(n)f(0)$$

$$= \sum_{k=0}^{n} f(k)f(n - k)$$

Convolution gives us a fairly direct method of solving this equation. Let $g(n) = \sum_{k=0}^{n} f(k)f(n - k)$ and note that g is the convolution of the sequence f with itself. It follows from Theorem 10.23 that the generating function of g is $(F(x))^2$, where F is the generating function of f. On taking the generating function of the left- and right-side of

$$f(n + 1) = \sum_{k=0}^{n} f(k)f(n - k) \tag{4}$$

we have

$$\frac{1}{x}[F(x) - 1] = (F(x))^2$$

Since $f(1) = 1$ (a single pair can be joined in only one way), and since from equation (4), $f(1) = f(0)f(0)$, we can assume that $f(0) = 1$. Thus, we have

$$\frac{1}{x}[F(x) - f(0)] = (F(x))^2$$

or, equivalently,

$$x(F(x))^2 - F(x) + 1 = 0 \tag{5}$$

We apply the quadratic formula to (5) to solve for $F(x)$ and obtain the two potential generating functions

$$F(x) = \frac{1 + \sqrt{1 - 4x}}{2x} \quad \text{and} \quad F(x) = \frac{1 - \sqrt{1 - 4x}}{2x} \tag{6}$$

It can be shown that the first of these functions is not a generating function of the desired sequence. It can also be shown that the second function F defined by (6) is the generating function of

$$f(n) = \frac{(2n)!}{n!(n + 1)!}$$

Thus, there are exactly $(2n)!/n!(n + 1)!$ ways to pair $2n$ points on a circle with nonintersecting chords. $\qquad\square$

10.7   **Problems**

In Problems 1–4 use Definition 10.22 to write the first four terms of the given products.

1. $(1 + x + x^2 + x^3 + x^4 + \cdots)(1 + 2x + 3x^2 + 4x^3 + 5x^4 + \cdots)$

2. $(1 - x + x^2 - x^3 + x^4 + \cdots)(1 + x + x^2 + x^3 + x^4 + \cdots)$

3. $(1 + x + x^2 + x^3 + x^4 + \cdots)(1 + x^2 + x^4 + x^6 + x^8 + \cdots)$

4. $\displaystyle\sum_{k=0}^{\infty} \frac{x^k}{k!} \cdot \sum_{k=0}^{\infty} \frac{(2x)^k}{k!}$

5. In how many ways can eight points on a circle be paired by using nonintersecting chords?

In Problems 6–9 compute the convolution of the sequences.

6. $f(n) = 1; g(n) = 1$ 7. $f(n) = n; g(n) = 1$

8. $f(n) = 1; g(n) = n^2$

9. $f(n) = g(n) = \begin{cases} n & \text{if } n = 0, 1, 2, 3 \\ 0 & \text{if } n > 3 \end{cases}$

10. What is the generating function for the convolution of the sequences $f(n) = \cos(n\pi/2)$ and $g(n) = \sin(n\pi/2)$?

11. Use the Convolution Theorem to find h where $X(h)(x) = 1/(1 - x)^3$.

12. If $f(n) = n$ and $X(g(n))(x) = X(f(n))(x) + x$, find g.

13. If $f(n) = 1$ and $X(g(n))(x) = X(f(n))(x) + 1 + x^2$, find g.

14. If $h(n) = c(1) + c(2) + \cdots + c(n)$, and $h(o) = 0$, then show that $X(h(n))(x) = X(c(n))(x)/(1 - x)$.

15. If $F(x) = (1 + x + x^2 + x^3)/(1 - x)$, find $f(n)$. (*Hint:* $1 + x + x^2 + x^3$ is the generating function of the sequence 1, 1, 1, 1, 0, 0, 0, ...)

16. If $X(g(n))(x) = X(f(n))(x) + a_0 + a_1 x + a_2 x^2$, determine the relationship between the sequences f and g.

17. Let $P(n)$ denote the number of ways of forming the product of n terms. For example, the product $x_1 x_2 x_3$ can be computed in two ways: $((x_1 x_2)x_3)$ and $(x_1(x_2 x_3))$, and therefore, $P(3) = 2$.
    (a) Show that $P(1) = 1$ and $P(4) = 5$.
    (b) Use the expression

$$((x_1 x_2 \cdots x_r)(x_{r+1} \cdots x_n))$$

to explain why

$$P(n) = \sum_{r=1}^{n-1} P(r)P(n - r)$$

   (c) Use the method of Example 4 to determine $P(n)$.

A COMBINATORIAL
10.8 VIEW OF GENERATING FUNCTIONS

Up to this point we have emphasized the role of generating functions in solving difference equations. In this and the next section we consider some examples that illustrate how to use generating functions to establish certain combinatorial results.

Consider the following problem. If $X(h)(x) = (x + x^2)^5(1 + x^3)^2$, what is h? By definition

$$X(h)(x) = h(0)x^0 + h(1)x^1 + h(2)x^2 + \cdots + h(n)x^n + \cdots$$

and thus we have

$$h(0)x^0 + h(1)x^1 + h(2)x^2 + \cdots + h(n)x^n + \cdots = (x + x^2)^5(1 + x^3)^2$$

It follows that for each nonnegative integer n, $h(n)$ is the coefficient of x^n in the product

$$(x + x^2)^5(1 + x^3)^2 = (x + x^2)(x + x^2)(x + x^2)(x + x^2)(x + x^2)$$
$$\times (1 + x^3)(1 + x^3) \tag{1}$$

Each term of the expanded form of this product is of the form

$$x^{e_1}x^{e_2}x^{e_3}x^{e_4}x^{e_5}x^{e_6}x^{e_7} \tag{2}$$

where the term x^{e_i} has been selected from the ith factor of the right-handed side of (1). Observe that for $1 \le i \le 5$, e_i is equal to either 1 or 2, while for $i = 6$ or $i = 7$, e_i is equal to either 0 or 3.

Note too that the term (2) will contribute to the coefficient $h(n)$ of x^n if and only if

$$e_1 + e_2 + \cdots + e_7 = n \tag{3}$$

Therefore, for each n, $h(n)$ is equal to the number of solutions to (3) where $e_1, e_2, \ldots, e_5 \in \{1, 2\}$ and $e_6, e_7 \in \{0, 3\}$. Thus, for example, $h(5) = 1$ since there is only one solution to

$$e_1 + e_2 + \cdots + e_7 = 5$$

that satisfies the given criteria. (The solution is $e_1 = e_2 = \cdots = e_5 = 1$ and $e_6 = e_7 = 0$.) Note also that $h(n) = 0$ for all $n \ge 17$. You may verify by multiplying out (1) that

$$X(h)(x) = x^{16} + 5x^{15} + 10x^{14} + 12x^{13} + 15x^{12} + 21x^{11} + 21x^{10}$$
$$+ 15x^9 + 12x^8 + 10x^7 + 5x^6 + x^5$$

where the coefficients appearing on the right-hand side define the function h.

Example 1   To describe the coefficient of x^n in the product $(x^3 + x)^4$ $(x + 1)^5(x^2 + x)$ note that each term of the product is of the form

$$x^{e_1}x^{e_2}x^{e_3}x^{e_4}x^{e_5}x^{e_6}x^{e_7}x^{e_8}x^{e_9}x^{e_{10}} \tag{4}$$

where for $1 \le i \le 10$, x^{e_i} has been selected from the ith factor of

$$(x^3 + x)(x^3 + x)(x^3 + x)(x^3 + x)(x + 1)(x + 1)$$
$$\times (x + 1)(x + 1)(x + 1)(x^2 + x)$$

Thus, we see that

$$e_i = 3 \quad \text{or} \quad e_i = 1 \qquad \text{if } i = 1, 2, 3, 4 \tag{5}$$
$$e_i = 1 \quad \text{or} \quad e_i = 0 \qquad \text{if } i = 5, 6, 7, 8, 9 \tag{6}$$
$$e_{10} = 2 \quad \text{or} \quad e_{10} = 1 \tag{7}$$

Observe that (4) will contribute to the term x^n if and only if

$$e_1 + e_2 + \cdots + e_{10} = n \tag{8}$$

It follows that the coefficient of x^n is the number of solutions of (8) that satisfy conditions (5), (6), and (7).   □

Reversing the process just discussed provides us with a way to compute the generating functions of certain combinatorial numbers.

Example 2   We use a generating function to find the sequence $\{a_n\}$ where for each n, a_n is the number of ways of placing n indistinguishable objects into three boxes so that the first box contains 2 or 3 of the objects, the second box 0, 1, or 2 objects, and the third box 4 or 5 objects.

Note that for each n, a_n is the number of solutions to the equation

$$e_1 + e_2 + e_3 = n$$

where e_i represents the number of objects placed in box i. That is, e_1 is equal to 2 or 3, e_2 is equal to 0, 1 or 2, and e_3 is equal to 4 or 5. It follows as in Example 1 that a_n is the coefficient of

$$x^n = x^{e_1 + e_2 + e_3} = x^{e_1}x^{e_2}x^{e_3}$$

in

$$F(x) = (x^2 + x^3)(x^0 + x^1 + x^2)(x^4 + x^5) \tag{9}$$

Thus, $F(x)$ is the generating function of the sequence $\{a_n\}$, where a_n is the number of ways of distributing the n indistinguishable objects as specified.

If we use elementary algebra to do the multiplication indicated in (9) we find that

$$F(x) = x^6 + 3x^7 + 4x^8 + 3x^9 + x^{10}$$

Thus,

$$a_n = 0, \text{ if } n = 0, 1, 2, 3, 4, 5, \text{ or } n \geq 11$$
$$a_6 = 1$$
$$a_7 = 3$$
$$a_8 = 4$$
$$a_9 = 3$$
$$a_{10} = 1$$ □

Example 3   For $n = 0, 1, \ldots$, let a_n denote the number of ways that n indistinguishable objects can be placed into two boxes where the first box is to have an even number of objects and the second box is to have 0, 1, 2, or 3 objects. We wish to find the generating function of the sequence $\{a_n\}$.

First note that the number a_n is also the number of solutions to $e_1 + e_2 = n$ where e_1 is a nonnegative even integer, and e_2 is equal to either 0, 1, 2, or 3. It follows that a_n is the coefficient of x^n in the product

$$(x^0 + x^2 + x^4 + x^6 + \cdots)(x^0 + x^1 + x^2 + x^3)$$

Thus

$$F(x) = (1 + x^2 + x^4 + x^6 + \cdots)(1 + x + x^2 + x^3) \qquad (10)$$

is the desired generating function. □

Example 4   We find the generating function of the sequence $\{a_n\}$ where a_n is the number of ways of distributing n indistinguishable objects into eight boxes so that each box contains at least one object. Note that for each n, a_n is the number of solutions of the equation

$$e_1 + e_2 + e_3 + \cdots + e_8 = n \quad \text{where } e_i \geq 1, \text{ for } i = 1, \ldots, 8$$

where e_i is the number of objects placed in ith box. Thus, a_n is the coefficient of x^n in the product $(x^1 + x^2 + \cdots + x^k + \cdots)^8$. In other words,

$$F(x) = (x + x^2 + \cdots + x^k + \cdots)^8$$

is the generating function of the sequence $\{a_n\}$. □

Example 5   We find the generating function of the sequence $\{a_n\}$ where, for each n, a_n is the number of ways that the sum n is obtained when five dice are rolled. Note that for each n, a_n is equal to the number of solutions to the equation

$$e_1 + e_2 + e_3 + e_4 + e_5 = n$$

where e_i represents the contribution of the ith die to the sum. Thus, for $i = 1, 2, 3, 4, 5$, e_i is an integer and $1 \leq e_i \leq 6$. It follows that a_n is the coefficient of x^n in the expansion of

$$F(x) = (x + x^2 + x^3 + x^4 + x^5 + x^6)^5$$

and that $F(x)$ is the desired generating function. □

In the next section we will study ways to invert generating functions such as those found in the last three examples. This will enable us to solve a number of difficult combinatorial problems.

10.8 Problems

In Problems 1–14 compute the generating function of the given sequence.

1. $\{h(n)\}$, where $h(n)$ is the number of integer solutions to $e_1 + e_2 + e_3 = n$, with $0 \leq e_1 \leq 2; 1 \leq e_2 \leq 4; e_3 = 4, 5$.

2. $\{g(n)\}$, where $g(n)$ is the number of integer solutions to $e_1 + e_2 + e_3 = n$, with $0 \leq e_1 \leq 3; e_2 = 1, 2, 5; 3 \leq e_3 \leq 5$.

3. $\{f(n)\}$, where $f(n)$ is the number of integer solutions to $e_1 + e_2 + e_3 + e_4 + e_5 = n$, with $-2 \leq e_1 \leq 0; e_2 = 3, 4; e_3 = -3, 0; 0 \leq e_4 \leq 4; 0 \leq e_5 \leq 2$.

4. $\{h(n)\}$, where $h(n)$ is the number of positive integer solutions to $e_1 + e_2 + e_3 + e_4 = n$.

5. $\{k(n)\}$, where $k(n)$ is the number of positive integer solutions to $e_1 + e_2 + e_3 + e_4 + e_5 = n$ where each e_i is even.

6. $\{f(n)\}$, where $f(n)$ is the number of nonnegative integer solutions to $e_1 + e_2 + e_3 + e_4 = n$, and e_1 and e_3 are odd, and e_2 and e_4 are even.

7. $\{g(n)\}$, where $g(n)$ is the number of ways of distributing n indistinguishable objects into five boxes so that each box contains an even number of objects.

8. $\{r(n)\}$, where $r(n)$ is the number of ways of distributing n indistinguishable objects into seven boxes so that each box contains at least two objects.

9. $\{h(n)\}$, where $h(n)$ is the number of ways of distributing n indistinguishable objects into five boxes so that the first box contains at most three objects and the remaining four are nonempty.

10. $\{f(n)\}$, where $f(n)$ is the number of ways of placing n indistinguishable objects into k boxes so that each box contains one or two objects.

11. $\{p(n)\}$, where $p(n)$ is the number of different ways to write the integer n as the sum of k positive integers where the order in which the integers are added together counts.

12. $\{f(n)\}$, where $f(n)$ is the number of code words of length n from the alphabet $\{a, b, c, d, e\}$ in which c appears an odd number of times.

13. $\{f(n)\}$, where $f(n)$ is the number of positive integers k such that $k < 10^n$ and the integers 2 and 3 appear at least once in k.

14. $\{h(n)\}$, where $h(n)$ is the number of ways the number n can be rolled using 60 standard dice.

10.9 SOME APPLICATIONS OF GENERATING FUNCTIONS TO COMBINATORIAL PROBLEMS

In the previous section we identified the generating functions associated with certain combinatorial sequences. In this section we shall use generating functions to solve a variety of applied combinatorial problems and to establish some combinatorial identities. We will make use of the following equalities.

$$\frac{1 - x^{k+1}}{1 - x} = 1 + x + x^2 + \cdots + x^k \qquad \text{if } x \neq 1 \qquad (1)$$

$$\frac{1}{1 - x} = 1 + x + x^2 + \cdots + x^n + \cdots \qquad \text{if } |x| < 1 \qquad (2)$$

$$(1 + x)^k = 1 + \binom{k}{1}x + \binom{k}{2}x^2 + \cdots + \binom{k}{n}x^n$$
$$+ \cdots + \binom{k}{k}x^k \qquad \text{if } k = 0, 1, 2, \cdots \qquad (3)$$

$$(1 - x)^{-k} = 1 + \binom{1 + k - 1}{1}x + \binom{2 + k - 1}{2}x^2$$
$$+ \cdots + \binom{n + k - 1}{n}x^n + \cdots \qquad \text{if } k = 0, 1, 2, \cdots \quad (4)$$

Equations (1)–(4) may be used as follows.

Example 1   Given $X(h)(x) = (x^2 + x^3 + x^4 + \cdots + x^n + \cdots)^3$, we compute $h(9)$.

On factoring we see that

$$X(h)(x) = x^6(1 + x + x^2 + \cdots)^3$$

and, therefore, by (2) we have

$$X(h)(x) = x^6\left(\frac{1}{1 - x}\right)^3 = x^6(1 - x)^{-3}$$

From (4), with $k = 3$ we obtain

$$X(h)(x) = x^6\left(1 + \binom{1 + 3 - 1}{1}x + \binom{2 + 3 - 1}{2}x^2 + \binom{3 + 3 - 1}{3}x^3 \right.$$
$$+ \cdots + \binom{n + 3 - 1}{n}x^n + \cdots\right)$$

and, hence, $h(9)$, the coefficient of x^9, is

$$\binom{3 + 3 - 1}{3} = \binom{5}{3} = 10 \qquad\qquad \square$$

Example 2   We   compute $h(n)$ if $X(h)(x) = (x + x^2 + \cdots + x^5)$ $(x^2 + x^3 + \cdots + x^n + \cdots)^3$. If we factor x from the first factor and x^2 from $x^2 + x^3 + \cdots + x^n + \cdots$, we have

$$X(h)(x) = x^7(1 + x + x^2 + x^3 + x^4)(1 + x + x^2 + \cdots)^3$$

From (1) and (2) we have

$$X(h)(x) = x^7\left(\frac{1 - x^5}{1 - x}\right)\left(\frac{1}{1 - x}\right)^3$$
$$= x^7(1 - x^5)(1 - x)^{-4}$$

It follows from (4) that

$$X(h)(x) = x^7(1 - x^5)\left[1 + \binom{1 + 4 - 1}{1}x + \binom{2 + 4 - 1}{2}x^2 \right.$$
$$\left. + \cdots + \binom{n + 4 - 1}{n}x^n + \cdots\right]$$

Observe now that terms involving x^n result from the products

$$x^7 \cdot 1 \cdot \binom{n - 7 + 4 - 1}{n - 7}x^{n - 7}$$

and

$$x^7 \cdot (-x^5) \cdot \binom{n - 12 + 4 - 1}{n - 12}x^{n - 12}$$

Thus, the coefficient $h(n)$ of x^n is given by

$$h(n) = \binom{n - 7 + 4 - 1}{n - 7} - \binom{n - 12 + 4 - 1}{n - 12}$$
$$= \binom{n - 4}{n - 7} - \binom{n - 9}{n - 12}$$

$\square$

The next examples illustrate the use of generating functions in solving a number of applied combinatorial problems.

Example 3   We compute the number of ways to distribute n indistinguishable objects into eight boxes so that each box contains at least one object.

Let $h(n)$ denote the number of ways of distributing the n objects as specified. As in Example 4 of the previous section, the generating function of $h(n)$ is

$$X(h)(x) = (x + x^2 + x^3 + \cdots)^8$$
$$= x^8(1 + x + x^2 + \cdots)^8$$

By (2), we have

$$X(h)(x) = x^8\left(\frac{1}{1-x}\right)^8 = x^8(1-x)^{-8}$$

Application of (4) gives

$$X(h)(x) = x^8\left(1 + \binom{1+8-1}{1}x + \binom{2+8-1}{2}x^2 \cdot\right.$$
$$\left. + \cdots + \binom{n+8-1}{n}x^n + \cdots\right)$$

Note that the only term involving x^n results from the product

$$x^8 \cdot \binom{n-8+8-1}{n-8}x^{n-8}$$

Thus,

$$h(n) = \binom{n-1}{n-8}$$

is the number of ways to distribute the n objects.   □

Example 4   We compute the number of ways to roll a total of 11 using five dice.

Let $h(n)$ denote the number of ways of obtaining a sum of n when the five dice are rolled. We have seen in Example 5 of Section 10.8 that the generating function of h is given by

$$X(h)(x) = (x + x^2 + x^3 + x^4 + x^5 + x^6)^5$$
$$= x^5(1 + x + x^2 + x^3 + x^4 + x^5)^5$$

By (1) we have

$$X(h)(x) = x^5\left(\frac{1-x^6}{1-x}\right)^5 = x^5(1-x^6)^5(1-x)^{-5}$$

and using (3) and (4) yields

$$X(h)(x) = x^5\left(1 + \binom{5}{1}(-x^6) + \binom{5}{2}(-x^6)^2 + \binom{5}{3}(-x^6)^3 + \binom{5}{4}(-x^6)^4\right.$$
$$\left. + \binom{5}{5}(-x^6)^5\right)\left(1 + \binom{1+5-1}{1}x + \binom{2+5-1}{2}x^2\right.$$
$$\left. + \cdots + \binom{n+5-1}{n}x^n + \cdots\right)$$

Note that the term involving x^{11} will result from the term involving $x^{11-5} = x^6$

in the product

$$\left(1 - \binom{5}{1}x^6 + \binom{5}{2}x^{12} - \binom{5}{3}x^{18} + \binom{5}{4}x^{24} - \binom{5}{5}x^{30}\right)$$

$$\times \left(1 + \binom{5}{1}x + \binom{6}{2}x^2 + \cdots + \binom{n+5-1}{n}x^n + \cdots\right)$$

This term is

$$1 \cdot \binom{10}{6}x^6 - \binom{5}{1}x^6 \cdot 1 = \left(\binom{10}{6} - \binom{5}{1}\right)x^6$$

Thus,

$$h(11) = \binom{10}{6} - \binom{5}{1}$$

is the number of ways of obtaining a sum of 11 on the roll of 5 dice. □

The techniques found in the previous examples can be used to establish a number of combinatorial identities as we illustrate in the following example.

Example 5   We use the identity

$$(1 + x)^m(1 + x)^n = (1 + x)^{m+n} \tag{5}$$

to establish Vandermonde's identity:

$$\binom{m}{0}\binom{n}{r} + \binom{m}{1}\binom{n}{r-1} + \cdots + \binom{m}{r}\binom{n}{0} = \binom{m+n}{r}$$

Substituting

$$(1 + x)^k = 1 + \binom{k}{1}x + \binom{k}{2}x^2 + \cdots + \binom{k}{k}x^k$$

in (5) for $(1 + x)^m$, $(1 + x)^n$, and $(1 + x)^{m+n}$, yields

$$\left(1 + \binom{m}{1}x + \binom{m}{2}x^2 + \cdots + \binom{m}{m}x^m\right)\left(1 + \binom{n}{1}x + \binom{n}{2}x^2 + \cdots + \binom{n}{n}x^n\right)$$

$$= 1 + \binom{m+n}{1}x + \binom{m+n}{2}x^2 + \cdots + \binom{m+n}{m+n}x^{m+n} \tag{6}$$

The coefficient of x^r on the right-hand side of equation (6) is $\binom{m+n}{r}$. The terms involving x^r on the left-hand side of the equation are:

$$1 \cdot \binom{n}{r}x^r + \binom{m}{1}x\binom{n}{r-1}x^{r-1} + \binom{m}{2}x^2\binom{n}{r-2}x^{r-2} + \cdots + \binom{m}{r}x^r \cdot 1$$

$$= \left[1\binom{n}{r} + \binom{m}{1}\binom{n}{r-1} + \binom{m}{2}\binom{n}{r-2} + \cdots + \binom{m}{r} \cdot 1\right]x^r$$

Since the coefficients of x^r on both sides must be equal, we have

$$\binom{m}{0}\binom{n}{r} + \binom{m}{1}\binom{n}{r-1} + \binom{m}{2}\binom{n}{r-2} + \cdots + \binom{m}{r}\binom{n}{0} = \binom{m+n}{r}$$

as we wanted to show.   □

In this chapter we have introduced generating functions and seen how they can be used to solve difference equations and certain combinatorial problems. Generating functions provide an excellent example of how a relatively abstract technique can be applied to very diverse situations.

10.9 Problems

1. If $X(h(n))(x) = (1 + x + x^2 + \cdots + x^n + \cdots)^4$, compute $h(7)$.

2. If $X(h(n))(x) = (1 + x + x^2 + \cdots + x^n + \cdots)^{10}$, compute $h(11)$.

3. If $X(g(n))(x) = (1 + x + x^2 + \cdots + x^{12})^5$, compute $g(10)$.

4. If $X(h(n))(x) = (1 + x + x^2 + \cdots + x^{10})^7$, compute $h(13)$.

5. If $X(h(n))(x) = (1 + x + x^2 + x^3)(1 + x + x^2 + \cdots + x^{12})^5$, compute $h(10)$.

6. If $X(k(n))(x) = (1 + x + x^2 + x^3)(1 + x^2 + x^4 + \cdots + x^{2n} + \cdots)^6$, compute $k(12)$.

7. If $X(g(n))(x) = (1 + x + x^3 + \cdots + x^n + \cdots)^3(x + x^2 + x^3)^5$, compute $g(10)$.

8. If $X(h(n))(x) = (1 + x + x^2 + \cdots + x^n + \cdots)^4(x + x^2 + x^3 + x^4)^7$, compute $h(11)$.

9. If $X(h(n))(x) = (1 + x + x^2 + \cdots + x^n + \cdots)^3$, compute $h(n)$.

10. If $X(r(n))(x) = (1 + x + x^2 + \cdots + x^n + \cdots)^5$, compute $r(n)$.

11. If $X(h(n))(x) = (1 + x + x^2 + x^3 + x^4)(1 + x + x^2 + \cdots + x^n + \cdots)^3$, compute $h(n)$.

12. If $X(g(n))(x) = (1 + x + x^2 + x^3)(1 + x + x^2 + \cdots + x^n + \cdots)^4$, compute $g(n)$.

13. If $X(h(n))(x) = (x + x^2 + x^3 + x^4)(1 + x + x^2 + \cdots + x^n + \cdots)^3$, compute $h(n)$.

14. If $X(p(n))(x) = (x + x^2 + \cdots + x^6)(1 + x + x^2 + \cdots + x^n + \cdots)^5$, compute $p(n)$.

15. If $X(h(n))(x) = (x + x^2)^k$, compute $h(n)$.

16. If $X(h(n))(x) = (x^2 + x^3 + x^4 + x^5)(x + x^2 + x^3 + \cdots + x^n + \cdots)^7$, compute $h(n)$.

17. If $X(k(n))(x) = (x^2 + x^3 + x^4 + \cdots + x^7)(x^2 + x^3 + x^4 + \cdots + x^n + \cdots)^6$, compute $k(n)$.

18. Compute the number of ways of distributing n indistinguishable objects into three boxes so that the first box contains from two to seven objects and the other two boxes are nonempty.

19. Compute the number of ways of distributing 20 indistinguishable objects into five boxes so that the first box contains at most three objects and the remaining four are nonempty.

20. Compute the number of positive integer solutions to $e_1 + e_2 + e_3 + e_4 + e_5 = 32$ where each e_i is even.

21. Compute the number of non-negative integer solutions to $e_1 + e_2 + e_3 + e_4 = n$, where e_1 and e_3 are odd, and e_2 and e_4 are even.

22. Compute the number of ways of distributing n indistinguishable objects into five boxes so that each box contains an even number of objects.

23. Compute the number of integers k, where $0 < k < 10^n$, such that the integers 2 and 3 appear at least once in k.

24. Compute the number of ways the number 100 can be rolled using 60 standard dice.

25. In how many positive integers smaller than 10^9 do both the integers 2 and 3 appear at least once?

26. Compute the number of ways of placing n indistinguishable objects into k boxes so that each box contains either one or two items.

Chapter 10   REVIEW

Concepts for Review

infinite sequence (p. 404)
upper bound (p. 405)
lower bound (p. 405)
bounded (p. 405)
least upper bound (p. 405)
greatest lower bound (p. 406)
Axiom of Completeness (p. 406)
limit of a sequence (p. 407)
convergence/divergence (p. 407)
increasing sequence (p. 409)
decreasing sequence (p. 409)
limit theorem (p. 411)
infinite sums (p. 415)
infinite series (p. 415)
sequence of partial sums (p. 415)
geometric series (p. 419)

generating function of a sequence (p. 423)
Shifting Theorem (p. 425)
partial fractions (p. 432)
inversion (p. 432)
convolution of two sequences (439)
Convolution Theorem (p. 439)

Review Problems

In Problems 1–4 compute the limit or state that it does not exist.

1. $\lim\limits_{n \to \infty} (1 - (-1)^n/n)$

2. $\lim\limits_{n \to \infty} \dfrac{2n^2 + 7n + 3}{n^2 - 200n}$

3. $\lim\limits_{n \to \infty} (-1)^{2n}$

4. $\lim\limits_{n \to \infty} \dfrac{n^3 + 3n^2 + 1}{n^4}$

5. Find a sequence $\{h(n)\}$ where $h(n)$ is the interest earned during the nth year on an initial investment of $100. Assume a simple interest at an annual rate of 10%.

In Problems 6–11 find the sum of the given series or state that the series is divergent.

6. $\sum\limits_{k=0}^{\infty} 7\left(\dfrac{2}{3}\right)^k$

7. $\sum\limits_{k=1}^{\infty} 3k2^{-k}$

8. $\sum\limits_{k=0}^{\infty} 2\left(\dfrac{5}{2}\right)^k$

9. $\sum\limits_{k=0}^{\infty} 3 \cdot 7^{-k} \cdot 2^{2k}$

10. $\sum\limits_{k=2}^{\infty} \dfrac{1 - k}{4^k}$

11. $\sum\limits_{k=1}^{\infty} \dfrac{k + 3(-1)^k}{5^k}$

In Problems 12–15 use Definition 10.16 to write $X(f)(x)$.

12. $f(n) = \begin{cases} 1 & n = 0, 1, 2, 4 \\ 0 & \text{otherwise} \end{cases}$

13. $f(5) = 7$; $f(n) = 0$ if $n \neq 5$

14. $f(n) = 1$

15. $f(n) = n^2$

16. Show that $X(f(n) + 1)(x) = X(f(n))(x) + 1/(1 - x)$.

In Problems 17–22 use Table 10.1 to write the generating function of the given sequence.

17. $f(n) = 3n^2$

18. $h(n) = -n$

19. $k(n) = 10^n$

20. $g(n) = \dbinom{n + 10}{n}$

21. $k(n) = 2^n - 7n3^n$

22. $f(n) = 2^{-n} \sin n$

23. Give the form of the partial fraction decomposition of

$$\frac{7x^5 - 32x + 1}{(x^2 + 2x + 1)^2(x + 3)^5(x^2 + 1)^2}$$

24. Find the partial fraction decomposition of $(5x - 9)/(3x^2 - x - 4)$.

In Problems 25–28 use generating functions to find the indicated particular solution to the difference equation.

25. $f(n + 2) - 6f(n + 1) + 9f(n) = 0; f(0) = 1, f(1) = 6$

26. $h(n + 2) - 3h(n + 1) - 4h(n) = 0; h(0) = 2, h(1) = 13$

27. $f(n + 1) - 10f(n) + 25f(n - 1) = 0; f(0) = 2, f(1) = 15$

28. $-2g(n + 2) + g(n + 1) + g(n) = -3(1 + 16 \cdot 3^n); g(0) = 1, g(1) = 4$

In Problems 29–30 compute the convolution of the sequences $f(n)$ and $g(n)$.

29. $f(n) = \begin{cases} 1 & n = 1, 2, 3 \\ 0 & \text{otherwise} \end{cases}; g(n) = \begin{cases} 2 & n = 2, 3, 4 \\ 0 & \text{otherwise} \end{cases}$

30. $f(n) = n; g(n) = (-1)^n$

In Problems 31–32 write the generating function of the given sequence.

31. $\{r(n)\}$ where $r(n)$ is the number of positive integer solutions to $e_1 + e_2 + e_3 = n$ such that e_1 is even, e_2 is odd, and e_3 is less than 10.

32. $\{g(n)\}$ where $g(n)$ is the number of ways of distributing n indistinguishable objects into eight boxes so that no box has less than three objects.

33. If $\quad X(g(n))(x) = (1 + x + x^2 + x^3 + x^4)(x + x^2 + x^3 + x^4 + \cdots)^4$, compute $g(n)$.

34. Use generating functions to compute the number of ways that k indistinguishable objects can be distributed into n boxes so that no box contains less than three objects.

11 Introduction to Automata and Formal Languages

A computer's ability to recognize specified patterns is critical in many applications. For example, text-editing software generally offers the user the option of replacing a given string of symbols with another string, and a compiler must scan the symbols of a computer program to locate certain key words. Both applications require that the computer be able to recognize specified patterns. As we shall see in this chapter, a wide class of pattern recognition problems can be solved through the use of machines called *automata*.

On a more theoretical level, mathematical machines, which include automata, have been suggested as models for simple computers. These models help researchers investigate the theoretical limits of computers.

11.1 AN INTRODUCTION TO MACHINES

The word "machine" in this chapter denotes a variety of mathematical models that abstract certain properties of computers. The machines we deal with all have a finite number of internal states and are assumed to "read" input from a tape containing symbols from an input alphabet. On reading an input symbol the machine moves in a prescribed way from one internal state to another, and prepares to read the next input symbol.

In Figure 11.1 we have a schematic illustration of a machine that, on reading the input symbol s, moves from internal state A to internal state B, and then prepares to read the next input symbol t. The horizontal arrow denotes the right-to-left movement of the tape.

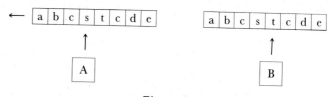

Figure 11.1

In general, the action of a machine can be specified by a table (called the *transition matrix* of the machine) that indicates how the machine moves from one state to another state upon reading a given symbol from an input tape. For instance, suppose that a machine has three possible internal states, A, B, and C, and an input alphabet 0, 1. Then the machine could be specified by a transition matrix such as that given in Table 11.1.

Present state	Input 0	1
A	A	B
B	C	B
C	C	C

Table 11.1

According to this table, if the machine is in state B and it reads an input of 0, then the machine assumes internal state C. If the machine is in state C, then either input (0 or 1) causes the machine to remain in this same state.

Although some types of machines can generate output or alter the input tape, we will restrict our attention to those machines that utilize a single input tape that is unmodified by the machine. In such cases the machine's job is essentially to recognize patterns: It reads a given a finite input tape containing a word and either accepts the word or rejects it. The word is accepted if, after reading the tape, the machine is in any one of a set of specifically designated "accepting states."

For instance, consider the machine described by Table 11.1. Suppose that this machine is initially in state A and suppose that B is the only "accepting" state among states A, B, and C. Figure 11.2 illustrates the action of this machine as it reads the word 00011.

Note that in Figure 11.2(e) the machine is in state B as it reads the last input letter. Since the machine is still in state B after it reads the last 1, the machine is in an accepting state (B) after having read the input word. Consequently, the word 00011 is accepted by this machine.

On the other hand, the word 0110 is not accepted since, as indicated in Figure 11.3, when the machine reads the last 0, it will move to internal state C, which is not an accepting state.

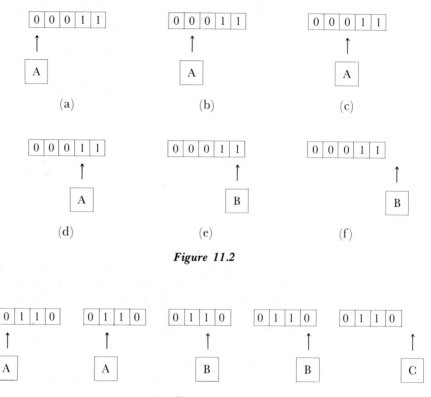

Figure 11.2

Figure 11.3

Machines of the type just described are called *deterministic finite automata*. Although these machines obviously have rather limited capabilities, there are good reasons for focusing our attention on them. Deterministic finite automata are the simplest machines of interest and as such they provide a natural introduction to the topic of machine theory. Furthermore, as noted in the introduction to this chapter, these machines can be used to solve a wide variety of pattern recognition problems.

More formally, a deterministic finite automaton (or DFA) is a machine M defined as follows:

    a. There is a finite set of internal states $\mathscr{S}$ that the machine can assume. This set is called the *state set*. Generally, we use capital letters to designate internal states, and we frequently refer to internal states simply as states.

    b. There is a finite set of symbols $\mathscr{I}$ that the machine will read from a tape containing a word. This set is called the *input alphabet*, and we generally designate its elements by integers or lowercase letters.

c. The machine has an *initial state*. This is the internal state (an element of $\mathscr{S}$) of the machine before it reads the first input symbol.

d. There is a set of *accepting states*. These are the internal states that determine whether or not a word will be accepted. If, after reading a word, the machine is in one of the designated accepting states, then the word is accepted; if the machine is not in such a state, then the word is not accepted. The set of all words accepted by machine M is designated by $ac(M)$.

e. There is a *next state function* $f: \mathscr{S} \times \mathscr{I} \rightarrow \mathscr{S}$. If $S \in \mathscr{S}$ and $a \in \mathscr{I}$, then $f(S, a)$ is the state that the machine will move to from state S upon reading a. Generally, we specify the next state function with a transition matrix such as Table 11.1.

A directed graph (with loops) provides a convenient way to describe a DFA. The vertices of the graph represent the internal states of the machine. The initial state is designated by an incoming arrow and vertices surrounded by a double circle represent accepting states. We label the edges of the digraph with letters of the input alphabet to indicate the appropriate transitions. For example, the directed graph given in Figure 11.4 describes the DFA specified by Table 11.1.

This machine has three internal states A, B, and C. The state A is the initial state (note the incoming arrow), and the state B is the only accepting state (only vertex B has a double circle around it). As determined by Table 11.1, if the machine is in state A when it reads 0, then the machine will remain in state A, but if it reads 1, then it makes a transition to state B. As long as the machine is in state B and it reads a 1, the machine remains in state B; when it reads a 0, the machine will make a transition from state B to state C. State C is known as a "dead state," because once the machine reaches state C, neither the input 0 nor the input 1 can change the state.

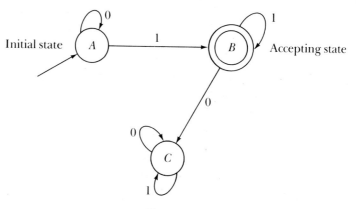

Figure 11.4

In this chapter we will consider the theoretical limits of DFA's and the problem of designing and simplifying DFA's that accomplish a specific recognition task. The next two examples introduce some types of patterns that a specific DFA can recognize.

Example 1   As you can verify, Figure 11.5 illustrates one possible DFA that will accept precisely those words that contain the sequence *abc* where the input alphabet is $\{a, b, c\}$.   □

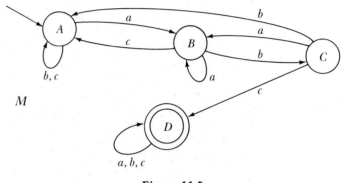

Figure 11.5

Note that in Example 1 we have used the convention that a single arrow may carry several labels. For example, if the machine is in state A, then reading either a b or a c will cause the machine to remain in state A.

Example 2   Figure 11.6 shows a DFA, M, for which $ac(M)$ is the set of words containing two successive 0's or two successive 1's. The input alphabet is $\{0, 1\}$.   □

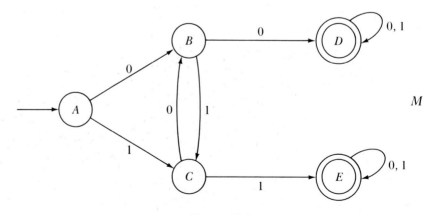

Figure 11.6

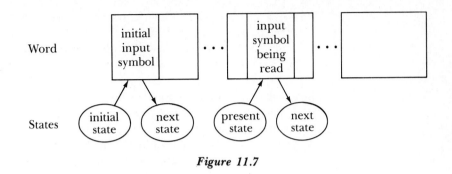

Figure 11.7

Note that in Example 2 it was simpler to use a machine with two accepting states, D and E. State D serves to accept any word with two or more successive 0's and the machine will be in accepting state E after reading any word containing two successive 1's.

It is important to note that the action of a DFA is always completely determined; that is, no matter what state the DFA is in, and no matter what input is read, the machine will move to one and only one next state (which may be the same as its present state). In terms of the directed graph representation of the machine, this means that each input symbol must appear on one and only one edge leaving each vertex.

One additional way to describe the action of an automaton is often useful. In Figure 11.7 we write below the word read by the machine the succession of states through which the machine moves. Using this notation, we can describe the action of the machine in Example 2 as it reads the tape 01010010 by

$$A \nearrow^{0} \searrow B \nearrow^{1} \searrow C \nearrow^{0} \searrow B \nearrow^{1} \searrow C \nearrow^{0} \searrow B \nearrow^{0} \searrow D \nearrow^{1} \searrow D \nearrow^{0} \searrow D$$

We are now in a position to consider a question of considerable theoretical and practical importance: What patterns can a DFA actually recognize? For example, a compiler must recognize identifiers, so if we wish to use a DFA in compiler design, we must use a class of identifiers that a DFA can recognize.

To see the significance of this question we first show that there are patterns that a DFA cannot recognize. Consider patterns consisting of a number of 0's followed by an equal number of 1's. Any DFA designed to recognize all such patterns would be required to accept 01, 0011, 000111, ..., $0^n 1^n$, ..., where for each n

$$0^n 1^n = \underbrace{00 \cdots 0}_{n \text{ 0's}} \underbrace{11 \cdots 1}_{n \text{ 1's}}$$

Suppose it is possible to design such a DFA and that it has m internal states. Then the DFA must reenter a previously held state on reading *any* string of more than m symbols.

Now consider the tape

$$\underbrace{00 \cdots 0}_{m + 1} \underbrace{11 \cdots 1}_{m + 1}$$

Since it is of the proper form, it must be accepted by our hypothetical machine. That is, after reading the last 1 the machine must be in an accepting state. Since the tape begins with $m + 1$ 0's, the DFA must enter the same state (call it R) at least twice during its reading of the first half of the tape. Therefore, the action of the machine can be described by Figure 11.8.

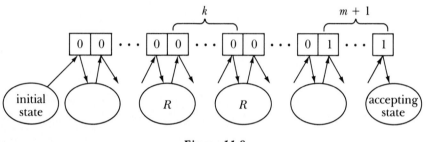

Figure 11.8

Here k 0's are read between the two times the DFA enters state R. Note that the insertion of k more zeros after the second occurrence of state R will not change the subsequent action of the machine. Thus, the action of the machine on the tape

$$00 \cdots 0 \quad \underbrace{11 \cdots 1}_{m + 1}$$
$$\underbrace{}_{m + k + 1}$$

is described by

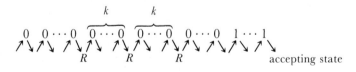

Since the automaton again assumes an accepting state after having read the tape, the automaton accepts the word

$$0^{m+k+1}1^{m+1} = \underbrace{00 \cdots 0}_{m + k + 1} \underbrace{11 \cdots 0}_{m + 1}$$

which is contrary to its design specifications.

In Section 11.4 we return to the problem of determining those sets of words that can be recognized by DFA's.

11.1   Problems

In Problems 1–4 draw the directed graph that describes the DFA with the given transition matrix.

1.

Present state \ Input	a	b
A	B	D
B	C	D
C	C	C
D	B	D

Initial state: *A*
Accepting state: *D*

2.

Present state \ Input	0	1
A	A	B
B	A	C
C	A	C

Initial state: *A*
Accepting state: *C*

3.

Present state \ Input	a	b	c
A	B	A	A
B	D	C	A
C	B	A	D
D	D	D	D

Initial state: *A*
Accepting state: *C*

4.

Present state \ Input	0	1	2
A	B	A	C
B	A	D	A
C	D	C	A
D	B	A	B

Initial state: *A*
Accepting states: *B, C*

5. Which of the following words is accepted by the DFA given in Problem 1?
   (a) *aabb*   (b) *abbb*
   (c) *bbab*   (d) *bbaabaa*

6. Which of the following words is accepted by the DFA given in Problem 2?
   (a) 0
   (b) 001
   (c) 001110
   (d) $001010 \cdots 10 = 00(10)^n$

7. Describe the set of words accepted by the DFA given in Problem 1.

8. Describe the set of words accepted by the DFA given in Problem 2.

9. Describe the set of words accepted by the DFA given in Problem 3.

In Problems 10–17 draw the directed graph of a DFA, M, that accepts precisely the given set $ac(M)$.

10. $ac(M)$ is the set of all words containing exactly one 0. The input alphabet is $\{0, 1, 2\}$.

11. $ac(M)$ is the set of all words containing at least one 0. The input alphabet is $\{0, 1, 2\}$.

12. $ac(M)$ is the set of all words containing the sequence 101. The input alphabet is $\{0, 1\}$.

13. $ac(M)$ is the set of all words starting with three successive 1's. The input alphabet is $\{0, 1, 2\}$.

14. $ac(M)$ is the set of all words ending with three successive a's. The input alphabet is $\{a, b\}$.

15. $ac(M)$ is the set of all words having an even number of 1's. The input alphabet is $\{0, 1\}$.

16. $ac(M)$ is the set of all words containing $3k$ b's, where k is an integer. The input alphabet is $\{a, b, c\}$.

17. $ac(M)$ is the set of all words in which every 0 is followed by a 1. The input alphabet is $\{0, 1\}$.

18. Show that there is no DFA that accepts all words containing only a prime number of 0's.

19. Show that there is no DFA, M, with $ac(M) = \{0^n 1^{2n} | n = 0, 1, 2, \ldots\}$.

11.2 NONDETERMINISTIC AUTOMATA

As we indicated in the previous section, a DFA is deterministic in the sense that for a given state and a given input, the next state of the machine is completely determined. Nondeterminism allows a machine to select arbitrarily from several possible responses to a given situation, including the possibility of selecting from several initial states. If one of the various responses to a word leaves the machine in an accepting state, then the word is said to be accepted.

More formally, a *nondeterministic finite automaton* (or NFA) is a machine M defined as follows:

a. There is a finite set of internal states $\mathscr{S}$ that the machine can assume. This set is called the *state set*. Generally, we use capital letters to designate internal states, and we frequently refer to internal states simply as states.

b. There is a finite set of symbols $\mathscr{I}$ that the machine will read from a tape. This set is called the *input alphabet*, and we generally designate its elements by integers or lowercase letters.

c. There is a subset of $\mathscr{S}$ called the *initial state set*.

d. There is a set of *accepting states*. This is a subset of $\mathscr{S}$ whose elements determine whether or not a word will be accepted. If, after reading the word, the machine M can be in one of the accepting states, then the word is accepted; otherwise, the word is not accepted. The set $ac(M)$ is the set of all words accepted by the machine M.

e. There is a *next state function* $f\colon \mathscr{S} \times \mathscr{I} \to \mathscr{P}(\mathscr{S})$. If $S \in \mathscr{S}$ and $a \in \mathscr{I}$, then $f(S, a)$ is the *set* of states to which the machine M can move from state S upon reading the input a.

Example 1   Let M be the NFA with state set $\mathscr{S} = \{A, B, C, D, E\}$, input alphabet $\mathscr{I} = \{0, 1\}$, initial states A and B, accepting state E, and next state function f given by

$$\begin{aligned}
f(A, 0) &= \{A, C\} & f(C, 0) &= \{E\} & f(E, 0) &= \{E\} \\
f(A, 1) &= \{A\} & f(C, 1) &= \varnothing & f(E, 1) &= \{E\} \\
f(B, 0) &= \{B\} & f(D, 0) &= \varnothing \\
f(B, 1) &= \{B, D\} & f(D, 1) &= \{E\}
\end{aligned}$$

In this case $f(A, 0) = \{A, C\}$ specifies that the machine, when in state A and presented with 0 as input, can assume either A or C as its next state. Note in addition that $f(C, 1) = \varnothing$ indicates that the machine, when in state C and presented with 1 as input, has no response.   □

As in the deterministic case, we can represent an NFA by a directed graph. Figure 11.9 illustrates the NFA defined in Example 1.

From Figure 11.9 we can easily trace the response of the machine to any input. For example, if the machine is in state A when it reads a 0, then the machine can exhibit two responses: It can either remain in state A or move to state C. Suppose that the machine does move to state C and then it reads a 1. At this point we see another aspect of nondeterminism: It may be the case that no response can be made. This presents no particular difficulty. It simply means that no word beginning with 01 can be accepted if the machine moves from state A to state C on reading the initial 0.

It can be seen that the NFA, M, described in Figure 11.9 accepts only those words containing two successive 0's or two successive 1's. Note that the

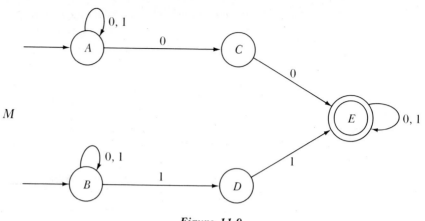

Figure 11.9

machine M will not accept the word 00 if it selects B as the initial state (or if it selects A initially but remains in state A upon reading the second 0). However, since there is a sequence of responses that will result in the machine's being in an accepting state after reading 00, the word 00 is accepted.

The transition matrix of an NFA allows both multiple entries and undefined transitions and thus reflects the nondeterminism of the machine. For example, Table 11.2 is the transition matrix of the machine given in Figure 11.9.

Present state	Input 0	1
A	A, C	A
B	B	B, D
C	E	—
D	—	E
E	E	E

Initial states: A, B
Accepting state: E

Table 11.2

There are several recognition tasks for which it is easier to design an NFA than a DFA. To describe one such task we adopt the following notation. If L is a set of words, then $r(L)$ will denote the set of all words formed by reversing the order of the words of L. Thus, for example, *pat* $\in r(L)$ if and only if *tap* $\in L$.

Let us now suppose that L is the set of words accepted by a DFA, M. That is, $ac(M) = L$. It is not difficult to find an NFA, $R(M)$, such that $ac(R(M)) = r(L)$. Thus, $R(M)$ recognizes exactly the words of $r(L)$. The *dual machine*, $R(M)$, is constructed as follows:

   a. The initial states of $R(M)$ are the final states of M.

   b. The accepting state of $R(M)$ is the initial state of M.

   c. All transitions are reversed. (For example, if on reading a, M makes a transition from state S to state T, then on reading a, $R(M)$ makes a transition from state T to state S.)

Graphically, we interchange the initial and terminal states and reverse the direction of each edge.

Example 2   We find the NFA $R(M)$ corresponding to the DFA, M, described in Example 1 of Section 11.1. Figure 11.10 shows once again the graph of M found in that example.

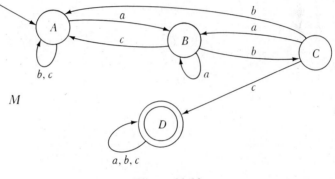

Figure 11.10

Recall that the machine, M, accepts only those words containing the triple abc. The dual machine, $R(M)$, illustrated in Figure 11.11, will accept precisely those

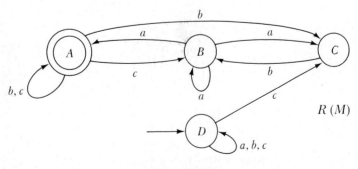

Figure 11.11

tapes containing the triple *cba*. Note that the resulting machine is nondeterministic. For example, if this machine is in state B, then the response to an a can be a transition to either state A, state B, or state C.   □

Next we introduce one more NFA associated with a DFA. The closure, $C(M)$, of a DFA, M, is the NFA that allows the same transitions from each of the accepting states of M as the DFA allows from the initial state of M. As we shall see, the closure of a DFA plays a central role in the design of automata. The next example illustrates the formation of $C(M)$.

Example 3   The DFA, M, illustrated in Figure 11.12 accepts only the word 11. To form the closure of M we must allow the same transitions from state C that are allowed from the initial state A. That is, we must allow the transition from C to B on reading a 1. The closure of M, $C(M)$, is shown in Figure 11.13. Note that $C(M)$ accepts any word consisting of $2k$ 1's where $k \geq 1$.   □

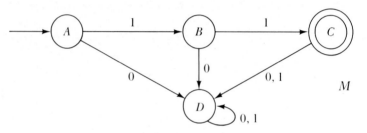

Figure 11.12

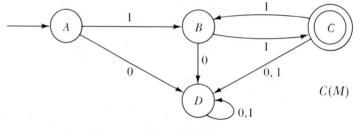

Figure 11.13

In Problem 12 you are asked to compare the set of words accepted by a machine M with the set of words accepted by the closure of M.

In this section we have seen how an NFA can be used to accomplish certain pattern recognition tasks. In the next section we will compare the pattern recognition capabilities of an NFA with those of a DFA.

11.2 **Problems**

In Problems 1–5 design an NFA that will accept precisely the given set of words. Use a directed graph to describe your NFA.

1. The set of words consists of those that contain either 111 or 000. The input alphabet is $\{0, 1\}$.

2. The set of words consists of those that contain either *dog* or *cat*. The input alphabet is $\{d, o, g, c, a, t\}$.

3. The set of words consists of those that are either a sequence of 0's followed by a 1, or a sequence of 1's followed by a 0. The input alphabet is $\{0, 1\}$.

4. The set of words consists of those that contain either an even number of 2's or an odd number of 1's. The input alphabet is $\{0, 1, 2\}$.

5. The set of words consists of all those words over the alphabet $\{a, b, c\}$ that begin with a and end with c.

In Problems 6–9:

(a) Diagram the dual, $R(M)$, of the given DFA, M, and give its transition matrix.

(b) Diagram the closure, $C(M)$, of the given DFA, M, and give its transition matrix.

6.

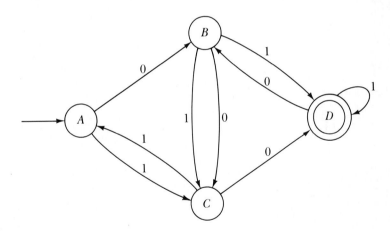

7.

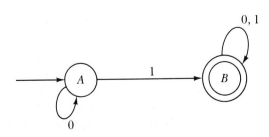

8.

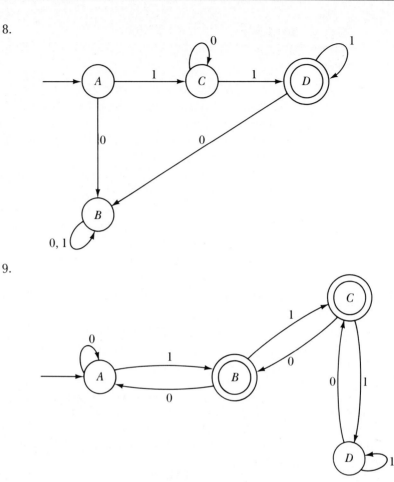

9.

10. Suppose that a DFA, M, accepts only the word *abc*. Describe the set of words accepted by the closure, $C(M)$.

11. Suppose that a DFA, M, accepts only the word *abc*. Describe the set of words accepted by $C(R(M))$.

12. Describe the relation between $ac(M)$ and $ac(C(M))$.

13. Let L_1 denote the set of words accepted by a DFA, M_1. Let L_2 denote the set of words accepted by a DFA, M_2. Explain how you would design an NFA that accepts exactly $L_1 \cup L_2$.

11.3 COMPARING PATTERN RECOGNITION CAPABILITIES OF NFA'S AND DFA'S

Since an NFA can respond in various ways to a specific input, it might be assumed that an NFA can perform more complex pattern recognition tasks than those performed by a DFA. In fact, this is not the case; any set of words recognized by an NFA can also be recognized by a DFA. To prove this we describe how to convert any NFA, M, into a DFA, $D(M)$, that accepts the same words as does M, that is, $ac(D(M)) = ac(M)$.

As you read the description of $D(M)$ you should keep in mind that the states of $D(M)$ are subsets (possibly including the empty set) of the set of states of M. We describe $D(M)$ as follows:

a. The initial state of $D(M)$ is the *set* of all initial states of M.

b. For any state $\{A_1, A_2, \ldots, A_k\}$ of $D(M)$ and any input x, the next state of $D(M)$ is the set of all states of M that can result as next states for M if M is in any of the states $A_1, A_2, \ldots, A_k$ when it reads x.

c. We apply **b.** as long as new states are introduced. Since M has only a finite number of states, this process will terminate.

d. The accepting states of $D(M)$ are those states that contain an accepting state of M.

Example 1 We construct $D(M)$ corresponding to the NFA, M, given in Figure 11.9 (p. 466). It follows from **a.** that the initial state of $D(M)$ is $\{A, B\}$. Note in Figure 11.9 that states A and B are carried to states A, B, and C by 0 and that states A and B are carried to states A, B, and D by 1. Thus, we obtain the diagram in Figure 11.14.

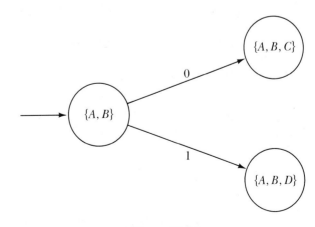

Figure 11.14

Next we apply **b.** to the states $\{A, B, C\}$ and $\{A, B, D\}$. The transition matrix of M (Table 11.2, p. 466) shows that the states A, B, and C are carried to states A, B, C, and E by 0; therefore, on reading 0, $D(M)$ will move from state $\{A, B, C\}$ to state $\{A, B, C, E\}$. Similarly, Table 11.2 indicates that when M reads a 1, it makes a transition from states A, B, and C to states A, B, and D, so it follows that on reading 1, $D(M)$ will move from state $\{A, B, C\}$ to state $\{A, B, D\}$. Continuing this process, we obtain the automaton $D(M)$ shown in Figure 11.15.

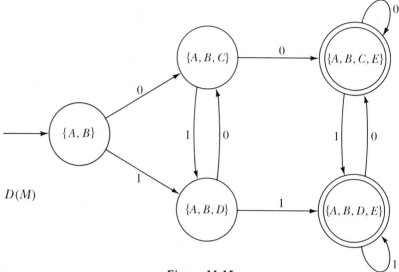

Figure 11.15

Note that $\{A, B, C, E\}$ and $\{A, B, D, E\}$ are the only accepting states of $D(M)$ since only these states contain an accepting state, E, of M. □

As shown in the next example the empty set may appear as a state of $D(M)$. Whenever this happens the empty set serves as a *dead state*, a state carried to itself by all inputs.

Example 2   The NFA with transition matrix

Present state	0	1
A	—	B
B	C	—
C	—	D
D	D	D

Initial state: A
Accepting state: D

will accept only those words beginning with 101. We construct $D(M)$.

Since A is the only initial state of M, the initial state of $D(M)$ is $\{A\}$. Since A is not carried to any other state by 0, the next state of $D(M)$ on reading 0 is the empty set, $\varnothing$. Similar remarks hold for states B and C. Figure 11.16 illustrates the resulting DFA.

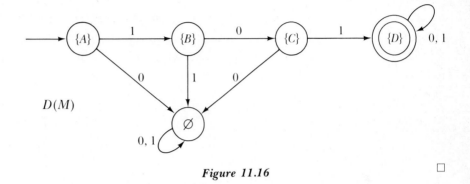

$D(M)$

Figure 11.16

Next we show that the DFA, $D(M)$, accepts precisely the same set of words as does M; that is, $ac(M) = ac(D(M))$. To see this, we first show that $ac(M) \subseteq ac(D(M))$. Suppose that $a_1 a_2 a_3 \cdots a_k \in ac(M)$ and that the action of M in accepting this word is described by

$$S_1 \nearrow^{a_1} \searrow S_2 \nearrow^{a_2} \searrow S_3 \cdots S_{k-1} \nearrow^{a_k} \searrow S_k \quad \text{(accepting state)}$$

Then S_1 is an initial state and S_k is an accepting state of M. Now suppose that the action of $D(M)$ on reading $a_1 a_2 \cdots a_k$ is

$$S'_1 \nearrow^{a_1} \searrow S'_2 \nearrow^{a_2} \searrow S'_3 \cdots S'_{k-1} \nearrow^{a_k} \searrow S'_k$$

Then by the construction of $D(M)$ we must have that $S_i \in S'_i$ for all i. In particular, we have $S_k \in S'_k$. Thus, since S_k is an accepting state of M, S'_k must be, by definition, an accepting state of $D(M)$. Consequently, $D(M)$ also accepts $a_1 a_2 \cdots a_k$, so any word accepted by M is accepted by $D(M)$. Hence, we have

$$ac(M) \subseteq ac(D(M)) \tag{1}$$

Next we show that $ac(D(M)) \subseteq ac(M)$. Suppose that $b_1 b_2 \cdots b_n \in ac(D(M))$ and that the action of $D(M)$ on accepting this word is described by

$$T'_1 \nearrow^{b_1} \searrow T'_2 \nearrow^{b_2} \searrow T'_3 \cdots T'_{n-2} \nearrow^{b_{n-1}} \searrow T'_{n-1} \nearrow^{b_n} \searrow T'_n \quad \text{(accepting state)}$$

We can now work backwards to describe an accepting action for the NFA, M.

Since T'_n is an accepting state of $D(M)$, there must be, by definition, an element of T'_n, call it T_n, that is an accepting state of M. Moreover, by the construction of $D(M)$ there must be some state in T'_{n-1}, call it T_{n-1}, that is carried by b_n to T_n. Thus, we have the following action of M:

$$T_{n-1} \quad \overset{b_n}{\nearrow \searrow} \quad T_n$$

Again it follows from the construction of $D(M)$ that there is some state in T'_{n-2}, say T_{n-2}, that is carried by b_{n-1} to T_{n-1}. We now have the following action of M:

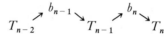

$$T_{n-2} \quad \overset{b_{n-1}}{\nearrow \searrow} \quad T_{n-1} \quad \overset{b_n}{\nearrow \searrow} \quad T_n$$

Continuing in this fashion we eventually obtain:

$$T_1 \quad \overset{b_1}{\nearrow \searrow} \quad T_2 \quad \overset{b_2}{\nearrow \searrow} \quad T_3 \cdots T_{n-1} \quad \overset{b_n}{\nearrow \searrow} \quad T_n$$

Since T'_1 consists of only initial states of M, T_1 must be such an initial state. Since T_1 is an initial state of M and T_n is an accepting state, M accepts $b_1 b_2 \cdots b_n$. Thus,

$$ac(D(M)) \subseteq ac(M) \tag{2}$$

and from (1) and (2) we obtain the following important result.

Theorem 11.1   If M is an NFA, then there is a DFA, D, such that $ac(D) = ac(M)$. That is, there is a DFA, D, that accepts precisely the set of words accepted by M.

According to Theorem 11.1, an NFA is no more powerful than a DFA. In view of this, one might well question the value of nondeterministic finite automata. As we shall see in the next sections, however, nondeterministic automata are particularly important in describing the kinds of sets that can be accepted by deterministic automata.

11.3   Problems

In Problems 1–6 use the procedure described in this section to convert the given NFA into a DFA that accepts precisely the same set. Draw the directed graph representation of the resulting DFA.

1.

Present state \ Input	0	1
A	B	—
B	—	A
C	—	D
D	C	—

Initial states: A, C
Accepting states: A, C

2.

Present state \ Input	0	1
A	B, C	—
B	—	A
C	C	D
D	—	—

Initial states: A, C
Accepting states: A, D

3.

Present state \ Input	a	b
A	A	B
B	—	—
C	D	C
D	—	—

Initial states: A, C
Accepting states: B, D

4.

Present state \ Input	a	b
A	A	B
B	—	C, D
C	D	C
D	—	—

Initial states: A, C
Accepting states: B, D

5.

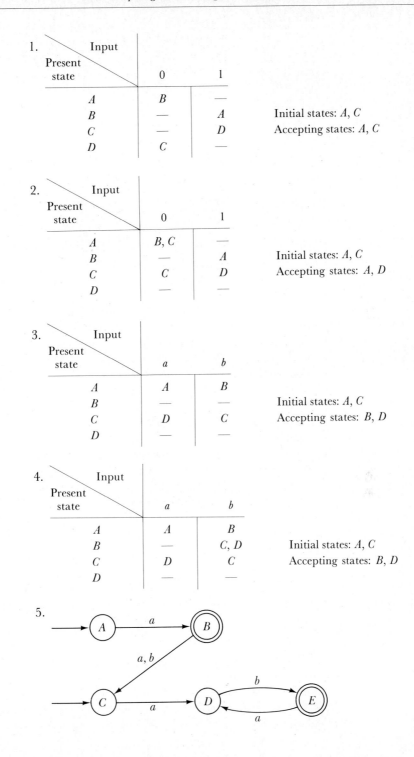

6.

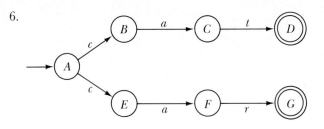

11.4   REGULAR SETS

We now set about the task of describing the sets that a DFA can recognize. We have seen that there are some sets (for example, $\{0^n 1^n | n = 1, 2, \ldots\}$) that cannot be recognized by a DFA. We have yet to describe, however, those patterns that are recognizable.

A little reflection should indicate the difficulty of this problem. At this point we lack even the ability to describe patterns, and without some method for doing so, we cannot expect to describe those patterns that can be recognized by a DFA. Some new ideas and notation are therefore necessary.

An *alphabet* Σ is a finite set of symbols, and a *word* over Σ is either a finite string of symbols from Σ or the "empty word", ε, consisting of no symbols. Thus, for example, $w_1 = abc$ and $w_2 = aabab$ are two words over the alphabet $\Sigma = \{a, b, c\}$. The *concatenation* of two words is formed by juxtaposing the symbols that form the words. For example, if $w_1 = abc$ and $w_2 = aabab$, then the concatenation of w_1 and w_2 is $w_1 w_2 = abcaabab$. Note that the concatenation of the empty word ε with any word w gives the same word; that is, $\varepsilon w = w \varepsilon = w$.

The idea of concatenation is extended to sets of words in a natural way. If L_1 and L_2 are sets of words, then the concatenation of L_1 and L_2, denoted by $L_1 L_2$, is the set of all words formed by concatenating a word from L_1 with a word from L_2; that is,

$$L_1 L_2 = \{w_1 w_2 | w_1 \in L_1 \text{ and } w_2 \in L_2\}$$

If, for example, $L_1 = \{a, aba, cab, \varepsilon\}$ and $L_2 = \{ca, cb\}$, then

$$L_1 L_2 = \{aca, acb, abaca, abacb, cabca, cabcb, ca, cb\}$$

Powers of L are used to designate the concatenation of L with itself the appropriate number of times. Thus, for instance, $L^2 = LL$ and $L^3 = LLL$. In addition, we will let $L^0 = \{\varepsilon\}$ and $L^1 = L$.

If, for example, $L = \{a, b\}$, then $L^0 = \{\varepsilon\}$, $L^1 = \{a, b\}$, $L^2 = LL = \{aa, ab, ba, bb\}$, and $L^3 = LLL = \{aaa, aab, aba, abb, baa, bab, bba, bbb\}$.

Definition 11.2   If L is a set of words, then L^*, the *Kleene closure of L*, is defined by

$$L^* = \bigcup_{i=0}^{\infty} L^i$$

The Kleene closure of L is the set of all words (including ε) that can be formed by concatenating words from L any number of times.

Example 1   If $\Sigma = \{0, 1\}$ and $L = \{0, 10\}$, then L^* consists of the empty word ε and all words that can be formed using 0 and the pair 10. Thus, L^* consists of ε and all words formed from 0's and 1's with the property that every 1 is followed by a 0.

We now give the most important definition of this section.

Definition 11.3   Let Σ be an alphabet. The *regular expressions* over Σ, and the sets they designate are defined as follows:

    a. If $a \in \Sigma$, then a is a regular expression designating the set $\{a\}$.

    b. If E and F are regular expressions designating the sets A and B, respectively, then:

    $E + F$ is a regular expression designating the set $A \cup B$.

    EF is a regular expression designating the set AB, the concatenation of A and B.

    E^* is a regular expression designating the set A^*, the Kleene closure of A.

As illustrated in the next example regular expressions give us an easy method of describing many sets.

Example 2   Let $\Sigma = \{a, b\}$. Then the following are regular expressions designating the indicated sets of words:

a	designates the set $\{a\}$
a^*	designates the set $\{a\}^* = \{\varepsilon, a, aa, aaa, \ldots\}$
b	designates the set $\{b\}$
ab	designates the set $\{a\}\{b\} = \{ab\}$
$a + b$	designates the set $\{a\} \cup \{b\} = \{a, b\}$
$(ab)^*$	designates the set $\{ab\}^* = \{\varepsilon, ab, abab, ababab, \ldots\}$

$a* + (ab)*$ designates the set $\{a\}* \cup \{ab\}* =$
$\{\varepsilon, a, aa, aaa, \ldots, ab, abab, ababab, \ldots\}$

$a*b$ designates the set $\{a\}*\{b\} = \{b, ab, aab, aaab, \ldots\}$

$b(ab)*$ designates the set $\{b\}\{ab\}* = \{b, bab, babab, \ldots\}$

$a*b(ab)*$ designates the set of all words that begin with any number (possibly zero) of a's followed by a single b, followed by any number (possibly zero) of pairs ab ☐

Definition 11.4   A *regular set* over an alphabet Σ is either

a. the empty set;

b. the set consisting of only the empty word; or

c. a set designated by some regular expression over Σ.

Example 3   Let $\Sigma = \{a, b\}$. By definition, $\varnothing$ and $\{\varepsilon\}$ are regular sets. In view of Example 2 the following sets are also regular:

$$\{a\}, \quad \{\varepsilon, a, aa, aaa, \ldots\}, \quad \{b\}, \quad \{a, b\},$$
$$\{\varepsilon, ab, abab, ababab, \ldots\}, \quad \{b, ab, aab, aaab, \ldots\}$$ ☐

In the next section, we study the relationship between regular sets and DFAs.

11.4   Problems

In Problems 1–10 explain why the given expressions are regular expressions over $\Sigma = \{a, b, c\}$ and describe the regular set designated by the expression.

1. $a + bc$ 2. $a + bc*$ 3. $a + (bc)*$

4. $(b + c)*$ 5. $(ab)*c*$ 6. $(ab)*(cb)*$

7. $(ab + c)*$ 8. $(aa)*$ 9. $(aaa)*$

10. $(abc)*$

In Problems 11–18 give a regular expression that designates the given set. In each case $\Sigma = \{a, b, c\}$.

11. $\{\varepsilon, b, a, a^2, a^3, \ldots\}$

12. $\{\varepsilon, ab, abab, ababab, \ldots\}$

13. $\{a, ab, ab^2, ab^3, \ldots\}$

14. $\{a, b, c\}$

15. $\{\varepsilon, a, b, c, bc, bcbc, bcbcbc, \ldots\}$

16. The set of words beginning with any number of pairs ab, followed by a c, followed by any number of triples abc

17. The set of all words containing the sequence *cba*

18. The set of all words containing an even number of *a*'s

11.5 KLEENE'S THEOREM

The importance of regular sets stems from the fact that for any regular set S, there is a DFA that accepts exactly S.

Theorem 11.5 If S is a regular set over Σ, then there is a DFA, M such that $ac(M) = S$.

Proof: We proceed through the three parts of Definition 11.4 and construct a DFA that accepts exactly the words arising from **a.**, **b.**, and **c.** of this definition.

  a. The automaton illustrated in Figure 11.17 accepts no words; that is, it accepts exactly the empty set.

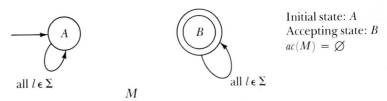

Initial state: A
Accepting state: B
$ac(M) = \emptyset$

all $l \in \Sigma$ all $l \in \Sigma$

M

Figure 11.17

  b. The DFA illustrated in Figure 11.18 accepts only the empty word ε.

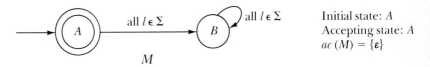

all $l \in \Sigma$ all $l \in \Sigma$

Initial state: A
Accepting state: A
$ac(M) = \{\varepsilon\}$

M

Figure 11.18

  c. To show that any set designated by a regular expression is accepted by some DFA, we proceed by induction on the length of the regular expression. (For example, the length of each of the expressions ab and $a*$ is two and the length of $a + b*$ is four.)

Suppose that the regular expression has length one. Then the regular expression is a for some $a \in \Sigma$. The DFA shown in Figure 11.19 accepts only the word a. Assume now that the theorem is true for all expressions of length n or less.

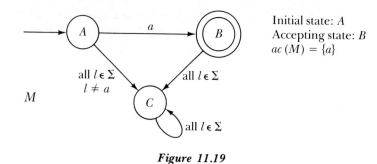

Initial state: A
Accepting state: B
$ac\,(M) = \{a\}$

Figure 11.19

Now consider the regular expressions $E + F$, EF, and E^*. Each of these expressions is longer than the regular expressions E or F. If the expression $E + F$ (or EF or E^*) has length $n + 1$, then the lengths of E and F are less than or equal to n, so we may assume that there are DFA's M_1 and M_2 that accept precisely the sets designated by E and F, respectively. That is, $ac(M_1) = E$ and $ac(M_2) = F$. We will designate M_1 and M_2 schematically as shown in Figure 11.20.

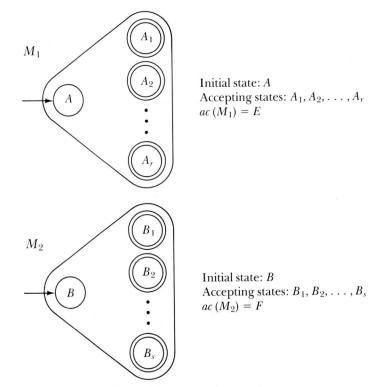

Initial state: A
Accepting states: $A_1, A_2, \ldots, A_r$
$ac\,(M_1) = E$

Initial state: B
Accepting states: $B_1, B_2, \ldots, B_s$
$ac\,(M_2) = F$

Figure 11.20

Figure 11.21 represents an NFA that accepts precisely the set designated by $E + F$, $ac(M_1) \cup ac(M_2)$. Note that since there are two initial states, the machine is indeed nondeterministic. However, by Theorem 11.1 there is a DFA that accepts precisely the same set.

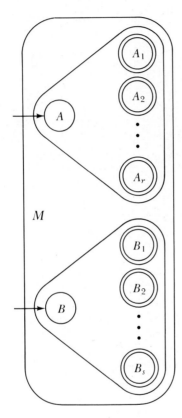

Initial states: A, B
Accepting states: $A_1, A_2, \ldots, A_r, B_1, B_2, \ldots, B_s$
$ac(M) = ac(M_1) \cup ac(M_2)$

Figure 11.21

Figure 11.22 illustrates a NFA that accepts exactly the set designated by EF. In that figure we have attached r copies of machine M_2 to the machine M_1 by using each accepting state of M_1 as the initial state for a copy of M_2. The resulting machine is an NFA. (Since the alphabets of E and F are not necessarily disjoint, the machine has no way of "knowing" when it has completed reading the E portion of the word; consequently, the machine is nondeterministic when in states (A_i, B).) Again, Theorem 11.1 establishes the existence of a DFA that accepts precisely the set $EF = ac(M_1)ac(M_2)$.

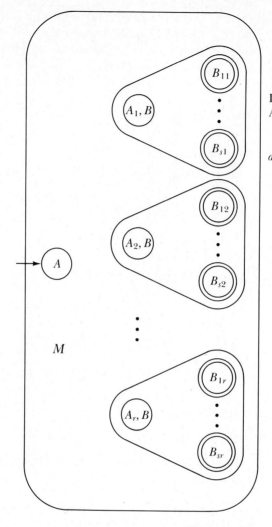

Initial State: A
Accepting states: $B_{11}, \ldots, B_{s1},$
$\qquad B_{12}, \ldots, B_{s2}, \ldots,$
$\qquad B_{1r}, \ldots, B_{sr}$
$ac\,(M) = ac\,(M_1)\,ac\,(M_2)$

M

Figure 11.22

Finally, we must obtain a DFA that accepts E^*. Since in part **b.** of the proof we found a DFA that accepts only the empty word $\{\varepsilon\}$, it suffices to find a DFA that accepts the words of E^* that are not empty. The closure of M_1, $C(M_1)$ (recall Example 3 of Section 11.2), accomplishes this task since after the machine reads a word in E, it is prepared to read another word; that is, $ac(C(M_1)) = (ac(M_1))^*$. As in the previous parts of this proof we can now use Theorem 11.1 to obtain a DFA that accepts the same set. This completes the inductive proof that every set represented by a regular expression is accepted by some automaton.   ■

The proof of Theorem 11.5 is particularly important because we can use the technique found there to construct a DFA that will accept a set designated by any regular expression E. This can be done as follows:

Step 1.   For each $a \in E$, construct a DFA, M, such that $ac(M) = \{a\}$. If ε is in the set designated by E, construct a DFA that accepts exactly the set $\{\varepsilon\}$ (see Figures 11.18 and 11.19).

Step 2.   Combine the DFA's obtained in step 1 to form the sum, concatenation, or Kleene closure as appropriate (see Figures 11.21 and 11.22, or form the closure of the DFA).

Step 3.   Continue using step 2 until the desired regular expression has been generated.

Step 4.   Construct a DFA, $D(M)$, from the NFA, M, formed in step 3.

Example 1   We construct a DFA that will accept the set designated by the regular expression $(ab)^* + b$.

The DFA's M_1 and M_2 shown in Figure 11.23 have $ac(M_1) = \{a\}$ and $ac(M_2) = \{b\}$.

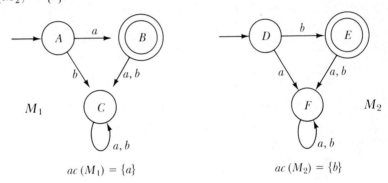

M_1 M_2

$ac(M_1) = \{a\}$ $ac(M_2) = \{b\}$

Figure 11.23

Figure 11.24 shows how M_1 and M_2 can be combined into an NFA, M, with $ac(M) = \{ab\}$.

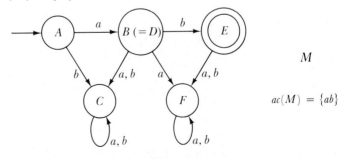

M

$ac(M) = \{ab\}$

Figure 11.24

Figure 11.25 illustrates an NFA that accepts the set designated by $(ab)^*$. Note that this NFA consists of the closure of the machine illustrated in Figure 11.24 and a DFA that accepts only the empty word.

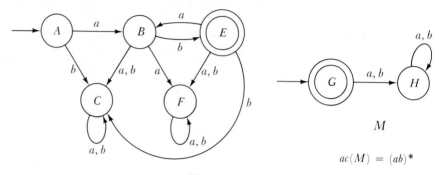

$$ac(M) = (ab)^*$$

Figure 11.25

In Figure 11.26 we combine the NFA of Figure 11.25 with a DFA that accepts only the word b. The result is a nondeterministic automaton M with $ac(M) = (ab)^* + b$.

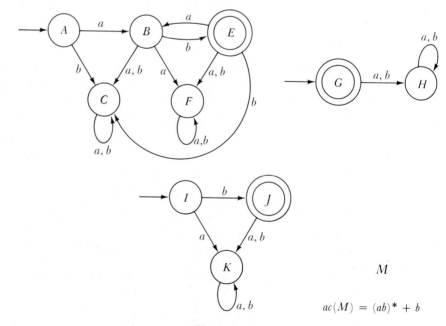

$$ac(M) = (ab)^* + b$$

Figure 11.26

Finally, in Figure 11.27 we have applied the procedure outlined in Section 11.3 to the NFA of Figure 11.26 to obtain the DFA that accepts precisely the set designated by the regular expression $(ab)^* + b$.   □

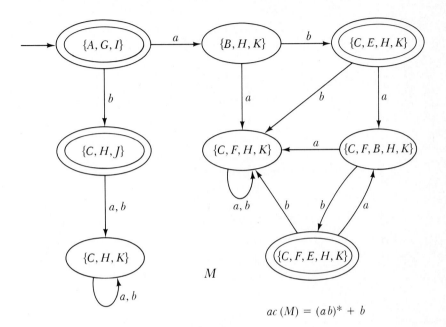

$$ac(M) = (ab)^* + b$$

Figure 11.27

The DFA that results from the procedure outlined in Example 1 is not necessarily a DFA with the smallest number of states that accepts the desired set. In the next section we will give a procedure that will "minimize" DFA's.

Induction can be used to prove the converse of Theorem 11.5, but since the proof is rather tedious it will be omitted. Theorem 11.5 and its converse give a convenient characterization of the sets that can be recognized by a DFA.

Theorem 11.6   *Kleene's Theorem.*   There is a DFA, M, such that $ac(M) = S$ if and only if S is a regular set.

11.5   Problems

In Problems 1–10 design a DFA, M, that accepts the regular set described by the given expression. In each problem the alphabet is $\Sigma = \{a, b, c\}$.

1. a 2. ab 3. a^*

4. $\cdot ba^*$ 5. $(ab)^*$ 6. $(ab)^* + c$

7. $(ab)*c$ 8. $(aa)*$ 9. $(aaa)*$

10. $a* + b*c$

11. Design a DFA that accepts exactly those words over $\Sigma = \{a, b, c\}$ that begin with an arbitrary number of a's and end with a b.

12. Design a DFA that accepts exactly those words over $\Sigma = \{a, b, c\}$ containing an even number of a's and no other symbols.

11.6 MINIMIZATION OF AUTOMATA

We begin this section by giving the solution to the problem alluded to at the end of the last section: For a given DFA, M, find a DFA with the smallest number of states that recognizes the same set of words as does M.

Recall that if M is a DFA, then $R(M)$ is the NFA that accepts the reverse of the tapes accepted by M. Recall also that if N is an NFA, then $D(N)$ is a DFA with $ac(N) = ac(D(N))$. The desired result is the following.

Theorem 11.7   If M is a DFA, then $D(R(D(R(M))))$ is a DFA with a minimal number of states that accepts the same set of words that M accepts.

We illustrate the use of Theorem 11.7 in the next example.

Example 1   The DFA given in Figure 11.28 accepts exactly those tapes that end in b. We use Theorem 11.7 to generate a DFA with a minimal number of states that accomplishes the same recognition task.

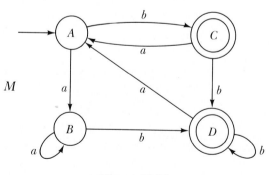

Figure 11.28

The dual automaton of M, $R(M)$, together with its transition matrix, is shown in Figure 11.29.

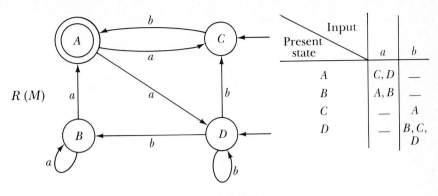

	Input		
Present state		a	b
A		C, D	—
B		A, B	—
C		—	A
D		—	B, C, D

Figure 11.29

Figure 11.30 describes $D(R(M))$.

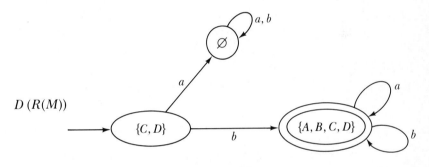

Figure 11.30

The NFA $R(D(R(M)))$ is shown in Figure 11.31. Note that the empty state may be omitted since no transition into that state is possible.

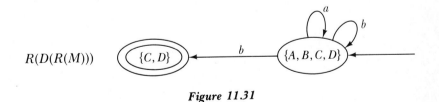

Figure 11.31

Since the set of words recognized by an automaton is not affected by the particular names assigned to the internal states of the machine, we let $X = \{C, D\}$ and $Y = \{A, B, C, D\}$ to simplify notation. We can then represent $R(D(R(M)))$ and its transition matrix as shown in Figure 11.32. In more complicated cases such a change can considerably reduce the complexity of the notation.

$R(D(R(M)))$

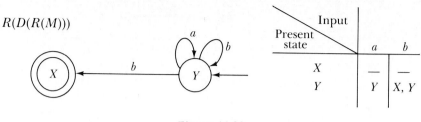

Figure 11.32

The DFA $D(R(D(R(M))))$ is shown in Figure 11.33. It is easy to see that $ac(D(R(D(R(M)))))$ is the set of words that end in b. Moreover, as guaranteed by Theorem 11.7, it is a DFA with the smallest number of states that will accomplish this task.   □

$D(R(D(R(M))))$

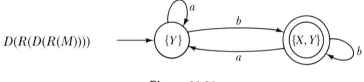

Figure 11.33

11.6 Problems

In Problems 1–4 use the procedure of this section to minimize the given DFA.

1.

Present state \ Input	0	1
A	B	D
B	C	B
C	C	B
D	D	E
E	D	D

Initial state: A
Accepting states: B, C

2.

Present state \ Input	a	b
A	B	A
B	A	C
C	A	B

Initial state: A
Accepting state: A

3.

Present state \ Input	a	b
A	C	B
B	G	B
C	E	D
D	C	F
E	F	E
F	F	E
G	G	B

Initial state: A
Accepting states: B, D, G

4.

Present state \ Input	0	1
A	C	B
B	E	B
C	E	D
D	A	B
E	C	D
F	E	F

Initial state: A
Accepting states: B, D, F

11.7 FORMAL LANGUAGES

In 1954 Noam Chomsky introduced the idea of a phrase structure grammar in an effort to describe the natural languages (English, Spanish, and so on). Phrase structure grammars also describe sets of words (formal languages) that are recognized by machines of various types. This is important in the design of compilers of programming languages.

A formal definition of a phrase structure grammar will follow shortly. At present our aim is to gain an intuitive understanding of this idea. A phrase structure grammar consists of a terminal alphabet whose symbols make up the resulting words; a nonterminal alphabet used to generate the patterns; a special member of the nonterminal alphabet called the "start symbol"; and a set of rules (called *productions*) that govern the generation of the pattern.

Example 1 We consider the phrase structure grammar G with:

Non–terminal alphabet: $\{S, A, B\}$
Terminal alphabet: $\{0, 1\}$
Start symbol: S
Productions: **a.** $S \to ASB$ **c.** $A \to 0$
  b. $S \to AB$ **d.** $B \to 1$

The productions indicate that the letter(s) to the left of the arrow can be replaced by the letter(s) to the right of the arrow. Thus, for instance, applying production **b.** to the word *AASBB* gives *AAABBB*. We denote this by

$$AASBB \xrightarrow{\textbf{b.}} AAABBB$$

The basic idea is to begin with the start symbol *S* and then successively apply productions to determine a word that contains only *terminal letters*. Any word obtainable in this way is said to be in the language generated by the grammar. For instance, the sequence of productions (called a *derivation*)

$$S \xrightarrow{\textbf{a.}} ASB \xrightarrow{\textbf{c.}} 0SB \xrightarrow{\textbf{b.}} 0ABB \xrightarrow{\textbf{d.}} 0AB1 \xrightarrow{\textbf{c.}} 00B1 \xrightarrow{\textbf{d.}} 0011$$

indicates that the word 0011 is in the language generated by the given grammar. Similarly, the derivation

$$S \xrightarrow{\textbf{b.}} AB \xrightarrow{\textbf{c.}} 0B \xrightarrow{\textbf{d.}} 01$$

shows that the word 01 is also in the language generated by the given grammar.

It is not difficult to see that the language generated by the grammar *G* consists of all words of the form

$$0^n1^n = \underbrace{00 \cdots 0}_{n}\underbrace{11 \cdots 1}_{n}$$

□

The following definitions more formally describe a phrase structure grammar and its language.

Definition 11.8   A *phrase structure grammar* (or simply, a *grammar*) *G* is a 4-tuple (N, T, S, P) where:

    a. *N* is the nonterminal alphabet.

    b. *T* is the terminal alphabet, $N \cap T = \varnothing$.

    c. $S \in N$ is the start symbol.

    d. *P* is a set of productions of the form $\alpha \to \beta$ where α is a string containing at least one nonterminal symbol.

Definition 11.9   The *language*, $L(G)$, generated by a phrase structure grammar *G*, is the set of all words in the terminal alphabet that can be derived from the start symbol by a sequence of productions.

Example 2   We describe $L(G)$ where $G = (N, T, S, P)$ is given by:

$$N = \{S, A, B\}; \qquad T = \{a, b\}; \qquad P = \{S \to AB, B \to bB, A \to a, B \to b\}$$

Note that every derivation must begin with $S \to AB$. Note further that since A is a nonterminal letter, it cannot appear in any word of $L(G)$. Thus, we must use the production $A \to a$ at some step of the derivation. Since the point at which we use that production is not important, we can begin the derivation of any word as:

$$S \to AB \to aB \to \cdots$$

We can use the production $B \to bB$ to replace B by any sequence of b's. For instance, we have

$$B \to b \quad \text{or} \quad B \to bB \to bb \quad \text{or} \quad B \to bB \to bbB \to bbb; \cdots$$

and, thus, derivations of the form

$$S \to AB \to aB \to \cdots \to abb \cdots b$$

yield all words of $L(G)$. Consequently, the language generated by the grammar G is the set of all words in the alphabet $\{a, b\}$ consisting of an a followed by a string of b's.   □

At times we will be interested in writing a grammar for a particular set L.

Example 3 We find a grammar that generates a language L consisting of all words of length 3, 6, 9, ... over the terminal alphabet $\{a\}$.

The objective is to generate the words *aaa, aaaaaa, aaaaaaaaa, ...* We could, for example, use the symbol A to generate the word *aaa* ($A \to aaa$) and use the production $A \to aaaA$ to produce any number of *aaa*'s. Then, introducing the start symbol S, we have

$$S \to A, \qquad A \to aaa, \qquad A \to aaaA$$

It follows that a grammar for L is given by $G = (N, T, S, P)$ where $N = \{A, S\}$, $T = \{a\}$, and $P = \{S \to A, A \to aaa, A \to aaaA\}$.   □

We can use other sets of productions to generate the same language. For instance, as you can easily verify, the set of productions

$$P_1 = \{S \to aaa, S \to aaaS\}$$

also generates the language given in Example 3.

In some circumstances a grammar will include *erasing productions*. These are productions of the form $\alpha \to \lambda$; λ is called the *null symbol* and the effect of such a production is to erase the entire string α.

Example 4 The grammar $G = (N, T, S, P)$ where $N = \{S, A, B, C, D\}$, $T = \{a\}$, and $P = \{S \to aA, A \to aB, B \to aC, C \to aB, C \to \lambda\}$ contains an erasing production that can be used to eliminate the nonterminal C. In fact, the erasing production $C \to \lambda$ must be used to obtain a word containing no nonterminals. You can verify that $L(G)$ is the language of Example 3.   □

A phrase structure grammar is too general to be of much use to us right now. By restricting the form that the productions can assume we are able to obtain more manageable classes of grammars.

Table 11.3 outlines Noam Chomsky's hierarchy of grammars, a listing in which grammars are given both a "type number" and a descriptive name. The Chomsky Hierarchy is indeed a hierarchy in the sense that every type 3 grammar is a type 2 grammar; every type 2 grammar is a type 1 grammar; and every type 1 grammar is a type 0 grammar (see Problem 16).

Chomsky Hierarchy

Type number	Descriptive name	Restrictions on productions $\alpha \to \beta$
0	phrase structure	α contains a nonterminal
1	context sensitive grammar	α contains a nonterminal and the length of α is less than or equal to the length of β
2	context free grammar	α consists of a single nonterminal
3	right linear grammar	only productions of the form $\alpha \to a\beta$ or $\alpha \to a$ can be used where α and β are single nonterminals and a is a terminal or λ

Table 11.3

The Chomsky Hierarchy can be applied to languages by using Definition 11.9.

Definition 11.9   A language is called a *type i* $(i = 0, 1, 2, 3)$ *language* if there is a type i grammar that generates the language.

Example 5   Let L be the language whose set of words consists of all words that use only the letter a and have lengths of 3, 6, 9, ... This language is generated by the type 3 grammar whose productions are $S \to aA$, $A \to aB$, $B \to aC$, $C \to aB$, $C \to \lambda$. Consequently, it is a type 3 language.   □

Observe that the same language can be generated by many different grammars. For instance, the language described in Example 5 is also generated by the Type 2 grammar whose productions are $S \to A$, $A \to aaaA$, $A \to aaa$. Although this latter grammar is not a Type 3 grammar, the language L it generates is a Type 3 language since there is some Type 3 grammar that generates L (see Example 5).

11.7   Problems

In Problems 1–10 describe the language generated by the given grammar.

1. $G = (N, T, S, P)$, where $N = \{S, A, B\}$, $T = \{a, b\}$, and
   $P = \{S \to ABS, A \to a, B \to b, S \to \lambda\}$

2. $G = (N, T, S, P)$, where $N = \{S, A\}$, $T = \{a\}$, and
   $P = \{S \to SA, A \to aa, S \to \lambda\}$

3. $G = (N, T, S, P)$, where $N = \{S\}$, $T = \{a, b\}$, and
   $P = \{S \to bSa, S \to b\}$

4. $G = (N, T, S, P)$, where $N = \{S, B, C\}$, $T = \{0\}$, and
   $P = \{S \to 0B, B \to 0C, C \to 0S, S \to \lambda\}$

5. $G = (N, T, S, P)$, where $N = \{S\}$, $T = \{1\}$, and
   $P = \{S \to 111S, S \to \lambda\}$

6. $G = (N, T, S, P)$, where $N = \{S, A, B\}$, $T = \{a, b, c\}$, and
   $P = \{S \to AB, A \to abA, B \to Bcb, A \to ab, B \to \lambda\}$

7. $G = (N, T, S, P)$, where $N = \{S, A, B\}$, $T = \{a, b\}$, and
   $P = \{S \to A, S \to B, A \to aA, B \to bB, A \to a, B \to b\}$

8. $G = (N, T, S, P)$, where $N = \{S, A, B\}$, $T = \{a, b\}$, and
   $P = \{S \to AB, A \to AB, B \to bB, A \to a, B \to b\}$

9. $G = (N, T, S, P)$, where $N = \{S, A, B\}$, $T = \{a, b, c\}$, and
   $P = \{S \to ASB, A \to acA, B \to b, A \to \lambda, S \to B\}$

10. $G = (N, T, S, P)$, where $N = \{S, A, B, C\}$, $T = \{a, b, c\}$, and
    $P = \{S \to ABSC, A \to a, B \to b, C \to cCa, C \to \lambda, S \to AB\}$

In Problems 11–15 give a grammar G that has the given set as its language, $L(G)$.

11. The set of all words over $\Sigma = \{a\}$ consisting of an even number of a's.

12. The set of all words over $\Sigma = \{0, 1\}$ consisting of n 0's followed by n 1's where $n = 1, 2, 3, \ldots$.

13. The set of all words over $\Sigma = \{a, b, c\}$ consisting of n a's followed by n b's followed by a single c, where $n = 1, 2, 3, \ldots$.

14. The set of all words over $\Sigma = \{a, b, c\}$ consisting of n a's followed by n b's followed by n c's, where $n = 1, 2, 3, \ldots$.

15. The set of all words over $\Sigma = \{a, b, c\}$ consisting of an even number of a's followed by an even number of b's followed by two c's.

16. Show that the Chomsky Hierarchy is a hierarchy in the sense that every type i grammar is a type $i - 1$ grammar for $i = 1, 2, 3$.

17. Give the type of each grammar found in Problems 1, 4, 7, 8, and 10.

18. Give the type of each grammar you found in Problems 11–15.

19. Give an example (other than the one described on page 492) of a type 2 grammar that is not a type 3 grammar and yet generates a type 3 language.

11.8   TYPE 3 LANGUAGES AND AUTOMATA

Type 3 languages form a small but important class of languages. This importance stems from two facts. First, although no computer language is regular, certain heavily used portions of computer languages such as identifiers are regular. In addition, the relative simplicity of type 3 languages provides a natural introduction to the study of the relation between languages and machines.

We begin this study by showing that a DFA can be constructed to recognize any type 3 language. Consider, for example, the language generated by the type 3 grammar $G = (N, T, S, P)$ where $N = \{S, A, B\}$, $T = \{0, 1\}$, and $P = \{S \to 0A, S \to 0, A \to 1B, B \to 0A, B \to 0\}$. It is not difficult to see that the language $L(G)$ generated by G is defined by the regular expression $0(10)^*$. In Figure 11.34 we illustrate an NFA, M, with $ac(M) = L(G)$. M is constructed by taking the nonterminals S, A, and B of G as its internal states and adding a special state F that serves as the only accepting state of M. The start symbol S of G is the only initial state of M.

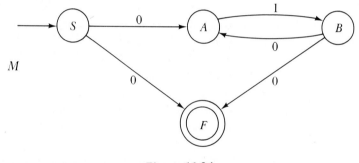

Figure 11.34

In general, for every production in P of the form $\alpha \to a\beta$ we define a transition in M from state α to state β whenever a is read. In other words we include

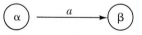

if $\alpha \to a\beta$ is a production of G.

For every production in P of the form $\alpha \to a$ we define a transition from state α to an accepting state F whenever a is read. In other words we include

if $\alpha \to a$ is a production of G.

Since G is a type 3 grammar, the nonerasing productions of G have one of the forms $\alpha \to a\beta$ or $\alpha \to a$. Thus, for each nonerasing production in G we have defined a corresponding transition in M. An erasing production, $\alpha \to \lambda$, results in the designation of α as an accepting state.

Example 1   Figure 11.35 illustrates the NFA, M, that corresponds to the type 3 language $L(G)$ generated by $G = (N, T, S, P)$, where $N = \{S, A, B, C\}$, $T = \{a, b\}$,   and $P = \{S \to aA, S \to bB, A \to aA, A \to \lambda, B \to b\}$.   □

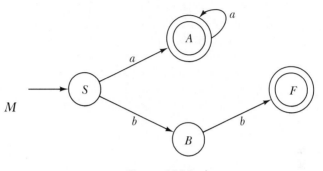

Figure 11.35

Note that in Example 1 the set of words accepted by M is exactly the language generated by G. That is, $ac(M) = L(G)$.

To see that this construction always results in a machine M such that $ac(M) = L(G)$ we observe:

a. A path in the directed graph representation of M beginning at S and ending at an accepting state corresponds to a sequence of productions in G beginning with S and ending in a string of terminal letters; and conversely,

b. a derivation in G beginning with S and ending in terminal letters corresponds to a path in the directed graph beginning at S and ending at an accepting state.

As an illustration, Figure 11.36 gives a derivation and the corresponding path in the directed graph graph representation of M.

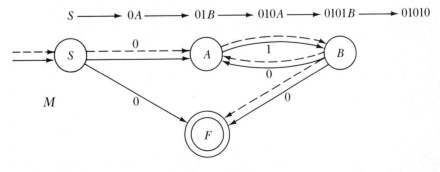

Figure 11.36

To find the DFA that recognizes a given type 3 language, we can apply the above procedure and then construct $D(M)$ using the procedure discussed in Section 3. The next example illustrates this method.

Example 2   We construct a DFA that accepts the language $L(G)$ where $G = (N, T, S, P)$ and $N = \{S, A, B, C, D\}$, $T = \{0, 1\}$,   and $P = \{S \to 0A, A \to 0B, B \to 0A, B \to \lambda, S \to 1C, C \to 1D, D \to 1C, D \to 1\}$.

An NFA, M, with $ac(M) = L(G)$ will have internal states S, A, B, C, D, and F. Productions of the form $S_1 \to tS_2$ result in a transition from state S_1 to state S_2 when a t is read. The transition $D \to 1$ results in a transition from state D to an accepting state F when a 1 is read. The erasing production $B \to \lambda$ results in the designation of B as an accepting state. On including the transitions resulting from all productions, we obtain the NFA, M, illustrated in Figure 11.37. Figure 11.38 illustrates $D(M)$. You may verify that $L(G) = ac(M)$ is the set designated by the regular expression $00(00)^* + 1(11)^*$.   □

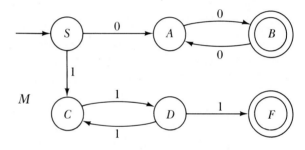

Figure 11.37

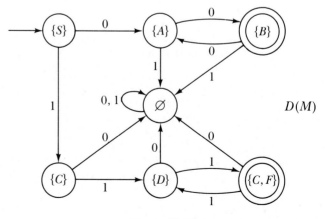

Figure 11.38

A reversal of the procedure in Example 2 will allow us to write a type 3 grammar for the set accepted by a DFA.

Example 3   We find a type 3 grammar G whose language $L(G)$ is the set of words accepted by the DFA, M, of Figure 11.39.

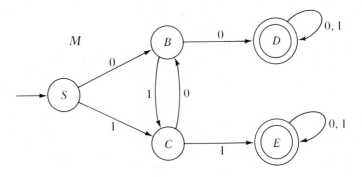

Figure 11.39

The grammar $G = (N, T, S, P)$ has $N = \{S, B, C, D, E\}$ and $T = \{0, 1\}$. Each transition results in a production. For example, the transition from S to B when 0 is read yields the production $S \rightarrow 0B$.

Each accepting state yields an erasing production of the form

$$(\text{accepting state}) \rightarrow \lambda$$

Thus, we must include both the productions $D \rightarrow \lambda$ and $E \rightarrow \lambda$. A complete description of P is $P = \{S \rightarrow 0B, B \rightarrow 1C, S \rightarrow 1C, C \rightarrow 0B, B \rightarrow 0D, C \rightarrow 1E, D \rightarrow 1D, D \rightarrow 0D, E \rightarrow 1E, E \rightarrow 0E, D \rightarrow \lambda, E \rightarrow \lambda\}$.

The language $L(G)$ is the set accepted by M. That is, $L(G)$ is the set of all words containing two successive 0's or two successive 1's.   □

Since it is always possible to design a DFA to accept any type 3 language, and since, conversely, it is always possible to describe the set of accepted words by a type 3 grammar, we have the following fundamental theorem.

Theorem 11.10   A set is recognizable by a DFA if and only if the set is a type 3 language.

This theorem reflects what has been a very fruitful area of research: finding a hierarchy of machines that corresponds to the Chomsky Hierarchy of languages. For example, it can be shown that if a DFA is supplied with a "stack" or "push-down" memory, the resulting machine (called a push-down acceptor) corresponds to the set of context free languages. In other words, if L is a context free language, then there is a push-down acceptor that accepts exactly the language L.

The hierarchy of machines, which begins with DFA's, ends with Turing Machines, the most powerful of mathematical machines. Although Turing Machines are rather simple, they can recognize the language generated by any phrase structure grammar. In fact, there are no algorithms that a Turing Machine cannot perform.

A Turing Machine is much like a DFA. On reading an input symbol, however, a Turing Machine can do any or all of the following (depending on the input symbol):

a. change its internal state;

b. print a new symbol on the tape to replace the one just read;

c. move its reading head either right *or* left.

More detailed considerations of Turing Machines and phrase structure grammars are contained in many texts in formal language theory.

11.8   Problems

In Problems 1–6 construct a DFA that accepts the language generated by the given grammar.

1. $G = (N, T, S, P)$, where $N = \{S, A, B\}$, $T = \{0\}$, and
   $P = \{S \to 0A, A \to 0B, B \to 0S, S \to \lambda\}$

2. $G = (N, T, S, P)$, where $N = \{S\}$, $T = \{1\}$, and $P = \{S \to 1S, S \to \lambda\}$

3. $G = (N, T, S, P)$, where $N = \{S, A, B\}$, $T = \{a, b\}$, and
   $P = \{S \to aA, S \to bB, A \to aA, B \to bB, A \to a, B \to b\}$

4. $G = (N, T, S, P)$, where $N = \{S\}$, $T = \{a, b\}$, and
   $P = \{S \to aS, S \to b\}$

5. $G = (N, T, S, P)$, where $N = \{S, A, B\}$, $T = \{a, b, c\}$, and
   $P = \{S \to a, S \to aA, S \to aS, S \to cB, B \to b, S \to \lambda, A \to a\}$

6. $G = (N, T, S, P)$, where $N = \{S, A, B, C\}$, $T = \{a, b, c\}$, and
   $P = \{S \to aA, A \to bB, B \to cC, A \to a, B \to b, C \to cC, C \to \lambda\}$

In Problems 7–10 give the grammar of the language accepted by the DFA whose transition matrix is given.

7.

Present state	a	b
A	B	D
B	C	D
C	C	C
D	B	D

Initial state: A
Accepting state: D

8.

Present state \ Input	0	1
A	A	B
B	A	C
C	A	C

Initial state: A
Accepting state: C

9.

Present state \ Input	a	b	c
A	B	A	A
B	D	C	A
C	B	A	D
D	D	D	D

Initial state: A
Accepting state: D

10.

Present state \ Input	0	1	2
A	B	A	C
B	A	D	A
C	D	C	A
D	B	A	B

Initial state: A
Accepting states: C, D

Chapter 11   REVIEW

Concepts for Review

transition matrix (p. 457)
deterministic finite automaton (DFA) (p. 458)
input alphabet (p. 458)
initial state (p. 459)
accepting states (p. 459)
next state function (p. 459)
nondeterministic finite automaton (NFA) (p. 464)
$R(M)$, where M is a DFA (p. 466)
$D(M)$, where M is a DFA (p. 471)
concatenation (p. 476)
regular expressions (p. 477)
Kleene closure (p. 477)
regular sets (p. 478)

phrase structure grammar (p. 490)
language generated by a grammar *G* (p. 490)
Chomsky hierarchy (p. 492)
type 3 languages (p. 494)
Turing Machine (p. 498)

Review Problems

1. Draw the directed graph that describes the DFA with the given transition matrix.

Present state	Input a	b
A	*B*	*A*
B	*B*	*C*
C	*B*	*D*
D	*D*	*D*

Initial state: *A*
Accepting state: *D*

2. Draw the directed graph that describes the DFA with the given transition matrix.

Present state	Input 0	1
A	*A*	*B*
B	*B*	*C*
C	*C*	*D*
D	*D*	*B*

Initial state: *A*
Accepting state: *D*

3. Use the notation of regular sets to describe the set of words accepted by the DFA in Problem 1.

4. Use the notation of regular sets to describe the set of words accepted by the DFA in Problem 2.

5. Design an NFA, *M*, with $\mathscr{I} = \{a, b, c\}$ and $ac(M) = a*bc$.

6. Design a DFA, *M*, with $\mathscr{I} = \{a, b, c\}$ and $ac(M) = ab + c$.

7. Design an NFA, *M*, with $\mathscr{I} = \{a, b, c\}$ and $ac(M) = ab + (ccc)*$.

8. Design a DFA, *M*, with $\mathscr{I} = \{0, 1\}$ and $ac(M) = (0 + 1)*$.

9. Give an example of a set that cannot be the set of words accepted by a DFA.

10. Explain why the concept of an NFA is a valuable one.

11. Use the procedure of Section 11.3 to convert the NFA, whose transition matrix is given, to a DFA.

Present state	Input 0	1	2
A	B	A, D	B, C
B	A, D	—	D
C	—	C	A
D	C	B	—

Initial states: B, C

Accepting state: A

12. Use the procedure of Section 11.2 to convert the NFA, whose transition matrix is given, to a DFA.

Present state	Input a	b
A	B	—
B	C	—
C	—	—
D	—	E
E	—	F
F	—	G
G	—	—

Initial states: A, D

Accepting states: C, G

13. Explain why $a + (bc)^*a^*$ is a regular expression. Describe the regular set designated by this expression.

14. Design a DFA, M, such that $ac(M) = a + (bc)^*a$.

15. Use the procedure of Section 11.6 to minimize the DFA whose transition matrix is given.

Present state	Input a	b
A	B	A
B	B	C
C	E	D
D	D	F
E	E	C
F	F	D

Initial state: A

Accepting states: D, F

16. Describe the language generated by the grammar $G = (N, T, S, P)$ where $N = \{S, A, B\}$, $T = \{a, b\}$, $P = \{S \rightarrow aAa, A \rightarrow bBb, B \rightarrow bB, B \rightarrow b\}$. Of what type is this grammar?

17. Describe the language generated by the grammar $G = (N, T, S, P)$ where $N = \{S\}$, $T = \{0, a, b\}$, $P = \{S \rightarrow aSb, S \rightarrow 0\}$. Of what type is this grammar?

18. Describe a grammar G that has $L(G) = (ab)^*c$.

19. Describe a grammar G where $L(G)$ consists of all words over $\Sigma = \{0, 1, 2\}$ consisting of two zeros followed by an even number of 1's followed by a single 2.

20. Construct an NFA that accepts the language generated by the grammar $G = (N, T, S, P)$ where $N = \{S, A, B, C\}$, $T = \{0, 1\}$, $P = \{S \rightarrow 1A, A \rightarrow 1B, B \rightarrow 0C, C \rightarrow 1S, S \rightarrow \lambda\}$.

21. Construct an NFA that accepts the language generated by the grammar $G = (N, T, S, P)$ where $N = \{S, A, B\}$, $T = \{a, b, c\}$, $P = \{S \rightarrow c, S \rightarrow aA, S \rightarrow bB, B \rightarrow bB, A \rightarrow aA, A \rightarrow \lambda, A \rightarrow c, B \rightarrow \lambda\}$.

22. Give the grammar of the language accepted by the DFA given in Problem 1.

23. Give the grammar of the language accepted by the DFA given in Problem 2.

24. Explain why the language accepted by an NFA is of type 3.

Appendix

A.1 SETS

In mathematics the notion of a "set" is generally left undefined. Although this is not a happy situation, it is unavoidable.* To create any logical structure one must begin with certain basic concepts that cannot be defined in terms of other concepts. It does little good to define idea A in terms of B, idea B in terms of C, and so on, since this process will either never end, or it will become circular.

Thus, for instance, to define a "set" of objects as a "collection" of objects is not particularly helpful since we are left with the problem of defining a "collection." Although the notion of a "set" is never really defined, it is still possible for us to have a good intuitive idea of what is meant by this term. For the purposes of this text, a set of objects, a class of things, a collection of items all represent the same idea.

We shall also assume that it is possible (at least theoretically) to determine whether a particular object is a member of, or belongs to, a given set. For instance, if A is the set of all U.S. cities, then Chicago belongs to this set while a toad does not. In general, if A represents a set and a is a member of the set we shall write

$$a \in A$$

where the symbol $\in$ can be read "is a member of," "is an element of," or "belongs to." Therefore, if A is a set consisting of the integers 1, 2, and 6, we can write

$$6 \in A$$

* Actually, in formal set theory, sets are defined in terms of what are called classes; however, classes themselves are undefined.

to indicate that 6 is an element of this set. We also write

$$4 \notin A$$

to indicate that 4 does not belong to A.

There are a number of ways that are commonly used to denote sets. In some cases it is desirable or necessary to simply list all the elements of the set. For instance, in the example just discussed we could write

$$A = \{1, 2, 6\}$$

Another possibility for describing the set A is

$$A = \{x | x = 1, 2, \text{ or } 6\}$$

where the vertical bar | is to be read "such that."

What if a set contains an infinite number of elements? Although for obvious reasons we could not list all of the elements of such a set, there are nevertheless a variety of ways to describe it. For instance, the set B of all positive even numbers could be written as:

$$B = \{2, 4, 6, 8, \ldots\}$$

or

$$B = \{x | x \text{ is a positive even number}\}$$

or

$$B = \{q | q = 2, 4, 6, 8, \ldots\}$$

In addition, if we let $\mathbf{Z}^+$ denote the set of all positive integers, then we can describe the set B by

$$B = \{2n | n \in \mathbf{Z}^+\}$$

At the other extreme, there is a set that contains no elements. This set is generally referred to as the *empty set*, and is denoted by $\emptyset$.

Finally, you should note that sets themselves may be elements of a given set. For instance, the set A defined by

$$A = \{4, 6, \{1, 4\}, \{4\}, 9\}$$

contains five elements: 4, 6, $\{1, 4\}$, $\{4\}$, and 9.

Definition A.1   A set B is a *subset* of a set A if every member of B is also a member of A.

If B is a subset of A, we write

$$B \subseteq A$$

If B is not a subset of A, we write

$$B \nsubseteq A$$

Example 1

(a) If $A = \{1, 2, -7, 6\}$ and $B = \{2, 6\}$, then since every element of B is an element of A, it follows that B is a subset of A, and we can write $B \subseteq A$.

(b) The set $B = \{1/2, -1, 4\}$ is not a subset of $A = \{3, -1, 4, 9, 11\}$ since $\frac{1}{2} \in B$, but $\frac{1}{2} \notin A$. Consequently, $B \nsubseteq A$. $\square$

Definition A.2   Two sets A and B are said to be *equal* if $A \subseteq B$ and $B \subseteq A$.

From Definition A.2 we see that two sets A and B are equal if every element of A is an element of B, and if every element of B is an element of A. Thus, two sets are equal if they contain precisely the same elements.

Example 2   The sets $\{1, -9, \pi\}$ and $\{-9, \pi, 1\}$ are equal, but the sets $\{1, \{-9\}, \pi\}$ and $\{-9, \pi, 1\}$ are not equal since the set $\{1, \{-9\}, \pi\}$ has $\{-9\}$ as an element while the set $\{-9, \pi, 1\}$ does not. $\square$

If we wish to emphasize that a set B is a subset of a set A, but is not equal to A, we write

$$B \subsetneqq A$$

Thus, for example,

$$\{2, 7\} \subsetneqq \{2, 3, 7, 10\}$$

It is important that you distinguish carefully between the notions "C is a subset of A" ($C \subseteq A$) and "C is an element of A" ($C \in A$).

Example 3   If

$$A = \{2, \{3, 4\}, 3, 6, -1\}$$

then all of the following are true:

$3 \in A$	$\{3, -1\} \subseteq A$
$\{3\} \subseteq A$	$4 \notin A$
$\{3, 4\} \in A$	$\{4\} \nsubseteq A$
$\{3, 4\} \nsubseteq A$	$\{3, -1, 2\} \subsetneqq A$

$\square$

There are a number of ways that new sets can be generated from given sets. Two such possibilities are defined next.

Definition A.3   Let A and B be sets. Then the *union* of A and B, $A \cup B$, is defined by

$$A \cup B = \{x|x \in A \quad \text{or} \quad x \in B\}.$$

The *intersection* of A and B, $A \cap B$, is defined by

$$A \cap B = \{x|x \in A \quad \text{and} \quad x \in B\}$$

A word here about the mathematical use of "or" and "and" is perhaps in order. As one might expect, to say that $x \in A$ *and* $x \in B$ means that x must belong to both of these sets. However, to say that $x \in A$ *or* $x \in B$ does not exclude the possibility that x is an element of both of these sets; that is, the word "or," when employed in a mathematical context, is not exclusive. For example, if $A = \{1, 5, 9\}$ and $B = \{2, 9, 5\}$, then $A \cup B = \{1, 2, 5, 9\}$. Note that 5 and 9 belong to both of the sets A and B.

Example 4

(a) If $A = \{-\frac{1}{2}, \Delta, *, 4, c\}$ and $B = \{\Delta, 8, \square, 4, C, \sqrt{2}\}$, then

$$A \cup B = \{-\tfrac{1}{2}, \Delta, *, 4, c, 8, \square, C, \sqrt{2}\}$$

and

$$A \cap B = \{\Delta, 4\}$$

(b) Let $A = \{x|1 \le x < 15 \text{ and } x \text{ is an odd integer}\}$ and let $B = \{x|1 \le x \le 18 \text{ and } x \text{ is a prime number}\}$. Then

$$A \cup B = \{1, 2, 3, 5, 7, 9, 11, 13, 17\}$$

and

$$A \cap B = \{3, 5, 7, 11, 13\}$$

(c) If $A = \{\sqrt{2}, \square, 3\}$ and $B = \{-1, 2\}$, then $A \cap B = \varnothing$.   □

As illustrated in the next example, set theoretic identities can be established using Definition A.2.

Example 5

We show that $A \cap (B \cup C) = (A \cap B) \cup (A \cap C)$. From Definition A.2 we must show that

$$A \cap (B \cup C) \subseteq (A \cap B) \cup (A \cap C) \tag{1}$$

and

$$(A \cap B) \cup (A \cap C) \subseteq A \cap (B \cup C) \tag{2}$$

To establish (1) we will assume that $x \in A \cap (B \cup C)$ and show that $x \in (A \cap B) \cup (A \cap C)$. If $x \in A \cap (B \cup C)$, then $x \in A$ and $x \in B \cup C$. Thus $x \in A$, and $x \in B$ or $x \in C$. Consequently $x \in A$ and $x \in B$, or $x \in A$ and $x \in C$;

that is, $x \in A \cap B$ or $x \in A \cap C$. It follows that $x \in (A \cap B) \cup (A \cap C)$ and (1) has been established.

To prove (2) we assume that $x \in (A \cap B) \cup (A \cap C)$ and show that $x \in A \cap (B \cup C)$. If $x \in (A \cap B) \cup (A \cap C)$, then $x \in A \cap B$ or $x \in A \cap C$. That is, $x \in A$ and $x \in B$, or $x \in A$ and $x \in C$. It follows that $x \in A$, and $x \in B$ or $x \in C$. Consequently $x \in A$ and $x \in B \cup C$; thus $x \in A \cap (B \cup C)$, so (2) has been proven. From (1) and (2) it follows that

$$A \cap (B \cup C) = (A \cap B) \cup (A \cap C) \qquad \square$$

Unions and intersections of more than two sets are defined in the obvious manner.

Definition A.4 If $A_m, A_{m+1}, \ldots, A_n$ are sets, then the *union* of these sets is defined by

$$\bigcup_{i=m}^{n} A_i = \{x \mid x \in A_i \quad \text{for some } i, m \leq i \leq n\}$$

The *intersection* of these sets is defined by

$$\bigcap_{i=m}^{n} A_i = \{x \mid x \in A_i \quad \text{for each } i, m \leq i \leq n\}$$

Example 6 If for each i, $1 \leq i \leq 5$, A_i is defined by

$$A_i = \{x \mid x \text{ is an integer} \quad \text{and} \quad i - 2 \leq x \leq 3i\}$$

then

$$\bigcup_{i=1}^{5} A_i = \{-1, 0, 1, 2, 3, 4, 5, 6, 7, 8, 9, 10, 11, 12, 13, 14, 15\}$$

and

$$\bigcap_{i=1}^{5} A_i = \{3\}$$

Note also that we have

$$\bigcap_{i=3}^{5} A_i = \{3, 4, 5, 6, 7, 8, 9\}$$

and

$$\bigcup_{i=2}^{4} A_i = \{0, 1, 2, 3, 4, 5, 6, 7, 8, 9, 10, 11, 12\} \qquad \square$$

A simple modification of Definition A.4 allows us to define the union and intersection of infinitely many sets.

Definition A.5 If $A_1, A_2, \ldots$ are sets, then the *union* of these sets is defined by

$$\bigcup_{i=1}^{\infty} A_i = \{x | x \in A_i \quad \text{for some } i, i = 1, 2, \ldots\}$$

The *intersection* of these sets is defined by

$$\bigcap_{i=1}^{\infty} A_i = \{x | x \in A_i \quad \text{for each } i, i = 1, 2, \ldots\}$$

Example 7
   (a) If for $i = 1, 2, \ldots, A_i = \{x | i \leq x \leq i + 1\}$, then

$$\bigcup_{i=1}^{\infty} A_i = \{x | x \geq 1\}$$

   (b) If for $i = 1, 2, \ldots, A_i = \{x | 1 - 1/i < x < 1 + 1/i\}$, then

$$\bigcap_{i=1}^{\infty} A_i = \{1\} \qquad \square$$

Definition A.6 If X and A are sets, then the *complement* of A in X, $X \backslash A$, is defined by

$$X \backslash A = \{x | x \in X \quad \text{and} \quad x \notin A\}$$

Example 8 If $X = \{-4, 2, 0, 6\}$ and $A = \{7, 2, 6, \pi\}$, then

$$X \backslash A = \{-4, 0\} \qquad \square$$

At times it is helpful to "picture" unions, intersections, and complements of sets with the aid of what are called Venn diagrams. In a Venn diagram, disks as indicated in Figure A.1 are used to represent sets.

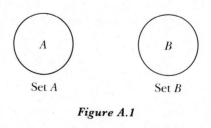

Set A Set B

Figure A.1

We can denote the intersection of sets A and B by shading in the area that is common to both of the sets, as in Figure A.2.

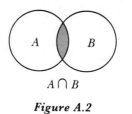

$A \cap B$

Figure A.2

In Figure A.3 the shading denotes $A\backslash B$, $A \cup B$, and $B\backslash A$.

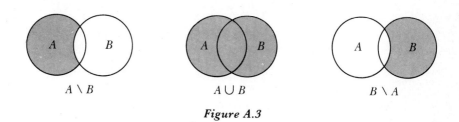

$A \backslash B$ $A \cup B$ $B \backslash A$

Figure A.3

Slightly more complicated combinations of unions, intersections, and complements are indicated in Figure A.4.

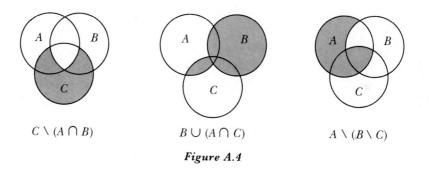

$C \backslash (A \cap B)$ $B \cup (A \cap C)$ $A \backslash (B \backslash C)$

Figure A.4

It is frequently useful to consider ordered pairs of elements. An *ordered pair* is a pair, (a, b), where a is the first coordinate of the pair and b is the second coordinate. Two pairs (a, b) and (c, d) are *equal* whenever $a = c$ and $b = d$. Thus, for example, the ordered pairs $(x + 2, -y/2)$ and $(1, 4)$ are equal if and only if $x + 2 = 1$ and $-y/2 = 4$, that is, if and only if $x = -1$ and $y = -8$. Note that the ordered pairs $(2, 5)$ and $(5, 2)$ are not equal.

The Cartesian product of two sets is defined in terms of ordered pairs.

Definition A.7 If A and B are sets, then the *Cartesian product of A and B*, $A \times B$, is defined by

$$A \times B = \{(a, b) | a \in A \quad \text{and} \quad b \in B\}$$

Example 9
   (a) If $A = \{2, -3\}$ and $B = \{0, -3, 4\}$ then

$$A \times B = \{(2, 0), (2, -3), (2, 4), (-3, 0), (-3, -3), (-3, 4)\}$$

Note that

$$B \times A = \{(0, 2), (0, -3), (-3, 2), (-3, -3), (4, 2), (4, -3)\}$$

and, hence $A \times B \neq B \times A$. In Problem 46 you are asked to establish conditions under which $A \times B = B \times A$.

   (b) If $\mathbf{R}^1$ denotes the set of real numbers, then

$$\mathbf{R}^1 \times \mathbf{R}^1 = \{(x, y) | x \in \mathbf{R}^1 \quad \text{and} \quad y \in \mathbf{R}^1\} \qquad \square$$

We frequently use the following notion.

Definition A.8 Let X be a finite set. The *power set* of X, $\mathscr{P}(X)$, is defined by

$$\mathscr{P}(X) = \{A | A \subseteq X\}$$

In other words, the power set of X is the set of all subsets of X.

Example 10
   (a) If $X = \{1, 2\}$, then $\mathscr{P}(X) = \{\varnothing, \{1\}, \{2\}, \{1, 2\}\}$.
   (b) If $X = \{a, b, c\}$, then

$$\mathscr{P}(X) = \{\varnothing, \{a\}, \{b\}, \{c\}, \{a, b\}, \{a, c\}, \{b, c\}, \{a, b, c\}\} \qquad \square$$

A.1   Problems

1. Use set notation to describe the following sets:
    (a) The set of all even integers between 3 and 105
    (b) The set of squares of all nonnegative integers
    (c) The set of integers divisible by 6
    (d) The set of odd integers between (and including) -9 and 541
2. Use set notation to describe the following sets:
    (a) The set of all integers that are the sum of two prime numbers
    (b) The set of all sets that contain precisely three elements

    (c) The set of all integers that can be written as the sum of two squared integers

    (d) The set of all integers whose squares lie between 3 and 1968

In Problems 3–18 indicate if the given statement is true or false.

3. $\{2, 5\} \subseteq \{2, 4, 6, 8\}$

4. $7 \notin \{n | n = 2k, k = 1, 2, 3, \ldots\}$

5. $\{a, b, c\} \in \{a, b, c, \{d, e\}\}$

6. $\{a, b, c, d\} \subseteq \{a, b, d, c, e\}$

7. $\{u, v\} \in \{\{u, v\}, u, w\}$

8. $\{u\} \in \{v, u\}$

9. $-1 \in \mathbf{Z}^+$

10. $\{1, 3, 6\} = \{1, 6, 3\}$

11. $\Delta \notin \{\{\Delta, \pi\}, u\}$

12. $\{3, \Delta\} \subsetneqq \{\{3\}, \Delta, 5\}$

13. $\{6, 7\} \in \{2, 3, 6, \{7\}\}$

14. $\{u, v, w, x\} \subsetneqq \{x, w, u, v\}$

15. $\{x, y, z\} = \{x, y, z, y\}$

16. $\{x, y, z\} \subseteq \{x, y, z, y\}$

17. $\{a, \Delta\} \subseteq \varnothing$

18. $a \notin \varnothing$

In problems 19–31 compute the indicated set.

19. $\{x, y, z\} \cup \{u, v, w\}$

20. $\{x, y, z\} \cap \{x, y, w\}$

21. $\{a, b, c, d\} \cap \{d, e, f\}$

22. $\{a, b, c\} \cap \{x\}$

23. $\{x, y\} \times \{1, 2, 3\}$

24. $\{r, s, t\} \backslash \{s, t\}$

25. $(\{s, t\} \backslash \{r, s, t\}) \cup \{r, t, u\}$

26. $\{r, s, t, x\} \cap \{x, y, z, s\}) \backslash \{y, s, r\}$

27. $\{1, 2\} \times \{1, 2, 3\}$

28. $\mathbf{Z}^+ \times \mathbf{Z}^+$

29. $\mathbf{Z}^+ \cap \varnothing$

30. $S \backslash \varnothing$ (S is any set)

31. $T \times \varnothing$ (T is any set)

32. Let: $A = \{n | 1 \leq n \leq 10$ and n is even$\}$
          $B = \{n | 1 \leq n \leq 10$ and n is prime$\}$

    Compute:

    (a) $A \cap B$
    (b) $A \times B$
    (c) $A \backslash B$

    (d) $B \backslash A$
    (e) $(A \cap B) \times B$
    (f) $(A \backslash B) \times B$

33. For each positive integer i, let $A_i = \{i, i + 1, i + 2, i + 3\}$.
    Compute:

    (a) $\displaystyle\bigcup_{i=1}^{4} A_i$
    (b) $\displaystyle\bigcup_{i=5}^{9} A_i$
    (c) $\displaystyle\bigcap_{i=1}^{3} A_i$

    (d) $\displaystyle\bigcap_{i=6}^{9} A_i$
    (e) $\displaystyle\bigcup_{i=3}^{6} (A_i \backslash A_{i+2})$
    (f) $\displaystyle\bigcup_{i=1}^{7} (A_i \cap A_{i+2})$

    (g) $\displaystyle\bigcap_{i=3}^{4} (A_i \cup A_{i+2})$

34. For $i = 1, 2, \ldots$, let $A_i = \{x | x > i\}$. Compute:

    (a) $\displaystyle\bigcup_{i=2}^{\infty} A_i$
    (b) $\displaystyle\bigcap_{i=1}^{\infty} A_i$

35. For each positive integer n, let $A_n = \{x| -1/n \le x < n + 1\}$. Compute:

(a) $\displaystyle\bigcup_{n=1}^{4} A_n$ 　　　　　　　　　　　　 (b) $\displaystyle\bigcap_{n=1}^{5} A_n$

(c) $\displaystyle\bigcap_{n=2}^{5} A_n$ 　　　　　　　　　　　　 (d) $A_7 \backslash A_6$

(e) $\displaystyle\bigcup_{n=1}^{4} (A_n \backslash A_{n+2})$ 　　　　　 (f) $A_4 \backslash \left(\displaystyle\bigcap_{n=1}^{3} A_n \right)$

(g) $\displaystyle\bigcap_{n=1}^{4} (A_n \backslash A_{n+2})$ 　　　　　 (h) $\displaystyle\bigcap_{n=1}^{\infty} A_n$

36. Use Venn diagrams to picture the following sets:
  (a) $(A \backslash B) \cup C$ 　　(b) $(A \backslash B) \backslash C$ 　　　(c) $(A \cup B) \backslash C$
  (d) $(A \cap B) \cup C$ 　　(e) $(A \cap B) \backslash C$ 　　　(f) $(A \cup B) \cap C$

37. Show that for any two sets A and B: $A \cap B \subseteq A \cup B$. (*Hint:* Show that if $x \in A \cap B$, then $x \in A \cup B$.)

In Problems 38–43 show the given equality is true for any three sets A, B, and C.

38. $A \cup (B \cap C) = (A \cup B) \cap (A \cup C)$
39. $A \times (B \cap C) = (A \times B) \cap (A \times C)$
40. $A \backslash (B \cap C) = (A \backslash B) \cup (A \backslash C)$
41. $(A \backslash B) \cap B = \varnothing$
42. $(A \cap B) \backslash C = (A \backslash C) \cap (B \backslash C)$
43. $A \backslash (B \cup C) = (A \backslash B) \cap (A \backslash C)$

44. Find a specific example to disprove: $A \cup B \subseteq A \cap B$. (Such an example is called a *counterexample*.)

45. Find a specific example to disprove $A \times B = B \times A$

46. Prove: If $A \times B = B \times A$ and $A \ne \varnothing$ and $B \ne \varnothing$, then $A = B$.

47. Compute the power sets of:
  (a) $X = \{a, 1, -2\}$
  (b) $\mathscr{P}(X)$, where $X = \{0, 1\}$

48. Use induction to show that if a set X has n elements, then $\mathscr{P}(X)$ has 2^n elements.

A.2　FUNCTIONS

Informally, a function from a set A to a set B is a "rule" that assigns to each element of A a unique element of B. For instance, if $A = \{1, 2, 4\}$ and $B = \{1, 2, 6, 8\}$, then the rule that assigns to the number 1 the number 6; to the number 2, the number 2; and to the number 4, the number 8 is a function. In a rather naive sense one might think of a function as a "computer" or "black box" that accepts elements of a given set as inputs and, corresponding to each input, produces an output consisting of a single element from another (or possibly the same) set.

To describe a function such as the one given above we could let f denote our "rule" and write

$$1 \xrightarrow{\; f \;} 6 \qquad 2 \xrightarrow{\; f \;} 2 \qquad 4 \xrightarrow{\; f \;} 8$$

More commonly, however, we use the notation

$$f(1) = 6 \qquad f(2) = 2 \qquad f(4) = 8 \tag{1}$$

to describe such a function, where, for example, $f(1)$ is read "f of 1." Letters other than f can be used as well. For instance, the function described by

$$g(1) = 6 \qquad g(2) = 2 \qquad g(4) = 8$$

is identical to the one defined by (1), since the rules of assignment are identical.

The set of all possible inputs for a function is called the *domain* of the function and the set of outputs is called the *range*. Thus, in our example, the domain of f is $\{1, 2, 4\}$ and the range is $\{2, 6, 8\}$.

Generally, if X is the domain of a function f and the range of f is contained in or is equal to a set Y, then we write

$$f : X \to Y$$

If $f : X \to Y$, we often say that f *maps* the set X into Y; in fact, functions are often referred to as *mappings*.

Frequently it will be possible to describe a function f in terms of a simple rule, instead of having to make a list as in (1). For example, if we wish to assign to each real number the square of the number, we can write

$$f(x) = x^2$$

where x can be any real number. This particular function could also be described by $g(y) = y^2$ or $H(z) = z^2$. Note that the domain of this function is $\mathbf{R}^1$ and the range is the set of all nonnegative numbers.

Example 1

(a) Let $X = \{a, b, c, d\}$ and $Y = \{1, 2, 3, 4\}$. We can define $f : X \to Y$ by $f(a) = 2$, $f(b) = 3$, $f(c) = 2$, $f(d) = 1$.

(b) Let X be a set. For each $A \in \mathscr{P}(X)$, let $|A|$ denote the number of elements in A. Then $g : \mathscr{P}(X) \to \mathbf{Z}^+ \cup \{0\}$ defined by

$$g(A) = |A|$$

is a function. Note that $g(\{a, b, c\}) = |\{a, b, c\}| = 3$, and $g(\varnothing) = |\varnothing| = 0$.

(c) Let $X = \mathbf{R}^1 \times \mathbf{R}^1$ and define $h: X \to \mathbf{R}^1$ by

$$h(a, b) = a^3 b^2 + 3ab + 7$$

Then h is a function that maps $\mathbf{R}^1 \times \mathbf{R}^1$ into the real numbers. Note that $h(-2, 1) = -7$ and $h(1, -2) = 5$.

(d) Define $f: \mathbf{R}^1 \to \mathbf{R}^1$ by

$$f(x) = \begin{cases} 2x + 1 & \text{if} \quad x \le 3 \\ x^2 & \text{if} \quad 3 < x \le 5 \\ 4 & \text{if} \quad x > 5 \end{cases}$$

Note that $f(-7) = -13$, $f(3.5) = 12.25$, and $f(1000) = 4$.

(e) Let X be a set, and let A be a subset of X. The *characteristic function* associated with A is the function $f: X \to \{0, 1\}$ defined by

$$f(x) = \begin{cases} 0 & \text{if} \quad x \notin A \\ 1 & \text{if} \quad x \in A \end{cases}$$

Note that the characteristic function describes which elements of X are in A.

$\square$

Functions can be categorized in a number of ways. One important class of functions is defined as follows.

Definition A.9 A function $f: X \to Y$ is said to be a *one-to-one* function if whenever $x_1 \ne x_2$, then $f(x_1) \ne f(x_2)$.

It is easy to verify that this idea is equivalent to the property that if $f(x_1) = f(x_2)$, then $x_1 = x_2$.

Example 2
(a) The function $f: \mathbf{R}^1 \to \mathbf{R}^1$ defined by

$$f(x) = 3x - 2$$

is one-to-one since if $3x_1 - 2 = 3x_2 - 2$, then $x_1 = x_2$.

(b) The function $f: \mathbf{R}^1 \to \mathbf{R}^1$ defined by

$$f(x) = x^2$$

is not one-to-one since, for example, $f(3) = f(-3)$ even though $3 \ne -3$. $\square$

Definition A.10 A function $f: X \to Y$ is an *onto* function if for each $y \in Y$ there is an $x \in X$ such that $f(x) = y$.

Example 3

(a) Let $X = \mathbf{R}^1$ and $Y = \{y \mid y \geq 0\}$ and define $f: X \to Y$ by $f(x) = x^2$. Then f is an onto function since for each $y \in Y$ there is an $x \in X$ (either $x = \sqrt{y}$ or $x = -\sqrt{y}$) such that $f(x) = y$.

(b) Let $X = \{1, 2, 3, 4\}$ and $Y = \{5, 9, 13, 17, 21\}$ and define $f: X \to Y$ by $f(x) = 4x + 1$. Then f is not onto since there is no $x \in X$ such that $f(x) = 21$.   □

Often it is useful to combine functions in the following way. Suppose that f is a function that maps a set X into a set Y, and that g is a function that maps the set Y into a set Z. Diagramatically we have

$$X \xrightarrow{\ f\ } Y \xrightarrow{\ g\ } Z$$

In this situation we can define a function from X to Z in a natural way. If $x \in X$, then $f(x) \in Y$. Since $g: Y \to Z$, we have that $g(f(x)) \in Z$. Thus, if we define a function

$$h: X \to Z$$

by

$$h(x) = g(f(x))$$

we have defined a function that combines the functions f and g and that maps the set X into the set Z. The function h is said to be the *composite function* of f by g, and it is usually denoted by $g \circ f$. Note that we have

$$X \xrightarrow[f]{} Y \xrightarrow[g]{} Z$$

with $g \circ f$ spanning from X to Z.

Example 4

(a) Let $f: \mathbf{R}^1 \to \mathbf{R}^1$ be defined by $f(x) = x^2$ and let $g: \mathbf{R}^1 \to \mathbf{R}^1$ be defined by $g(x) = 3x + 1$. Then $g \circ f(x) = g(f(x)) = g(x^2) = 3x^2 + 1$.

(b) Let $f: \mathbf{R}^2 \to \mathbf{R}^1$ be defined by $f(x, y) = x^3y^2$ and let $g: \mathbf{R}^1 \to \mathbf{R}^1$ be defined by

$$g(x) = \frac{x}{1 + |x|}$$

where $|x|$ denotes the absolute value of x. Then

$$g \circ f(x, y) = g(f(x, y)) = g(x^3y^2) = \frac{x^3y^2}{1 + |x^3y^2|}$$ □

Observe that in part (a) of Example 3 we could also define the composite function $f \circ g$, where $f \circ g: \mathbf{R}^1 \to \mathbf{R}^1$ is defined by

$$f \circ g(x) = f(g(x)) = f(3x + 1) = (3x + 1)^2$$

Note that $f \circ g(x) \neq g \circ f(x)$.

A.2 Problems

1. Suppose a function f is defined by $f(x) = x + 1/x$, for all possible real numbers x. Find:

   (a) The domain of f (b) The range of f

   (c) $f(2)$ (d) $f(-1)$

   (e) $f(a + b)$ (f) $f(1/a)$

   (g) $\dfrac{1}{f(b)}$

2. Suppose a function f is defined by $f(x) = 3 - x^2$ for all possible real numbers x. Find:

   (a) The domain of f (b) The range of f

   (c) $f(-3)$ (d) $f(0)$

   (e) $f(c^2)$ (f) $f(x - 2x^2)$

3. Find the domain and range of the function $f(x) = x/|x|$, where x denotes the absolute value of x.

4. If $f : \mathbf{R}^1 \to \mathbf{R}^1$ is defined by

$$f(x) = \begin{cases} -x^3 & \text{if } x \le 3 \\ 6 & \text{if } 3 < x < 6 \\ 2x - 1 & \text{if } x \ge 6 \end{cases}$$

   find:

   (a) $f(6)$ (b) $f(-3)$

   (c) $f(100)$ (d) $f(4.5)$

5. Suppose

$$f(x) = \begin{cases} 2x & \text{if } x > 3 \\ 4 & \text{if } -1 \le x < 2 \\ x^2 & \text{if } x < -1 \end{cases}$$

   Find:

   (a) The domain of f (b) The range of f

   (c) $f(0)$ (d) $f(-1)$

   (e) $f(4.6)$ (f) $f(-5)$

6. Give the characteristic function of $A \subseteq X$ where $A = \{1, 3, 5\}$ and $X = \{0, 1, 2, 3, 4, 5, 6\}$.

7. Give the characteristic function of $A \subseteq X$ where $A = \{a, b, c, d\}$ and $X = \{a, b, c, d, e, f, g, h\}$.

8. The function $f : X \to \{0, 1\}$ is defined by

$$f(1) = f(3) = f(5) = f(6) = 1$$
$$f(2) = f(4) = f(7) = f(8) = 0$$

   Of what set is f the characteristic function?

9. Compute:
   (a) $|\{1, 2, 5\}|$
   (b) $|\varnothing|$
   (c) $|\{2\}|$
   (d) $|\{\{1, 2\}, \{3\}\}|$
   (e) $|\{\{a, b, c\}\}|$

10. Which of the following functions are one-to-one? Unless otherwise specified, the domain of each function is the set of all real numbers.
    (a) $f(x) = 3x - 4$
    (b) $f(x) = \sqrt{x}, x \geq 0$
    (c) $g(x) = 7x^2$
    (d) $h(x) = |x - 1|$
    (e) $f(x) = 7x^2, x \geq 0$
    (f) $f(x) = x^4, x \in \{-1, 1, 2\}$

11. Which of the following functions are one-to-one? Unless otherwise specified, the domain of each function is the set of all real numbers.
    (a) $f(x) = x^3$
    (b) $f(x) = |x - 3|, x \leq 1$
    (c) $f(x) = x^2 + x^3$
    (d) $g(x) = 5$
    (e) $l(x) = x^4, x \in \{1, 2, 3\}$
    (f) $g(x) = 7x - 1, x > 2$

12. Which of the following functions $f : X \to Y$ are onto?
    (a) $X = \mathbf{R}^1, Y = \mathbf{R}^1$, and $f(x) = 2x + 4$
    (b) $X = \{1, 7, 6\}, Y = \{b, d\}$, and $f(1) = d, f(7) = d$, and $f(6) = b$
    (c) $X = \{x | x \geq 0\}, Y = \{y | y \geq 0\}$, and $f(x) = \sqrt{x + 2}$

13. Which of the following functions $f : X \to Y$ are onto?
    (a) $X = \mathbf{R}^2, Y = \mathbf{R}^1, f(a, b) = a - 2b$
    (b) $X = \mathbf{R}^1, Y = \{y | y \geq 0\}, f(x) = x^2/3$
    (c) $X = \{x | x > 0\}, Y = \{y | y > 0\}, f(x) = \sqrt{x + 9}$
    (d) $X = \{x | x \in \mathbf{R}^1 \text{ and } x \text{ is rational}\}, Y = \{y | y \in \mathbf{R}^1 \text{ and } y \text{ is irrational}\}$,
    $f(x) = \sqrt{2}x$

14. (a) If $f(x) = 2x - 3$ and $g(x) = x^2 + 1$, find $f \circ g(2)$ and $g \circ f(2)$.
    (b) If $h(x) = 1/x$ and $k(z) = 5 - z$, find $h \circ k(4), h \circ k(5), k \circ h(4)$, and $k \circ h(5)$.

15. If $f : \mathbf{R}^1 \times \mathbf{R}^1 \to \mathbf{R}^1$ is defined by $f(x, y) = x^2 + xy$ and $g : \mathbf{R}^1 \to \mathbf{R}^1$ is defined by $g(t) = 4 + 6t$, find $g \circ f(2, 3)$ and $g \circ f(q, r)$. What can be said about $f \circ g(-10)$?

16. A function $f : \mathbf{R}^1 \to \mathbf{R}^1$ is said to be an *even function* if $f(x) = f(-x)$ for each $x \in \mathbf{R}^1$; f is said to be an *odd function* if $f(-x) = -f(x)$ for each $x \in \mathbf{R}^1$.
    (a) Show that $f(x) = x^2$ is an even function and that $f(x) = x^3$ is an odd function. Can you generalize this example?
    (b) Find a function $g : \mathbf{R}^1 \to \mathbf{R}^1$ that is neither even nor odd.

17. Determine whether the following functions are odd, even, or neither odd nor even:
    (a) $f(x) = 3x^2 + 2x^4$
    (b) $f(x) = 3x^2 + 2x^3$
    (c) $g(x) = \sqrt{4 - x^2}$
    (d) $k(x) = (x^2 - 1)(3x + 2)$
    (e) $g(x) = (x^2 + 1)^2 x^3$
    (f) $g(x) = |x| - 4$
    (g) $f(x) = -16$
    (h) $h(x) = 2^x$

18. Suppose that $f: \mathbf{R}^1 \to \mathbf{R}^1$ is an even function and that $g: \mathbf{R}^1 \to \mathbf{R}^1$ is an odd function.
    (a) What can be said about the function $h: \mathbf{R}^1 \to \mathbf{R}^1$ defined by $h(x) = f(x) + g(x)$?
    (b) What can be said about the function $k: \mathbf{R}^1 \to \mathbf{R}^1$ defined by $k(x) = f(x) \cdot g(x)$?

19. Show that every function $f: \mathbf{R}^1 \to \mathbf{R}^1$ can be written as the sum of an even function and an odd fuction (*Hint*: Consider the functions $g(x) = f(x) + f(-x)$ and $h(x) = f(x) - f(-x)$.)

A.3   MATRICES AND MATRIX OPERATIONS

One of the most useful constructs in mathematics is a matrix. Matrices arise naturally in a wide variety of contexts including engineering, physics, economics, and computer science. They provide not only a convenient notational device, but also serve as a backdrop for much applied and theoretical work in mathematics and related disciplines.

In this section we consider those basic properties of matrices and matrix operations used in this text.

Definition A.11   An $m \times n$ *matrix* is a rectangular array of numbers that consists of m rows and n columns. The numbers found in a matrix are the *entries* of the matrix. A matrix with m rows and n columns is said to be of order $m \times n$.

Example 1
    (a) The rectangular array

$$\begin{bmatrix} 1 & 0 & -\frac{1}{2} \\ \pi & 3 & 1 \\ -\frac{1}{3} & \sqrt{2} & 2 \\ 4 & 4 & 7 \end{bmatrix}$$

is a 4×3 matrix since it consists of 4 rows and 3 columns of numbers. Entries of this matrix include 1, $\sqrt{2}$, and 4.

    (b) The rectangular array

$$\begin{bmatrix} 1 & -\frac{4}{7} \\ 4 & \frac{1}{2} \end{bmatrix}$$

is a 2×2 matrix; matrices having the same number of rows as columns are called *square* matrices.

(c)   A *zero matrix* is a matrix whose entries are all 0. For example,

$$\begin{bmatrix} 0 & 0 & 0 \\ 0 & 0 & 0 \end{bmatrix} \quad \text{and} \quad \begin{bmatrix} 0 & 0 \\ 0 & 0 \end{bmatrix}$$

are zero matrices.   □

It will be convenient to adopt the standard matrix notation

$$A = [a_{ij}]_{m \times n} \tag{1}$$

to denote an $m \times n$ matrix. The symbol a_{ij} represents the (i, j) th entry—the entry occurring in the ith row and jth column of the matrix A. Thus (1) is simply a "shorthand" notation to describe

$$A = \begin{bmatrix} a_{11} & a_{12} & \cdots & a_{1j} & \cdots & a_{1n} \\ a_{21} & a_{22} & \cdots & a_{2j} & \cdots & a_{2n} \\ \vdots & & & & & \\ a_{i1} & a_{i2} & \cdots & a_{ij} & \cdots & a_{in} \\ \vdots & & & & & \\ a_{m1} & a_{m2} & \cdots & a_{mj} & \cdots & a_{mn} \end{bmatrix}$$

Example 2   If

$$A = \begin{bmatrix} 1 & 4 & 2 & 1 \\ 0 & 0 & -1 & 5 \\ \frac{3}{2} & 1 & 6 & 2 \end{bmatrix}$$

then $a_{13} = 2$, since the entry found in the first row and third column is 2. Similarly, we find that $a_{24} = 5$ and $a_{31} = \frac{3}{2}$.   □

Definition A.12   Two matrices $A = [a_{ij}]_{m \times n}$ and $B = [b_{ij}]_{p \times q}$ are *equal* if they are of the same order and if their corresponding entries are equal: that is, $A = B$ if $m = p$, $n = q$, and $a_{ij} = b_{ij}$ for each i, j.

We now turn to the basic matrix operations of addition, subtraction, scalar multiplication, and multiplication. Addition (subtraction) of two $m \times n$ matrices is done in the obvious way: Corresponding entries of the two matrices are added (subtracted). Only matrices of the same order may be added or subtracted.

Definition A.13   If $A = [a_{ij}]_{m \times n}$ and $B = [b_{ij}]_{m \times n}$ are two $m \times n$ matrices, then

$$A + B = [a_{ij}]_{m \times n} + [b_{ij}]_{m \times n} = [a_{ij} + b_{ij}]_{m \times n} \tag{2}$$

and

$$A - B = [a_{ij}]_{m \times n} - [b_{ij}]_{m \times n} = [a_{ij} - b_{ij}]_{m \times n} \qquad (3)$$

Example 3
    (a) If

$$A = \begin{bmatrix} 1 & 2 & -3 \\ 1 & 4 & 6 \end{bmatrix} \quad \text{and} \quad B = \begin{bmatrix} \frac{1}{2} & -2 & 6 \\ 1 & \pi & 0 \end{bmatrix}$$

then

$$A + B = \begin{bmatrix} 1 & 2 & -3 \\ 1 & 4 & 6 \end{bmatrix} + \begin{bmatrix} \frac{1}{2} & -2 & 6 \\ 1 & \pi & 0 \end{bmatrix} = \begin{bmatrix} \frac{3}{2} & 0 & 3 \\ 2 & 4 + \pi & 6 \end{bmatrix}$$

and

$$A - B = \begin{bmatrix} 1 & 2 & -3 \\ 1 & 4 & 6 \end{bmatrix} - \begin{bmatrix} \frac{1}{2} & -2 & 6 \\ 1 & \pi & 0 \end{bmatrix} = \begin{bmatrix} \frac{1}{2} & 4 & -9 \\ 0 & 4 - \pi & 6 \end{bmatrix}$$

    (b) The matrices

$$A = \begin{bmatrix} 1 & 2 & -3 \\ 1 & 4 & 6 \end{bmatrix} \quad \text{and} \quad B = \begin{bmatrix} \frac{1}{2} & 1 \\ -2 & \pi \\ 6 & 0 \end{bmatrix}$$

cannot be added or subtracted since they are not of the same order.   □

A matrix may be multiplied by a real number. Such multiplication is called *scalar multiplication* and is defined as follows.

Definition A.14   If $A = [a_{ij}]_{m \times n}$ and $\alpha \in \mathbf{R}^1$, then

$$\alpha A = [\alpha a_{ij}]_{m \times n}$$

Thus, to multiply a matrix A by a real number, or "scalar," α we multiply each entry of A by α.

Example 4   If

$$A = \begin{bmatrix} 4 & 2 & -10 & 1 \\ -1 & 3 & 15 & 1 \end{bmatrix}$$

and $\alpha = -6$, then

$$\alpha A = -6A = \begin{bmatrix} -24 & -12 & 60 & -6 \\ 6 & -18 & -90 & -6 \end{bmatrix}$$
  □

Multiplication of two matrices is somewhat more complicated. In view of the definitions of matrix addition and subtraction, you might expect that if

$$A = [a_{ij}]_{m \times n} \quad \text{and} \quad B = [b_{ij}]_{m \times n}$$

are two $m \times n$ matrices, then the product of A and B would be defined by

$$AB = [a_{ij} \cdot b_{ij}]_{m \times n}$$

Although this product does make mathematical sense, it has limited use and is rarely, if ever, employed. The following notion turns out to be far more useful, though at first glance it appears to be somewhat contrived.

Definition A.15 Let $A = [a_{ij}]_{m \times q}$ and $B = [b_{ij}]_{q \times n}$ be matrices of order $m \times q$ and $q \times n$, respectively. The *product AB* is defined by

$$AB = [c_{ij}]_{m \times n}$$

where for each i, j

$$c_{ij} = a_{i1}b_{1j} + a_{i2}b_{2j} + \cdots + a_{iq}b_{qj} \tag{4}$$

Note that this definition applies *only* to matrices A and B with the property that the number of columns of A is equal to the number of rows of B.

This rather peculiar definition requires some elucidation. Mechanically, it works as follows. To obtain c_{ij}, the (i, j)th entry of the product AB, we "multiply" the ith row of A by the jth column of B. This is accomplished by multiplying each entry in the ith row of A by the corresponding entry in the jth column of B and then forming the sum of these products. (See Figure A.5.)

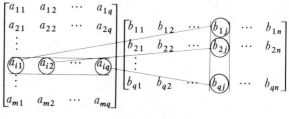

Figure A.5

The following examples should help clarify matters.

Example 5   Let

$$A = \begin{bmatrix} 1 & 3 & 2 \\ 2 & 1 & 5 \end{bmatrix}_{2 \times 3} \quad \text{and} \quad B = \begin{bmatrix} -1 & 0 & 4 & 3 \\ 2 & -4 & 0 & 4 \\ 1 & 3 & 1 & 3 \end{bmatrix}_{3 \times 4}$$

Note that the number of columns of A is equal to the number of rows of B. We have

$$AB = \begin{bmatrix} 1 & 3 & 2 \\ 2 & 1 & 5 \end{bmatrix} \begin{bmatrix} -1 & 0 & 4 & 3 \\ 2 & -4 & 0 & 4 \\ 1 & 3 & 1 & 3 \end{bmatrix} = [c_{ij}]_{2 \times 4}$$

and from (4) we find that

$$c_{11} = (1)(-1) + (3)(2) + (2)(1) = 7$$
$$c_{12} = (1)(0) + (3)(-4) + (2)(3) = -6$$
$$c_{13} = (1)(4) + (3)(0) + (2)(1) = 6$$
$$c_{14} = (1)(3) + (3)(4) + (2)(3) = 21$$
$$c_{21} = (2)(-1) + (1)(2) + (5)(1) = 5$$
$$c_{22} = (2)(0) + (1)(-4) + (5)(3) = 11$$
$$c_{23} = (2)(4) + (1)(0) + (5)(1) = 13$$
$$c_{24} = (2)(3) + (1)(4) + (5)(3) = 25$$

Consequently, we have

$$AB = [c_{ij}] = \begin{bmatrix} 7 & -6 & 6 & 21 \\ 5 & 11 & 13 & 25 \end{bmatrix} \qquad \square$$

Example 6   Let

$$A = \begin{bmatrix} -2 & 0 & 1 \\ 5 & -1 & 2 \end{bmatrix} \quad \text{and} \quad B = \begin{bmatrix} 6 & 2 \\ 3 & -2 \\ 0 & 4 \end{bmatrix}$$

Then you may verify that

$$AB = \begin{bmatrix} -12 & 0 \\ 27 & 20 \end{bmatrix} \qquad \square$$

Note in Example 6 that since the number of columns of the matrix A is equal to the number of rows of the matrix B, we can also form the product BA, which will be a 3×3 matrix. Straightforward calculations show that

$$BA = \begin{bmatrix} -2 & -2 & 10 \\ -16 & 2 & -1 \\ 20 & -4 & 8 \end{bmatrix}$$

Clearly, then, $AB \neq BA$; indeed, this will almost always be the case—matrix multiplication is not commutative. Of course, as we can see from Example 5, it may happen that although the matrix product AB is defined, the product BA is not defined.

Example 7   The $n \times n$ *identity matrix* is the matrix I defined by

$$I = [c_{ij}]_{n \times n}$$

where

$$c_{ij} = \begin{cases} 1 & \text{if} \quad i = j \\ 0 & \text{if} \quad i \neq j \end{cases}$$

For instance, the 3×3 identity matrix is

$$I = \begin{bmatrix} 1 & 0 & 0 \\ 0 & 1 & 0 \\ 0 & 0 & 1 \end{bmatrix} \qquad \square$$

In the context of matrix multiplication, an identity matrix plays the same role as does the number 1 in the multiplication of real numbers. In Problem 14 you are asked to show that if A is an $m \times n$ matrix, and I is the $n \times n$ identity matrix, then $AI = A$, and if I is the $m \times m$ identity matrix, then $IA = A$.

Although, as we have seen, matrix multiplication is not commutative, it does turn out to be associative, that is, if A is an $m \times q$ matrix, B is a $q \times r$ matrix, and C is an $r \times n$ matrix, then the $m \times n$ matrices $A(BC)$ and $(AB)C$ are defined and

$$A(BC) = (AB)C \tag{5}$$

The proof of this fact is based on a somewhat tedious though straightforward application of the definition of matrix multiplication (see Problem 11).

Example 8
(a) If

$$A = \begin{bmatrix} 1 & 0 \\ -1 & 1 \end{bmatrix}, \qquad B = \begin{bmatrix} 2 & 3 & 4 \\ 1 & 0 & 1 \end{bmatrix}, \quad \text{and} \quad C = \begin{bmatrix} 1 \\ 2 \\ -1 \end{bmatrix}$$

then

$$A(BC) = \begin{bmatrix} 1 & 0 \\ -1 & 1 \end{bmatrix} \left(\begin{bmatrix} 2 & 3 & 4 \\ 1 & 0 & 1 \end{bmatrix} \begin{bmatrix} 1 \\ 2 \\ -1 \end{bmatrix} \right) = \begin{bmatrix} 1 & 0 \\ -1 & 1 \end{bmatrix} \begin{bmatrix} 4 \\ 0 \end{bmatrix} = \begin{bmatrix} 4 \\ -4 \end{bmatrix}$$

and

$$(AB)C = \left(\begin{bmatrix} 1 & 0 \\ -1 & 1 \end{bmatrix} \begin{bmatrix} 2 & 3 & 4 \\ 1 & 0 & 1 \end{bmatrix} \right) \begin{bmatrix} 1 \\ 2 \\ -1 \end{bmatrix} = \begin{bmatrix} 2 & 3 & 4 \\ -1 & -3 & -3 \end{bmatrix} \begin{bmatrix} 1 \\ 2 \\ -1 \end{bmatrix}$$

$$= \begin{bmatrix} 4 \\ -4 \end{bmatrix}$$

(b) If A is a square matrix, then it follows from (5) that all powers of A (that is, A^2, A^3, ...) are well defined; for instance, $A^3 = (AA)A = A(AA)$. If

$$A = \begin{bmatrix} 1 & 0 \\ -2 & 1 \end{bmatrix}$$

then you can easily verify that

$$A^2 = \begin{bmatrix} 1 & 0 \\ -4 & 1 \end{bmatrix} \quad \text{and} \quad A^3 = \begin{bmatrix} 1 & 0 \\ -6 & 1 \end{bmatrix} \qquad \square$$

Matrices also obey a distributive rule in the following sense. If A is an $m \times q$ matrix and B and C are $q \times n$ matrices, then

$$A(B + C) = AB + AC \tag{6}$$

and

$$A(B - C) = AB - AC \tag{7}$$

Again the proofs of these properties of matrices follow directly from the definition of matrix multiplication, addition, and subtraction (see Problem 12).

Example 9   We show by direct calculation that $A(B + C) = AB + AC$ for the matrices

$$A = \begin{bmatrix} 2 & 1 \\ 0 & -1 \\ 3 & 3 \end{bmatrix}, \quad B = \begin{bmatrix} 1 & 0 & 1 & -1 \\ 4 & 2 & 1 & 3 \end{bmatrix}, \quad \text{and} \quad C = \begin{bmatrix} 0 & -1 & 1 & 1 \\ 2 & 1 & 3 & 4 \end{bmatrix}$$

We first compute $A(B + C)$.

$$A(B + C) = \begin{bmatrix} 2 & 1 \\ 0 & -1 \\ 3 & 3 \end{bmatrix} \left(\begin{bmatrix} 1 & 0 & 1 & -1 \\ 4 & 2 & 1 & 3 \end{bmatrix} + \begin{bmatrix} 0 & -1 & 1 & 1 \\ 2 & 1 & 3 & 4 \end{bmatrix} \right)$$

$$= \begin{bmatrix} 2 & 1 \\ 0 & -1 \\ 3 & 3 \end{bmatrix} \begin{bmatrix} 1 & -1 & 2 & 0 \\ 6 & 3 & 4 & 7 \end{bmatrix}$$

$$= \begin{bmatrix} 8 & 1 & 8 & 7 \\ -6 & -3 & -4 & -7 \\ 21 & 6 & 18 & 21 \end{bmatrix}$$

We now compute $AB + AC$.

$$AB + AC = \begin{bmatrix} 2 & 1 \\ 0 & -1 \\ 3 & 3 \end{bmatrix} \begin{bmatrix} 1 & 0 & 1 & -1 \\ 4 & 2 & 1 & 3 \end{bmatrix} + \begin{bmatrix} 2 & 1 \\ 0 & -1 \\ 3 & 3 \end{bmatrix} \begin{bmatrix} 0 & -1 & 1 & 1 \\ 2 & 1 & 3 & 4 \end{bmatrix}$$

$$
= \begin{bmatrix} 6 & 2 & 3 & 1 \\ -4 & -2 & -1 & -3 \\ 15 & 6 & 6 & 6 \end{bmatrix} + \begin{bmatrix} 2 & -1 & 5 & 6 \\ -2 & -1 & -3 & -4 \\ 6 & 0 & 12 & 15 \end{bmatrix}
$$

$$
= \begin{bmatrix} 8 & 1 & 8 & 7 \\ -6 & -3 & -4 & -7 \\ 21 & 6 & 18 & 21 \end{bmatrix}
\qquad \square
$$

We summarize a number of rules of matrix arithmetic in the following theorem. Proofs of these rules for 2×2 matrices are left as exercises (see Problem 13).

Theorem A.16   Assume that the orders of the matrices A, B, C, and I are such that the following operations are defined. Then

(a) $A(BC) = (AB)C$

(b) $A(B + C) = AB + AC$

(c) $A + (B + C) = (A + B) + C$

(d) $\alpha(A + B) = \alpha A + \alpha B$

(e) $(\alpha + \beta)A = \alpha A + \beta A$

(f) $\alpha(AB) = (\alpha A)B = A(\alpha B)$

(g) $IA = AI = A$

To conclude this section we introduce the idea of the determinant of a square matrix. Although it is possible to define the determinant of any square matrix, it is sufficient for our purposes to evaluate 2×2 and 3×3 determinants.

Definition A.17   (a) The *determinant* of the matrix

$$
\begin{bmatrix} a_{11} & a_{12} \\ a_{21} & a_{22} \end{bmatrix}
$$

is defined by

$$
\begin{vmatrix} a_{11} & a_{12} \\ a_{21} & a_{22} \end{vmatrix} = a_{11}a_{22} - a_{12}a_{21}
$$

(b) The *determinant* of the matrix

$$
\begin{bmatrix} a_{11} & a_{12} & a_{13} \\ a_{21} & a_{22} & a_{23} \\ a_{31} & a_{32} & a_{33} \end{bmatrix}
$$

is defined in terms of 2×2 determinants by

$$\begin{vmatrix} a_{11} & a_{12} & a_{13} \\ a_{21} & a_{22} & a_{23} \\ a_{31} & a_{32} & a_{33} \end{vmatrix} = a_{11} \begin{vmatrix} a_{22} & a_{23} \\ a_{32} & a_{33} \end{vmatrix} - a_{12} \begin{vmatrix} a_{21} & a_{23} \\ a_{31} & a_{33} \end{vmatrix} + a_{13} \begin{vmatrix} a_{21} & a_{22} \\ a_{31} & a_{32} \end{vmatrix}$$

Example 10
  (a) If

$$A = \begin{bmatrix} 1 & 2 \\ -1 & 4 \end{bmatrix}$$

then

$$\begin{vmatrix} 1 & 2 \\ -1 & 4 \end{vmatrix} = 1(4) - 2(-1) = 6$$

  (b) If

$$A = \begin{bmatrix} 3 & 2 & 4 \\ 2 & -1 & 3 \\ 4 & -2 & -3 \end{bmatrix}$$

then

$$\begin{vmatrix} 3 & 2 & 4 \\ 2 & -1 & 3 \\ 4 & -2 & -3 \end{vmatrix} = 3 \begin{vmatrix} -1 & 3 \\ -2 & -3 \end{vmatrix} - 2 \begin{vmatrix} 2 & 3 \\ 4 & -3 \end{vmatrix} + 4 \begin{vmatrix} 2 & -1 \\ 4 & -2 \end{vmatrix}$$

$$= 3(3 + 6) - 2(-6 - 12) + 4(-4 + 4) = 63 \qquad \square$$

A.3 Problems

1. If

$$A = [a_{ij}]_{3 \times 5} = \begin{bmatrix} 1 & 2 & 3 & -4 & 6 \\ 2 & 1 & 0 & 5 & \frac{1}{2} \\ 1 & -1 & \pi & 0 & 0 \end{bmatrix}$$

find a_{33}, a_{12}, a_{21}, and a_{34}.

2. If

$$A = \begin{bmatrix} 2 & 1 \\ 4 & -1 \\ 0 & 2 \\ 1 & 1 \end{bmatrix}, \quad B = \begin{bmatrix} 3 & 0 \\ 0 & 2 \\ -1 & 6 \\ 4 & 5 \end{bmatrix}, \quad C = \begin{bmatrix} 1 & 3 & 2 \\ 4 & 1 & 5 \end{bmatrix},$$

$$D = \begin{bmatrix} -1 & 2 & 3 \\ 0 & 0 & 4 \end{bmatrix}, \quad \text{and} \quad E = \begin{bmatrix} 1 & -2 & 0 & 1 \\ 3 & 1 & 2 & 1 \\ -1 & 1 & -1 & 1 \end{bmatrix}$$

find the following or state that they do not exist:

(a) $A + B$ (b) AB (c) $D - C$
(d) $A + D$ (e) $6A$ (f) $A(C + D)$
(g) $B(C - D)$ (h) $B(CE)$ (i) $(BC)E$
(j) ED (k) DE (f) $A(C + D)$

3. (a) If

$$A = \begin{bmatrix} 1 & 2 \\ -1 & 3 \end{bmatrix}$$

find $A^2 + 3A$.

(b) If

$$A = \begin{bmatrix} 0 & -1 & 3 \\ 0 & 0 & 2 \\ 0 & 0 & 0 \end{bmatrix}$$

find A^7.

(c) If

$$A = \begin{bmatrix} 0 & -1 \\ -1 & 0 \end{bmatrix}$$

find A^{14}.

4. Let

$$A = \begin{bmatrix} 0 & 1 \\ 0 & 2 \end{bmatrix}, \quad B = \begin{bmatrix} 2 & 7 \\ 3 & 4 \end{bmatrix}, \quad \text{and} \quad C = \begin{bmatrix} -4 & 6 \\ 3 & 4 \end{bmatrix}$$

show that $AB = AC$, but that $B \neq C$.

5. (a) Show that if A and B are square matrices and $AB = BA$, then

$$(A - B)(A + B) = A^2 - B^2 \tag{8}$$

(b) Is (8) true if $AB \neq BA$?

6. True or false? If A is an $n \times n$ matrix and A^2 is the $n \times n$ zero matrix, then A is the $n \times n$ zero matrix.

7. Suppose that a matrix A has a row of zeros. Show that if B is any matrix such that AB is defined, then AB has a row of zeros.

8. Let A_i be the ith row of a matrix A. Show that if B is a matrix such that AB is defined, then A_iB is the ith row of AB. Find (and prove) an analogous result for the jth column of AB.

9. Show that $A - B = A + (-1)B$.

10. Use induction to show that for each positive integer n,

$$\begin{bmatrix} a & 0 \\ b & a \end{bmatrix}^n = \begin{bmatrix} a^n & 0 \\ na^{n-1}b & a^n \end{bmatrix}$$

11. Show that if A, B, and C are 2×2 matrices, then

$$A(BC) = (AB)C$$

12. Show that if A is a 2×3 matrix, and if B and C are 3×4 matrices, then

$$A(B + C) = AB + AC$$

13. Prove Theorem A.16 in the case that each of the matrices is a 2×2 matrix.

14. Show that if A is an $m \times n$ matrix, and I is the $n \times n$ identity matrix, then

$$AI = A$$

Show also that if I is the $m \times m$ identity matrix, then

$$IA = A$$

15. Compute the determinants:

(a) $\begin{vmatrix} 1 & -2 \\ -3 & 4 \end{vmatrix}$

(b) $\begin{vmatrix} 5 & -2 \\ -3 & 6 \end{vmatrix}$

(c) $\begin{vmatrix} 1 & 2 & 3 \\ 4 & 5 & -2 \\ 1 & -6 & 2 \end{vmatrix}$

(d) $\begin{vmatrix} -1 & 1 & -1 \\ 1 & 1 & 0 \\ 1 & 0 & 1 \end{vmatrix}$

A.4   SIGMA NOTATION

From time to time it is convenient to adopt a compact notation to describe sums of numbers. We can denote the sum

$$a_m + a_{m+2} + \cdots + a_n$$

by

$$\sum_{i=m}^{n} a_i \tag{1}$$

Thus, for example,

$$\sum_{i=3}^{8} c_i$$

denotes the sum

$$c_3 + c_4 + c_5 + c_6 + c_7 + c_8$$

Frequently, the values of the summands are specified. For instance,

$$\sum_{j=1}^{5} j^2$$

denotes the sum

$$1^2 + 2^2 + 3^2 + 4^2 + 5^2 \qquad (2)$$

Observe that

$$\sum_{q=1}^{5} q^2$$

also denotes the sum found in (2); the particular letters used in describing (2) are not important.

For another example of the use of sigma notation, you can check easily that

$$\sum_{k=0}^{n} (n - k)$$

is the shorthand representation of the sum

$$n + (n - 1) + (n - 2) + \cdots + 1 + 0$$

For any real number c

$$\sum_{i=1}^{n} c$$

indicates the number c is to be added to itself n times. Thus, for example,

$$\sum_{n=1}^{6} c = c + c + c + c + c + c = 6c$$

Similarly,

$$\sum_{n=4}^{8} c = c + c + c + c + c = 5c$$

You can readily verify that the following theorem follows immediately from basic properties of real numbers.

Theorem A.18   For any real numbers k, a_i, and b_i

(a) $\displaystyle\sum_{i=j}^{n} (a_i + b_i) = \sum_{i=j}^{n} a_i + \sum_{i=j}^{n} b_i$

(b) $\displaystyle\sum_{i=j}^{n} k a_i = k \sum_{i=j}^{n} a_i$

Example 1   From Theorem A.18 we have

$$\sum_{q=4}^{9} (q^2 - 3q + 4) = \sum_{q=4}^{9} q^2 - 3 \sum_{q=4}^{9} q + 24 \qquad \square$$

A.4   Problems

1. Write out the following sums:

    (a) $\displaystyle\sum_{j=1}^{5} a_j$
    (b) $\displaystyle\sum_{p=3}^{6} c_p^2$
    (c) $\displaystyle\sum_{k=1}^{4} k^3$

2. Compute the following sums:

    (a) $\displaystyle\sum_{k=1}^{5} 4k$
    (b) $\displaystyle\sum_{j=1}^{4} (3j + 2)$

    (c) $\displaystyle\sum_{n=1}^{5} (-1)^n n^2$
    (d) $\displaystyle\sum_{k=3}^{6} k!$

    (Recall that for any positive integer k, $k! = k(k-1)(k-2) \cdots 2 \cdot 1$)

3. Compute the following sums:

    (a) $\displaystyle\sum_{k=4}^{7} \frac{1}{k}$
    (b) $\displaystyle\sum_{i=2}^{5} (i^2 + 3i - 2)$

    (c) $\displaystyle\sum_{n=0}^{5} 6$
    (d) $\displaystyle\sum_{n=1}^{5} (-1)^{n+1} \frac{1}{n^3}$

4. Use sigma notation to describe the following sums:
    (a) $2 + 2^2 + 2^3 + 2^4 + 2^5$
    (b) $1 + 3 + 5 + 7 + 9 + 11$
    (c) $1 + 8 + 27 + 64 + 125$
    (d) $1 + 2 + 6 + 24 + 120 + 720$

5. Use sigma notation to describe the following sums:
    (a) $1 - 2 + 3 - 4 + 5 - 6 + 7$
    (b) $5^2 + 6^2 + 7^2 + 8^2$
    (c) $3 + \left(\dfrac{3}{2}\right)^2 + \left(\dfrac{3}{3}\right)^2 + \left(\dfrac{3}{4}\right)^2 + \left(\dfrac{3}{5}\right)^2 + \left(\dfrac{3}{6}\right)^2$
    (d) $\dfrac{2}{4} + \dfrac{4}{9} + \dfrac{8}{16} + \dfrac{16}{25} + \dfrac{32}{36} + \dfrac{64}{49}$

A.5 RELATIONS

Relations arise in contexts ranging from computer arithmetic to comparisons of numbers, sets, and geometric objects. Relations are defined in terms of ordered pairs.

Definition A.19 A *relation* R on a set X is a subset of $X \times X$. If $(x, y) \in R$, we write $x R y$ (read as "x is related to y").

Example 1

(a) If $X = \mathbf{R}^1$, then $R = \{(x, y) | (x, y) \in \mathbf{R}^1 \times \mathbf{R}^1 \text{ and } x \le y\}$ is a relation on $\mathbf{R}^1$, and we write $x R y$ if $x \le y$.

(b) If $X = \mathbf{Z}$, the set of integers, then $R = \{(x, y) | (x, y) \in X \times X \text{ and } x - y \text{ is divisible by } 4\}$ is a relation on $\mathbf{Z}$, and we write $x R y$ if $x - y$ is divisible by 4. (Recall that an integer d is divisible by an integer c if there is an integer q such that $d = cq$.)

(c) If $X = \mathbf{R}^1$, then $R = \{(x, y) | (x, y) \in \mathbf{R}^1 \times \mathbf{R}^1 \text{ and } x - y \ne 0\}$ is a relation on $\mathbf{R}^1$, and we write $x R y$ if $x - y \ne 0$.

(d) Let Y be any nonempty set and let $X = \mathscr{P}(Y)$ be the power set of Y. Then $R = \{(C, D) | (C, D) \in X \times X \text{ and } C \subseteq D\}$ is a relation on X, and we write $C R D$ if $C \subseteq D$.

(e) If X is the set of inhabitants of the earth, then $R = \{(x, y) | (x, y) \in X \times X \text{ and } x \text{ is in love with } y\}$ is a relation on X, and we write $x R y$ if x is in love with y. $\square$

We can classify relations as follows.

Definition A.20

(a) A relation R on a set X is *reflexive* if $x R x$ for each $x \in X$.

(b) A relation R on a set X is *symmetric* if whenever $x R y$, then $y R x$.

(c) A relation R on a set X is *transitive* if whenever $x R y$ and $y R z$, then $x R z$.

Example 2

(a) The relation given in (a) of Example 1 is reflexive (since $x \le x$), transitive (since if $x \le y$ and $y \le z$, then $x \le z$), but not symmetric (since $3 \le 5$, but $5 \nleq 3$).

(b) The relation given in (c) of Example 1 is symmetric (since if $x - y \neq 0$, then $y - x \neq 0$), but neither reflexive (since $x - x = 0$) nor transitive (since $2 - 3 \neq 0$ and $3 - 2 \neq 0$ do not imply $2 - 2 \neq 0$).

(c) What can you say about (e) of Example 1?   □

The relation given in (b) of Example 1 is reflexive, symmetric, *and* transitive. This is a consequence of the following result.

Theorem A.21   Let m be a positive integer. The relation R on **Z** defined by $R = \{(x, y) \mid (x, y) \in$ **Z** $\times$ **Z** and m divides $x - y\}$ is reflexive, symmetric, and transitive.

> ***Proof:***   Since for each $x \in$ **Z**, m divides $x - x$, it follows that R is reflexive.
> The relation R is symmetric since if m divides $x - y$, then m divides $y - x$.
> To show that R is transitive, suppose that m divides both $x - y$ and $y - z$. We need to show that m divides $x - z$. Since m divides $x - y$, there is an integer c such that $x - y = mc$. Since m divides $y - z$, there is an integer d such that $y - z = md$. Then $x - z = x - y + y - z = mc + md = m(c + d)$, from which it follows that m divides $x - z$. Hence, R is transitive.   ∎

Relations that satisfy the three conditions, reflexivity, symmetry and transitivity, are called *equivalence relations*. These relations have a wide variety of applications in computer science and almost all branches of mathematics.

We make a few observations about equivalence relations. Note that in the case of (b) of Example 1, the integers $\{\ldots, -8, -4, 0, 4, 8, \ldots\}$ are all equivalent, as are the integers $\{\ldots, -7, -3, 1, 5, \ldots\}$, $\{\ldots, -6, -2, 2, 6, \ldots\}$ and $\{\ldots, -5, -1, 3, 7, \ldots\}$. Moreover, it is clear that every integer falls into one and only one of these four sets of integers, so the equivalence relation R has generated a "partition" of the set **Z** (see Problem 12). Such a partitioning is characteristic of equivalence relations, as we see in the next theorem.

Theorem A.22   Suppose that R is an equivalence relation on a set X. For each $x \in X$, let $[x] = \{y \mid x R y\}$. Then,

a. Every element of X belongs to some set $[x]$, and

b. If $x \in X$ and $y \in X$, then either $[x] = [y]$ or $[x] \cap [y] = \varnothing$.

> ***Proof:***   **a.** Since R is reflexive we have that $x \in [x]$ for each $x \in X$, and hence each number of X belongs to at least one equivalence class.
> **b.** We show that if $[x] \cap [y] \neq \varnothing$, then $[x] = [y]$. If $[x] \cap [y] \neq \varnothing$, then $z \in [x] \cap [y]$ for some $z \in X$. We first show that $[x] \subseteq [y]$. Suppose that $w \in [x]$. Then $x R w$ and, since $z \in [x] \cap [y]$, we also have $x R z$ and $y R z$.

Using the fact that R is symmetric, we obtain zRx and, applying transitivity twice to yRz, zRx, and xRw, we obtain yRw, which implies that $w \in [y]$. Therefore, $[x] \subseteq [y]$.

To see that $[y] \subseteq [x]$ suppose that $q \in [y]$. Then yRq, and since (by the symmetry of R) we have xRz, zRy, and yRq, it follows from the transitivity of R that xRq. Thus, $q \in [x]$. Therefore, $[y] \subseteq [x]$ and this completes the proof. ∎

The sets $[x]$ are called the *equivalence classes* associated with the equivalence relation R. The upshot of Theorem A.22 is that an equivalence relation partitions a set X into disjoint equivalence classes.

Example 3

(a) Let $X = \mathbf{Z}$ and let m be a positive integer. Then by Theorem A.21 the relation R defined by xRy if m divides $x - y$ is an equivalence relation. Note that in this case we have disjoint equivalence classes $[0], [1], [2], \ldots [m-1]$.

(b) Let X be the set of all animals, and define a relation R on X by xRy if x and y have the same number of legs. Then R is an equivalence relation that partitions the set of animals into those having one leg, those having two legs, and so on.

(c) Let $X = \mathbf{Z}$ and define a relation R on X by xRy if $x + y$ is an even integer. Then R is an equivalence relation that partitions $\mathbf{Z}$ into two equivalence classes, the even integers and the odd integers.

(d) If X is any set, then the relation R defined by xRy if $x = y$ is an equivalence relation. Note in this case that the equivalence classes correspond to the elements of X. □

Order relations (also called partial orders) have many applications in both theoretical and applied mathematics. To define an order relation we need the following concept.

Definition A.23 A relation R on a set X is said to be *antisymmetric* if whenever xRy and yRx, then $x = y$.

Example 4 If $X = \{x \mid x \in \mathbf{R}^1$ and $x > 0\}$ and R is defined by xRy if x divides y, then R is antisymmetric. □

Definition A.24 A relation R on a set X is said to be an *order relation* (or a *partial order*) if R is reflexive, antisymmetric, and transitive.

Example 5

(a) If $X = \mathbf{R}^1$ and a relation R on X is defined by $x\,R\,y$ if $x \leq y$, then R is an order relation (or a partial order) on X. (See Problem 17.)

(b) Let Y be any nonempty set. Then the relation R defined on the power set $X = \mathscr{P}(Y)$ by $A\,R\,B$ if $A \subseteq B$ is an order relation on X. (See Problem 18.)

(c) If $X = \mathbf{Z}^+$, then the relation R on X defined by $a\,R\,b$ if a divides b is a partial order on X. (See Problem 19.)

(d) (Lexicographic ordering) Let $X = \mathbf{R}^1 \times \mathbf{R}^1$ and define a relation R on X by $(a, b)\,R\,(c, d)$ if $a < c$, or, if $a = c$, then $b \leq d$. Then R is a partial order on X. (See Problem 20.) ☐

Parts (b), (c), and (d) of Example 5 show that the term *partial order* is appropriate since in each of these cases there are elements a and b in X for which neither $a\,R\,b$ nor $b\,R\,a$; that is, a and b cannot be compared or ordered with respect to one another. In the case of a set X that is partially ordered by a relation R, we will often replace R with $\preceq_R$ (or $\preceq$) and write $x \preceq_R y$ (or $x \preceq y$) in place of $x\,R\,y$. If $x \preceq y$ and $x \neq y$, then we write $x \prec y$.

A.5 Problems

In Problems 1–4 determine whether the given relation R is reflexive, whether it is symmetric, whether it is transitive, and whether it is antisymmetric.

1. X is the set of integers; R is defined by $x\,R\,y$ if x is a multiple of y.
2. X is the set of real numbers; R is defined by $x\,R\,y$ if $|x - y| < 2$
3. X is the set of lines in the plane; R is defined by $x\,R\,y$ if x is perpendicular to y.
4. X is the set of all men; R is defined by $x\,R\,y$ if x is a brother of y.
5. Let $X = \{a, b, c\}$. Which of the following relations are
    (a) reflexive?
    (b) symmetric?
    (c) transitive?
    (d) antisymmetric?

$$R_1 = \{(a, a), (b, b), (a, c), (c, b), (b, c)\}$$
$$R_2 = \{(a, a), (b, b), (c, c), (a, c), (c, a)\}$$
$$R_3 = \{(b, b), (a, b), (b, a), (a, c), (b, c), (c, b)\}$$
$$R_4 = \{(a, a), (b, b), (c, c)\}$$
$$R_5 = \{(a, b), (a, c), (b, c)\}$$

6. Give your own examples of:
    (a) A relation that is reflexive and symmetric but not transitive.

(b) A relation that is reflexive and transitive but not symmetric.

(c) A relation that is symmetric and transitive but not reflexive.

7. Find a relation that is neither reflexive, symmetric, nor transitive.

8. What is wrong with the following proof that "shows" that symmetry and transitivity imply reflexivity? Let R be a relation that is symmetric and transitive. Suppose that $a\,R\,b$. Since R is symmetric we have $b\,R\,a$; consequently, it follows from the transitivity of R that $a\,R\,b$ and $b\,R\,a$ imply $a\,R\,a$. Therefore, R is also reflexive.

9. Suppose that R and S are relations on a given set X, and that X has at least three elements. Prove or give a counterexample:

(a) If R and S are symmetric, then so is $R \cap S$ (recall that R, S, and $R \cap S$ are all subsets of $X \times X$).

(b) If R and S are symmetric, then so is $R \cup S$.

(c) If R and S are transitive, then so is $R \cap S$.

(d) If R and S are reflexive, then so is $R \cap S$.

(e) If R is reflexive and S is symmetric, then $R \cap S$ is reflexive and transitive.

10. Which of the following equivalence classes in $\mathbf{Z_4}$ are equal?

$$[2], \quad [12], \quad [-3], \quad [5], \quad [45], \quad [0], \quad [-5], \quad [134], \quad [-67]$$

11. Which of the following equivalence classes in $\mathbf{Z_{12}}$ are equal?

$$[0], \quad [6], \quad [12], \quad [15], \quad [-3], \quad [17], \quad [-5], \quad [24], \quad [162], \quad [-29].$$

12. Let X be a set and suppose $A_1, A_2, \ldots, A_n$ are mutually disjoint subsets of X such that $X = \bigcup_{i=1}^n A_i$ (such a family is called a *partition* of X). Define a relation on X by $x\,R\,y \Leftrightarrow x$ and y belong to the same subset A_i. Show that R is an equivalence relation.

13. Which of the relations in Problems 1–4 are equivalence relations?

14. Which of the relations in Problems 5 are equivalence relations?

15. Which of the relations given in Problems 1–4 are partial orders?

16. Which of the relations given in Problem 5 are partial orders?

17. Show that (a) of Example 5 is a partial order.

18. Show that (b) of Example 5 is a partial order.

19. Show that (c) of Example 5 is a partial order.

20. Show that (d) of Example 5 is a partial order.

21. Let X be the set of integers. Define a relation R on $X \times X$ by $(a, b)\,R\,(c, d)$ if $a \le c$ and $b \le d$. Is R a partial order?

22. Suppose that R and S are order relations on a set X. Prove or give a counterexample:

(a) $R \cap S$ is an order relation.

(b) $R \cup S$ is an order relation.

Answers to Selected Odd-Numbered Problems

1.1 Problems

3. False: If $x = 3/2$, then $[2x] = 3$ and $2[x] = 2$.

7.

Step	F	L	K	$A(K)$
1	1	10	5	CF
2	6	10	8	EK
3	9	10	9	GR
4	10	10	10	HI
5	11	10		

Since $F > L$, the output is "NOT FOUND."

9. Z is the maximum of X and Y.

11. CT is the number of positive numbers in the list $A(1), A(2), \ldots, A(N)$.

19. (a) 8   (b) 40   (c) 16

1.2 Problems

	Initial Order	Pass 1					Pass 2			
5. J	—	1	1	1	1	1	2	2	2	2
I	—	1	2	3	4	5	1	2	3	4
$A(1)$	10	10	10	10	10	10	10	10	10	10
$A(2)$	9	9	9	9	9	9	9	9	9	9
$A(3)$	6	6	6	8	8	8	8	8	8	8
$A(4)$	8	8	8	6	6	6	6	6	6	6
$A(5)$	5	5	5	5	5	5	5	5	5	5
$A(6)$	4	4	4	4	4	4	4	4	4	4

Since no changes were made in the last pass, you may terminate the algorithm.

	Initial Order	Pass 1					Pass 2			
7. J	—	1	1	1	1	1	2	2	2	2
I	—	1	2	3	4	5	1	2	3	4
$A(1)$	9	9	9	9	9	10	10	10	10	10
$A(2)$	10	10	10	10	10	9	9	9	9	9
$A(3)$	8	8	8	8	8	8	8	8	8	8
$A(4)$	3	3	5	5	5	5	5	5	5	5
$A(5)$	4	5	3	3	3	3	4	4	4	4
$A(6)$	5	4	4	4	4	4	3	3	3	3

Since no changes were made in the last pass, you may terminate the algorithm.

1.4 Problems

1.

i	$A(i)$	Applic. 1	Applic. 2	Applic. 3
1	5	8	8	8
2	8	7	7	7
3	3	5	5	5
4	1	3	3	3
5	7	1	1	2
6	2	2	2	1

3.

i	$A(i)$	Applic. 1	Applic. 2	Applic. 3	Applic. 4
1	8	8	8	8	8
2	2	2	4	6	6
3	1	1	6	4	4
4	4	4	3	3	3
5	6	6	2	2	2
6	0	0	1	1	1
7	3	3	0	0	0

1.5 Problems

1. $B = \{1, 3, 4, 6, 7\}$; $TW = 18$; $TV = 33$
3. $B = \{2, 6, 3, 1\}$; $TW = 31$; $TV = 28$ (using the order $v_2, v_6, v_3, v_1, v_4, v_5, v_8, v_7$)
5. $x_1 x_6 x_4 x_2 x_3 x_1$; Cost: 33

1.6 Problems

1. k	$MINNET$	$f(1)$	$f(2)$	$f(3)$	$f(4)$	$f(5)$	$p(2)$	$p(3)$	$p(4)$	$p(5)$
3	$\langle 1, 3\rangle, \langle 3, 2\rangle, \langle 2, 5\rangle$	3	5	2	1	5	3	1	—	2
3	$\langle 1, 3\rangle, \langle 3, 2\rangle, \langle 2, 5\rangle$	3	5	3	1	5	3	1	—	2
3	$\langle 1, 3\rangle, \langle 3, 2\rangle, \langle 2, 5\rangle$	3	5	4	1	5	3	1	—	2
4	$\langle 1, 3\rangle, \langle 3, 2\rangle, \langle 2, 5\rangle, \langle 3, 4\rangle$	3	5	4	1	5	3	1	3	2

OUTPUT: $\langle 1, 3\rangle, \langle 3, 2\rangle, \langle 2, 5\rangle, \langle 3, 4\rangle$

3. OUTPUT: $\langle 1, 2\rangle, \langle 2, 5\rangle, \langle 1, 4\rangle, \langle 4, 3\rangle, \langle 1, 6\rangle$

1.7 Problems

5. True   7. True   9. True

11. True   13. False   23. N^2 times

Chapter 1 Review

1. S is the sum of the squares of the first N positive integers.

3.
Step	F	L	K	$A(K)$
1	1	11	6	*HAT*
2	7	11	9	*MAR*
3	10	11	10	*RAS*

5.
LARGE	K	M
−6	2	1
4	2	2
4	3	2
16	4	4
16	5	4
16	6	4

7.
i	Original list	Pass 1	Pass 2	Pass 3	Pass 4	Pass 5
1	3	8	8	8	8	8
2	8	3	7	7	7	7
3	−4	−4	−4	5	5	5
4	−6	−6	−6	−6	3	3
5	5	5	5	−4	−4	−4
6	7	7	3	3	−6	−6

9.
I	—	1	1	1	1	1	2	2	2	2	3	3	3	4	4
J	—	1	2	3	4	5	1	2	3	4	1	2	3	1	2
$A(1)$	3	3	3	3	3	4	4	4	4	4	4	4	4	4	4
$A(2)$	−4	−4	−4	−4	4	3	3	3	3	3	3	3	3	3	3
$A(3)$	4	4	4	4	−4	−4	−4	−4	2	2	2	2	2	2	2
$A(4)$	2	2	2	2	2	2	2	2	−4	−4	−4	1	1	1	1
$A(5)$	1	1	1	1	1	1	1	1	1	1	1	−4	−4	0	0
$A(6)$	0	0	0	0	0	0	0	0	0	0	0	0	0	−4	−4

13.

I	$A(I)$	1	2
1	5	6	7
2	6	7	6
3	4	5	5
4	7	4	4
5	1	1	1

17. Algorithm 1.7: $B = \{2, 3, 5, 6, 7\}$; $TW = 25$; $TV = 27$
   Algorithm 1.8: $B = \{2, 6, 4, 5, 7, 3\}$; $F = 8$; $TV = 28.8$

2.1 Problems

1. $1132 = 3205_7$
7. 1906

3. $20{,}976 = 50760_8$
9. 16,288

5. $1{,}027{,}565 = FADED_{16}$
11. 360

2.2 Problems

1. $x(x(2x - 3) + 4) - \frac{1}{2}$
3. $x(x(x(3x - 1) + 0) + 1) + 0$
5. Problem 1 of this section:

$$S \leftarrow 2$$
$K = 1, \quad S \leftarrow 4(2) - 3 = 5$
$K = 2, \quad S \leftarrow 4(5) + 4 = 24$
$K = 3, \quad S \leftarrow 4(24) - \frac{1}{2} = 95.5$

Thus, $f(4) = 95.5$

Problem 3 of this section:

$$S \leftarrow 3$$
$K = 1, \quad S \leftarrow 4(3) - 1 = 11$
$K = 2, \quad S \leftarrow 4(11) + 0 = 44$
$K = 3, \quad S \leftarrow 4(44) + 1 = 177$
$K = 4 \quad S \leftarrow 4(177) + 0 = 708$

Thus, $f(4) = 708$

7. 19
9. 964
11. 228

2.3 Problems

1. $946 = 1662_8$
7. $4792 = 11270_8$
13. 111100001_2
19. 1372_{16}
25. 45462_8
31. $CO3_{16}$

3. $1096 = 448_{16}$
9. 266_8
15. 1000000001_2
21. 100111000010_2
27. 30302_8

5. $87 = 10020_3$
11. 100_8
17. 275_{16}
23. 1100101011111110_2
29. $7E_{16}$

2.4 Problems

1. (a) 11111   (b) 101111100   (c) 1110010   (d) 1001.0111
3. (a) $a* = 94{,}306$; $a** = 94{,}307$; 32,111
   (b) $a* = 9{,}994{,}016$; $a** = 9{,}994{,}017$; 998,584
   (c) $a* = 990{,}000$; $a** = 990{,}001$; 668,235

2.5 Problems

1. (a) False   (b) True   (c) True   (d) True
3. (a) $[0] = \{\ldots -12, -6, 0, 6, 12, \ldots\}$
      $[1] = \{\ldots -11, -5, 1, 7, 13, \ldots\}$

$$[2] = \{\ldots -10, -4, 2, 8, 14, \ldots\}$$
$$[3] = \{\ldots -9, -3, 3, 9, 15, \ldots\}$$
$$[4] = \{\ldots -8, -2, 4, 10, 16, \ldots\}$$
$$[5] = \{\ldots -7, -1, 5, 11, 17, \ldots\}$$
(b) $[0] = \{\ldots -6, -4, -2, 0, 2, 4, 6, \ldots\}$
$$[1] = \{\ldots -5, -3, -1, 1, 3, 5, \ldots\}$$
9. (a) [3]   (b) [0]   (c) [3]   (d) [0]
13. [3] does not have an inverse in $\mathbf{Z}_6$.

2.6 Problems

1. (a) 4   (b) 8   (c) 1
3. (a) [760], [717], [843]   (b) [676], [442], [126]

Chapter 2 Review

1. 26
3. 3686
5. $x(x(x(x(x(5x + 0) + 0) + 3) - 1) + 0) + 8$
7. 51
9. 47,964
11. 10420_5
13. $3BD_{16}$
15. $1777_8; 3FF_{16}$
17. $111111110010011_2; 7F93_{16}$
19. (a) 110100   (b) 101100
21. [2]
23. 4, 8, 3, 1, 9, 10, 6, 2, 5
25. 3
26. (a) [283], [792]

3.1 Problems

1.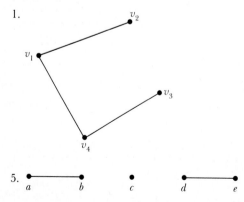

3.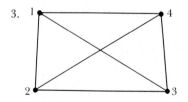

5.

7.

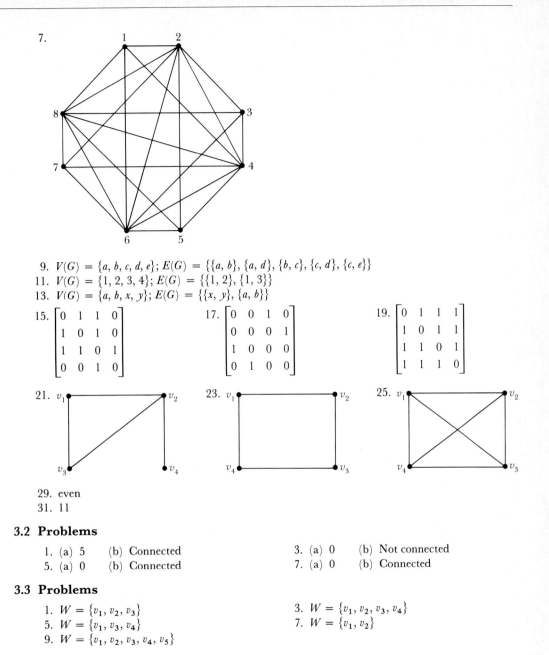

9. $V(G) = \{a, b, c, d, e\}$; $E(G) = \{\{a, b\}, \{a, d\}, \{b, c\}, \{c, d\}, \{c, e\}\}$
11. $V(G) = \{1, 2, 3, 4\}$; $E(G) = \{\{1, 2\}, \{1, 3\}\}$
13. $V(G) = \{a, b, x, y\}$; $E(G) = \{\{x, y\}, \{a, b\}\}$

15. $\begin{bmatrix} 0 & 1 & 1 & 0 \\ 1 & 0 & 1 & 0 \\ 1 & 1 & 0 & 1 \\ 0 & 0 & 1 & 0 \end{bmatrix}$

17. $\begin{bmatrix} 0 & 0 & 1 & 0 \\ 0 & 0 & 0 & 1 \\ 1 & 0 & 0 & 0 \\ 0 & 1 & 0 & 0 \end{bmatrix}$

19. $\begin{bmatrix} 0 & 1 & 1 & 1 \\ 1 & 0 & 1 & 1 \\ 1 & 1 & 0 & 1 \\ 1 & 1 & 1 & 0 \end{bmatrix}$

29. even
31. 11

3.2 Problems

1. (a) 5   (b) Connected
3. (a) 0   (b) Not connected
5. (a) 0   (b) Connected
7. (a) 0   (b) Connected

3.3 Problems

1. $W = \{v_1, v_2, v_3\}$
3. $W = \{v_1, v_2, v_3, v_4\}$
5. $W = \{v_1, v_3, v_4\}$
7. $W = \{v_1, v_2\}$
9. $W = \{v_1, v_2, v_3, v_4, v_5\}$

3.4 Problems

1. No; the sum of the degrees of the vertices is odd.
3. No; $\Sigma_{i=1}^{8} \delta(v_i) = 16 = 2|E(G)|$, and therefore, the number of edges of G is 8. But $|V(G)| = 8$, and if G is a tree, then $|E(G)| = |V(G)| - 1$.
7. False.

3.5 Problems

1.

Edge considered	$N(v_1)$	$N(v_2)$	$N(v_3)$	$N(v_4)$	$N(v_5)$	$N(v_6)$	$N(v_7)$	T	k
	1	2	3	4	5	6	7	$\varnothing$	0
e_1	1	1	3	4	5	6	7	$\{e_1\}$	1
e_2	1	1	1	4	5	6	7	$\{e_1, e_2\}$	2
e_3	1	1	1	4	5	6	7	$\{e_1, e_2\}$	2
e_4	1	1	1	1	5	6	7	$\{e_1, e_2, e_4\}$	3
e_5	1	1	1	1	5	6	7	$\{e_1, e_2, e_4\}$	3
e_6	1	1	1	1	1	6	7	$\{e_1, e_2, e_4, e_6\}$	4
e_7	1	1	1	1	1	1	7	$\{e_1, e_2, e_4, e_6, e_7\}$	5
e_8	1	1	1	1	1	1	1	$\{e_1, e_2, e_4, e_6, e_7, e_8\}$	6

$E(T) = \{e_1, e_2, e_4, e_6, e_7, e_8\}$

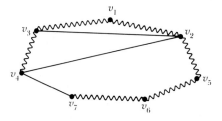

3.

Edge considered	$N(v_1)$	$N(v_2)$	$N(v_3)$	$N(v_4)$	$N(v_5)$	$N(v_6)$	$N(v_7)$	T	k
	1	2	3	4	5	6	7	$\varnothing$	0
e_1	1	1	3	4	5	6	7	$\{e_1\}$	1
e_2	1	1	3	4	1	6	7	$\{e_1, e_2\}$	2
e_3	1	1	1	4	1	6	7	$\{e_1, e_2, e_3\}$	3
e_4	1	1	1	1	1	6	7	$\{e_1, e_2, e_3, e_4\}$	4
e_5	1	1	1	1	1	6	7	$\{e_1, e_2, e_3, e_4\}$	4
e_6	1	1	1	1	1	1	7	$\{e_1, e_2, e_3, e_4, e_6\}$	5
e_7	1	1	1	1	1	1	1	$\{e_1, e_2, e_3, e_4, e_6, e_7\}$	6

$E(T) = \{e_1, e_2, e_3, e_4, e_6, e_7\}$

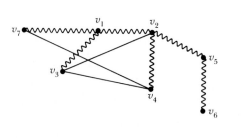

5.

Edge considered	$N(v_1)$	$N(v_2)$	$N(v_3)$	$N(v_4)$	$N(v_5)$	$N(v_6)$	$N(v_7)$	$N(v_8)$	$N(v_9)$	T	k
	1	2	3	4	5	6	7	8	9	$\varnothing$	0
e_1	1	2	3	4	5	6	7	8	8	$\{e_1\}$	1
e_2	1	2	3	4	5	6	6	8	8	$\{e_1, e_2\}$	2
e_3	1	2	3	4	5	6	6	5	5	$\{e_1, e_2, e_3\}$	3
e_4	1	2	3	4	5	5	5	5	5	$\{e_1, e_2, e_3, e_4\}$	4
e_5	1	2	3	4	4	4	4	4	4	$\{e_1, e_2, e_3, e_4, e_5\}$	5
e_6	1	2	3	3	3	3	3	3	3	$\{e_1, e_2, e_3, e_4, e_5, e_6\}$	6
e_7	1	2	3	3	3	3	3	3	3	$\{e_1, e_2, e_3, e_4, e_5, e_6\}$	6
e_8	1	2	2	2	2	2	2	2	2	$\{e_1, e_2, e_3, e_4, e_5, e_6, e_8\}$	7
e_9	1	2	2	2	2	2	2	2	2	$\{e_1, e_2, e_3, e_4, e_5, e_6, e_8\}$	7
e_{10}	1	1	1	1	1	1	1	1	1	$\{e_1, e_2, e_3, e_4, e_5, e_6, e_8, e_{10}\}$	8

$E(T) = \{e_1, e_2, e_3, e_4, e_5, e_6, e_8, e_{10}\}$

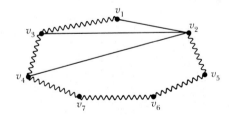

9. This algorithm, when applied to the graph in Problem 1 of this section, yields the spanning tree indicated by the wavy lines:

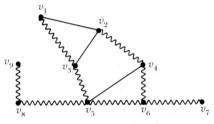

This algorithm, when applied to the graph in Problem 3 of this section, yields the spanning tree indicated by the wavy lines:

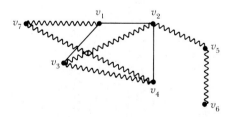

This algorithm, when applied to the graph in Problem 5 of this section, yields the spanning tree indicated by the wavy lines:

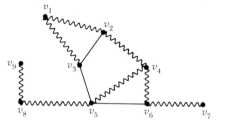

3.6 Problems

1. $W(H) = 8$

3. $\begin{bmatrix} 0 & 1 & \infty & \infty & 3 & \infty \\ 1 & 0 & 1 & \infty & 0 & \infty \\ \infty & 1 & 0 & 2 & \infty & 4 \\ \infty & \infty & 2 & 0 & \infty & 3 \\ 3 & 0 & \infty & \infty & 0 & 1 \\ \infty & \infty & 4 & 3 & 1 & 0 \end{bmatrix}$

5.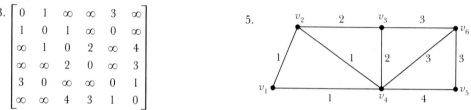

7. Edge set of the minimal spanning tree, T: $\{\{1, 2\}, \{1, 4\}, \{2, 3\}, \{3, 6\}, \{5, 6\}\}$; $W(T) = 10$
9. Edge set of the minimal spanning tree, T: $\{\{1, 2\}, \{3, 5\}, \{5, 6\}, \{2, 3\}, \{3, 4\}\}$; $W(T) = 6$
11. Edge set of the minimal spanning tree, T: $\{\{1, 2\}, \{1, 3\}, \{5, 6\}, \{1, 4\}, \{4, 6\}, \{3, 7\}, \{7, 8\}\}$; $W(T) = 17$
13. Edge set of the minimal spanning tree, T: $\{\{v_1, v_2\}, \{v_2, v_3\}, \{v_3, v_4\}, \{v_4, v_5\}\}$; $W(T) = 11$
15. Edge set of the minimal spanning tree, T: $\{\{v_1, v_2\}, \{v_1, v_5\}, \{v_2, v_3\}, \{v_2, v_4\}\}$; $W(T) = 7$
17. Edge set of the minimal spanning tree, T: $\{\{v_1, v_2\}, \{v_2, v_6\}, \{v_6, v_5\}, \{v_2, v_3\}, \{v_3, v_4\}\}$; $W(T) = 8$
19. (c) Edge set of the minimal spanning tree, T (for Problem 7 of this section): $\{\{1, 2\}, \{1, 4\}, \{2, 3\}, \{3, 6\}, \{5, 6\}\}$; $W(T) = 10$
    Edge set of the minimal spanning tree, T (for Problem 9 of this section): $\{\{1, 2\}, \{2, 3\}, \{3, 5\}, \{5, 6\}, \{3, 4\}\}$; $W(T) = 6$
    Edge set of the minimal spanning tree, T (for Problem 11 of this section): $\{\{1, 2\}, \{1, 3\}, \{1, 4\}, \{4, 6\}, \{6, 5\}, \{3, 7\}, \{7, 8\}\}$; $W(T) = 17$

3.8 Problems

1. $C = v_1, v_2, v_4, v_3, v_5, v_1$; $W(C) = 14$
3. $C = v_1, v_6, v_5, v_3, v_4, v_2, v_1$; $W(C) = 18$
5. (For Problem 1 of this section) $C = v_1, v_5, v_3, v_4, v_2, v_1$; $W(C) = 14$
   (For Problem 3 of this section) $C = v_1, v_2, v_3, v_5, v_6, v_4, v_1$; $W(C) = 15$
7. No
9. $C = v_1, v_3, v_5, v_2, v_6, v_4, v_1$; $W(C) = 18$
11. $C = v_1, v_6, v_5, v_3, v_4, v_2, v_1$; $W(C) = 16$
13. $C = v_1, v_4, v_5, v_6, v_2, v_3, v_1$; $W(C) = 14$
15. $w(\{v_2, v_1\}) + w(\{v_1, v_4\}) = 1 + 1 = 2 \not\geq 5 = w(\{v_2, v_4\})$

17. Using $w_a = 24$, which corresponds to the cycle $v_1, v_6, v_2, v_3, v_4, v_5, v_1$,

$$w_m \le 24 \le 2w_m \qquad \text{or, equivalently,} \qquad 12 \le w_m \le 24$$

19. (For Problem 17 of this section) Using $w_a = 20$, which corresponds to the cycle $v_1, v_2, v_6, v_5, v_4, v_3, v_1$,

$$w_m \le 20 \le 3w_m + \tfrac{1}{2} \qquad \text{or, equivalently,} \qquad \tfrac{39}{2} \le w_m \le 20$$

27. (a) $(0,0), (2,3), (2,6), (2,4), (1,4), (1,2), (1,1), (0,0);$ Approximate distance: 16
    (b) $(0,0), (1,1), (1,2), (1,4), (2,4), (2,3), (2,6), (0,0);$ Approximate distance: 18
    (c) (For (a)) $8 \le w_m \le 16$ (For (b)) $\tfrac{35}{6} \le w_m \le 18$

3.9 Problems

1.

3.

5. $V(G) = \{u, v, w\}; E(G) = \{(u, v), (u, w), (w, u), (w, v)\}$
7. $V(G) = \{v_1, v_2, v_3, v_4, v_5\}$
   $E(G) = \{(v_1, v_2), (v_2, v_4), (v_2, v_5), (v_3, v_1), (v_4, v_3), (v_5, v_4)\}$

9. $\begin{bmatrix} 0 & 1 & 1 \\ 1 & 0 & 0 \\ 0 & 0 & 0 \end{bmatrix}$

11. $\begin{bmatrix} 0 & 1 & 1 & 0 \\ 0 & 0 & 0 & 0 \\ 0 & 0 & 0 & 1 \\ 1 & 0 & 0 & 0 \end{bmatrix}$

13. (a) Number of edges with terminal vertex v_j (b) Number of edges with initial vertex v_i

17. $\begin{bmatrix} 0 & 1 & \infty & \infty \\ \infty & 0 & 2 & \infty \\ \infty & \infty & 0 & 7 \\ 5 & 3 & \infty & \infty \end{bmatrix}$

19. $\begin{bmatrix} 0 & 2 & 1 & \infty & \infty \\ \infty & 0 & \infty & \infty & \infty \\ \infty & \infty & 0 & 6 & 2 \\ \infty & 1 & \infty & 0 & 3 \\ \infty & \infty & \infty & \infty & 0 \end{bmatrix}$

21. (a)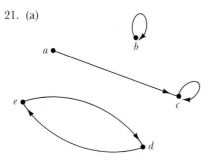

(b)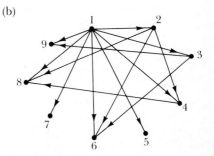

25.

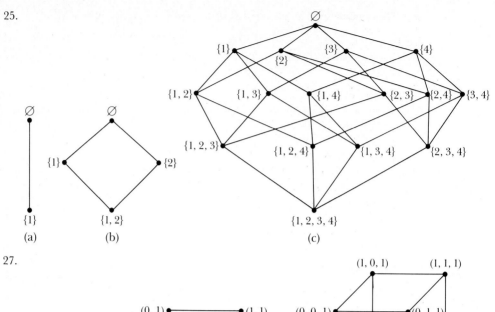

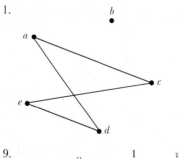

(a)   (b)   (c)

27.

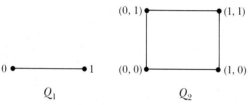

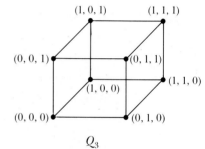

Q_1 Q_2 Q_3

Chapter 3 Review

1.

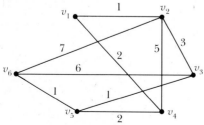

3. 14

5. $\{v_1, v_6, v_2, v_7\}$

7. The spanning tree has edge set
$\{e_1, e_2, e_3, e_4, e_6\}$.

9.

11. The minimal spanning tree has edge set $\{\{v_1, v_2\}, \{v_3, v_4\}, \{v_1, v_6\}, \{v_2, v_5\}, \{v_4, v_6\}\}$; $W = 8$

13. $C = v_1, v_6, v_5, v_4, v_3, v_2, v_1$; $\qquad W(C) = 16$

17. $\frac{31}{6} \leq w_m \leq 16$

19.

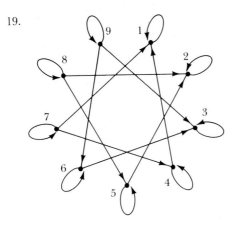

4.1 Problems

1.

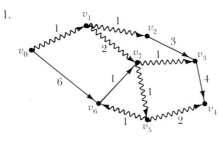

3.

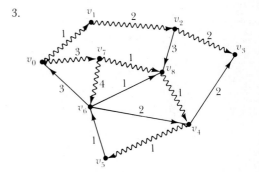

5.

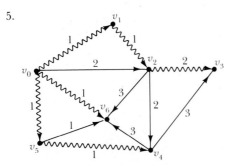

4.2 Problems

1.

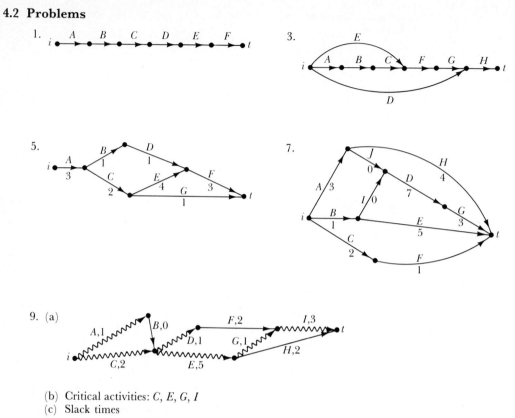

3.

5.

7.

9. (a)

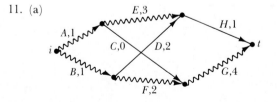

(b) Critical activities: *C, E, G, I*
(c) Slack times

Activity	Slack time
A	0
B	1
C	0
D	0
E	0
F	3
G	0
H	2
I	0

11. (a)

(b) Critical activities: B, F, G

(c) Slack times

Activity	Slack time
A	0
B	0
C	2
D	1
E	0
F	0
G	0
H	2

13. Critical activities: purchase machine C; install machine C; test machine C; produce parts a, b, c; assemble parts a, b, c; test final product; ship product.

Slack Times

Activity	Slack time	Activity	Slack time
purchase machine A	0	modify machine D	25
purchase machine B	0	test machine A	5
purchase machine C	0	test machine B	6
install machine A	0	test machine C	0
install machine B	0	produce parts a, b, c	0
install machine C	0	assemble parts a, b, c	0
purchase test equipment	0	test final product	0
install test equipment	19	ship	0

4.3 Problems

1. Graph (a) and Graph (c) have Euler paths; none have Euler tours.

4.4 Problems

1. 1111000010011010

3. 32

5. 111110000010001100111010010010110

4.5 Problems

1.

(a)
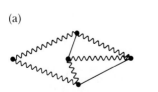

(c)
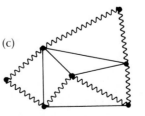

5.

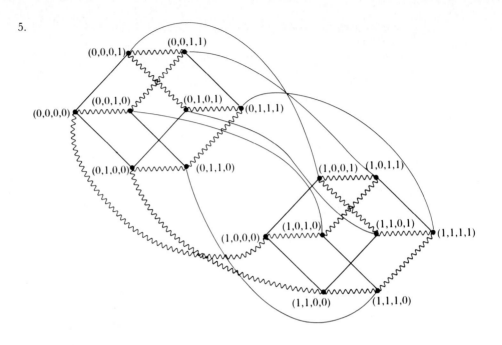

7. 32

9.

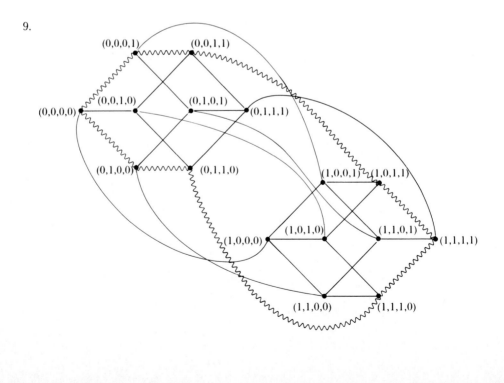

11. If $L_1 = 0000, 0001, 0011, 0010, 0110, 0111, 0101, 0100, 1100, 1101, 1111, 1110, 1010, 1011,$
1001, 1000
then $L_4 = 00000, 00010, 00110, 00100, 01100, 01110, 01010, 01000, 11000, 11010, 11110,$
11100, 10100, 10110, 10010, 10000, 10001, 10011, 10111, 10101, 11101, 11111, 11011, 11001,
01001, 01011, 01111, 01101, 00101, 00111, 00011, 00001

Chapter 4 Review

1.

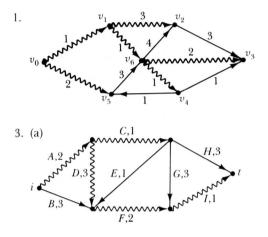

3. (a)

(b) Critical activities: A, D, F, I

(c)
Activity	Slack time
A	0
B	2
C	0
D	0
E	1
F	0
G	1
H	2
I	0

5. Graphs (a) and (b) have Euler paths.
Graph (b) has an Euler tour.

7. The Euler tour consisting of the edges 1111, 1110, 1100, 1000, 0000, 0001, 0010, 0100, 1001,
0011, 0110, 1101, 1010, 0101, 1011, 0111 yields the de Bruijn sequence 1111000010011010.

5.1 Problems

1. (b)
a	b	$a \cdot b$	$b \cdot a$
0	0	0	0
0	1	0	0
1	0	0	0
1	1	1	1

(c)

a	b	c	$(a + b) + c$	$a + (b + c)$
0	0	0	0	0
0	0	1	1	1
0	1	0	1	1
1	0	0	1	1
0	1	1	1	1
1	0	1	1	1
1	1	0	1	1
1	1	1	1	1

(d)

a	b	c	$(a \cdot b) \cdot c$	$a \cdot (b \cdot c)$
0	0	0	0	0
0	0	1	0	0
0	1	0	0	0
1	0	0	0	0
0	1	1	0	0
1	0	1	0	0
1	1	0	0	0
1	1	1	1	1

3. Property (e) is not satisfied by the real numbers.

5. (f) $A \cap (B \cup C) = (A \cap B) \cup (A \cap C)$ since

$$x \in A \cap (B \cup C)$$

if and only if

$$x \in A \quad \text{and} \quad (x \in B \text{ or } x \in C)$$

if and only if

$$x \in A \cap B \quad \text{or} \quad x \in A \cap C$$

if and only if

$$x \in (A \cap B) \cup (A \cap C)$$

Parts (g) $A \cup \emptyset = A$, (h) $A \cap X = A$, (i) $A \cup (X \backslash A) = X$, and (j) $A \cap (X \backslash A) = \emptyset$ are obvious.

7. (a)

a	b	a	$ab + ab'$
0	0	0	0
0	1	0	0
1	0	1	1
1	1	1	1

(b)

a	b	c	$ac + a'b$	$(a + b)(a' + c)(b + c)$
0	0	0	0	0
0	0	1	0	0
0	1	0	1	1
1	0	0	0	0
0	1	1	1	1
1	0	1	1	1
1	1	0	0	0
1	1	1	1	1

5.2 Problems

1. $x'y + xy$

3. $x'yz' + x'y'z + xyz' + x'yz$

5.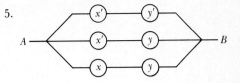

7.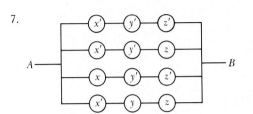

9.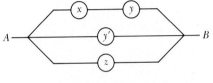

11.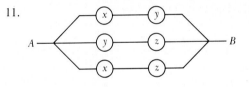

13. $xy + xz + y'z$

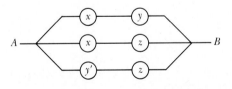

15. $x'yx + x'yyz + zx + zyz + x'z' + yz' = x'yz + zx + zy + x'z' + yz'$

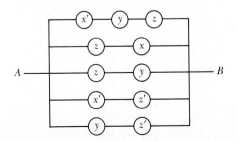

17. $x + y + z'$

19. $xz + z(xy + z') + yz$

5.3 Problems

1. (a) Variables: x, y, z
   Literals: x, y, y', z'

3.

x	y	z	$(x + y)(x + z)$	$x + yz$
0	0	0	0	0
0	0	1	0	0
0	1	0	0	0
1	0	0	1	1
0	1	1	1	1
1	0	1	1	1
1	1	0	1	1
1	1	1	1	1

5.

x	y	z	$xx' + yz$	yz
0	0	0	0	0
0	0	1	0	0
0	1	0	0	0
1	0	0	0	0
0	1	1	1	1
1	0	1	0	0
1	1	0	0	0
1	1	1	1	1

9. $x(y + z) = xy + xz$

11. $xx' + yz = 0 + yz = yz$

5.4 Problems

1. The product terms wxy and $wx'y$ are adjacent.
3. The product terms wxy and wxy' are adjacent.
5. $wx'yz + w'x'yz + wx'yz' + w'x'yz' = (w + w')(x'yz) + (w + w')(x'yz')$
   $= x'yz + x'yz' = (x'y)(z + z') = x'y$

7.

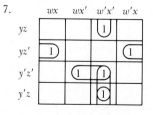

9.

11.

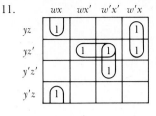

	wx	wx'	$w'x'$	$w'x$
yz	1			1
yz'		1	1	1
$y'z'$			1	
$y'z$	1			

$wxz + x'yz' + w'x'z' + w'xy$

13.

	wx	wx'	$w'x'$	$w'x$
yz	1			
yz'	1	1	1	1
$y'z'$		1	1	
$y'z$	1	1	1	1

$yz' + y'z + wxy + x'y'z'$

17. (a) $vwxy + wxyz' + w'x'yz' + x'y' + v'wy'z$
(b) $wxyz + vx'yz' + vw'y' + w'y'z + v'w'x'y'$
(c) $x'yz + wxz' + vw'x' + w'x'y'z'$

5.5 Problems

1. Yes, f implies g.
3. Yes, f implies g.
7. $x'y'$ and $y'z'$ are prime implicants of $x'y' + xz' + yz$; xy' is not.
9. x and yz' are prime implicants of $xyz' + x'yz' + xz + xy'z'$; y and xyz' are not.

5.6 Problems

1. $x_1 x_2' x_3$ and $x_1 x_2 x_3$
3. $xy + xy' + x'y$
5. $xy + x'y$
7. $xyz + xy'z + xy'z' + x'y'z + x'y'z'$
9. $yz', x'y, xz'$
11. $xy'z, xyz'$
13. $xy, w'x'z, w'yz, wxz'$

5.8 Problems

1. $xz' + x'y$
3. $xy' + z'$
5. $w'z' + x'y'z' + w'xy + wx'y'$
7. $wx'z + w'xz' + xyz + wx'y' + wyz$
9. $wx' + w'x + x'z'$

Chapter 5 Review

1. $x'y'z' + x'y'z + xy'z' + xy'z = y'$

3. (a)

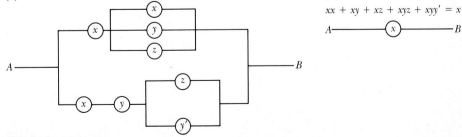

(b)

$xx + xy + xz + xyz + xyy' = x$

$A \longrightarrow \!\!\bigcirc\!x\!\bigcirc\!\longrightarrow B$

5.

x	y	z	y + z	x + y	(y + z)(x + y)	y + xz
0	0	0	0	0	0	0
0	0	1	1	0	0	0
0	1	0	1	1	1	1
0	1	1	1	1	1	1
1	0	0	0	1	0	0
1	0	1	1	1	1	1
1	1	0	1	1	1	1
1	1	1	1	1	1	1

7. $wx' + xz$

9. $xyz + xyz' + xy'z' + xy'z$

11. $x'z' + wx'y + wyz' + w'y'z'$

13. $y'z' + wxy'$

6.1 Problems

1.

p	q	$p \lor q$	$\sim p$	$\sim q$	$\sim(p \lor q)$	$(\sim p) \land (\sim q)$
T	T	T	F	F	F	F
T	F	T	F	T	F	F
F	T	T	T	F	F	F
F	F	F	T	T	T	T

3.

p	q	r	$p \lor q$	$q \lor r$	$(p \lor q) \lor r$	$p \lor (q \lor r)$
T	T	T	T	T	T	T
T	T	F	T	T	T	T
T	F	T	T	T	T	T
T	F	F	T	F	T	T
F	T	T	T	T	T	T
F	T	F	T	T	T	T
F	F	T	F	T	T	T
F	F	F	F	F	F	F

5.

p	q	r	$p \land q$	$q \land r$	$(p \land q) \land r$	$p \land (q \land r)$
T	T	T	T	T	T	T
T	T	F	T	F	F	F
T	F	T	F	F	F	F
T	F	F	F	F	F	F
F	T	T	F	T	F	F
F	T	F	F	F	F	F
F	F	T	F	F	F	F
F	F	F	F	F	F	F

7.

p	q	r	$q \wedge r$	$p \vee q$	$p \vee r$	$p \vee (q \wedge r)$	$(p \vee q) \wedge (p \vee r)$
T	T	T	T	T	T	T	T
T	T	F	F	T	T	T	T
T	F	T	F	T	T	T	T
T	F	F	F	T	T	T	T
F	T	T	T	T	T	T	T
F	T	F	F	T	F	F	F
F	F	T	F	F	T	F	F
F	F	F	F	F	F	F	F

9.

p	q	r	$p \vee q$	$\sim(p \vee q)$	$\sim p$	$(\sim p) \wedge q$	$\sim(p \vee q) \vee [(\sim p) \wedge q] \vee p$
T	T	T	T	F	F	F	T
T	T	F	T	F	F	F	T
T	F	T	T	F	F	F	T
T	F	F	T	F	F	F	T
F	T	T	T	F	T	T	T
F	T	F	T	F	T	T	T
F	F	T	F	T	T	F	T
F	F	F	F	T	T	F	T

6.2 Problems

1. F   3. T

5. T   7. T

9.

p	q	$p \Rightarrow q$	$(\sim p) \vee q$
T	T	T	T
T	F	F	F
F	T	T	T
F	F	T	T

11.

p	q	r	$(q \vee r)$	$p \Rightarrow (q \vee r)$	$(p \Rightarrow q) \vee (p \Rightarrow r)$
T	T	T	T	T	T
T	T	F	T	T	T
T	F	T	T	T	T
T	F	F	F	F	F
F	T	T	T	T	T
F	T	F	T	T	T
F	F	T	T	T	T
F	F	F	F	T	T

13.

p	q	$p \vee q$	$p \Rightarrow (p \vee q)$
T	T	T	T
T	F	T	T
F	T	T	T
F	F	F	T

15.

p	q	$p \wedge q$	$(p \wedge q) \Rightarrow p$
T	T	T	T
T	F	F	T
F	T	F	T
F	F	F	T

17.

p	q	$p \wedge q$	$p \vee q$	$(p \wedge q) \Rightarrow (p \vee q)$
T	T	T	T	T
T	F	F	T	T
F	T	F	T	T
F	F	F	F	T

19.

p	q	$\sim q$	$p \Rightarrow q$	$(\sim q) \wedge (p \Rightarrow q)$	$[(\sim q) \wedge (p \Rightarrow q)] \Rightarrow \sim p$
T	T	F	T	F	T
T	F	T	F	F	T
F	T	F	T	F	T
F	F	T	T	T	T

23. (a) If x is an odd integer, then $x = 6$. (F)
    (b) If x is not an odd integer, then $x \neq 6$. (F)
    (c) If $x \neq 6$, then x is not an odd integer. (F)

25.

p	q	$p \Rightarrow q$	$q \Rightarrow p$	$p \Leftrightarrow q$
T	T	T	T	T
T	F	F	T	F
F	T	T	F	F
F	F	T	T	T

27.

p	q	r	$p \wedge r$	$q \wedge (\sim r)$	$(p \wedge r) \vee (q \wedge (\sim r))$	$(\sim p) \wedge r$	$(\sim q) \wedge (\sim r)$
T	T	T	T	F	T	F	F
T	T	F	F	T	T	F	F
T	F	T	T	F	T	F	F
T	F	F	F	F	F	F	T
F	T	T	F	F	F	T	F
F	T	F	F	T	T	F	F
F	F	T	F	F	F	T	F
F	F	F	F	F	F	F	T

$(p \wedge r) \vee (q \wedge (\sim r))$	$((\sim p) \wedge r) \vee ((\sim q) \wedge (\sim r))$	$((p \wedge r) \vee (q \wedge (\sim r))) \Leftrightarrow (((\sim p) \wedge r) \vee ((\sim q) \wedge (\sim r)))$
T	F	F
T	F	F
T	F	F
F	T	F
F	T	F
T	F	F
F	T	F
F	T	F

6.3 **Problems**

1.

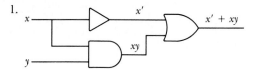

3.

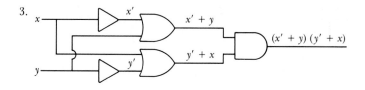

5.

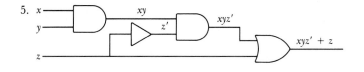

7. $((x + y)z)'$

9. $(x + y + z)'$

11.

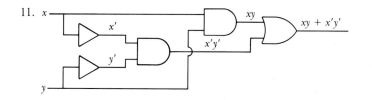

13.

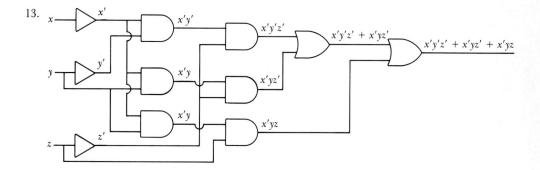

15.

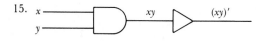

17. (a)

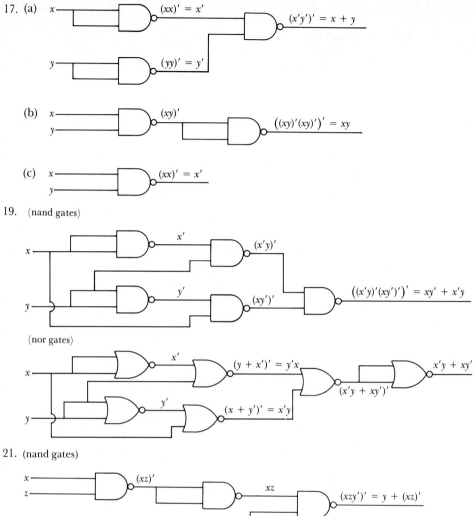

19. (nand gates)

(nor gates)

21. (nand gates)

(nor gates)

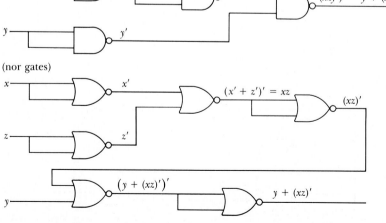

6.4 Problems

1.

x	y	$x'y + xy'$	$(x + y)(xy)'$
0	0	0	0
0	1	1	1
1	0	1	1
1	1	0	0

3.

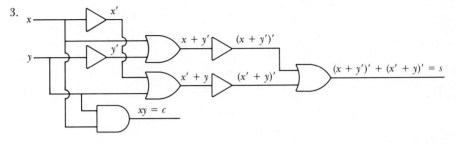

x	y	$s = (x + y')' + (x' + y)'$	$c = xy$
0	0	0	0
0	1	1	0
1	0	1	0
1	1	0	1

5.

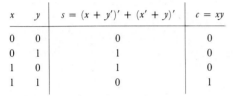

x	y	$s = (xy + x'y')'$	$c = xy$
0	0	0	0
0	1	1	0
1	0	1	0
1	1	0	1

7.

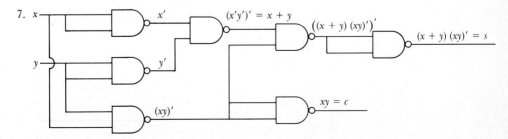

9. For the difference

$$\begin{aligned} & x_4\,x_3\,x_2\,x_1 \\ - & y_4\,y_3\,y_2\,y_1 \\ \hline & s_4\,s_3\,s_2\,s_1 \end{aligned}$$

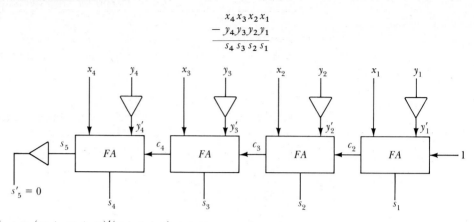

11. $s = (xy + yz + xz)'(x + y + z) + xyz;\ c = xy + yz + xz$

Chapter 6 Review

1.

p	q	s	$q \wedge s$	$\sim(p \vee (q \wedge s))$	$(\sim p) \wedge (\sim q)$	$(\sim p) \wedge (\sim s)$	$((\sim p) \wedge (\sim q)) \vee ((\sim p) \wedge (\sim s))$
T	T	T	T	F	F	F	F
T	T	F	F	F	F	F	F
T	F	T	F	F	F	F	F
T	F	F	F	F	F	F	F
F	T	T	T	F	F	F	F
F	T	F	F	T	F	T	T
F	F	T	F	T	T	F	T
F	F	F	F	T	T	T	T

3.

p	q	$p \vee q$	$\sim(p \vee q)$	$\sim p$	$\sim(p \vee q) \Rightarrow \sim p$
F	F	F	T	T	T
F	T	T	F	T	T
T	F	T	F	F	T
T	T	T	F	F	T

5. Statement: $p \Rightarrow q$ Contrapositive: $(\sim q) \Rightarrow (\sim p)$

p	q	$p \Rightarrow q$
F	F	T
F	T	T
T	F	F
T	T	T

p	q	$\sim q$	$\sim p$	$(\sim q) \Rightarrow (\sim p)$
F	F	T	T	T
F	T	F	T	T
T	F	T	F	F
T	T	F	F	T

7. (a) If $2x - 10 = 0$, then $x = 5$; true
   (b) If $2x - 10 \neq 0$, then $x \neq 5$; true
   (c) If $x \neq 5$, then $2x - 10 \neq 10$; true

9.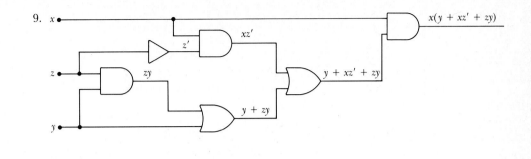

7.1 Problems

1. $3(6 \cdot 2^{n-1} + 4) - 2(6 \cdot 2^{n-2} + 4) = 6 \cdot 2^n + 4$
3. $2^{n+2} + (n+2)2^{n+2} - 4(2^{n+1} + (n+1)2^{n+1}) + 4(2^n + n2^n) = 0$
5. $(n+2)(-1)^{n+2} + 2(n+1)(-1)^{n+1} + n(-1)^n = 0$
7. $(n+2)! - (n+1)! - (n+1)^2 n! = 0$
11. $S(n) = 200S(n-1) + 225S(n-2) + 50S(n-3)$

13. $S(n) = S(n-1) + \sin \dfrac{n\pi}{2}$

15. $S(n) = S(n-1) + n^2$
17. $S(n) = S(n-1) + 3n^{5/2}$
19. $S(n) = S(n-1) + S(n-2)$
21. $f(n) = f(n-2) + f(n-1)$
23. $f(n) = f(n-1) + f(n-2)$
25. $S(n) = 1.1S(n-1) + 50n$

7.2 Problems

1. $\alpha_1 + \alpha_2(-2)^n$, for any α_1, α_2
3. $\alpha_1(3/2)^n + \alpha_2(-3/2)^n$, for any α_1, α_2
5. $\alpha_1 3^n + \alpha_2(-3)^n$, for any α_1, α_2
7. $\alpha_1 6^n + \alpha_2(-3)^n$, for any α_1, α_2
9. $\alpha_1(-3)^n + \alpha_2(-5)^n + \alpha_3$, for any $\alpha_1, \alpha_2, \alpha_3$
11. $\alpha_1(1/3)^n + \alpha_2(-7/3)^n$, for any α_1, α_2
13. $\alpha 2^n$, for any α
15. α, for any α

7.3 Problems

1. General solution: $\alpha_1(-1)^n + \alpha_2 2^n$
   Particular solution: $(-1/3)(-1)^n + (1/3)2^n$
3. General solution: $\alpha 7^n$
   Particular solution: $2(7^n)$
5. General solution: $\alpha_1(1/2)^n + \alpha_2(-2)^n$
   Particular solution: $(8/15)(1/2)^n + (2/15)(-2)^n$
7. General solution: $\alpha_1(1/2)^n + \alpha_2(-1)^n + \alpha_3(-2)^n$
9. $\begin{vmatrix} 3^n & n3^n \\ 3^{n-1} & (n-1)3^{n-1} \end{vmatrix} = -3^{2n-1} \neq 0$, and direct substitution shows that 3^n and $n3^n$ are solutions
   of $f(n+2) - 6f(n+1) + 9f(n) = 0$

11. $\begin{vmatrix} 1 & n \\ 1 & n-1 \end{vmatrix} = -1 \neq 0$, and direct substitution shows that 1 and n are solutions of $f(n) - 2f(n-1) + f(n-2) = 0$

13. $\begin{vmatrix} \cos\dfrac{n\pi}{4} & \sin\dfrac{n\pi}{4} \\ \cos\dfrac{(n-1)\pi}{4} & \sin\dfrac{(n-1)\pi}{4} \end{vmatrix} \neq 0$, and direct substitution shows that $\cos(n\pi/4)$ and $\sin(n\pi/4)$

are solutions of $f(n+2) - \sqrt{2}f(n+1) + f(n) = 0$.

15. $f(n) = \alpha_1 \left(\dfrac{1 + \sqrt{5}}{2}\right)^n + \alpha_2 \left(\dfrac{1 - \sqrt{5}}{2}\right)^n$, where $\alpha_1 = \dfrac{3 + \sqrt{5}}{5 + \sqrt{5}}$ and $\alpha_2 = \dfrac{\sqrt{5} - 1}{2\sqrt{5}}$

17. $f(n) = 2(5^{n-1}); \; 1.2207(10^{10})$

19. $f(n) = \alpha_1(2 + \sqrt{6})^n + \alpha_2(2 - \sqrt{6})^n$, where $\alpha_1 = \dfrac{2 + \sqrt{6}}{2\sqrt{6}}$ and $\alpha_2 = \dfrac{5 - 2\sqrt{6}}{6 - 2\sqrt{6}}; \; 54{,}692{,}416$

7.4 Problems

1. General solution: $\alpha_1(-2)^n + \alpha_2 n(-2)^n$

3. General solution: $\alpha_1 2^n + \alpha_2 n 2^n$

5. General solution: $\alpha_1 + \alpha_2 n$
   Particular solution: $2 - 2n$

7. General solution: $\alpha_1(1/3)^n + \alpha_2 n(1/3)^n$

9. General solution: $\alpha_1\left(\dfrac{1 + \sqrt{5}}{2}\right)^n + \alpha_2\left(\dfrac{1 - \sqrt{5}}{2}\right)^n + \alpha_3$

11. General solution: $\alpha_1(-7)^n + \alpha_2 n(-7)^n$
    Particular solution: 0

13. General solution: $\alpha_1(-7)^n + \alpha_2 n(-7)^n$
    Particular solution: $(-7)^n - (9/7)n(-7)^n$

15. General solution: $\alpha_1 + \alpha_2 n + \alpha_3 n^2$
    Particular solution: $(9/2)n - (5/2)n^2$

17. General solution: $\alpha_1(-1)^n + \alpha_2 n(-1)^n + \alpha_3 2^n + \alpha_4 n 2^n$

19. Difference equation: $f(n) = 2f(n-1) - f(n-2)$
    Expression for nth term: $f(n) = -3 + 5n$

7.5 Problems

1. $\alpha_1 2^{n/2} \sin\dfrac{n\pi}{2} + \alpha_2 2^{n/2} \cos\dfrac{n\pi}{2}$

3. General solution: $\alpha_1(2\sqrt{2})^n \sin\dfrac{3\pi n}{4} + \alpha_2(2\sqrt{2})^n \cos\dfrac{3\pi n}{4}$

   Particular solution: $(3/2)(2\sqrt{2})^n \sin\dfrac{3\pi n}{4} + 2(2\sqrt{2})^n \cos\dfrac{3\pi n}{4}$

5. $2^n\left(\alpha_1 \sin\dfrac{2\pi n}{3} + \alpha_2 \cos\dfrac{2\pi n}{3}\right)$ 7. $\alpha_1 + \alpha_2 \sin\dfrac{n\pi}{2} + \alpha_3 \cos\dfrac{n\pi}{2}$

9. $\alpha_1(-4)^n + \alpha_2 \sin\dfrac{n\pi}{2} + \alpha_3 \cos\dfrac{n\pi}{2}$

11. $2^n\left(\alpha_1 \sin\dfrac{n\pi}{2} + \alpha_2 \cos\dfrac{n\pi}{2} + \alpha_3 \sin\dfrac{2\pi n}{3} + \alpha_4 \cos\dfrac{2\pi n}{3}\right)$

13. $f(n) = 2^{n/2}\left(\alpha_1 \sin \dfrac{n\pi}{4} + \alpha_2 \cos \dfrac{n\pi}{4}\right)$

Special case: $f(n) = 2^{n/2}\left(25 \sin \dfrac{n\pi}{4} + 100 \cos \dfrac{n\pi}{4}\right)$

In year 5, the number of prey is 500 below the base level.

7.6 Problems

1. $\alpha_1 + \alpha_2 2^n - n$
3. General solution: $\alpha_1 + \alpha_2 2^n + (1/2)3^n$
   Particular solution: $5/2 - (2)2^n + (1/2)3^n$
5. General solution: $\alpha_1 2^n + \alpha_2(-2)^n + (n/8)2^n$
   Particular solution: $(11/16)2^n - (11/16)(-2)^n + (n/8)2^n$
7. $\alpha_1 5^n + \alpha_2 n 5^n + (1/9)2^{n+2}$
9. $\alpha_1 + \alpha_2 2^n + n 2^{n-1}$
11. General solution: $\alpha_1 \sin \dfrac{n\pi}{2} + \alpha_2 \cos \dfrac{n\pi}{2} + \dfrac{n}{2}$

   Particular solution: $-\dfrac{1}{2}\sin \dfrac{n\pi}{2} + \cos \dfrac{n\pi}{2} + \dfrac{n}{2}$

13. General solution: $\alpha + n$
    Particular solution: n
15. $\alpha_1(-3)^n + \alpha_2(-2)^n + \dfrac{2}{21}4^n$

7.7 Problems

1. $\alpha_1(-1)^n + \alpha_2 2^n + (1/4)3^n$
3. $\alpha_1 + \alpha_2 3^n + 2n$
5. General solution: $\alpha_1 + \alpha_2 n - 2n^2$
   Particular solution: $1 + 3n - 2n^2$
7. $\alpha_1 3^n + \alpha_2(-3)^n - (1/8)n^2 - (1/16)n + 35/64$
9. General solution: $\alpha_1(-4)^n + \alpha_2 2^n + n - 2$
   Particular solution: $(2/3)(-4)^n + (4/3)2^n + n - 2$
11. $\alpha_1 2^n + \alpha_2(-2)^n + 2n^2 - 3 - (1/4)n2^n$
13. $S(n) = \dfrac{n(n+1)}{2}$
15. $S(n) = (\frac{1}{2})(-1)^n n + (1/4)(-1)^n - 1/4$
17. $S(n) = 2^{n+1}(n-1) + 2$
21. $e(n) = e(n-1) + 3^{n-1}$
    $e(n) = \frac{1}{2}(3^n + 1)$

7.8 Problems

7. (b) $(1/2)(n^2 + n) + 1$
9. $(5/6)2^n + (2/3)(-1)^n$

Chapter 7 Review

3. General solution: $\alpha_1 5^n + \alpha_2(-2)^n$
   Particular solution: $f(n) = (3/7)5^n + (4/7)(-2)^n$
5. $\alpha_1(-5)^n + \alpha_2$

7. General solution: $\alpha_1(-3/2)^n + \alpha_2 n(-3/2)^n$ 9. $\alpha_1 5^{n/2} \sin \dfrac{2\pi n}{3} + \alpha_2 5^{n/2} \cos \dfrac{2\pi n}{3}$

Particular solution: $f(n) = 0$

11. $\alpha_1 3^n + \alpha_2(-2)^n - (1/4)2^n$

13. $\alpha_1(-1)^n + \alpha_2 n(-1)^n + 1/2$

19. $f(n) = 2f(n-1) + 2f(n-2); f(1) = 3, f(0) = 1$

Solution: $f(n) = \dfrac{2 + \sqrt{3}}{2\sqrt{3}} (1 + \sqrt{3})^n + \dfrac{\sqrt{3} - 2}{2\sqrt{3}} (1 - \sqrt{3})^n$

8.1 Problems

1. 63	3. 650	5. 1296
7. 5,153,632	9. 488	11. 24
13. 16	15. 30	17. mn

8.2 Problems

1.

abc	bac	cab	dab
acb	bca	cba	dba
abd	bad	cad	dac
adb	bda	cda	dca
acd	bcd	cbd	dbc
adc	bdc	cdb	dcb

3. 720

5. 20

7. 24

9. $(n!)^2$

11. $\dfrac{n!}{r(n-r)!}$

8.3 Problems

1. (a) 21 (c) 1 (e) 1	3. 2,598,960
5. 5148	7. 6,589,440
9. 2.22×10^{32}	11. 72,072
13. 1.03×10^8	15. 24,024
17. 720	19. $\dbinom{n}{k} 2^{n-k}$

8.4 Problems

1. 1716	5. 2520

8.6 Problems

1. $\{1, 1, 1, 1\}, \{1, 1, 1, 2\}, \{1, 1, 2, 2\}, \{1, 2, 2, 2\}, \{2, 2, 2, 2\}$.

3. 6

5.

Box No.	1	2	3	4	5	6	7	8
■	**			***			*	* ■

7. 11,440

9. 362,880

11. 1,860,480

15. 220

8.7 Problems

1. 19
5. 21

3. 7
9. 12

Chapter 8 Review

1. 17
5. 720
9. 247,104
15. $\sum_{i=k}^{n} \binom{n}{i} 9^{n-i}$
19. 792

3. 4,626,984,729
7. 277,200
11. 15,120
17. 286
21. 14,684,570

9.1 Problems

1. $\{(x_1, x_2, x_3, x_4, x_5) | x_i \in \{L, R, S\}, 1 \le i \le 5\}$
3. the set of all ordered triples, where each triple consists of three of the five people
7. (a) {(red, black), (red, blue), (black, red), (blue, red)}
   (b) {(red, red), (red, black), (red, blue), (black, red), (blue, red)}
   (c) {(red, black), (red, blue), (black, blue), (blue, black), (black, red), (blue, red)}
9. 5/16
11. 1/3
13. (a) 2/9   (b) 1/2

9.2 Problems

1. 46/91; 3/13
3. (a) 1/6   (b) 5/12
5. 11/12
9. (a) 0.3412   (b) 0.2995   (c) 0.9583
11. $1 - \left(\dfrac{(365!)/(265!)}{365^{100}} \right)$
13. (a) 0.1   (b) 0.49   (c) 0.3   (d) 0.11

9.3 Problems

1. $P(A|B) = 5/8; P(B|A) = 5/14$
3. If $P(A) = P(B)$
5. 2/5
7. 1/2
11. 5/9
15. (a) 1/256   (b) 1/32   (c) 1/70

9.4 Problems

1. No
3. $(q - q^2)/(1 - q^2)$
7. No, since $P(A|B) \ne P(A)$
9. (a) 1/18   (b) 5/9   (c) 7/18

9.5 Problems

1. (a) 15/64   (b) 57/64
3. 0.03279
5. (a) 12/1296   (b) 72/1296   (c) 625/1296
7. (a) 0.03534   (b) 0.17095   (c) 0.45096
9. (a) 0.0810   (b) $(5/6)^{10}$ (c) 0.5236

9.6 Problems

1. 0.5385   3. 0.5475   5. 0.1770; .9204

9.7 Problems

1. (a) $X(H, H, H, H) = 12$, $\dot{X}(H, H, H, T) = 7$, $X(H, H, T, T) = 2$, $X(H, T, T, T) = -3$
   $X(T, T, T, T) = -8$
   (b)
   $$f(x) = \begin{cases} 1/16 & \text{if } x = 12 \\ 1/4 & \text{if } x = 7 \\ 3/8 & \text{if } x = 2 \\ 1/4 & \text{if } x = -3 \\ 1/16 & \text{if } x = -8 \\ 0 & \text{otherwise} \end{cases}$$

3. (a) $X(T, T, T) = 0, X(H, A, A) = 1, X(T, H, A) = 2, X(T, T, H) = 3$ (A can be either H or T.)
   (b)
   $$f(x) = \begin{cases} 1/8 & \text{if } x = 0 \\ 1/2 & \text{if } x = 1 \\ 1/4 & \text{if } x = 2 \\ 1/8 & \text{if } x = 3 \\ 0 & \text{otherwise} \end{cases}$$

5.
   $$f(x) = \begin{cases} 1/36 & \text{if } x = 2 \text{ or } x = 12 \\ 2/36 & \text{if } x = 3 \text{ or } x = 11 \\ 3/36 & \text{if } x = 4 \text{ or } x = 10 \\ 4/36 & \text{if } x = 5 \text{ or } x = 9 \\ 5/36 & \text{if } x = 6 \text{ or } x = 8 \\ 6/36 & \text{if } x = 7 \\ 0 & \text{otherwise} \end{cases}$$

11. (a) .032   (b) .9997   (c) .3984   (d) .608
13. (a) $X = 0, 1, 2, \ldots, 20$
   (b)
   $$f(x) = \begin{cases} \binom{20}{x}(1/5)^x(4/5)^{20-x} & \text{if } x = 0, 1, 2, \ldots, 20 \\ 0 & \text{otherwise} \end{cases}$$

9.8 Problems

1. 13/12
5. 2
7. 2
9. Yes (1445/35 ¢ expected gain per game)
11. (a) $\dfrac{x/2}{1 - x/2}$ (b) $1.00   (c) The expected value is infinite.
13. 7/2
15. $0.51n + 0.49$

9.9 Problems

3. (a) e^{-2} (b) $\frac{19}{3}e^{-2}$ (c) $1 - 3e^{-2}$
5. (a) 0.1966   (b) 0.1108   (c) 0.1687
7. (a) 0.1353   (b) 0.3068   (c) 0.3679   (d) 0.1465
9. 0.2381

9.10 Problems

1. (a) 1.6 minutes   (b) 3.2 cars   (c) 0.0524
3. .2843
5. 7/4 days; 3.2 people
11. (a) 0.1595 hours   (b) 9.0467   (c) .0788   (d) 0.2261 hours

Chapter 9 Review

1. The set of all ordered pairs (x, y), where $x = 1, 2, \ldots, 6$ and $y = 1, 2, \ldots, 6$; x is the number up on the green die and y is the number up on the red die.
3. (a) 1/3   (b) 1/6   (c) 21/36
5. $P(A) = 1/4, P(B) = 1/13, P(A \cup B) = 4/13$
7. 4/13
9. Events A and B are not independent.
   Events A, B, and C are not independent.
11. $-\$.20$
13. $\dfrac{200^{100}e^{-200}}{100!} \approx 1.8797 \times 10^{-15}$
15. (a) 1/30 hour   (b) 4/15   (c) 192/5000
19. 11.6721 people/hour

10.1 Problems

1. (a) None exists   (b) None exists   (c) None exists   (d) None exists
3. (a) 1, 150   (b) 1   (c) 0, -16 (d) 1/2
5. (a) 2, 3   (b) 2   (c) 0, -1 (d) 0
7. (a) 2, 7   (b) 3/2   (c) 0, 1   (d) 1
9. (a) None exists   (b) None exists   (c) -5, 1/2   (d) 2
11. $f(n) = 10(.6)^n, n = 1, 2, 3, \ldots$
13. Problem 1 of this section: Does not exist
    Problem 3 of this section: 1
    Problem 5 of this section: Does not exist
    Problem 7 of this section: 1
    Problem 9 of this section: Does not exist
19. (a) N can be any positive integer   (b) $N = 2$ (c) $N = 4$ (d) $N = [1/\varepsilon]$

10.2 Problems

1. 3/2
3. 0
5. Does not exist   7. Does not exist
9. Does not exist

10.3 Problems

1. $15/2$
3. Divergent
5. $-7/3$
7. 2
9. $5/12$
11. $9/4$
13. $-11/36$
15. 1
17. Divergent
25. (b) 2

10.4 Problems

1. $1 + x + x^2 + x^3 + x^4 + x^5$

3. $5 + 4x + 3x^2 + 2x^3 + x^4$

5. $\dfrac{1 - x^{21}}{1 - x}$

7. $\dfrac{1}{1 - 2x} + \dfrac{3x}{(1 - 3x)^2}$

9. $\dfrac{x \sin \pi/2}{1 + x^2 - 2x \cos \pi/2} = \dfrac{x}{1 + x^2}$

11. $\dfrac{x^2 + x}{(1 - x)^3}$

13. $(1 - x)^{-4}$

15. $\dfrac{1 - (1/3)x \cos \pi/4}{1 + (1/9)x^2 - (2/3)x \cos \pi/4}$

10.5 Problems

1. $f(n) = n2^n$
3. $g(n) = n(-1)^n$
5. $f(n) = 0$
7. $k(n) = 1$
9. $f(n) = (2)5^n$
11. $f(n) = 6^n$
13. $g(n) = n3^n$
15. $f(n) = (-3)^n$
17. $f(n) = 2n$
19. $f(n) = 2^n + n2^n$

10.6 Problems

1. $\dfrac{a}{x + 7} + \dfrac{b}{x - 3} + \dfrac{c}{x - 2}$

3. $\dfrac{a}{x} + \dfrac{b}{x - 3} + \dfrac{c}{x - 4} + \dfrac{d}{(x - 4)^2}$

5. $\dfrac{ax + b}{x^2 - 2x + 2} + \dfrac{c}{x - 1} + \dfrac{d}{(x - 1)^2}$

7. $\dfrac{ax + b}{x^2 + 2x + 2} + \dfrac{cx + d}{(x^2 + 2x + 2)^2} + \dfrac{e}{x - 1} + \dfrac{f}{(x - 1)^2} + \dfrac{g}{(x - 1)^3}$

9. $\dfrac{2}{x - 3} + \dfrac{1}{x + 5}$

11. $\dfrac{1}{x} - \dfrac{3}{x + 2}$

13. $\dfrac{-7}{x - 4} + \dfrac{4}{x - 5}$

15. $\dfrac{2x - 1}{x^2 + 3x + 5} - \dfrac{1}{x + 2}$

17. $f(n) = 3^n + 2(-2)^n$

19. $f(n) = (-2/5)(-2)^n + (17/5)3^n$

21. $f(n) = (4/5)3^n + (1/5)(-2)^n$

23. $f(n) = (3/2)2^n + (-2)^{n-1}$

25. $f(n) = 2^{n-2}$ for $n \geq 2$; $f(0) = f(1) = 0$

10.7 Problems

1. $1 + 3x + 6x^2 + 10x^3$
3. $1 + x + 2x^2 + 2x^3$
5. 14
7. $n(n + 1)/2$
9. $h(n) = \begin{cases} 0 & \text{if} \quad n = 0, 1 \\ 1 & \text{if} \quad n = 2 \\ 4 & \text{if} \quad n = 3 \\ 10 & \text{if} \quad n = 4 \\ 12 & \text{if} \quad n = 5 \\ 9 & \text{if} \quad n = 6 \\ 0 & \text{if} \quad n > 6 \end{cases}$

11. $h(n) = \dfrac{(n + 1)(n + 2)}{2}$

13. $g(h) = \begin{cases} 2 & \text{if} \quad n = 0 \\ 1 & \text{if} \quad n = 1 \\ 2 & \text{if} \quad n = 2 \\ 1 & \text{if} \quad n > 2 \end{cases}$

15. $f(n) = \begin{cases} n + 1 & \text{if} \quad 0 \le n \le 3 \\ 4 & \text{if} \quad n > 3 \end{cases}$

10.8 Problems

1. $(1 + x + x^2)(x + x^2 + x^3 + x^4)(x^4 + x^5)$
3. $(x^{-2} + x^{-1} + 1)(x^3 + x^4)(x^{-3} + 1)(1 + x + x^2 + x^3 + x^4)(1 + x + x^2)$
5. $(x^2 + x^4 + x^6 + \cdots)^5$
7. $(1 + x^2 + x^4 + x^6 + \cdots)^5$
9. $(1 + x + x^2 + x^3)(x + x^2 + x^3 + x^4 + \cdots)^4$
11. $(1 + x + x^2 + x^3 + \cdots)^k$
13. $\sum_{n=0}^{\infty} (10^n - 8^n)x^n$

10.9 Problems

1. $h(7) = 120$
3. $g(10) = 1001$
5. $h(10) = 2541$
7. $g(10) = 792$
9. $h(n) = \dbinom{n + 2}{n}$

11. $h(n) = \dbinom{n + 3}{n} - \dbinom{n - 2}{n - 5}$

13. $h(n) = \dbinom{n + 2}{n - 1} - \dbinom{n - 2}{n - 5}$

15. $h(n) = \dbinom{k}{n - k}, k \le n \le 2k; h(n) = 0,$ otherwise

17. $k(n) = 0, n \le 13; k(n) = \dbinom{n - 9}{n - 14}, 14 \le n \le 19; k(n) = \dbinom{n - 9}{n - 14} - \dbinom{n - 15}{n - 20}, n \ge 20$

19. $\dbinom{20}{20 - 4} - \dbinom{20 - 4}{20 - 8} = 3025$

21. $h(n) = \begin{cases} 0 & \text{if} \quad n \text{ is odd} \\ \dbinom{\dfrac{n - 2}{2} + 3}{\dfrac{n - 2}{2}} & \text{if} \quad n \text{ is even} \end{cases}$

23. $10^n - 8^n$
25. $10^9 - 8^9$

Chapter 10 Review

1. 1
3. 1
5. $10(1.1)^{n-1}$, $n = 1, 2, \ldots$
7. 6
9. 7
11. $-3/16$
13. $7x^5$
15. $x + 4x^2 + 9x^3 + \cdots + n^2 x^n + \cdots$
17. $\dfrac{3(x^2 + x)}{(1 - x)^3}$
19. $1/(1 - 10x)$
21. $\dfrac{1}{1 - 2x} - \dfrac{21x}{(1 - 3x)^2}$
23. $\dfrac{a}{x + 1} + \dfrac{b}{(x + 1)^2} + \dfrac{c}{(x + 1)^3} + \dfrac{d}{(x + 1)^4} + \dfrac{e}{x + 3} + \dfrac{f}{(x + 3)^2} + \dfrac{g}{(x + 3)^3} + \dfrac{h}{(x + 3)^4}$
$\quad + \dfrac{i}{(x + 3)^5} + \dfrac{jx + k}{x^2 + 1} + \dfrac{lx + m}{(x^2 + 1)^2}$
25. $3^n + n3^n$
27. $(2)5^n + n5^n$
29.
$$h(n) = \begin{cases} 0 & \text{if } n = 0, 1, 2 \text{ or } n \geq 8 \\ 2 & \text{if } n = 3 \text{ or } n = 7 \\ 4 & \text{if } n = 4 \text{ or } n = 6 \\ 6 & \text{if } n = 5 \end{cases}$$
31. $X(r)(x) = (x^2 + x^4 + x^6 + \cdots)(x + x^3 + x^5 + \cdots)(x + x^2 + x^3 + \cdots + x^9)$
33. $g(n) = 0$, $n \leq 3$; $g(n) = \dbinom{n}{n - 4}$, $4 \leq n \leq 8$; $g(n) = \dbinom{n}{n - 4} - \dbinom{n - 5}{n - 9}$, $n \geq 9$

11.1 Problems

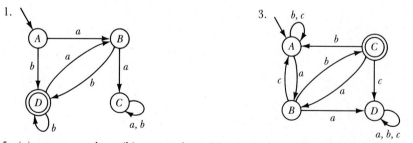

1.

3.

5. (a) not accepted   (b) accepted   (c) accepted   (d) not accepted
7. The DFA in Problem 1 accepts only words that end in b, and do not have two or more consecutive a's.
9. The DFA in Problem 3 accepts any word that ends in ab and does not have either two or more consecutive a's or an abc.

11.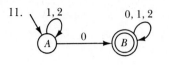

13.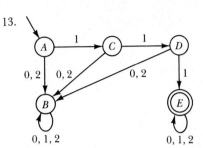

15.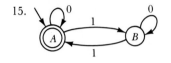

17.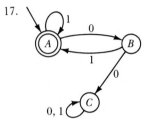

11.2 Problems

1.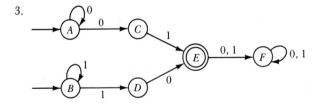

3.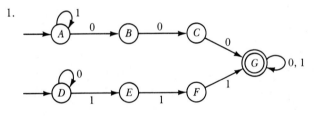

5.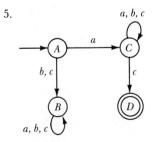

7. (a)

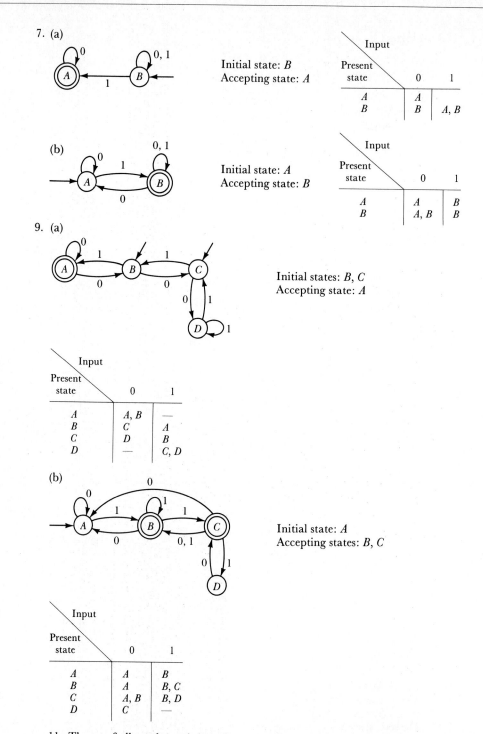

Initial state: *B*
Accepting state: *A*

Present state	Input	
	0	1
A	*A*	
B	*B*	*A, B*

(b)

Initial state: *A*
Accepting state: *B*

Present state	Input	
	0	1
A	*A*	*B*
B	*A, B*	*B*

9. (a)

Initial states: *B, C*
Accepting state: *A*

Present state	Input	
	0	1
A	*A, B*	—
B	*C*	*A*
C	*D*	*B*
D	—	*C, D*

(b)

Initial state: *A*
Accepting states: *B, C*

Present state	Input	
	0	1
A	*A*	*B*
B	*A*	*B, C*
C	*A, B*	*B, D*
D	*C*	—

11. The set of all words consisting of any number of repetitions of *cba*.

11.3 Problems

1.

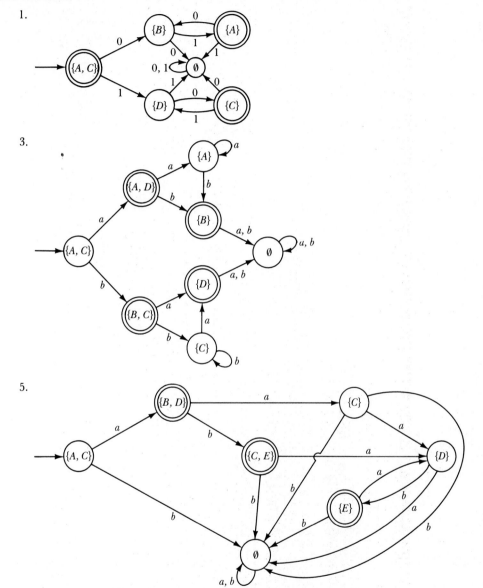

3.

5.

11.4 Problems

1. $\{a, bc\}$
3. $\{a, \varepsilon, bc, bcbc, bcbcbc, \dots\}$
5. The set consisting of the empty word, ε, and all words formed from any number of ab's followed by any number of c's.
7. The set consisting of the empty word, ε, and all words formed from any number of ab's and any number of c's.

9. The set consisting of the empty word, ε, and all words consisting of k a's, $k = 0, 3, 6, 9, \ldots$

11. $b + a*$ 13. $ab*$ 15. $a + b + c + (bc)*$

17. $(a + b + c)*(cba)(a + b + c)*$

11.5 Problems

1.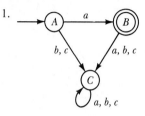

3.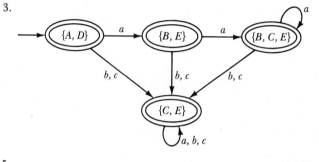

5.

7.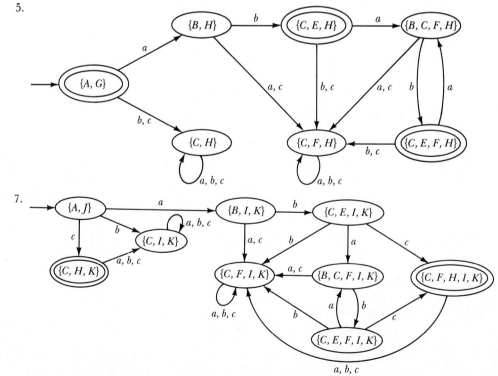

9.

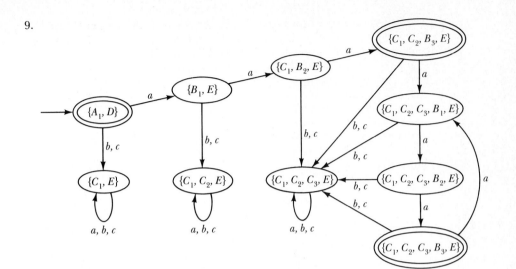

11.

11.6 Problems

1.

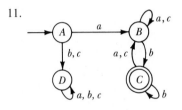

3.

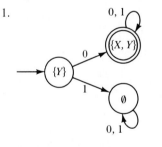

11.7 Problems

1. $\{ab\}^*$
3. The set of all words of the form $b^{n+1}a^n$, where $n = 0, 1, 2, \ldots$.
5. $\{111\}^*$
7. $aa^* + bb^*$
9. $\{ac\}^*\{b\}^*b$
11. $G = (N, T, S, P)$, where $N = \{S\}$, $T = \{a\}$, $P = \{S \rightarrow aaS, S \rightarrow \lambda\}$

13. $G = \{N, T, S, P\}$, where $N = \{S, A, B, C, D\}$, $T = \{a, b, c\}$,
    $P = \{S \rightarrow ADBc, D \rightarrow ADB, D \rightarrow \lambda, A \rightarrow a, B \rightarrow b\}$
15. $G = (N, T, S, P)$, where $N = \{S, A, B\}$, $T = \{a, b, c\}$,
    $P = \{S \rightarrow ABcc, A \rightarrow aaA, B \rightarrow bbB, A \rightarrow \lambda, B \rightarrow \lambda\}$
17. Problem 1 of this section: Type 2
    Problem 4 of this section: Type 3
    Problem 7 of this section: Type 3
    Problem 8 of this section: Type 2
    Problem 10 of this section: Type 2

11.8 Problems

1.

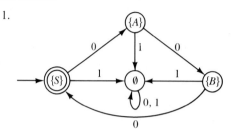

3.

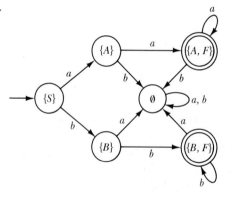

5.

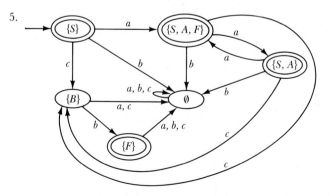

7. $G = (N, T, S, P)$, where $N = \{S, B, C, D\}$, $T = \{a, b\}$, $P = \{S \to aB, S \to bD, B \to bD, D \to aB, D \to \lambda, B \to aC, C \to aC, C \to bC, D \to bD\}$

9. $G = (N, T, S, P)$, where $N = \{S, B, C, D\}$, $T = \{a, b, c\}$, $P = \{S \to bS, S \to cS, S \to aB, B \to cS, B \to bC, B \to aD, C \to bS, C \to aB, C \to cD, D \to aD, D \to bD, D \to cD, D \to \lambda\}$

Chapter 11 Review

1.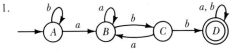

3. $b*a(a*(ba)*)*bb(a + b)*$

5.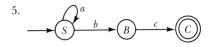

7.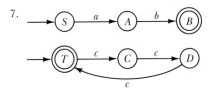

11.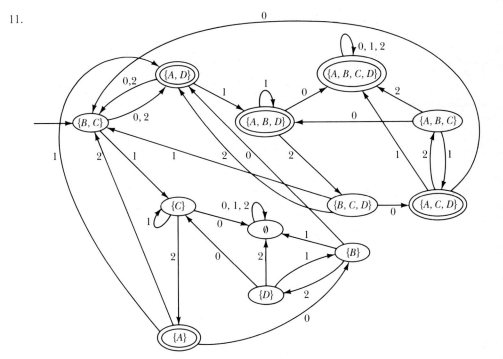

13. a, b, and c are regular expressions; bc is a regular expression; $(bc)*$ and $a*$ are regular expressions; $(bc)*a*$ is a regular expression; $a + (bc)*a*$ is a regular expression

15.

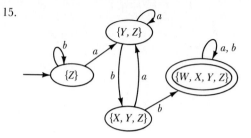

17. The language consists of all words of the form $a^n 0 b^n$, for $n = 1, 2, \ldots$. This is a Type 2 grammar.

19. $G = (N, T, S, P)$, where $N = \{S, A\}$, $T = \{0, 1, 2\}$, $P = \{S \to 00A2, A \to 11A, A \to \lambda\}$

21.

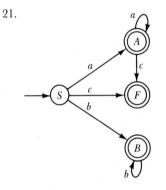

23. $G = (N, T, S, P)$, where $N = \{S, B, C, D\}$, $T = \{0, 1\}$, $P = \{S \to 0S, S \to 1B, B \to 0B, B \to 1C, C \to 0C, C \to 1D, D \to 0D, D \to 1B, D \to \lambda\}$

A.1 Problems

1. (a) $\{x \mid x \text{ is an even integer and } 3 < x < 105\}$
   (b) $\{z \mid z = n^2, n = 0, 1, 2, \ldots\}$
   (c) $\{y \mid y \text{ is an integer and } 6 \text{ divides } y\}$
   (d) $\{x \mid x = 2n + 1, -5 \le n \le 270 \text{ and } n \text{ is an integer}\}$
3. False
5. False
7. True
9. False
11. True
13. False
15. True
17. False
19. $\{x, y, z, u, v, w\}$
21. $\{d\}$
23. $\{(x, 1), (x, 2), (x, 3), (y, 1), (y, 2), (y, 3)\}$
25. $\{r, t, u\}$

27. $\{(1, 1), (1, 2), (1, 3), (2, 1), (2, 2), (2, 3)\}$
29. $\varnothing$
31. $\varnothing$
33. (a) $\{1, 2, 3, 4, 5, 6, 7\}$ (c) $\{3, 4\}$ (e) $\{3, 4, 5, 6, 7\}$ (g) $\{3, 4, 5, 6, 7, 8, 9\}$
35. (a) $\{x | -1 \le x < 5\}$ (c) $\{x | -1/5 \le x < 3\}$ (e) $\{x | -1 \le x < -1/6\}$ (g) $\varnothing$
47. (a) $\{\varnothing, \{a\}, \{1\}, \{-2\}, \{a, 1\}, \{a, -2\}, \{1, -2\}, \{a, 1, -2\}\}$
    (b) $\{\varnothing, \{\varnothing\}, \{\{0\}\}, \{\{1\}\}, \{\{0, 1\}\}, \{\varnothing, \{0\}\}, \{\varnothing, \{1\}\}, \{\varnothing, \{0, 1\}\}, \{\{0\}, \{1\}\}, \{\{0\}, \{0, 1\}\},$
    $\{\{1\}, \{0, 1\}\}, \{\varnothing, \{0\}, \{1\}\}, \{\varnothing, \{0\}, \{0, 1\}\}, \{\varnothing, \{1\}, \{0, 1\}\}, \{\{0\}, \{1\}, \{0, 1\}\}, \{\varnothing, \{0\},$
    $\{1\}, \{0, 1\}\}\}$

A.2 Problems

1. (a) $\{x | x \ne -1\}$ (b) $\{y | y \ne 0\}$ (c) $1/3$ (d) not defined   (e) $\dfrac{1}{a + b + 1}$

   (f) $\dfrac{1}{(1/a) + 1}$ (g) $b + 1$

3. Domain: $\{x | x \ne 0\}$ Range: $\{-1, 1\}$
5. (a) $\{x | x < 2\} \cup \{x | x > 3\}$ (b) $\{y | y > 1\}$ (c) 4 (d) 4 (e) 9.2 (f) 25

7. $f(x) = \begin{cases} 0 & \text{if } x = e, f, g, h \\ 1 & \text{if } x = a, b, c, d \end{cases}$

9. (a) 3 (b) 0 (c) 1 (d) 2 (e) 1
11. (a) $f \circ g(2) = 7; g \circ f(2) = 2$
    (b) $h \circ k(4) = 1; h \circ k(5)$ is not defined; $k \circ h(4) = 4\frac{3}{4}; k \circ h(5) = 4\frac{4}{5}$

A.3 Problems

1. $a_{33} = \pi; a_{12} = 2; a_{21} = 2; a_{34} = 0$

2. (a) $\begin{bmatrix} 5 & 1 \\ 4 & 1 \\ -1 & 8 \\ 5 & 6 \end{bmatrix}$

   (c) $\begin{bmatrix} -2 & -1 & 1 \\ -4 & -1 & -1 \end{bmatrix}$

   (e) $\begin{bmatrix} 6 & 18 & 12 \\ 24 & 6 & 30 \end{bmatrix}$

   (g) $\begin{bmatrix} 6 & 3 & -3 \\ 8 & 2 & 2 \\ 22 & 5 & 7 \\ 28 & 9 & 1 \end{bmatrix}$

   (i) $\begin{bmatrix} 24 & 9 & 12 & 15 \\ 4 & -4 & -6 & 20 \\ 4 & -15 & -22 & 54 \\ 42 & 2 & 1 & 74 \end{bmatrix}$

   (k) $\begin{bmatrix} 2 & 7 & 1 & 4 \\ -4 & 4 & -4 & 4 \end{bmatrix}$

3. (a) $\begin{bmatrix} 2 & 14 \\ -7 & 16 \end{bmatrix}$

   (b) $\begin{bmatrix} 0 & 0 & 0 \\ 0 & 0 & 0 \\ 0 & 0 & 0 \end{bmatrix}$

   (c) $\begin{bmatrix} 1 & 0 \\ 0 & 1 \end{bmatrix}$

15. (a) 10 (c) -109

A.4 Problems

1. (a) $a_1 + a_2 + a_3 + a_4 + a_5$ (b) $c_3^2 + c_4^2 + c_5^2 + c_6^2$ (c) $1^3 + 2^3 + 3^3 + 4^3$
3. (a) 319/840   (b) 88   (c) 36   (d) 0.9044
5. (a) $\sum_{n=1}^{7} (-1)^{n+1} n$ (b) $\sum_{k=5}^{8} k^2$ (c) $\sum_{k=1}^{6} 3(k^2)$ (d) $\sum_{k=1}^{6} 2^k/(k+1)^2$

A.5 Problems

1. Reflexive, transitive
3. Symmetric
5. (a) R_2, R_4 (b) $R_2, R_4,$ (c) R_2, R_4, R_5 (d) R_4, R_5
11. $[0] = [12] = [24]; [6] = [162]; [-5] = [-29]$
13. None are equivalent relations.
15. (1)
21. Yes

Index

A

Absolute value, 515
Accepting state, 457, 459, 465
Accessible vertex, 134
$ac(M)$, 459, 465
Addition modulo m, 74
Adjacency matrix of a graph, 88
Adjacent product terms in a Karnaugh map, 207
Adjacent vertices, 86
Algorithm, 1
    average-case analysis, 48, 384
    backtracking, 35
    BINARY SEARCH, 5, 45, 263, 296
    BUBBLESORT, 10, 299
    complexity of, 44
    divide-and-conquer, 263, 297
    Division, 63
    greedy, 28
    HEAPSORT, 47
    Horner's, 59
    Kruskal's, 115
    MERGESORT, 297
    minimal path tree, 154
    Prim's, 120
    QUICKSORT, 21
    Quine–McCluskey, 224
    recursive, 21
    Russian peasant, 67
    SEQUENTIAL SEARCH, 2, 45, 384
    sorting, 8
    spanning cycle, 125
    spanning tree, 109
    to generate Gray codes, 183
    to generate permutations in lexicographic order, 310
    TRINARY SEARCH, 264, 296
    worst-case analysis, 45
Alphabet, 476
Alternate form of the Principle of Induction, 21
And gate, 247
Antisymmetric relation, 138, 533
Arrival process, 393
Automata, 456
Average-case analysis, 384
Axiom of Completeness, 406

B

Backpack problem, 28
Backtracking algorithm, 35
Base of a number system, 56
Basic identities involving binomial coefficients, 322
Bayes' Theorem, 374
Bernoulli, 366
Bernoulli process, 366
Bernoulli trial, 366
Big O notation, 48
Binary channel, 366
Binary number system, 55
Binary operation, 191
BINARY SEARCH, 5, 45, 263, 296

Binary symmetric channel, 366
Binary tree, 107
Binomial coefficient, 319, 378
Binomial distribution, 378
Binomial probability, 368
Binomial process, 368
Binomial Theorem, 319
Bit (binary digit), 56
Boole, 187
Boolean algebra, 187, 191
Boolean expression, 194
Boolean function, 194
Bounded sequence, 405
BUBBLESORT, 10, 299

C

Canonical sum, 224
Cardano, 342
Carry bit, 252
Cartesian product, 510
Casorati's determinant, 271
Characteristic equation, 267
Characteristic function, 514
Chomsky, 489
Chomsky's Hierarchy, 492
Circuit, 246
Closure, 468
$C(M)$, 468
$C(n, r)$, 313
Combination, 313
Combinatorial proof, 322
Complement gate, 248
Complement of a graph, 94
Complement of a set, 508
Complementary switch, 189
Complete bipartite graph, 307
Complete graph, 123
Completion, 228
Complexity of an algorithm, 44
Component of a graph, 99
Composite function, 515
Compound interest, 261, 297
Concatenation, 476
Conclusion, 243
Conditional probability, 356
Conjunction, 239

Connected graph, 95
Context free grammar, 492
Context sensitive grammar, 492
Contradiction (logical), 241
Contradiction (proof by), 244
Contrapositive, 243
Convergent sequence, 407
Converse, 243
Conversion from decimal to other number
    systems, 62
Convolution, 438
Convolution Theorem, 439
Counterexample, 512
Counting with repetitions, 328
Cover, 221
Covering problem, 230
Critical activity, 159
Critical path, 160
Critical path method, 157
Cube, 143, 190
Cycle, 103
    Hamiltonian, 180
    of spread, 2, 182

D

Dead state, 459, 472
de Bruijn sequence, 175
Decreasing sequence, 409
Degree of a vertex, 90
Derangement, 337
Derivation, 490
Determinant of a matrix, 525
Deterministic finite automaton, 458
DFA, 458
Diagram of a graph, 86
Difference equation, 260
Digraph, 131
Directed graph, 131
    edge of, 132
    of a DFA, 459
    of an NFA, 465
    of a relation, 136
    path in, 133
    unilaterally connected, 143
    vertex of, 132
Directed path, 133

Disconnected graph, 95
Disjunction, 239
Distribution, 378
    binomial, 378
    geometric, 380
    Poisson, 389
    uniform, 378
Divergent sequence, 407
Divide-and-conquer algorithm, 263, 297
Division algorithm, 63
$D(M)$, 471
Domain of a function, 513
Dual machine, 467

E

Edge of a digraph, 132
Edge of a graph, 85
Efficiency of an algorithm, 6, 44
Empty set, 504
Entry of a matrix, 518
Enumeration, 302
Equal matrices, 519
Equal sets, 505
Equivalence class, 533,
Equivalence relation, 532
Equivalent Boolean expressions, 203
Equivalent statements, 240
Erasing production, 491
Essential prime implicant, 232
Euler path, 168
Eulerian graph, 173
Euler's Theorem, 169
Euler tour, 169
Even function, 517
Event, 344
Expected value, 382
Exponential complexity, 49

F

Fibonacci numbers, 22
Forest, 107
Formal languages, 456, 489
Four Queens Problem, 40
Full adder, 254

Function, 512
    Boolean, 194
    characteristic, 514
    composite, 515
    domain of, 513
    even, 517
    next-state, 459, 465
    odd, 517
    one-to-one, 514
    onto, 514
    range of, 513

G

Gambler's ruin problem, 277, 351
Gate
    and, 247,
    inverter, 248
    nand, 251
    nor, 251
    or, 247
Gate implementation, 247
Generalized product principle, 304
General law of compound probability, 359
General solution, 271
Generating function of a sequence, 423, 429,
        444, 448
Generator, 76
Geometric distribution, 380
Geometric series, 419
Goldbach Conjecture, 13
Good diagram, 176
Grammar, 490
    context free, 492
    context sensitive, 492
    nonterminal alphabet in, 489
    phrase structure, 489
    production in, 489
    right linear, 492
    start symbol in, 489, 490
    terminal alphabet in, 489, 490
    type 1, 492
    type 2, 492
    type 3, 492
Graph, 85
    adjacency matrix of, 88
    complete, 123

complete bipartite, 307
connected, 95
cycle in, 103, 180, 182
directed, 131
disconnected, 95
edge in, 5
loop in, 132
subgraph of, 88, 133
vertex in, 85, 132
weighted, 113
Gray code, 178, 181
Greatest common divisor, 7, 79
Greatest integer function, 3
Greatest lower bound, 406
Greedy algorithm, 28

H

Half adder, 253
Hamiltonian cycle, 180
Hamiltonian graph, 131, 180
Hasse diagram, 140
HEAPSORT, 47
Hexadecimal number system, 57
Homogeneous difference equation, 266
Horner's algorithm, 59
Hypothesis, 243

I

Identity matrix, 520
If and only if, 246
Immediate predecessor in a partial order, 139
Implicant, 219
Implication, 242
Implies, 219
Improper fraction, 432
Increasing sequence, 409
Indegree of a vertex, 172
Independent events, 361, 364
Induction, 12
Infinite sequence, 404
Infinite series, 415
Infinite sum, 415
Initial condition, 274
Initial state, 459, 465

Input alphabet, 458, 465
Internal vertices of a tree, 107
Intersection of sets, 506–508
Inverse, 243
Inverse in $\mathbf{Z}_m$, 78
Inversion, 432
Inverter gate, 248

K

Karnaugh map, 206
in four variables, 206
in five variables, 218
in three variables, 218
Kleene closure, 477
Kleene's theorem, 485
Königsberg bridge problem, 168
Kruskal's algorithm, 115, 120

L

Language, 490
and automata, 494
formal, 489
generated by a grammar, 490
natural, 489
type 3, 494
type i, 492
Laplace, 342
Least integer function, 6
Least upper bound, 405
Leaves of a tree, 107
Length of a directed path, 133
Length of a path, 94
Lexicographic order, 309, 534
Limit of a sequence, 407
Limit theorem, 411
Linear difference equation, 266
Literal, 204
Logic circuit, 246
Logic gate, 246
Logical equivalence, 240
Loop in a digraph, 132
Lower bound, 405

M

Machine, 456
Magic square, 43
Mapping, 513
Mathematical induction, 12
Matrix, 518
    addition of, 518
    adjacency, 88
    determinant of, 525
    entry of, 518
    identity, 523
    order of, 518
    product of, 521
    square, 518
    symmetric, 89
    transition, 457
    zero, 519
Matrix operations, 518
Maximal path tree, 149
Maximal shift register sequence, 175
Member of a set, 503
MERGESORT, 297
Method of undetermined coefficients, 290
Minimal connector problem, 115
Minimal counterexample, 169
Minimal path tree, 149
Minimal path tree algorithm, 154
Minimal representation of a Boolean
        function, 204
Minimal spanning tree, 113, 115
Minimization of automata, 486
Minimum weight spanning cycle, 122
Modular arithmetic, 55, 73
Modular code, 78
Multinomial theorem, 321
Multiplication of matrices, 521

N

Nand gate, 251
Natural language, 489
n–cube, 143, 180
Negation, 239
Next-state function, 459, 465
NFA, 464
Nine's complement, 69

Nondeterministic finite automaton, 464
Nonterminal alphabet, 490
Nor gate, 251
Null symbol, 491
Number system, 55
    base of, 56
    binary, 55
    hexadecimal, 57
    octal, 57

O

Octal number system, 57
Odd function, 517
$O(f)$, 48
One's complement, 71
One-to-one function, 514
Onto function, 514
Operations with binary numbers, 69
Or gate, 247
Order (Big O notation), 44, 58
Ordered pair, 509
Order of a difference equation, 266
Order relation, 533
Outcome, 343
Outdegree of a vertex, 172

P

$P(A|B)$, 356
Parallel connection, 188
Partial fraction decomposition, 433
Partial fractions, 432
Partial order, 138, 533
Partial sum, 415
Partition of a set, 532, 535
Pascal, 342
Pascal's triangle, 324
Path, 94
    directed, 133
Pattern recognition problem, 456
p d f, 377
Permutation, 307
    of n objects taken r at a time, 311
Phrase structure grammar, 490
    derivation in, 490

nonterminal alphabet in, 489
production in, 489
terminal alphabet in, 489
type i, 492
$P(n, r)$, 311
Poisson distribution, 389
Poisson parameter, 388
Polynomial complexity, 49
Polynomial of degree n, 59
Power set, 510
Predator–prey problem, 283
Prime implicant, 222
Prim's algorithm, 120
Principle of inclusion and exclusion, 335
Principle of Induction, 13
Probability density function, 377
Probability measure, 344
Probability space, 344
Probability theory, 343
Production, 489
Product of matrices, 521
Product of series, 438
Product principle, 303
Proof by contradiction, 244
Proper fraction, 432
Pseudorandom number, 76
$\mathscr{P}(X)$ (power set of X), 510

Q

Queue discipline, 393
Queueing theory, 393
QUICKSORT, 21
Quine–McCluskey algorithm, 224

R

Random experiment, 343
Random variable, 376
Range of a function, 513
Rational function, 432
r-combination, 313
Recursive algorithm, 21
Recursive definition, 194
Redundant r-combination, 332
Redundant r-permutation, 329

Reflexive relation, 138, 531
Regular expression, 477
Regular set, 478
Relation, 531
    antisymmetric, 138, 533
    equivalence, 532
    order, 533
    partial order, 138, 533
    reflexive, 138, 531
    symmetric, 531
    transitive, 138, 531
Relatively prime integers, 79
Right linear grammar, 492
$r(L)$, 466
$R(M)$, 466
Root of a tree, 107
Rooted tree, 148
r-permutation, 311
Russian peasant algorithm, 67

S

Sample space, 343
Scalar multiplication, 520
Seed, 76
SEQUENTIAL SEARCH, 2, 45, 384
Series connection, 188
Service process, 393
Set, 503
    Cartesian product of, 510
    complement, 508
    equal, 505
    intersection, 506–508
    power, 510
    subset of, 504
    union, 506–508
Shifting theorem, 425
Sigma notation, 528
Simple path, 95
Slack of an edge, 152
Sorting algorithms, 8
    BUBBLESORT, 10, 299
    HEAPSORT, 47
    MERGESORT, 297
    QUICKSORT, 21
Spanning cycle, 123
Spanning subgraph, 107

Spanning tree, 107
Square matrix, 518
Standard product term, 224
Start symbol, 489
State set, 458, 465
Statement, 238
Sterling number, 328
Strongly connected graph, 135, 172
Subgraph of a digraph, 133
Subgraph of a graph, 88
Subset, 504
Sum in canonical form, 224
Sum of matrices, 519
Sum of products, 196, 204
Sum principle, 304
Switch, 187
Switching system, 187
Syllogism, 245
Symbolic logic, 238
Symmetric matrix, 89
Symmetric relation, 531

T

Tautology, 241
Ten's complement, 69
Terminal alphabet, 490
Terminal letter, 490
Tour, 169
Transition matrix, 457
Transitive relation, 138, 531
Traveling salesman problem, 30
Tree, 103
    minimal spanning, 113, 115
Tree rooted at v_0, 148
Triangle inequality, 126
TRINARY SEARCH, 264, 296
Truth table, 239
Turing machine, 498

Two's complement, 71
Type i grammar, 492
Type 3 grammar, 494

U

Unary operation, 191
Undetermined coefficients, 290
Uniform distribution, 378
Unilaterally connected digraph, 143
Union of sets, 506–508
Upper bound, 405

V

Vandermonde's identity, 328, 451
Venn diagram, 508
Vertex, 85
    adjacent, 86
    degree of, 90
    indegree of, 172
    of a digraph, 135
    of a graph, 85
    outdegree of, 172

W

Weighted digraph, 135
Weighted graph, 113
Weight matrix, 113
    of a weighted digraph, 135
Worst-case analysis, 45
Worst-case complexity, 45

Z

Zero matrix, 519

A 5
B 6
C 7
D 8
E 9
F 0
G 1
H 2
I 3
J 4